Aeration Control
System Design

Aeration Control System Design

A Practical Guide to Energy and Process Optimization

Thomas E. Jenkins, PE
JenTech Inc.

WILEY

Published by John Wiley & Sons, Inc., Hoboken, New Jersey
Published simultaneously in Canada

***Library of Congress Cataloging-in-Publication Data*:**

Jenkins, Thomas E., 1950-

Aeration control system design : a practical guide to energy and process optimization / Thomas E. Jenkins, PE, JenTech Inc. – First edition.

pages cm

Includes index.

ISBN 978-1-118-38998-0 (hardback)

1. Sewage–Purification–Aeration. 2. Water–Aeration. 3. Supervisory control systems. 4. Sewage disposal plants–Energy conservation. I. Title.

TD758.J47 2013

628.1′65–dc23

2013020446

Printed in the United States of America

10 9 8 7 6 5 4 3 2 1

Contents

Preface xi

Acknowledgments xiii

List of Figures xv

List of Tables xxi

1 Introduction 1

1.1 Basic Concepts and Objectives / 2
1.2 Safety / 9
1.3 The Importance of an Integrated Approach / 10
1.4 Importance of Operator Involvement / 13
1.5 The Benefits of Successful Aeration Process Automation / 14
1.5.1 Energy Cost Reduction / 14
1.5.2 Treatment Performance / 18
1.5.3 Improved Equipment Life / 18
Example Problems / 19

2 Initial System Assessment 21

2.1 Define Current Operations / 24
2.1.1 Energy Cost / 25
2.1.2 Energy Consumption Patterns / 29

2.1.3 Influent and Effluent Process Parameters / 35
2.1.4 Treatment Performance / 36
2.2 Evaluate Process and Equipment / 37
2.3 Benchmark Performance / 40
2.4 Estimate Potential Energy Savings and Performance Improvement / 42
2.5 Prepare Report / 45
Example Problems / 47

3 Aeration Processes **49**

3.1 Process Fundamentals / 49
3.1.1 Peripheral Equipment and Processes / 55
3.1.2 BOD Removal / 62
3.1.3 Nitrification / 66
3.1.4 Denitrification / 67
3.2 Loading Variations and Their Implications / 68
3.3 Process Limitations and Their Impact on Control Systems / 70
Example Problems / 74

4 Mechanical and Diffused Aeration Systems **77**

4.1 Oxygen Transfer Basics / 78
4.2 Types of Aerators / 87
4.2.1 Mechanical Aerators / 88
4.2.2 Mechanical Aeration Control Techniques / 90
4.2.3 Diffused Aeration / 95
4.2.4 Diffused Aeration Control Techniques / 103
4.3 Savings Determinations / 106
Example Problems / 111

5 Blowers and Blower Control **113**

5.1 Common Application and Selection Concerns / 114
5.1.1 Properties of Air / 114
5.1.2 Effect of Humidity / 119
5.1.3 Pressure Effects / 123
5.1.4 Common Performance Characteristics / 125
5.2 Positive Displacement Blowers and Control Characteristics / 134
5.2.1 Types and Characteristics / 134
5.2.2 Lobe Type PD Blowers / 134

5.2.3 Screw Blowers / 138
5.2.4 Control and Equipment Protection Considerations / 141
5.3 Dynamic Blowers / 143
5.3.1 Types and Characteristics / 144
5.3.2 Multistage Centrifugal Blowers / 150
5.3.3 Geared Single Stage Centrifugal Blowers / 152
5.3.4 Turbo Blowers / 154
5.3.5 Control and Protection Considerations / 155
Example Problems / 157

6 Piping Systems **161**

6.1 Design Considerations / 162
6.1.1 Layout / 162
6.1.2 Pipe Size / 172
6.1.3 Pipe Material / 174
6.2 Pressure Drop / 178
6.3 Control Valve Selection / 182
Example Problems / 187

7 Instrumentation **191**

7.1 Common Characteristics and Electrical Design Considerations / 192
7.2 Pressure / 202
7.3 Temperature / 205
7.4 Flow / 209
7.5 Analytic Instruments / 216
7.5.1 Dissolved Oxygen / 217
7.5.2 Offgas Analysis / 221
7.5.3 pH and ORP / 224
7.6 Motor Monitoring and Electrical Measurements / 224
7.7 Miscellaneous / 226
Example Problems / 230

8 Final Control Elements **233**

8.1 Valve Operators / 234
8.2 Guide Vanes / 238
8.3 Motor Basics / 239
8.4 Motor Control / 247
8.5 Variable Frequency Drives / 251
Example Problems / 259

9 Control Loops and Algorithms **261**

9.1 Control Fundamentals / 264
9.1.1 Discrete Controls / 264
9.1.2 Analog Control / 267
9.1.3 Proportional-Integral-Derivative / 271
9.1.4 Deadband Controllers / 274
9.1.5 Floating Control / 276
9.2 Dissolved Oxygen Control / 280
9.3 Aeration Basin Air Flow Control / 287
9.4 Pressure Control / 288
9.5 Most-Open-Valve Control / 291
9.6 Blower Control and Coordination / 293
9.7 Control Loop Timing Considerations / 302
9.8 Miscellaneous Controls / 303
Example Problems / 305

10 Control Components **309**

10.1 Programmable Logic Controllers / 310
10.1.1 System Architecture / 314
10.1.2 Program Structure / 315
10.1.3 Communications Networks / 318
10.1.4 Accommodating Instrument Inaccuracy and Failure / 322
10.2 Distributed Control Systems / 323
10.3 Human Machine Interfaces / 323
10.3.1 Supervisory Control and Data Acquisition / 325
10.3.2 Touchscreens / 327
10.4 Control Panel Design Considerations / 328
Example Problems / 330

11 Documentation **333**

11.1 Specification Considerations / 335
11.2 Data Lists / 338
11.3 Process and Instrumentation Diagrams / 341
11.4 Ladder and Loop Diagrams / 342
11.5 One-Line Diagrams / 344
11.6 Installation Drawings / 345
11.7 Loop Descriptions / 347
11.8 Operation and Maintenance Manuals / 348
Example Problems / 349

12 Commissioning 351

12.1 Inspection / 354
12.2 Testing / 357
12.3 Tuning / 361
12.4 Training / 365
12.5 Measurement and Verification of Results / 368
Example Problems / 369

13 Summary 371

13.1 Review of Integrated Design Procedure / 371
13.2 Potential Problem Areas / 374
13.3 Benefits / 375
Example Problems / 375

Appendix A: Example Problem Solutions 377

Appendix B: List of Equations and Variables 447

Bibliography 485

Index 487

Preface

This is an engineering manual.

There are a lot of excellent resources available for energy conservation in wastewater treatment facilities. They contain a great deal of useful information on developing and implementing energy conservation programs. Most of them discuss aeration in general and automated aeration control in particular. Many of them include case histories identifying successful implementations of aeration control and showing the resulting savings. A few identify unsuccessful attempts and how to avoid problems.

To the best of my knowledge none of the available resources provide the detailed engineering procedures required to design, commission, and test an aeration control system. To the best of my knowledge none of the available resources provide detailed guidance on applying the multiple engineering disciplines—mechanical, electrical, and environmental—necessary for successful system design.

The information in this book is the result of over 30 years experience in analyzing energy consumption in wastewater treatment plants. It includes the lessons learned in the design of over 200 aeration control systems. This is hard-won knowledge, and was gained by personal experience in all of the tasks needed to develop concepts, sell management on the cost-effectiveness, get operator buy-in, and work through the inevitable start-up issues. The intent is to explain the nuts and bolts details of what to do—and what not to do—in designing aeration control systems.

There are two concepts that appear repeatedly in this book. First, there are no hard and fast rules. Steady-state operation is virtually nonexistent in wastewater treatment. This leads to the second concept. Whenever possible the assistance and input of equipment manufacturers should be obtained in order to obtain the highest level of precision possible in calculations.

Unfortunately, equipment manufacturers may not be responsive enough to meet the tight deadlines associated with many energy conservation and control designs. Worse, in some cases the information obtained may not be pertinent or even accurate. A thorough knowledge of the many aspects of aeration system operation is necessary to filter good information from bad and to make independent evaluations when outside sources fail. My intent is to provide that knowledge.

Another theme that appears throughout this text is "reasonable accuracy." One unfortunate side effect of the wide availability of computers, math software, and advanced modeling programs is the tendency to create elaborate analyses of various alternatives. The results, calculated to 10 or 20 digits, are reassuring and intellectually satisfying. There is a tendency to forget that the elaborate calculations are all based on initial assumptions that are only correct to two or three significant digits!

There is a time for detailed and precise calculations, and sophisticated modeling can provide insight into relationships between variables that would be difficult to achieve in any other way. However, when the early stages of the design process require deciding between multiple alternatives, it is generally adequate to use what used to be called "slide rule accuracy." Then, when the problem has been narrowed and the needed supporting data gathered, more advanced calculation methods can be used for the final analysis and design confirmation.

This book will show in detail how to predict savings, how to design systems that meet the mechanical, electrical, and process requirements, and how to commission the systems to secure successful operation. It is wide ranging in scope, but focused on providing practical guidance to creating aeration control systems that the operators will feel comfortable leaving in automatic.

Acknowledgments

First and foremost the credit for making this book possible goes to my wife, Ginny. I know—everybody says that. In this case Ginny has earned the credit a hundred times over. She gave me support and encouragement, of course. But more than that she backed me 30 years ago when I had "an idea," supported me when I began a business to commercialize that idea, figured out how to feed our family through the long lean years getting started, and for many years handled the finances for the business purely as an act of love. "Thanks" is inadequate, but all I have.

I also need to thank our employees. Over the years they contributed their hard work, ideas, and expertise to improving and implementing aeration control systems. All of them were important, but a special "thank you" goes to Tim Hilgart. He stuck it out through the tough times and was in it for the long haul. In my long career I have met no better engineer or finer man.

Finally, I want to thank my brother Paul and all of the plant operators I have worked with over the years. These professionals contributed insights, ideas, and encouragement. Some also contributed harsh and much needed criticism because they knew and understood reality in a way that no engineer really can. I hope this book repays their efforts by helping engineers design solutions — not problems!

THOMAS E. JENKINS

Milwaukee, WI

List of Figures

Figure 1.1	Simplified biological process	2
Figure 1.2	Typical activated sludge treatment processes	3
Figure 1.3	Effect of power factor	7
Figure 1.4	Typical diurnal hydraulic load variation	12
Figure 1.5	Typical WWTP energy use	15
Figure 1.6	Problem 1.2 pump curve	20
Figure 2.1	Initial assessment steps	22
Figure 2.2	Sample energy rates	28
Figure 2.3	Sample of benchmark energy versus flow	41
Figure 2.4	Comparison of simple payback and present value	44
Figure 2.5	EPA approach to ECM implementation	46
Figure 3.1	Plug flow basin	53
Figure 3.2	Plug flow basin with serpentine flow path	53
Figure 3.3	Oxidation ditch	54
Figure 3.4	Sample SBR cycles	54
Figure 3.5	Circular clarifier	56
Figure 3.6	Rectangular clarifier	56

Figure 3.7 Airlift pump schematic 58
Figure 3.8 Airlift pump air requirements 58
Figure 3.9 MLE process 68
Figure 4.1 Effect of DO concentration on oxygen transfer requirements 80
Figure 4.2 Effect of altitude on DO saturation 81
Figure 4.3 Effect of salinity on DO saturation 82
Figure 4.4 Types of mechanical aerators 88
Figure 4.5 Sample horizontal aerator 89
Figure 4.6 Sample vertical shaft aerator 90
Figure 4.7 Sample relationship of brush aerator power versus speed 91
Figure 4.8 Sample relationship of brush aerator oxygen transfer versus speed 91
Figure 4.9 Sample relationship of brush aerator aeration efficiency versus speed 92
Figure 4.10 Example of brush OTR versus DO concentration 94
Figure 4.11 Types of diffusers 95
Figure 4.12 Example tubular coarse bubble diffuser 96
Figure 4.13 Example tubular diffuser SOTE 97
Figure 4.14 Example tubular diffuser pressure drop 97
Figure 4.15 Example membrane fine-pore diffuser 98
Figure 4.16 Example membrane fine-pore SOTE 98
Figure 4.17 Variation of oxygen demand along tank length 99
Figure 4.18 Example membrane fine-pore pressure drop 102
Figure 4.19 Example of membrane fine-pore OTR versus DO concentration 105
Figure 4.20 Example weekday hourly power cost 108
Figure 4.21 Example weekend hourly power cost 108
Figure 4.22 Five point diurnal analysis 110
Figure 5.1 Reading a psychrometric chart 121
Figure 5.2 Example system curve 125
Figure 5.3 Example mass flow rate versus power consumption 130

Figure 5.4 Blower categories 133
Figure 5.5 Simplified diagram of lobe type PD blower 135
Figure 5.6 Example screw blower rotor profiles 139
Figure 5.7 Simplified diagram of screw blower 139
Figure 5.8 Example variation of screw blower performance with discharge pressure 140
Figure 5.9 Example variation of screw blower performance with speed 140
Figure 5.10 Effect of inlet throttling on dynamic blowers 147
Figure 5.11 Effect of variable speed on dynamic blowers 148
Figure 5.12 Simplified diagram of multistage centrifugal blower 151
Figure 5.13 Simplified diagram of geared single stage centrifugal blower 152
Figure 5.14 Simplified diagram of turbo blower package 154
Figure 5.15 Problem 5.3 plant air flow trend 158
Figure 5.16 Problem 5.3 blower performance curves 159
Figure 6.1 Piping layout example 163
Figure 6.2 Typical "Y-wall" configuration 169
Figure 6.3 Vertical goose neck 171
Figure 6.4 Horizontal takeoff 171
Figure 6.5 System pressure gradient 180
Figure 6.6 Typical butterfly valve C_v 185
Figure 6.7 Four inch (4 in.) butterfly valve throttling performance 187
Figure 6.8 Problem 6.1 piping layout 188
Figure 7.1 Contact arrangements 194
Figure 7.2 Typical solid-state switches 194
Figure 7.3 Transmitter wiring 196
Figure 7.4 Ground loop 199
Figure 7.5 Example transmitter output 201
Figure 7.6 Typical differential pressure transmitter 205
Figure 7.7 Typical temperature transmitter 208
Figure 7.8 Example orifice meter installation 215

Figure 7.9 Example DO probe installation 218
Figure 7.10 Offgas analysis 222
Figure 7.11 Problem 7.1 sample transmitter load limits 230
Figure 7.12 Problem 7.6 transmitter wiring 231
Figure 8.1 Typical valve positioner schematic 236
Figure 8.2 Typical direct valve control schematic 236
Figure 8.3 Motor torque and current variation with speed 242
Figure 8.4 Variation of power factor and amps with motor load 246
Figure 8.5 Example induction motor efficiencies 247
Figure 8.6 Example induction motor power factors 250
Figure 8.7 PWM output voltage and motor current 251
Figure 8.8 Example PWM VFD configuration 253
Figure 9.1 Feedback control 262
Figure 9.2 ISA automatic reset alarm sequence 265
Figure 9.3 ISA manual reset alarm sequence 266
Figure 9.4 Process measurement transformations 268
Figure 9.5 Effect of digital filtering 271
Figure 9.6 Example PID tuning 273
Figure 9.7 Deadband pressure control for blow-off valve 275
Figure 9.8 Floating control with deadband logic 277
Figure 9.9 Sample floating control with deadband 277
Figure 9.10 Proportional speed floating control 279
Figure 9.11 Cascaded DO and air flow control loops 280
Figure 9.12 DO control logic sequence 282
Figure 9.13 Typical pressure control system 290
Figure 9.14 Excess pressure with pressure control 291
Figure 9.15 Example savings with MOV control 292
Figure 9.16 Blower control and protection requirements 294
Figure 10.1 Sample of ladder logic 311
Figure 10.2 Typical ethernet system 321

Figure 10.3 Trend with multiple parameters 324
Figure 10.4 Sample SCADA screen shot 326
Figure 11.1 Example I/O point list 339
Figure 11.2 Example bill of material information 339
Figure 11.3 Example alarm list 340
Figure 11.4 Example setpoint list 341
Figure 11.5 Simple process and instrumentation diagram 342
Figure 11.6 Ladder diagram 343
Figure 11.7 Simplified loop diagram 344
Figure 11.8 Simplified one-line diagram 345
Figure 12.1 Start-up activities 353
Figure 12.2 Inspection sequence 354
Figure 12.3 Testing sequence 358
Figure 12.4 Tuning floating control 362
Figure 12.5 Initial DO control tuning estimate 364
Figure 12.6 Problem 12.2 sample DO control performance 370
Figure 13.1 Design procedure outline 372
Figure A.1 Problem 1.2 pump curve 379
Figure A.2 Problem 4.3 OTR vs. DO 399
Figure A.3 Problem 5.3 blower performance with IGV 409
Figure A.4 Problem 5.3 blower performance with VFD 411
Figure A.5 Problem 9.7 flow control valve response 438
Figure A.6 Problem 10.2 debounce timer 440

List of Tables

Table 1.1 Operating Wastewater Treatment Plants in the United States (EPA, 2004) 15

Table 2.1 Example Electrical Energy Benchmarks for Activated Sludge 41

Table 4.1 Variation of DO Saturation with Barometric Pressure 80

Table 4.2 Five Point Diurnal Estimate 110

Table 5.1 Composition of Dry Air 115

Table 5.2 Saturation Water Vapor Pressure 120

Table 5.3 Approximate Wire-to-Air Efficiency of Various Blowers at Mid-range Flow (for Preliminary Estimating Use Only) 129

Table 6.1 Nominal Pipe Support Spacing, Air Service, feet 167

Table 6.2 Typical Linear Coefficients of Thermal Expansion and Young's Modulus (Properties will vary with Temperature and Composition) 168

Table 6.3 Typical Distribution Piping Air Velocities 172

Table 6.4 Typical Steel and Stainless Steel Pipe Dimensions 173

Table 6.5 Typical Temperature Ratings of Common Elastomers 177

Table 6.6 Equivalent Length of Pipe Fittings 179

Table 7.1 Typical Motor Temperature Limits 225

Table 8.1 Partial Listing of Three-Phase Motor Locked Rotor Current 245

Table 8.2 Partial Listing of NEMA Starter Sizes 248

Table 8.3 Partial Listing of IEC Starter Sizes 249

Table 8.4 Typical Speed Limits for Induction Motors 254

Table 11.1 Nominal Wire Sizing 346

Chapter 1

Introduction

The activated sludge process is the most common method of secondary wastewater treatment throughout the world. In most wastewater treatment plants, it is the process that is most critical to meeting treatment objectives, and receives the most operator attention. Supplying oxygen to the activated sludge basins typically consumes more than 50% of the total energy used in wastewater treatment. This represents a significant portion of the plant's operating budget.

It is therefore surprising to realize how poorly automated the activated sludge process is in most wastewater treatment plants.

This book has two objectives. The first, and most important, is to improve and sustain process performance and enable operators to consistently meet treatment objectives. The second objective is to provide the tools and techniques required to achieve the first objective while minimizing the energy requirements of the process. Fortunately, the two objectives are complementary—it isn't necessary to sacrifice process performance and stability in order to optimize energy use.

There are many unit processes employed in wastewater treatment of both municipal waste from primarily residential sources and industrial waste. The principal concern of this text is the automation and energy optimization of suspended growth activated sludge facilities.

Aeration Control System Design: A Practical Guide to Energy and Process Optimization,
First Edition. Thomas E. Jenkins.

1.1 BASIC CONCEPTS AND OBJECTIVES

The primary objective of the automation system is maintaining the required process performance. This objective must be kept in mind during all phases of the evaluation and design procedure. They don't build wastewater treatment plants to save energy—they build them to remove pollutants from wastewater. A treatment plant doesn't become a headline in the morning paper by using the same amount of electricity at the treatment plant that they did last year! Furthermore, in the United States and most of the world, a government body issues permits establishing the required level of treatment. Failure to meet the permit requirements can result in fines and other penalties.

The activated sludge process is an *aerobic biological* process. In the basic *activated sludge* process, organic pollutants in the influent waste stream are absorbed by microorganisms. In an aeration basin, the microbes are suspended in the wastewater, and this combination of biology, water, and pollutants is referred to as *mixed liquor*. The microbes are not suspended as individual organisms, but they form clumps or *floc* particles. The microorganisms utilize oxygen dissolved in the wastewater to metabolize the pollutants (Figure 1.1). After being carried through the aeration basin by the wastewater flow, the microorganisms pass into the secondary clarifier. In the clarifier, the floc settles to the bottom and the microbes are pumped back to the aeration basin for another cycle of removing pollutants. The settled microbes are called *sludge*. The treated wastewater passes to disinfection and then into the receiving water (Figure 1.2).

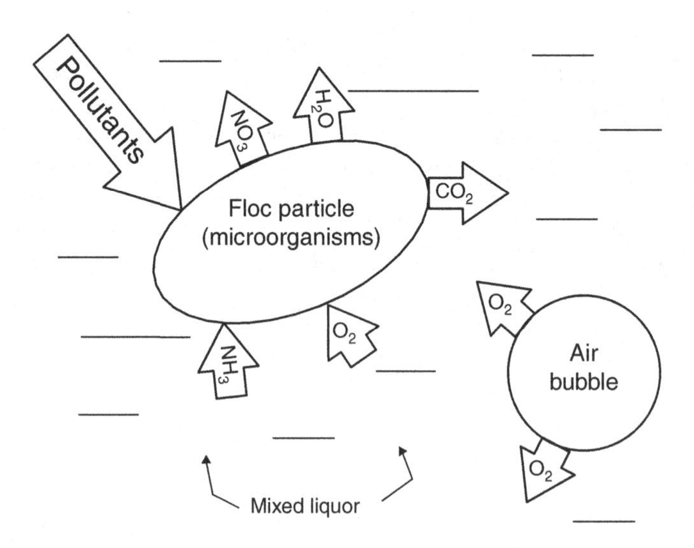

FIGURE 1.1 *Simplified biological process*

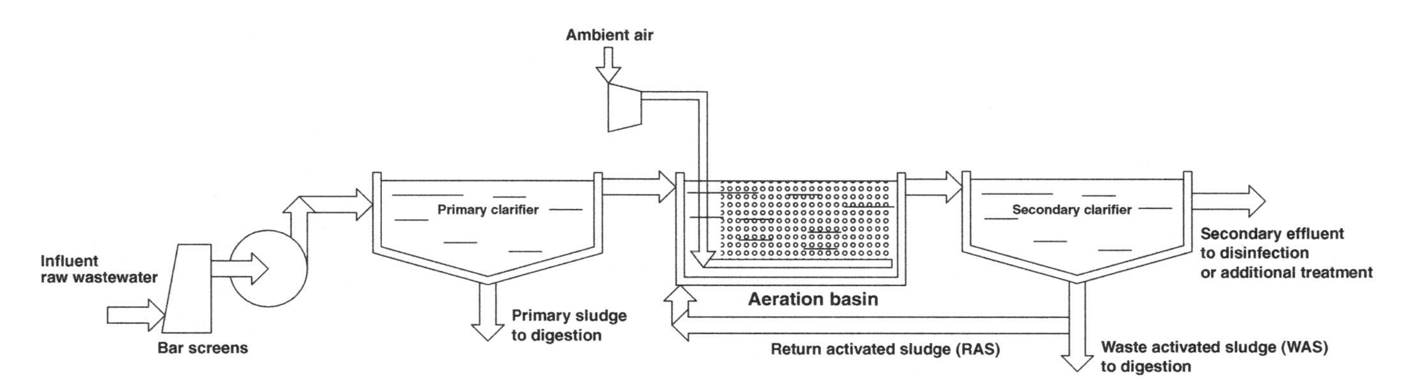

FIGURE 1.2 *Typical activated sludge treatment processes*

The most prevalent pollutants fall into one of two categories. The first category consists of *carbonaceous* compounds—essentially the organic compounds that make up human and industrial waste. These are most often measured and reported as *biochemical oxygen demand* (BOD). The second principal category consists of *nitrogenous* compounds, the most significant of which is ammonia (NH_3). Conversion of the ammonia to nitrate (NO_3) is called nitrification. Phosphorus is a compound that is regulated in some permits, and its removal requires a high level of process control but little additional energy. Toxic compounds are sometimes a concern, as are odors, unsightly appearance, and pathogens.

The most significant effect of both carbonaceous and nitrogenous compounds is depletion of the oxygen in the receiving water to a level that is detrimental to aquatic life. In some respects, the purpose of the treatment facility is to supply the pollutants' oxygen demand in a controlled fashion so that the demand is not exerted on the receiving water.

The flow into a treatment process or facility is referred to as *influent*. The flow out of a process or facility is referred to as *effluent*. The load to the treatment plant is further categorized as *hydraulic load*, which is the volumetric flow rate, and *organic load*, which is the combination of carbonaceous and nitrogenous pollutants that must be treated. Both types of load influence the design and performance of the treatment processes. The concentration of pollutants varies from one facility to another and also varies with time within a given facility. In some instances, the organic load dominates treatment process design and performance, and in other instances, the hydraulic load is more significant.

Once the primary objective of pollution abatement is achieved, it is possible to move on to optimizing the operation of the treatment plant—providing the required treatment at the lowest cost. This can include many aspects of running the treatment plant, including optimal staffing, extending equipment life, and reducing chemical use. Automation can help optimize all of these. One of the most desirable goals of optimization is reducing energy cost, and automation is critical to its successful achievement.

Aeration is an energy-intensive process. Approximately half of all energy used in a typical wastewater treatment plant is used to provide oxygen to the microorganisms in the aeration basins. The usual source of oxygen is simply ambient air. The air is either mechanically mixed into the wastewater using surface aerators or diffused into the wastewater by bubbling air under pressure into the bottom of the aeration basin.

Note that the stated goal is reducing *energy cost*, not reducing *energy consumption*. When discussing electrical energy, the terms *energy*, *cost*, and *power* tend to be used interchangeably, at least in casual conversation. However, as professionals, it is important that we distinguish between these terms. They are related, but definitely distinct.

Energy is the ability to do work, or the amount of work done. It theoretically consumes a fixed amount of energy to move a given mass of water uphill a given distance, for example. Units of measure for energy are the same as those for work and include Joules, Watt-hours, and foot-pounds. In mechanics, energy is calculated as force $\times$ distance.

Equation 1.1

$$E = F \cdot d$$

where

E = energy, ft lb_f
F = force, lb_f
d = distance, ft

Energy is available in many forms—heat, electricity, kinetic energy, potential energy, and so on. Aeration systems rely primarily on electricity, and it is the principal focus of aeration energy conservation efforts and aeration control system design.

Mechanical, heat, and electrical energy can be converted to one another. One British thermal unit (BTU), a common measurement of heat energy, is the equivalent of 778.2 ft lb or 0.0002931 kWh. Analyzing the conversion and transmission of energy consumes a great deal of engineering effort. In some cases, the conversion is intentional. For example, boilers and steam turbines convert heat energy to mechanical energy, and generators convert mechanical energy to electricity. In other cases, the conversion is unwanted, such as the heat generated by friction or the resistance in electrical circuits. The heat created by electricity is proportional to the resistance (R) and the square of the current (I) passing through it, and is often referred to as "I squared R" or "I^2R" losses.

Power is the rate of energy use or the rate of work. Power includes a time dependency. To move a given amount of water the same distance in 1 second takes more power than to do it in 1 hour, although the energy used is the same in both cases. Units of measure for power include Watts and horsepower. In mechanics, power is calculated as work divided by time:

Equation 1.2

$$\text{horsepower} = \frac{550 \text{ ft lb}_f}{\text{s}}$$

where

s = time, seconds

Pumping water is a common energy use in wastewater conveyance and treatment. Pumping power is a function of two variables. One is the flow rate—the weight of water being pumped. The other is the head—the pressure of the system, generally measured as the equivalent height to which the water is being lifted.

Equation 1.3

$$P_W = \frac{Q_W \cdot h \cdot \text{SG}}{3960}$$

where

P_w = water power, hp
h = pump head, ft
Q_w = flow rate, gpm
SG = specific gravity, water = 1.0, dimensionless

In electrical systems, power is a function of voltage and current. The determination in direct current (DC) circuits is straightforward. For simple DC circuits:

Equation 1.4

$$P = V \cdot I$$

where

P = power, W
V = voltage or electromotive force (emf), V
I = current, A

In alternating current (AC) circuits, additional considerations regarding phase and power factor must be considered:

Equation 1.5 (Single-phase AC only)

$$P = V \cdot I \cdot \mathrm{PF}$$

Equation 1.6 (Three-phase AC only)

$$P = V \cdot I \cdot \sqrt{3} \cdot \mathrm{PF}$$

where

P = power, W
V = voltage or electromotive force (emf), V
I = current, A
PF = power factor, decimal

The basic unit of electrical power is the Watt, but that is too small a unit of measure for many applications. The kilowatt (kW), 1000 W, is most often used when discussing power requirements for motors and similar loads in a wastewater treatment plant.

Power requirements for pumps, blowers, and other process equipment are often given as brake horsepower (bhp). This is the power required at the shaft of the driven equipment. This may be converted to pump motor power in kilowatt:

Equation 1.7

$$P = \frac{\mathrm{bhp} \cdot 0.746}{\eta_{\mathrm{m}} \cdot \eta_{\mathrm{vfd}}}$$

where

P = power, kW
bhp = shaft power, hp
η_{m} = efficiency of motor, decimal
η_{vfd} = efficiency of variable-frequency drive if used, decimal

Power factor is a concept that is commonly misunderstood. Power factor in an AC system is the ratio between a system's real power (used to perform work) and apparent power (VA or kVA). For three-phase loads, the apparent power is

Equation 1.8

$$S = V \cdot I \cdot \sqrt{3} = \frac{P}{\text{PF}}$$

where

S = apparent power, VA

In an AC system with inductive loads such as motors or transformers, the flow of current lags behind the rising voltage because the current flow must create a magnetic field in the inductor. Energy is stored in the inductor, and as the voltage drops, the magnetic field collapses, releasing current into the circuit. The real power waveform also lags behind the voltage waveform, and real power may be negative—in effect actually sending power back into the source.

The power factor is the cosine of the angle of phase difference between the voltage and the current waveforms (Figure 1.3). It is usually expressed as a decimal between 0 and 1, but may also be expressed as a percentage between 0 and 100%. The greater

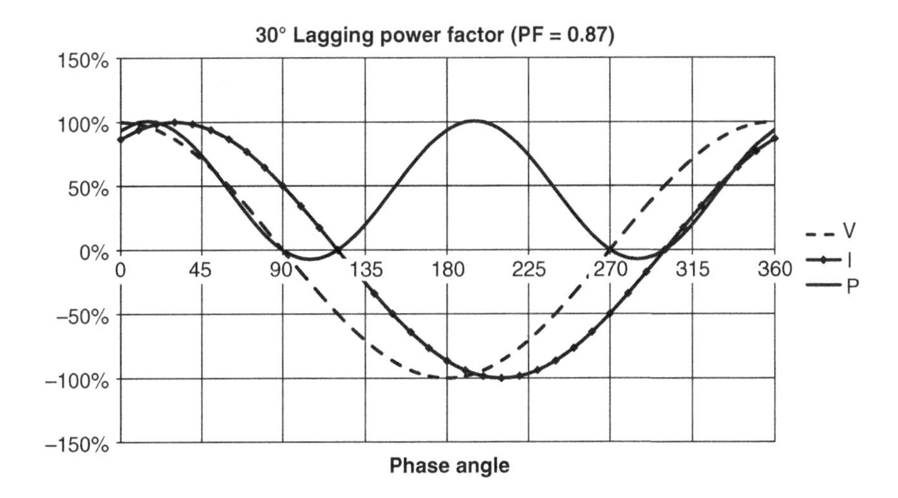

FIGURE 1.3 *Effect of power factor*

the phase shift, the lower the power factor is. In a circuit with purely resistive loads, the power factor is 1.0.

A system with a power factor less than 1 requires higher current from the electric utility to deliver the same real power to the load. This increases the I^2R losses in the transmission system, increasing the cost to the utility to deliver the useful power. Low power factors also require increasing the size of wires and switchgear, resulting in higher costs for transmission and distribution equipment.

Real power is the rate of energy usage. In electrical systems, energy consumption itself is usually measured and expressed in kilowatt-hours (kWh):

Equation 1.9

$$E = P \cdot t$$

where

E = energy, Wh or kWh
P = power, W or kW
t = time, h

Cost is an economic consideration. Consumption is one basis for the cost of energy, but far from the only one. Other factors that affect the cost of energy include time of day and rate of use. It's possible to reduce energy cost without any change in energy consumption. In most systems, the objective is to reduce the cost of energy, so the utility billing structure is just as important as the energy consumption itself.

This leads to consideration of *efficiency*, a term too commonly misused and abused. The concept is simple enough: Efficiency is the ratio of desired output to total input, usually expressed as a percent. For a pump, for example, efficiency is the ratio of "water power" to "shaft power." This is straightforward, and the terms are well defined. If we want to consider energy cost, however, we need to include the losses for the pump's motor, and if a variable speed drive is used, its losses must be accounted for. In other words, *system efficiency* has to be determined, and the pump efficiency alone isn't sufficient to determine the consumption of electricity.

A common term in pumping is "wire to water efficiency." This takes into account not only the efficiency of the pump itself but also the electric motor and variable-frequency drives (VFD):

Equation 1.10

$$\eta_{\mathrm{ww}} = \frac{(Q \cdot h/3960) \cdot 0.746}{P_{\mathrm{e}}} \cdot 100$$

where

η_{ww} = system wire to water efficiency, %
Q = flow rate, gpm
h = pump head, ft
P_{e} = total electrical input power to the pump system, kW

For blowers and compressors, it's more complicated—the blower efficiency used can be referred to as adiabatic, isentropic, polytropic, or isothermal. Further complicating the use of this term is that for any given device the efficiency varies depending on the load and other factors. Manufacturers of pumps and blowers like to quote their best efficiency point (BEP) value or the efficiency at the design point, but in fact, the equipment seldom operates at the exact flow and pressure corresponding to these points.

Even using system efficiency correctly in the analysis may not accurately reflect the actual energy use or cost. If a pump is throttled to reduce flow, the discharge pressure will increase. The pump and motor efficiencies may both decrease. The flow rate will also decrease, and consequently, the pump shaft power demand will also decrease. The drop in efficiency is more than offset by the drop in power, and throttling the pump will reduce the energy consumption and the energy cost.

Misuse of the term efficiency gets worse when we look at the general public. "Energy efficiency" is frequently used as a synonym for "energy conservation." A vehicle's "fuel efficiency" is often expressed in miles per gallon, although technically it should be the ratio of the engine output energy to the energy contained in the fuel consumed.

In this text, we will minimize the use of the term "efficiency" and take pains to clearly define it when it is used.

1.2 SAFETY

Electricity kills.

Some mistakes you only make once—then you're dead.

The first and most important task in development of any system is the integration of safety into design and commissioning procedures. An important task in any design process is verification that proper safety features are included in the design. The last task in system commissioning is training the operators in proper safety procedures and precautions.

Hazards in wastewater aeration systems come in many forms. Air under pressure in piping systems creates hazards. The pressure may be released suddenly and explosively. Most electrical equipment operates at dangerous voltage levels; lockout/tag out procedures, personal protective equipment (PPE), and proper test equipment and procedures must be used. Physical injury can result from rotating equipment starting automatically and unexpectedly. Belt drives can fail suddenly and catastrophically, sending belt fragments flying across a building; adequate guards and precautions must be taken. Confined spaces in wastewater treatment plants contain poisonous and explosive gases, many of them undetectable by humans; proper entry procedures and gas detection devices should be used. Aeration tanks are generally so turbulent that they represent an extreme drowning hazard; proper railings, lifelines, and a companion should be available. Sewage contains a variety of pathogens; appropriate inoculations should be obtained, and washing and disinfection procedures should be followed after exposure. Every construction site has open trenches, and many instrumentation and control system components are installed in elevated locations; observation of surroundings, vigilance, and using appropriate procedures are essential.

This is far from a comprehensive list of hazards and threats found in wastewater treatment facilities. The intent of identifying potential hazards isn't to frighten the reader away from the field. The intent is to ensure that proper precautions and safety procedures are always followed, so the reader can enjoy a long career in the field!

The Occupational Safety and Health Administration (OSHA), the National Electric Code (NEC), and Underwriters Laboratories Inc. (UL) are among the many organizations and standards intended to promote safety in the workplace. Conformance with these standards is often a legal requirement, either because they are directly part of the law or because they have been incorporated into building codes by the local jurisdictions.

Safety procedures can, on occasion, seem burdensome. Sometimes the requirements seem to have been developed by someone who has never performed fieldwork. It doesn't matter. Safety precautions and safe work practices can never be ignored! Safety should be integrated into every part of automation and control system design and implementation.

1.3 THE IMPORTANCE OF AN INTEGRATED APPROACH

The successful operation of an activated sludge process requires knowledge in a variety of technical areas. As with any automation project, the successful aeration system design also involves integration of many aspects of instrumentation and control. Some familiarity with fluid mechanics, biology, chemistry, and electricity is also needed just to maintain minimum process performance. It shouldn't be a surprise that designing an aeration control system demands integration of many engineering disciplines.

There are many terms that are used interchangeably in common use, but should be distinguished during the design process. This text will make the following distinctions:

- *Measurement* is sensing and quantifying a physical parameter or real-world phenomenon.
- *Indication* is the display of measurements in a manner observable and useable by an operator.
- *Monitoring* is tracking and/or recording measurements on a regular or continuous basis.
- *Data acquisition* is a special category of monitoring where a computer is used to indicate and archive monitored measurements.
- *Control* is manipulating devices or equipment to change the performance of the process. Control may be manual or automatic.

- *Equipment protection* is a category of control where the objective is to prevent damage to process equipment by stopping it or shutting it down when operation falls outside defined safe parameters.
- *Process optimization* is a category of control where the intent is to maximize the performance of at least one aspect of the process.

The need to be proficient in instrumentation is an obvious starting point for designing an automatic aeration control system. You can't control what you can't measure. Many of the common instruments for process control—flow, pressure, level, and temperature—are found in aeration control systems. There are also a variety of analytic instruments not commonly found in other industries. For example, mixed liquor dissolved oxygen (DO) measurement is a critical parameter for aeration system control. The instrumentation is integrated into the control strategy as inputs to the control system. This is typical for any process automation, not just aeration systems.

Once the critical process measurements are obtained, the next step is to use them to automatically optimize process performance. The required control logic can be executed in a variety of ways. The simplest devices, such as relay logic and single-loop PID controllers, are still used. They are usually not up to the task of any but the most rudimentary control, and so they are seldom found in aeration control systems. On the opposite end of the complexity scale are full-blown distributed control systems (DCS). These are most often found in large facilities that were early adopters of process automation. Because of the cost and proprietary nature of the programming, use of DCS is decreasing in most wastewater treatment facilities.

The most common system architecture for wastewater treatment process automation is programmable logic controller (PLC) based. Because of the physical separation of process equipment, a number of PLCs are employed with each one dedicated to a process or piece of equipment. The individual PLCs are connected by a communications network. This allows sharing the equipment status and process parameters between PLCs, and coordinating the operation of the separate parts of the process and individual equipment. In most systems, the communications network also links the PLCs to a personal computer (PC). Human machine interface (HMI) or supervisory control and data acquisition (SCADA) software is used to display the measured process parameters, tune the system operation, trend critical data, and log alarms. In addition to the centralized SCADA system, most systems also have localized HMI panels for monitoring the process close to the equipment.

After the control logic is executed, it must act on the process through manipulation of final control elements. The final control elements for implementing the control logic aren't unique to water and wastewater. They include motor starters and variable-frequency drives for pumps, blowers, and general motor control. Valve operators are employed to modulate the flow of process fluids, and motor operators open and close slide gates in process channels.

Integrating the instrumentation, controllers, and final control elements is common to every process automation or monitoring project. Many other aspects of system integration are commonly applied in aeration control. Equipment protection, for example, is typically included in process control systems. This makes economic sense, since the incremental cost of adding equipment protection to the control system is usually insignificant. The primary sensors are generally part of the control strategy, and the incremental sensors needed to provide equipment protection are a small percentage of total cost.

In addition to simple shutdown and equipment protection, incorporating protection and control into the system provides benefits to operators and owners. Alarm messages, trending of critical variables, and time and date stamps provide diagnostic tools for identifying the initial cause of the equipment failure. Including protection and control in a single system also enables start-up of standby machinery immediately after failure of the primary equipment. This ability obviously reduces operator aggravation, but even more importantly, it reduces or eliminates process upsets and maintains treatment levels.

The requirements for equipment protection are fairly straightforward. Allowable bearing temperatures, maximum motor currents, allowable range of flow, and so on are readily understood and usually well documented in the manufacturer's operating manuals. An additional level of integration is required to incorporate the operating characteristics of the process equipment into the system. This is where the requirement for a cross-disciplinary approach to system design begins.

Process optimization requires a more in-depth understanding of the equipment's performance. What is the BEP? How does it vary as the controlled devices are adjusted? How does the change of one parameter affect other aspects of the equipment operation? What is the response time of the equipment to control changes? These questions have implications on the control strategy.

An understanding of the variations in process loadings is necessary for developing the control logic. Unlike many industrial processes, loads to the treatment plant vary continuously and uncontrollably. Both hydraulic and organic loadings are affected by the daily fluctuations in population activity. These *diurnal* variations (Figure 1.4) are part of the challenge in developing an aeration control strategy. There are further

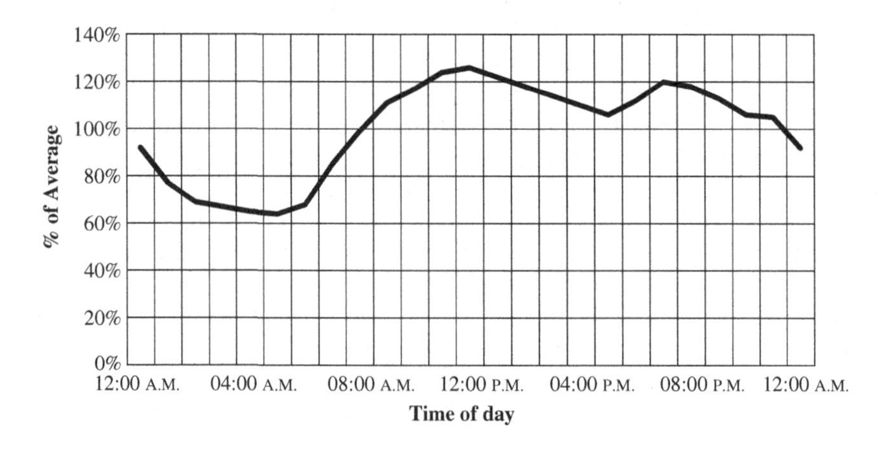

FIGURE 1.4 *Typical diurnal hydraulic load variation*

challenges from rain events, industrial slug loads, plant internal sidestreams, and seasonal temperature changes. These must all be accommodated in the control strategy.

A more demanding consideration is the process and energy implications of modulating process equipment. The operation of the equipment has an impact on the process—otherwise it wouldn't be needed—and on energy cost. Unlike many processes, activated sludge has a variety of complex interactions. Changes in one parameter, such as air flow rate, have a cascading effect. The oxygen transfer efficiency of the diffusers will change, the DO concentration will change, and the biology's rate of metabolism may change. This can affect settleability in the clarifiers, altering the concentration of the microbes in the return sludge to the aeration basins, further modifying process performance. Some of these reactions are very rapid; others may take hours or even days to occur.

1.4 IMPORTANCE OF OPERATOR INVOLVEMENT

There is generally no one who understands the operation of a treatment plant and the process interactions better than the plant operators. Every treatment plant is a unique combination of process equipment, loadings, and performance requirements—no two plants are the same. A "green field" plant, starting from scratch with a blank slate, is a rarity. It is absolutely essential to obtain the input of the operators and their experience with the facility and the process during the design stage and throughout the implementation and commissioning.

A lot of lip service is paid to the concept of operator involvement. Unfortunately, when operators express views that contradict the designer's preconceived ideas or don't match a "standard" design, the operator's opinion is often discounted or ignored. This is almost always a mistake. The plant staff may not be able to articulate the explanation for a potential problem, and they may not know the scientific basis for the phenomenon they observed. However, if an operator makes a statement about the advisability of a particular device or strategy, it behooves the designer to dig until he understands the basis for the statement. Then appropriate measures can be taken to either eliminate the cause or accommodate the occurrence in the system design or the control strategy.

When an operator makes cryptic statements such as "We tried that and it didn't work," it's understandable that the result is frustration for the designer. Remember, however, that the operator is also frustrated to see past mistakes being repeated! The designer's normal response is to ignore the comment and proceed based on the designer's past experience, whether or not this conflicts with the operator's opinion. The designer instead should investigate the details, determine the circumstances, and come to understand the causes and science behind the statement. Then changes can be made to either eliminate the causes or accommodate the realities in the control strategy. This not only improves the performance of the system but also creates a valuable ally in the operations staff! Above all, the designer must avoid clinging to theoretical purity in the face of pragmatic reality.

It is a truism that if an operator is convinced a particular system or device won't work, it won't. This doesn't imply sabotage or negligence by the plant staff! In general wastewater treatment plant operators are well above average workers in their conscientiousness, capability, and professionalism. This statement is simply an acknowledgment of human nature. Commissioning a control system, debugging the program, and maintaining analytic instruments involve a lot of effort. Most treatment plants are minimally staffed, and many tasks compete for the operator's time and attention. It's to be expected that a system forced on an unwilling and skeptical staff won't receive extra effort and attention. The inevitable result is a self-fulfilling prophecy: "It won't work."

The configuration of HMI screens and SCADA systems are particularly important areas for operator involvement. They are the operator's window into the process. The operator must be able to quickly read the screens, turn data into information, and make decisions about how the process and the equipment are performing. Colors, grouping of displays, units of measure, trending, and alarm messages must be configured in ways that make sense to the operator, not necessarily to the designer. Navigation from screen to screen should be as intuitive as possible.

During commissioning, the operator's opinions should be solicited and accommodated. This creates "buy in" by the staff. If they can see that the intent is to provide a tool to make their job easier and the plant run better, they will devote the extra effort and attention needed to make the control system successful.

Above all the designer and start-up personnel must remember that it is the operator's plant, and they will live with the system for years. Acknowledging their needs and expertise isn't just good engineering—it's giving fellow professionals the respect they merit.

1.5 THE BENEFITS OF SUCCESSFUL AERATION PROCESS AUTOMATION

When it's done right, an aeration control system is a win for all stakeholders. The supplier obtains a commercial success by providing a valuable service. The operating staff gets a useful tool for doing their job better and more efficiently. The owner and rate payers have a reduction in expense and operating costs for a necessary community service. The public in general benefits from a better environment. This all speaks to the triple bottom line: public, planet, and profit.

1.5.1 Energy Cost Reduction

The justification for aeration controls is usually based on energy cost reduction. This can have a significant impact on a community's budget. Water treatment and wastewater treatment combined represent 3–4% of the total US energy consumption, with approximately equal amounts for each. For a typical municipality, these two treatment plants consume 30–60% of all the energy used by the municipality.

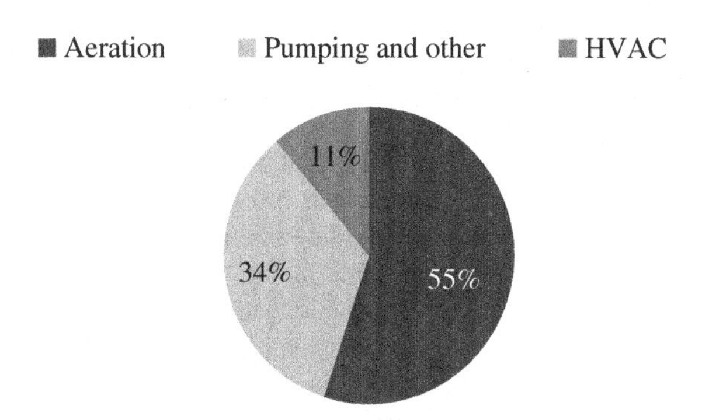

FIGURE 1.5 *Typical WWTP energy use*

Inside the treatment plant, the aeration process is nearly always the largest single energy use, usually accounting for more than half of all the electric power used (Figure 1.5).

Clearly, the aeration process represents the first and best target for energy conservation and energy cost reduction. All indications show that energy cost will continue to escalate, and the potential benefit of aeration control will also increase.

The amount of energy savings obtainable from automation varies significantly from plant to plant. A conservative value for most plants is 25% reduction in energy cost for the aeration process by changing from manual control to automatic aeration system control. Most systems provide a simple payback of 3–5 years for the control system investment.

There is a wide variation in the savings and cost of aeration controls. Plant size, expressed in terms of hydraulic load, makes an obvious difference in the total value of the savings available. Note that small and large treatment plants generally have the same unit processes—large plants just have more and larger equipment. This is fortunate, as there are many more small plants than large ones (see Table 1.1). The

TABLE 1.1 Operating Wastewater Treatment Plants in the United States (EPA, 2004)

Design Flow Range, mgd	Number of Facilities	Total Existing Flow, mgd	% Facilities
0–0.1	6,830	298	41.2%
>0.1–1.0	6,431	2,327	38.8%
>1.0–10.0	2,771	8,766	16.7%
>10.0–100.0	503	13,233	3.0%
>100.0	41	9,033	0.2%
Other	7	–	–
Total	16,583	33,657	100.0%

percentage savings available in a small facility may equal or even exceed the *percentage* savings available in a large plant.

A common metric for judging energy efficiency in wastewater treatment facilities is the amount of energy consumed per million gallons treated, kWh/mil gal. A typical activated sludge treatment plant uses between 2000 and 3000 kWh/mil gal treated, but there is a huge variation in that value. Some facilities use half the energy; others use several times that amount. Plant size is a factor, and in general, large plants use less energy per million gallons treated. Organic loading is another factor that affects the potential savings, and in many cases, this is more significant than hydraulic load. Plants that have a significant load from industrial facilities such as food processors may have organic loading larger than plants treating predominantly domestic wastewater. In these facilities, energy per pound of BOD removed, kWh/lb BOD, is a better metric for evaluating savings.

The level and type of treatment required by the plant's discharge permit changes the energy requirements. Adding nitrification, for example, can easily double the aeration system's power requirement. If the process design includes denitrification, the energy requirement will decrease because some oxygen is recaptured by the denitrification process.

The source and delivery method of the oxygen provided to the aeration process are other significant variables. In diffused aeration systems, the efficiency of the blowers and the oxygen transfer efficiency of the diffusers have an impact. In mechanical aeration systems, the oxygen transfer rate is a key parameter. Solids handling and sidestreams will also change the organic loading to the aeration basin.

Operator process control strategy will also affect energy use by the aeration system. Higher solids inventory (larger microbe population) directly affects the oxygen uptake rate. If the DO concentration in the mixed liquor is higher than necessary, it will result in increased air flow and aeration power. Operating unneeded aeration tanks at minimum air flow increases energy use without benefiting the process. Each of these factors affects others, and all must be integrated into the control system design.

One of the critical steps in developing an aeration control system is projecting the savings in energy costs. This establishes the size and complexity of the system, since the cost of the system must be offset by the savings in a reasonable time. If the savings are overestimated, the project will be a commercial failure, whether or not it is a technical success. Making accurate projections requires a detailed understanding of the process interactions.

There are two methods commonly used to determine if a proposed control system is cost-effective or to determine the best selection among alternates. Simple payback is commonly used for quick analysis, evaluation over short time frames, or for comparing alternative systems. The other, more sophisticated analysis is, based on present value or present worth. It includes consideration of the time value of money, potential earnings from other investments, and inflation. Simple payback is more common in evaluating control system additions to existing facilities, and present worth analysis is more common in evaluating large projects or comparing aspects that are part of a major facility modification.

Equation 1.11

$$P = \frac{C}{S}$$

where

P = simple payback period, the time required for the savings to offset the investment, years
C = net total initial cost for equipment and installation
S = net annual savings (energy and maintenance), \$/year

Equation 1.12

$$\text{NPW} = (\text{PWF} \cdot S) - C$$

Equation 1.13

$$\text{PWF} = \frac{(1+r)^n - 1}{r \cdot (1+r)^n}$$

Equation 1.14

$$r = \frac{R - I}{1 + I}$$

where

NPW = net present worth of investment, \$, NPW > 0 means the investment is economically justified
S = annual savings, \$
C = capital cost for equipment and installation \$,
n = total evaluation period, years
I = annual inflation rate, decimal
R = annual discount rate, decimal, typically the interest rate or rate of return on alternate investments
r = effective annual discount rate, decimal
PWF = present worth factor, dimensionless

Regardless of which method of economic analysis is used, there is a considerable level of uncertainty involved. The savings and equipment costs are typically known with a fair level of confidence, but will vary significantly with the accuracy of the initial assumptions and the analysis methodology. Greater levels of uncertainty are associated with estimated inflation rates and discount rates, as these are influenced by factors in the general economy outside the control of the system designer and which must be projected into the future. Many engineers assume that the more complex present worth analysis will provide more accurate results. The longer the evaluation period, however, the greater the uncertainty associated with this method. The shorter the evaluation period, the smaller the difference in the results obtained from the two methods. For most projects, the simple payback method is recommended.

The ultimate decision on the project design must include intangible factors as well as economic evaluation. These intangibles include customer preferences, reliability, and experience. Engineering judgment should take precedence over blind adherence to predictions of economic effects, regardless of the calculation method.

1.5.2 Treatment Performance

From an operator's perspective, the most important benefit of a properly functioning aeration control system may well be improved process performance. They recognize the advantages of reduced energy cost, of course, but process performance is always the first priority. The operating reductions in energy expense justify the cost of the system, but the process improvement and stability provide a substantial impetus for adoption.

One guarantee of process failure is operating the aeration basin *oxygen limited*—not providing sufficient oxygen to the mixed liquor to support the metabolism of the organic load by the microorganisms. Even more significant, if the process operates at an oxygen deficit for an extended period the microbe population will change, with the population of less desirable microorganisms increasing. This has long-term negative implications for treatment performance. Proper aeration control can prevent this.

There are other process benefits derived from a successful aeration control system. Marginal DO concentrations in the mixed liquor may result in growth of undesirable filamentous organisms. Even if the required BOD removal or nitrification occurs, these filamentous organisms may not settle properly in the secondary clarifier. This can result in a permit violation for excess total suspended solids (TSS) in the effluent, may inhibit solids dewatering and may interfere with solids digestion. The same deleterious effect may result from excess aeration breaking up the floc and reducing settling or from filamentous organisms that grow after long-term operation at very high DO concentrations. The sludge volume index (SVI) is a measure of solids settleability, and experience has shown that SVI improves when DO control is operating.

The cost reduction associated with process improvement is difficult to quantify and using it in economic justification is not recommended. Furthermore, since process considerations override cost considerations, the impact of aeration control on treatment may be sufficient justification in itself, regardless of savings. This is particularly true in advanced processes, where traditional manual control methods may prove inadequate in providing the required level of treatment.

1.5.3 Improved Equipment Life

A wastewater treatment plant represents a significant capital expense, and is often one of the largest investments for a municipality. A large part of this investment is for structures such as buildings and tanks, but a significant part of the investment is for equipment. Ensuring that the process equipment reaches the 20-year design life is an important part of the operators' responsibility. Machinery preventive maintenance and repair is a major expense and manpower requirement. Improvement in equipment life and reduction in maintenance expenses are benefits of most automation systems, and this is particularly true for aeration control.

The equipment protection functions are responsible for much of the improvement. By monitoring equipment operation and performance, the system can shut down machinery as soon as unsafe operating conditions are detected. This eliminates "run to failure" operation and reduces the scope and complexity of repairs. The early warning of operational problems will also provide diagnostic information to simplify and accelerate troubleshooting procedures.

The process equipment in a treatment plant must run 24 hours a day, 7 days a week, 365 days per year. It is simply not possible to stop operations because of a failure. Publicly owned treatment works are consequently required to have standby equipment for all critical functions. In order to keep all pieces of equipment operational, it must be run periodically. A machine that sits idle for months or years and is then started will not run long before failure occurs. By simplifying alternation of the running machinery or by automatic alternation, the aeration control system can provide uniform operating time for all of the process equipment.

Clearly, aeration control systems provide a great many benefits. Some of these are quantifiable as reductions in operating cost, and some are less easily defined. Obtaining these benefits requires integration of several areas of technology and a thorough understanding of the process and the process machinery. The analysis and design procedures that follow provide the guidance needed to successfully implement aeration control.

EXAMPLE PROBLEMS

Problem 1.1

(a) Plant staff reports a pump motor current measurement of 223 A. From the plant electric bill, it is determined that the nominal plant supply voltage is 480 VAC three phase and the average plant power factor is 83%. What is the apparent power of the motor? What is the estimated power consumption?

(b) Subsequent measurements with a power meter at the motor terminals show the actual motor voltage to be 469 V, and the measured motor power factor is 87.8%. What is the apparent and actual motor power with the revised data?

Problem 1.2

The influent lift station at a treatment plant has a design capacity of 8 mgd. The station has three identical pumps with characteristic curve shown in Figure 1.6. The

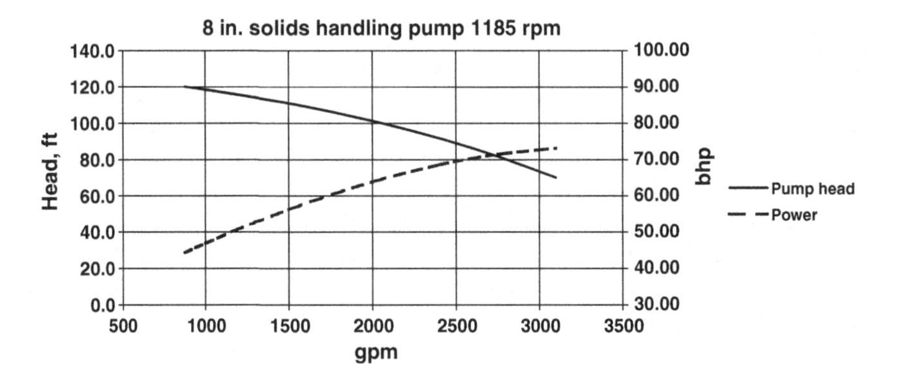

FIGURE 1.6 *Problem 1.2 pump curve*

average dry weather flow for the plant is 2 mgd, the reported discharge head is 80 ft, and the power consumption is measured as 58 kW using a true power meter at the motor control center.

(a) What is the operating flow rate of the pump?
(b) What is the corresponding pump shaft power?
(c) What is the wire to water efficiency?
(d) What is the pump efficiency?

Problem 1.3

What is the operating cost of the pump if the cost of electricity is $0.08/kWh?

Problem 1.4

It is proposed that 75 hp variable-frequency drives be provided for the pump system in Problem 1.2. The installed cost of each drive is estimated at $7500 each.

(a) Is this cost-effective if the billing is based on energy consumption only at $0.08/kWh?
(b) Is this cost-effective if the billing is based on energy consumption at $0.06/kWh and a demand charge of $9.00/kW?

Problem 1.5

List at least three benefits obtained from automating the aeration process.

Chapter 2

Initial System Assessment

As the old saying goes, "If you don't know where you are going any path will get you there."

The first step to any successful Energy Conservation Measure (ECM) program in an existing installation, including aeration controls, is establishing baseline data. This defines the starting point for energy and process performance. The second step is to benchmark the performance of the plant based on the process equipment on site and at similar installations. This establishes reasonable expectations for the available savings. The third step is to define goals for the project based on the baseline performance compared to the benchmark. The initial assessment should include these three steps and a report of their results to the plant staff.

The initial assessment is often referred to as an "energy audit." This term is too limited, since there is a great deal more involved than just examining the energy consumption of the facility. The assessment must also include the treatment process and the treatment equipment, since a successful system must reduce energy cost while maintaining adequate treatment levels. Changes in energy cost and process performance are integrally tied together—consideration of one must include the other. The initial assessment should also include a preliminary analysis of potential ECMs and the report to the owner should include recommendations on ECM implementation (see Figure 2.1).

The goals defined in the report should include specific energy cost reduction targets, since this is critical in developing the economic justification for any proposed system changes. The report should also include the limiting economic criteria such as

Aeration Control System Design: A Practical Guide to Energy and Process Optimization,
First Edition. Thomas E. Jenkins.

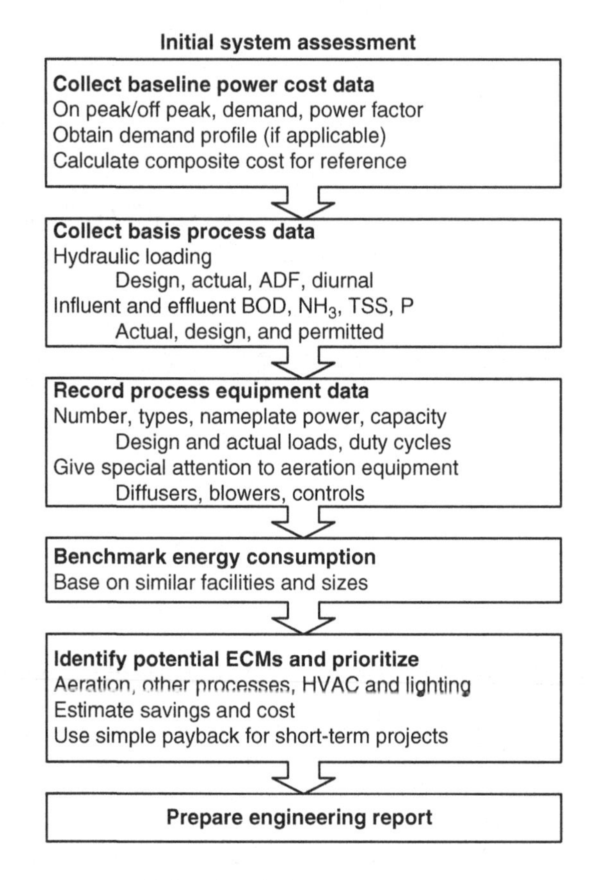

FIGURE 2.1 *Initial assessment steps*

maximum payback period. Needed process improvements should be included, since these can provide justification of expenditures if energy cost reductions alone don't provide sufficient economic benefit. If some ECMs can be implemented by simple operational changes without capital expenditure, they should be identified and the anticipated energy cost reductions estimated.

If the control system design is for a wastewater treatment plant that is under construction and therefore doesn't have operating data, the sequence is modified slightly. This is also the case if the changes to an existing facility are part of a significant upgrade, so that the existing operating data is not comparable to the final process configuration. In both cases, baseline energy data can be examined and recorded, but aren't necessarily useful in the economic justification of control system cost. Baseline process data, particularly influent characterization, is important as the basis of process calculations to determine operating requirements.

Benchmarking is still useful for new construction, since it provides at least a basis for establishing the minimum level of expected performance—a reality check if you will. The energy cost savings and process performance will be calculated based on operation with and without the proposed system. In that case either manual operation or a less sophisticated automatic control system serves as the baseline. Once the baseline and benchmark are established, setting goals will include the same considerations as when designing ECMs for an existing plant.

A concept that receives a considerable amount of attention is the "triple bottom line." This refers to the evaluation of projects based on their impact on "people, planet, and profit" rather than relying on traditional evaluations of economic impact alone. The evaluation of triple bottom line includes the social impact (people) and sustainability implications such as greenhouse gases and resource depletion (planet). The evaluation of profit embodies the traditional aspects of economic impact, cost, savings, and payback.

This text will concentrate on the economic aspects of aeration control system evaluation. That isn't intended to imply that the other factors aren't important. However, they are to some degree intangible and are certainly hard to quantify. The author's expertise doesn't include evaluation of the social factors. Ultimately the go/no-go decision on any given project will generally rely on the economic evaluation. Consideration of social affects and sustainability are unlikely to be adequate on their own to justify a significant expenditure for upgrading a control system. A system that provides justification through the energy cost savings will be implemented unless the negative social and sustainability evaluation demonstrates severe negative consequences—an unlikely occurrence. If two alternate proposals are both similar in cost-effectiveness, it is possible that differences in social impact or sustainability can influence which alternate is selected.

One exception to the above is a case where the implementation of automatic controls is expected to take a plant from violation of permit requirements to compliance. To some extent this is an economic consideration, since fines, penalties, and mandated capital improvement expenditures will result from consistent or chronic permit violation. The primary objective of any facility is meeting treatment requirements, so the economic evaluation is of secondary importance if the process is failing. If automatic controls will improve effluent quality enough to ensure permit compliance, then any benefit in reduced operating cost is inconsequential. At most the economic analysis will consist of comparison of capital or present worth costs of alternative systems.

Pumping is not the major focus of the aeration control system, and pumping is generally a much smaller portion of treatment energy than aeration. However, it is often possible to implement pumping-related ECMs at a small incremental cost, and the inclusion of pumping in the initial assessment requires very little additional effort. A basic pumping evaluation and pump-related ECMs should generally be included as part of the plant aeration system evaluation. Influent and effluent pumps generally have higher horsepower requirements than return activated sludge (RAS) and waste activated sludge (WAS) pumps, but all pumps should be evaluated.

Many wastewater treatment facilities also include remote pump stations as part of the conveyance system. Pump ECMs are applicable to in-plant pumps and pump stations, but remote pump stations usually have less complex billing rates and the economic evaluation must consider this. The available literature, which has ample material on techniques for reducing pumping energy costs, should be consulted for assessment and analysis.

Calculation of potential savings for some ECMs is straightforward and not time-consuming. For example, an operational change that shifts a load to off-peak times is simply the energy used times the rate difference plus the power times the demand charge. Other ECMs, such as replacing aeration blowers, are much more complex and require expertise in process, control systems, and mechanical equipment.

2.1 DEFINE CURRENT OPERATIONS

Establishing a baseline requires definition of current operating conditions. On the face of it, this seems like a fairly straightforward task. It is, instead, often one that requires a lot of effort, judgment, and assumptions. The information available varies from one installation to the next, but it is rare to have all the pertinent data readily available.

A recent study indicates that only 10% of treatment plant operators actually see their energy bills regularly. Of those who do see their energy bills, fewer than 10% of them feel that they understand them! Getting accurate energy information is clearly a challenge in the initial assessment.

The availability of information on the process operation is somewhat better, but still not ideal. Influent organic and hydraulic loads are almost always measured and reported to the appropriate regulatory authority, so this data is readily available. The final effluent permit defines final pollutant levels and is also included in reports, so this data is also available. Unfortunately, this data alone is insufficient to provide accurate estimates of the loading to the aeration system and the energy performance of the aeration process.

One example of needed information that may not be available is the impact of nitrification. Many smaller facilities are only required to meet discharge limits on BOD and total suspended solids (TSS). Concentrations of ammonia in the influent and nitrate in the effluent are not measured, therefore this information is not available. If the plant is hydraulically under loaded, though, nitrification is likely to occur. This can double the oxygen demand and the air flow rate needed to maintain DO concentrations. If the operators aren't monitoring nitrification, the accuracy of the assessment is impaired and the impact of control system improvements can't be accurately estimated.

Conventional energy audit procedures will not give good results in wastewater treatment applications. A simple walk-through, counting motors and adding horse power, won't suffice. Unlike heating ventilating and air conditioning (HVAC) or general industrial applications, wastewater treatment plants are legally required to have standby equipment. Determining the actual duty cycle of equipment is required. The process equipment is usually designed for anticipated growth over a 20-year design life, and is sized to ensure permit compliance at worst-case operating conditions. The Environmental Protection Agency (EPA) estimates that most municipal treatment plants are actually operating at one third of the design capacity.

Actual output and power draw may be much lower than rated capacity and motor nameplate rating.

The excess capacity presents both a challenge and an opportunity. Turndown limits on equipment may prevent matching supply to demand. On the other hand, taking some equipment out of service can lower energy use with no capital investment. Replacing oversized process equipment such as blowers with smaller units sized to meet current loadings can result in dramatic savings and excellent payback.

The items to be included in establishing the baseline energy expense include

- energy cost,
- energy consumption patterns,
- influent and effluent process parameters, and
- treatment performance.

It goes without saying that all of the data collected should be recorded in a consistent and retrievable manner. Extensive note-taking is encouraged. Copies of actual energy bills, discharge monitoring reports (DMR), and process data should be obtained. Measurements of pipe sizes are useful. Copies of "as-built" drawings of the facilities are very valuable. The sizes and quantities of tanks and major equipment should be recorded, along with the number of each that are in and out of service. Photographs of the facility and equipment are extremely useful.

The plant altitude should be obtained and noted. This provides the average site barometric pressure, which is necessary for subsequent process and aeration system calculations. The altitude is usually available on the plant's as-built drawings. If these are not accessible, the data can be obtained from organizations such as American Society of Heating, Refrigerating and Air Conditioning Engineers (ASH-RAE) or National Oceanic and Atmospheric Administration (NOAA).

2.1.1 Energy Cost

The objective of aeration control is to optimize the cost of energy while improving or maintaining process performance. An understanding of the utility cost structure is necessary for evaluating the impact of the system on operating cost. One of the first steps in initial assessment and establishing the baseline is obtaining the actual utility bills.

In most cases, the aeration equipment is powered by electricity. Some blower and pumping equipment uses internal combustion engines, gas turbines, or steam turbines, but these are the exception. Alternate energy sources are usually powered by digester gas or natural gas. In both cases, the cost structure of the energy source is straightforward, and primarily a function of the energy consumption alone.

The distinction between energy consumption and energy cost should always be borne in mind. High consumption means high cost, of course, but this is generally not the only consideration, and may not even be the most important one! Electric bills are generally complex, reflecting the complex nature of power generation and usage patterns. The first hurdle to overcome in the assessment process is determining the cost structure of the energy used.

The electric utility uses billing to recover the cost of generation and transmission. Factors that increase the utility's cost result in higher charges. The utilities also use billing to influence customers to change their patterns of energy use in ways that reduce the utility's cost or maximize power availability. In the United States, most utilities are investor-owned regulated monopolies. Rates and profit levels are generally overseen and regulated by governmental agencies.

An often-overlooked resource in reducing energy cost is the electric company's own engineers or customer service representatives. They are knowledgeable about the various rate structures and are able to explain the impact of various billing components on a treatment plant's energy cost. The utility representatives have access to detailed and accurate usage data for each customer that can be invaluable in establishing baseline data and identifying cost reduction opportunities. The representatives are anxious to help customers reduce power demand. They are also the first point of contact in establishing the availability of incentives or rebates for implementing ECMs.

It may seem counterintuitive for the utility to assist customers in reducing their energy costs. The obvious question is "Why would they spend money so they can make less money themselves?" There are several reasons for this. First, it is more efficient and profitable for the utility to generate power at a uniform rate and eliminate large fluctuations in consumption. This is the goal of many "demand side management" programs established by various utilities. By providing power at a uniform rate, the transmission losses are minimized and use of lower efficiency peaking generators is minimized.

Second, adding generation and transmission capacity is very costly and time-consuming. The permitting process can be cumbersome. Public resistance can delay or kill projects regardless of economic or technical justifications. The utility would much prefer to maximize the use of existing investments.

Finally, many states have legislation or regulatory requirements that mandate energy conservation programs. These are seen as a benefit to society at large, and often have wide political support. The mandates often include a charge to customers that is designated specifically for conservation and so-called green energy initiatives. This provides the pool of capital used to finance ECMs.

The utility's engineers may not be familiar with wastewater treatment processes and potential cost reductions related to treatment systems and equipment. They can clarify details of the plant's current rate structure and identify alternate rate structures that would reduce the wastewater treatment plant's costs. They are also able to identify aspects of the current electric energy usage pattern that are most significant and offer potential for cost reductions.

There are several components of electric billing that are common to virtually all wastewater plants. The items in a simple rate structure include

- energy consumption charges (typically $/kWh) for energy used,
- facility charges, typically a fixed monthly fee, and

- miscellaneous charges for
  - fuel cost adjustment,
  - energy conservation funding, and
  - taxes.

The basis of simple consumption based billing is obvious—the utility needs to recoup the cost of generation and capital investment and return a profit to the shareholders. Fixed fees such as facility charges also cover the cost of maintaining accounts and billing. Fuel surcharges offset the variability in the cost of fuel used to generate the electricity. Many jurisdictions have implemented mandatory conservation programs financed by the utilities, and the cost of these programs is funded by a charge added to the energy bills.

For lift stations and very small facilities, these may be the only charges for electric energy. More complex billing requires more complex (and expensive) power meters so that the utility can track the energy consumption more accurately. In general, as the total monthly power consumption for a facility increases, the complexity of the billing structure also increases. This allows the electric utility to reflect the cost of generation and distribution in the rates. The more complex rates generally include all of the charges in a simple rate structure, but may add the following:

- Time of day rates for energy consumption
  - on peak (highest rates for usage during times of peak demand)
    - typically 9:00 A.M. to 9:00 P.M. weekdays
  - off peak (lower rates for all times not included in on peak)
- Demand charges (charges for the highest power usage during a month)
  - $/kW
  - usually based on the highest averages over a sliding window (typically 15 minutes)
- Power factor charge
  - a penalty assessed if the power factor is less than a minimum value (typically 85%)
  - usually billed against the peak demand use
- Ratchet charge
  - the demand charge is billed for the highest peak over a 12 month period.

A sample of energy charges is shown in Figure 2.2. Note that charges vary widely from one utility to another.

The demand for electric power varies with the time of day and day of the week. It is impractical to store electric power, so it must be used as it is generated. During periods of high demand, typically daytime on weekdays, so-called "peaking generators" are used to meet the highest demand for power. These peaking generators use natural gas or oil, and are less efficient than the large generators used for meeting

Charge Type	Usage	Units	Rate	Charge	% of Total
Service				$500.00	2
On-peak energy	190,000	kWh	$0.06	$11,400.00	33
Off-peak energy	320,000	kWh	$0.03	$9,600.00	28
Demand (summer)	775	kW	$15.00	$11,625.00	34
Taxes			3%	$993.75	3
Total				$34,118.75	

FIGURE 2.2 *Sample energy rates*

the base electric load. Transmission lines and transformers must be sized to meet the highest anticipated current requirement, so high peak loads add to capital expense for the utility.

To offset the higher cost, and to encourage users to shift power use to times of lower demand, the utilities charge higher rates for power used during the times of peak demand. This is the basis for higher "on-peak" rates, peak demand charges, and ratchet charges.

Demand is usually calculated as the highest average power (kW) used over a 15 minute interval during the billing period. The power used is typically averaged over a "sliding window" so that the highest usage for any occurrence during the billing period is captured. Many utilities also implement a "ratchet charge" on demand billing. The highest peak demand charge is maintained for the month of occurrence and for 11 successive months, unless a still higher demand is incurred.

Power factor charges offset the loss of transmission efficiency when a low power factor results in high I^2R losses. The power factor penalty is usually based on peak demand. Charges usually result when the plant's power factor falls below a minimum value, typically 85%.

It is sometimes useful to use a "composite rate" for power cost during preliminary ECM analysis and comparisons. This isn't an actual charge from the utility, but rather a convenient approximation of the average energy cost. It is obtained by dividing the total of all charges for the month by the total energy used in the month, resulting in an average cost per kilowatt-hour of energy. Time of day, demand, and other charges are lumped together in a single value. The composite rate should not be used for final detailed savings analysis, however, since the true savings will depend on the actual billing structure for the wastewater treatment plant.

Because of the complexity of energy billing, part of the initial assessment should include contact with a utility representative. This will generally require the participation of plant management or an employee of the owner who is authorized to access financial information. The utility representative can provide copies of actual energy bills so that accurate evaluations can be made.

Another important service that the electric utility's representative can provide is a comprehensive review of the current rate structure. In many cases, plants experience changes in operation or differences between projected and actual electric power usage. Most utilities offer a variety of rate structures, generally based on total energy consumption and/or peak demand. The utility representative can review the rate structure to identify potential cost savings that may be available simply by changing rate structures. For example, many utilities offer reduced rates if the treatment plant will shed loads to a minimal level on request. This can usually be implemented with little or no impact on process performance. Generally the plant is allowed to operate at a minimum load to cover critical process equipment, and the maximum duration of the shedding is identified in the agreement.

The initial assessment should determine which aspect of the utility bill is most significant. For a typical utility billing structure, there are 60 hours per week of on-peak rates and 108 hours per week of off-peak rates, but the energy charges for the two rates are usually nearly equal. The demand charges are approximately equal to the on-peak and the off-peak energy charges. This is a fairly common occurrence, and a typical municipal treatment facility's energy cost is composed of roughly one third on-peak energy charges, one third off-peak energy charges, and one third demand charges. If the initial assessment shows significant departure from these proportions, the cause should be investigated.

2.1.2 Energy Consumption Patterns

The literature for energy conservation in wastewater treatment plants recommends that lighting and HVAC energy consumption be examined for possible ECMs. Because building management loads are prevalent in almost any commercial or municipal activity, a lot of energy auditors are familiar with these loads. There are a lot of off-the-shelf ECMs available, so improvements in this area may be very cost-effective. Data on lighting and HVAC energy should always be collected in the initial assessment and analyzed in the report.

Another area that receives significant attention in the literature is pumping energy. This is usually the second largest use of energy in a treatment plant, and techniques for analyzing and improving pump efficiency are also widely available. It is very important to include duty cycle information in the pump data collected. Some pumps in wastewater treatment plants operate 24/7, while some only operate intermittently. It is usually most cost-effective to concentrate on the pumps with the most operating time in order to maximize payback. Similarly, data on actual pressures and flow should be obtained so that realistic savings estimates can be calculated. Design values are usually based on a worst-case situation and are generally encountered rarely in actual operation.

Treatment plants with anaerobic digestion for sludge produce digester gas as a by-product. Digester gas is typically composed of 60–70% methane, with the balance primarily CO_2. Although in many plants methane is simply flared, it can be used for heating or for fueling generators. Digester gas also contains traces of some troublesome compounds such as hydrogen sulfide (H_2S) and siloxanes. These must usually be removed before using the digester gas as fuel. The high cost of the generation equipment coupled with the cost of cleaning the gas prior to use may make beneficial use of the gas uneconomical. Nonetheless, in large facilities, the initial assessment should look at the potential for converting the digester gas to useful energy.

The most fruitful area for energy conservation is almost always secondary treatment in general and aeration in particular. First, this is usually the most energy intensive process. Second, the potential savings are significant, generally in the 25–50% range. The capital investment for many aeration ECMs may also be high, which obviously affects payback. Nonetheless, it is appropriate to devote a high proportion of the initial survey to collecting complete and accurate data on the operation of the aeration system.

Energy consumption for wastewater treatment is constantly fluctuating as a direct result of process requirements. In general, the process loading is outside the capability of the operator to control. These variations in the process come from a variety of sources and across a wide range of time periods:

- diurnal variations in plant influent hydraulic loads,
- diurnal variations in influent organic loads due to both hydraulic fluctuations and changes in waste concentrations,
- slug loads from industrial facilities discharging into the sanitary sewers,
- slug loads from internal sidestreams such as digesters and belt presses,
- storm events flushing high concentrations of organic contaminants from the sewers,
- storm events producing high flows of very dilute wastewater,
- weekend decreases in hydraulic and organic loads,
- seasonal fluctuations in wastewater temperatures and loading,
- seasonal variations in discharge permit requirements.

Many individuals with experience in energy conservation in other fields are stymied by the challenges of implementing ECMs in wastewater treatment plants. Controlling the aeration system is particularly difficult because of the complexity of the process and equipment.

Building HVAC energy requirements correlate with heating and cooling degree days. Wastewater treatment energy in general and aeration in particular don't have the same level of correlation between energy use and process loads. The loading is more complex and involves more factors than building energy management.

Establishing the fluctuations in process loads often represents the greatest challenge to establishing the baseline for evaluation. Operators have a tendency to respond to process fluctuations based on experience, without determining the root cause of the process change. By design, wastewater treatment processes tolerate normal fluctuations without violating permit requirements. Internal sidestreams are not normally monitored or recorded, although they can significantly affect energy demands. The only data recorded may be the information required to be reported by regulatory agencies, particularly in small plants.

The widespread implementation of SCADA systems has improved record keeping to some extent, but challenges remain. In many cases data displayed is not archived, or is not archived in a convenient format. Finding a "typical" day's trend in a year's archived data can be problematic. If seasonal variations are significant, it may be necessary to find a typical day's trends for each season.

SCADA operations are more often than not dictated by process control considerations. The process variables needed for energy evaluations are not necessarily those needed for process control. For example, final effluent DO is commonly monitored, but individual aeration basin DO may not be. Air flows and pressure may not be recorded. It is uncommon to have separate power metering on even the largest process equipment.

Rain events can increase or decrease the demands on the process. Rain increases the hydraulic load to the treatment plant through infiltration and inflow (I & I). Infiltration is groundwater that enters the sewer system through leaks in pipes or manholes. Because rain increases the amount and level of groundwater, infiltration increases during and after a rainstorm. Inflow is rainwater that enters the sewer system from roof gutters, sump pumps, and improper connections between drainage and sanitary sewer systems. These sources also increase hydraulic load during rain events. In older communities, the problem of increased hydraulic load is further exacerbated by combined sewer systems, where sanitary and storm waters are transported to the treatment plant in a common system.

The impact of rain events on organic loading is more complex. In the early period of the event, the "first flush" occurs. This refers to the increased hydraulic load and higher than normal water velocity flushing accumulated debris and organic material that has settled in the sewer. This can create an organic load to the process much higher than normal. However, once the accumulation is washed through the sewers, the I & I results in dilution of the influent wastewater. The rain and groundwater may also have a high concentration of dissolved oxygen, further reducing the aeration system demand.

The high hydraulic loading may cause other process problems for the treatment plant, particularly with solids retention. The high flows reduce the settling capability of the clarifiers, and may result in diluting the RAS or washing the microbes out of the plant. The lost of microbiology reduces treatment efficiency during the storm and after the high flows end. Many operators change operating procedures during storms to minimize the long-term disruption of the process. These changes may include bypassing peak flows around the secondary treatment process or taking measures to retain microbial solids.

In assessing the operation of the aeration system, it is important to determine the frequency and severity of storm events and their impact on the treatment plant operation. The necessary response to these events may limit the ability to take excess operating capacity offline, and it may require modification of the normal control strategy to incorporate storm response measures.

Equalization (EQ) basins are a method used to attenuate the fluctuations in plant loads. This has obvious benefits for limiting peak loads during storm events. Experience has shown that equalization can also reduce energy costs during routine operations. Diverting diurnal flow peaks to the equalization basin and then pumping them to the treatment process during off-peak hours can significantly reduce power cost by shifting the process demand to periods of lower power rates.

Equalization does require some additional energy, which will offset some of the savings. Aeration is generally required to maintain mixing of the wastewater, to keep solids in suspension, and to eliminate septicity and odors. The air for the EQ basin is often tapped from the main aeration blower system. Because of the lower static head and pressure required in the EQ basin, it may be more economical to have a dedicated blower system. The contents of the EQ basin must be pumped to the treatment processes, which adds to the total energy requirement. However, because the pumping can take place during less expensive off-peak periods, the cost of the energy is lower. The initial assessment must include aeration and EQ basins energy requirements to accurately assess cost impacts.

If the treatment plant is subject to peak demand charges, a demand profile for several typical months or weeks should be obtained. The demand profile can be used both to evaluate the impact of ECMs on energy cost and to provide guidance for potential ECMs. One disadvantage that treatment facilities have in managing demand is that the highest loading to the process usually coincides with the highest demand periods for electric power. However, observation of the demand profile can identify events, such as alternation of equipment, which may be shifted to lower demand periods to reduce cost.

Inrush current from starting process equipment can add to peak demand charges. The plant survey should include the operator's pattern of motor starting and equipment alternation. It is possible to reduce demand charges simply by shifting some equipment starting to off-peak times. This usually isn't a major operational issue, since the first shift at most facilities begins well before the time when higher on-peak rates begin. Automatic alternation that occurs off-peak is another way to reduce demand charges.

The impact of motor starting on demand is generally greatly exaggerated. The current inrush lasts at most for a few seconds. Averaged over the typical 15 minute demand period, inrush doesn't materially increase the demand charge. In reality it is running two pieces of equipment when only one is needed that pushes the peak demand charge higher. It's common practice for operators to start the lag equipment, let it run, and only turn off the previous lead equipment after a warm-up period. This adds to energy cost without improving process performance. Operating procedures should be examined and, if possible, modified to stop previous lead equipment before or immediately after starting the lag unit so only one is operating at a time. This has a negligible effect on process, but will dramatically reduce demand charges.

Submetering refers to providing separate power meters for major pieces of process equipment such as aeration blowers or raw wastewater pumps. The data on power consumption can identify patterns of usage as well as the equipment with the highest power usage. This data is particularly useful if it is transmitted to a SCADA system and available as trends showing daily or weekly usage. In addition to providing valuable data for the initial assessment, submetering is useful for measurement and verification of savings after ECM implementation.

Because aeration is the single largest energy consumption in most facilities, particular attention should be paid to this area. Aeration electric power usage is certainly valuable data, and this is a critical area for submetering. If the treatment plant is not equipped with permanent submetering, portable recording power measurement equipment may be installed on a temporary basis to monitor power over a typical week or month. "Snapshot" data at a single point in time is not nearly as useful as extended data and trends.

It may be possible to determine the patterns of usage from air flow and pressure data for the blowers and aeration basins. From this information and the blower manufacturer's performance curves, it is possible to calculate power consumption. If the performance curves are not available, the power may be estimated from Equation 2.1:

Equation 2.1

$$kW_b = 0.01151 \cdot \frac{Q_i \cdot p_i \cdot X}{\eta_b \cdot \eta_m \cdot \eta_{vfd}}$$

where

kW_b = blower electrical power, kilowatts
Q_i = blower inlet flow rate, inlet cubic feet per minute (ICFM)
p_i = blower inlet pressure, psia
$\eta_{b,m,vfd}$ = blower, motor, and VFD efficiency, decimal
X = adiabatic factor, dimensionless (see Equation 2.2)

Equation 2.2

$$X = \left(\frac{p_d}{p_i}\right)^{0.283} - 1$$

where

p_d = blower discharge pressure, psia

Blower efficiency is not a constant. It varies with the blower design, inlet conditions, the actual flow and pressure relative to those at the best efficiency point (BEP), and the flow control method used. Similarly motor efficiency varies with motor type, percentage of full load, and actual voltage. For quick approximations in the absence of more precise data, the following may be used:

- motor efficiency: 0.93
- positive displacement blower efficiency: 0.64
- multistage centrifugal blower efficiency (throttled): 0.67
- multistage centrifugal blower efficiency (variable speed): 0.70

- single-stage centrifugal blower efficiency (inlet guide vane): 0.73
- single-stage centrifugal blower efficiency (variable speed): 0.75

Actual measurements of system discharge pressure should be used. It is common for the design pressure and system nominal pressure to exceed actual operating conditions by 1 psig or more. This leads to overestimating power consumption if the design pressure is used.

Measurements of DO at each aeration basin should be recorded along with the actual location of the DO probe within the basin. Many plants have DO transmitters connected to their SCADA system, and trends of this data can be very useful for identifying excess aeration and potential savings. If this data isn't available then manual readings taken at several times of the day, including nighttime and early morning, can be used instead. As with blower power, trends over time are more useful than a single snapshot.

The location of the probe relative to the hydraulic flow is important. If the tank is configured as a complete mix system, the location should be noted but may not be critical unless it is immediately adjacent to an influent or effluent weir. In the more common plug flow configuration, the location is always critical. For example, it is common to install a single DO transmitter in each basin close to the effluent weir. This is not typically the location that provides the most accurate information about the relationship between air flow supplied and the process demand. Because the hydraulic retention time generally exceeds the process requirement, the biological process is complete well before the basin effluent. As a result the effluent DO concentration is elevated, and may be consistently close to saturation. The DO concentration in other basins locations may be fluctuating as process load changes but air flow is held constant.

If possible a DO profile along the length of the basin should be obtained, with DO readings taken at 0, 25, 50, 75, and 100% of the distance from influent to effluent. This data should be collected at several times during the day. The concentrations at peak flow and average flow are particularly useful.

In addition to the aeration basins for secondary treatment and EQ basins, there are a many other aeration applications in most treatment plants. These include post aeration, channel aeration, aerobic digesters, and sludge holding tanks. The source of this air, the flow rates, and the actual pressure required at worst case and average conditions should be noted. Diurnal variations in demand and water levels should also be noted.

Digester decanting is an important sidestream in many plants, since the supernatant may contain very high concentrations of BOD, ammonia, and hydrogen sulfide. This can exert a sudden and significant increase in the process demand for aeration. Both aerobic and anaerobic supernatant increase process loads. The schedule for decanting, the volume of supernatant, and concentrations of contaminants should be documented. The method of decanting and the point where the sidestream reenters the process should also be noted.

Sludge processing results in a variety of other sidestreams. Dewatering equipment such as centrifuges and belt presses produce sidestreams that may be similar in composition and impact to digester supernatant. These sidestreams may be significant in determining fluctuations in air demand in the secondary treatment system.

Many plants receive waste trucked in from septic tanks (septage), holding tanks, or landfill leachate. Landfill leachate is generated principally from rainwater that passes through the waste within the landfill. The trucked waste consists of different organic and inorganic compounds that may be either dissolved or suspended. These wastes seldom impact hydraulic loading to the facility, but they may constitute a large portion of the organic load that must be treated. These loads usually vary seasonally and diurnally. The assessment must include these loads and their impact.

2.1.3 Influent and Effluent Process Parameters

Since the primary purpose of the plant is to treat wastewater and remove the pollutants, any assessment must include the process performance. There are two sets of parameters: influent loading to the plant and levels of pollutants in the plant effluent. Both may be further categorized into two sets. First are design levels—those regulated by the appropriate jurisdiction or authority and used as the basis of process design. Second are actual levels—those loads currently being treated by the plant.

There is typically a wide disparity between design influent and actual loads. This is partially due to the plant design being based on projected future flows. Most plants have a design life of 20 years or more, and the design capacity is determined based on estimated population growth. There is also an inherent conservatism in the design process, and equipment and structures are sized to accommodate worst-case conditions. The EPA estimates that most municipal treatment plants operate at one third of their design capacity.

Industrial treatment plants are much less likely to have excess capacity. Wastewater treatment in most industrial facilities is considered an overhead expense, and the investment in equipment is minimized as much as possible. Furthermore, many industrial facilities discharge into municipal sewer systems rather than directly into receiving waters. They are generally charged by the municipal facility for the amount of load they place on the discharge to the sewer, but they are not subject to the fines and other penalties that are incurred by facilities that discharge directly to receiving waters.

The influent to the plant, the basis of the process demand, is not controlled but it is subject to scrutiny by the regulatory agencies. In most jurisdictions, if the plant influent approaches the design value, the regulatory agency will require that the owner initiate an upgrade or expansion.

Most treatment plants monitor the following influent parameters:

- Hydraulic loading (flow), mgd
- BOD_5, mg/l or ppm
- NH_3, mg/l or ppm
- TSS, mg/l or ppm

The hydraulic loading is a parameter that is commonly used to gauge the relative size of treatment facilities. Design and actual flows are usually expressed as average daily flow (ADF). In some treatment plants, the waste is more concentrated and the organic load is more important than the hydraulic load. There is usually a large diurnal variation in hydraulic load, even in dry weather. It is quite common to have the peak dry weather flow 20–50% higher than the ADF, and the peak flow can easily be twice the minimum flow.

For municipal treatment facilities, the effluent standards are usually set by a state regulatory agency. Most plants discharge to some sort of surface body of water—typically a river or stream. The discharge limits are based on limiting the impact of pollutants on the receiving water body.

Discharge requirements frequently vary by season, with one set of limits for winter and another for summer. This is particularly true of limits on nutrients such as nitrogen or phosphorus. A minimum DO concentration is often required to avoid reducing the average DO of the receiving water by dilution with low DO plant effluent. Discharge permits may also accommodate storm events, allowing flow to bypass secondary treatment during periods of severe hydraulic loading. It is important to incorporate all the permit requirements into the data collected during the initial assessment.

Treatment plants may monitor and report the following effluent parameters:

- Hydraulic flow, including separate bypass flow, mgd
- BOD_5, mg/l
- NH_3, mg/l
- NO_3, mg/l
- Phosphorus, mg/l
- TSS, mg/l
- DO, mg/l

The parameters monitored are dictated by the plant permit.

2.1.4 Treatment Performance

Permit requirements define the upper limits of regulated parameters. Most facilities routinely produce effluent well below the maximum limits. This is partly a result of conservatism—allowing a margin of safety so that upsets, slug loads, and equipment failures don't push the effluent over the regulated limits. In some cases, the "excess" level of treatment is the result of the design capacity exceeding actual loads coupled with a design that limits the turndown capability of the process equipment. This may result from operational procedures that don't take advantage of the turndown available.

In some cases, the plant operators take pride in the excess treatment, since they rightfully consider themselves environmental stewards. The higher level of treatment is considered evidence of going beyond minimum requirements and improving environmental quality. This attitude has some justification, but it neglects considerations of energy and sustainability. Small incremental improvements in pollutant removals are often obtained at the cost of large increases in energy consumption. The increase in carbon footprint and the depletion of energy resources may offset any improvement in water quality. And, of course, the high cost of energy and the impact on treatment cost are a consideration.

It is important to obtain actual effluent data and compare it to permit levels. This may identify potential ECMs, many of which can be implemented with little or no capital expenditure.

2.2 EVALUATE PROCESS AND EQUIPMENT

The biological treatment process is fairly robust, and the microorganisms can typically adapt to a range of operating conditions and still meet treatment objectives. The operating range is not unlimited, and the evaluation of the impact of ECMs on process performance must consider the inherent limitations.

In many cases, the process equipment has less flexibility than the process itself. There are discrete numbers of clarifiers and aeration tanks, so taking them off-line may result in large incremental changes in capacity. Most regulatory agencies require multiple process units and standby equipment to ensure continuous operation. Unfortunately, some design engineers attempt to minimize construction and equipment cost by using a few larger units instead of a greater number of smaller units. While this practice admittedly results in some reduction in initial cost, it limits the operator's flexibility. In most cases, the increase in operating cost over the long term exceeds the initial savings, as well as increasing the operator's challenges in maintaining efficient treatment. This is particularly true in the early years of plant operation, when actual loads are much lower than the ultimate design capacity.

Process equipment such as pumps, blowers, and diffusers have a maximum capacity. This may be greater than the "design" point. It is common practice in design to specify based on the anticipated worst-case conditions—the highest pressure, the highest flow rate, and so on. Motors are often specified to be "non-overloading"—that is, under these worst-case conditions the power draw will be less than the motor's nominal power. This means that there may be additional capacity beyond the design point, and taking advantage of this additional capacity may allow meeting process demand with one blower instead of two, for example.

Process equipment is also restricted on minimum capacity—in other words, turndown is limited. Turndown is usually expressed as a percentage of maximum flow or capacity:

Equation 2.3

$$\text{Turndown\%} = \frac{Q_{\text{max}} - Q_{\text{min}}}{Q_{\text{max}}} \cdot 100$$

where

$Q_{max,\ min}$ = maximum and minimum safe flow rates, SCFM

The turndown limit often creates a problem for operations and for energy conservation. Running a pump or blower below the safe range can result in equipment damage. Insufficient air flow to a basin can result in solids deposition and septicity. Oversized valves make flow control difficult. Many flow meters drop out—produce a zero flow or below range error—at less than 10% of max range. Velocities in RAS or wastewater piping that are too low result in solids settling out in the pipe.

A common situation is a plant that is producing a nitrified effluent when only BOD removal is required by permit. This is typically a result of long residence times—operating with more hydraulic capacity than needed. The additional air flow required to maintain nitrification represents an increased energy cost.

In making the initial assessment, it is important to identify the minimum and maximum capacity of major process equipment, including tanks and clarifiers. For example, if aeration basins are operating at very high DO concentrations during much of the day, an obvious ECM is installation of automatic DO control. However, if the blower is operating at the minimum safe flow rate, or the basin air flow is at the mixing limit, the DO control system can't provide any reduction in energy cost.

Nameplate data for all motors on major process equipment should be recorded, and motor data sheets copied. Copies of performance curves for pumps and blowers should be obtained. This information is usually found in the plant's Operation and Maintenance Manuals. These manuals should be investigated for other useful information such as the operating limits of the plant design. As-built drawings of the plant and equipment are also valuable sources of information on the facility.

One aspect of the data collection process that is often overlooked is recording piping system layout and design. Particularly critical is obtaining the sizes of flow control valves, since accurate flow control depends on using a satisfactory valve travel range. If the existing valves are oversized, it may be necessary to replace the valves and some piping, a measure that will certainly affect payback.

Examination of the existing instrumentation is another area that occasionally receives insufficient attention. Many operators don't distinguish between an instrument with local indication only and a transmitter that can provide a signal to SCADA or control systems. Maintenance and calibration histories should be obtained. An instrument that isn't critical to operations may be left inoperable after failure because the cost of maintenance isn't justified by the benefits.

Transmitter ranges should be verified and compared to actual operations instead of design or worst-case conditions. In some cases the primary element, such as a pitot tube, may be reused with a new transmitter. Existing transmitters may have range adjustments and can therefore be recalibrated for improved accuracy at actual and anticipated operating conditions. A common example of this is DO transmitters, which may be scaled to 0–20 ppm even though saturation is 10 ppm and the preferred operating concentration may be 1.5 ppm.

The devices in the instrument's signal loops and the output signal type should be determined. A 4–20 mA output signal is the most common in the United States, but in other areas or in some older systems, voltage outputs are used. The number and nature of other devices in the loop must be recorded. For example, if the signal runs to an RTU, PLC, or chart recorder, it may not be possible to add additional elements without creating a ground loop (see Chapter 7). Signal isolators may be required.

Communications networks, serial or Ethernet, are commonly found in existing SCADA or control systems tying remote Inputs and Outputs (I/O) or PLCs together. This may make it possible to obtain readings from existing instrumentation by tying in to the communications system. The type of communications signal, system architecture, and communications protocol should be examined. The initial survey should also identify the individual responsible for the system maintenance and programming. This may be a local integrator, a member of the plant staff, or a consultant.

Flow instrumentation is a particularly problematic area in most aeration systems. The two most common problems are installation errors (fittings or valves too close to the primary flow element) and sizing errors (the transmitter output scaled much higher than actual flows).

Valves and valve operators also require special attention. As noted above, proper sizing is critical to achieving accurate flow control. Additional problems with valves and operators are common. Valves that are not exercised regularly can lock up. Electric operators may be damaged by moisture and corrosion. Positioners may need calibration. Seats and seals exposed to hot air for years may lose resilience and prevent full closure of the valve.

Diffusers are always high on the list of potential ECMs in any diffused aeration plant. Unless the diffusers are quite new, it is likely to be cost-effective to replace them with newer units that have higher oxygen transfer efficiency (OTE). The most common retrofit is replacement of coarse bubble diffusers with fine-pore diffusers. Another common retrofit is upgrading fouled or degraded ceramic diffusers with new ceramic inserts or flexible membranes. Older flexible fine-pore membranes may have degraded as the elastomer hardens or the pores foul with biological or mineral deposits. In some cases, with either ceramic or flexible membrane diffusers, draining the tank then using chemical and/or mechanical cleaning may be sufficient to restore OTE. Diffuser upgrades require significant capital investment, so payback analysis is definitely in order.

Regardless of the condition or type of the existing diffusers, the depth of submergence must be obtained. The height of water above the air release point determines the static pressure of the aeration system—the minimum pressure required in the air piping before any air can be pushed through the diffusers. This parameter is critical for the analysis of blowers and for design of the control system.

If the plant uses mechanical aerators, the critical performance parameter is the actual aerator efficiency (AE) of the aerators, usually expressed in pounds O_2 per horsepower-hour. Mechanical aerator performance may also be reported as actual oxygen transfer rate (OTR) expressed in pounds of O_2 transferred per hour. Quantities, locations, and nameplate horsepower should be recorded, and actual power draw of the aerators should be measured. Most systems are designed so that the aerators do not use the full motor nameplate horsepower, so accurate comparison of alternate aeration systems cannot be based on nominal motor power alone.

Mechanical aeration systems in general have less flexibility in modulating oxygen transfer than diffused aeration systems. They are generally designed to meet peak loading. DO concentrations in the aeration basins should be measured at various times during the day and night, since the aeration basins may be at a reasonable level during peak loading but approach saturation during evening and nighttime operation. If some method of modulating the aerators, such as two speed motors or variable speed is available, the operator's operating procedures for aerator control should be examined.

Photographs are a convenient and useful tool for recording details about a plant configuration. With the ready availability of digital cameras the cost is inconsequential, and the ability to preserve plant information and incorporate it into reports and other documents can be invaluable. Every site survey should include extensive use of this technology.

The initial energy survey and process assessment can be time-consuming and even tedious. The above recommendations may seem a bit overwhelming. However, the success of all subsequent steps in the implementation and design of aeration control—and all other ECMs—depends on the accuracy of the initial assessment and obtaining accurate performance data. This is the foundation for the selection of alternatives and for proper system design.

It may not be possible to get all of the appropriate information for all of the areas identified above for every site survey. In general, however, the more the information obtained, the better the analysis and projections will be. In some cases an experienced engineer may be able to do the survey in two stages, with the first stage consisting of a walk-through used to highlight the most probable areas for energy cost reductions, followed by a more detailed survey of these areas.

2.3 BENCHMARK PERFORMANCE

An energy benchmark is a basis of comparison for a specific plant's performance relative to other facilities with similar processes and treatment levels. Comparison to the appropriate benchmark can provide an initial indication on the relative magnitude of potential energy cost reductions.

The term "benchmark" should not be confused with the term "baseline." Baseline data reflects the actual energy usage and process loading of the facility at the time of the survey. The baseline is used as the basis of comparison for energy cost before and after the implementation of ECMs. It is the starting point for calculating and verifying energy cost savings.

A benchmark, on the other hand, is a level of energy consumption that can be expected for a specific facility. A benchmark is derived from plants similar in size and configuration. It is useful for gauging the potential savings at the plant being

TABLE 2.1 Example Electrical Energy Benchmarks for Activated Sludge

Flow, mgd	Typical kWh/mil gal	High Efficiency kWh/mil gal
0–1	5500	3000
1–5	2500	1700
>5	2300	1500

surveyed and for setting target cost reductions. A benchmark is not a standard, but rather it is one means of judging the energy performance of a facility by comparing it to similar plants.

Comparison to benchmarks should not be taken as an absolute judgment on the energy efficiency of a given facility. Every treatment plant is different, with different influent and effluent characteristics, different unit processes, and different climatic conditions. The likelihood of finding another facility that is identical to the one being surveyed is extremely remote. The EPA and others have developed techniques for calculating the benchmark for common treatment processes. The appropriate benchmark for the plant being surveyed should be developed and used for comparisons. Some typical values are shown in Table 2.1 and Figure 2.3.

Benchmark data is available from a variety of sources, and a number of metrics, or key performance indicators, are used. The most common is kilowatt-hours per million gallons treated (kWh/mil gal). If the plant organic load is predominantly high strength industrial waste, using kilowatt-hours per pound BOD may be more appropriate.

Benchmarking should be used to identify the potential for savings and the areas for improvement. If the energy consumption for a plant being analyzed is below the benchmark value, it doesn't mean that there are no savings available. If the energy consumption is above the benchmark, it doesn't mean that the plant is poorly operated or that it wastes energy. The assessment should examine the operation and the energy use, process performance, and aeration equipment to determine what is

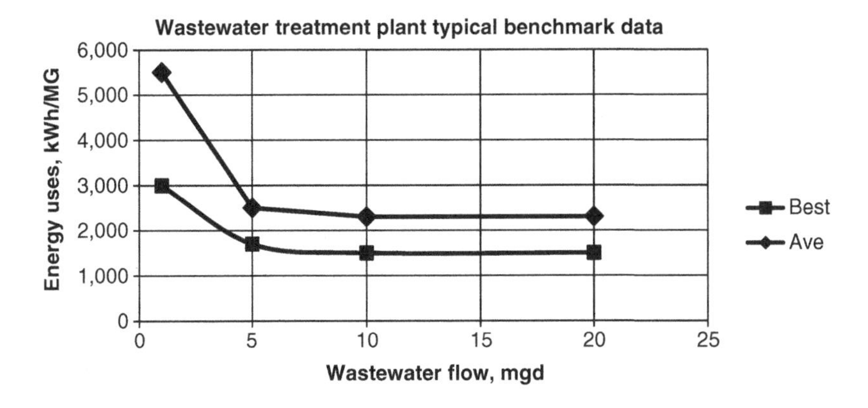

FIGURE 2.3 Sample of benchmark energy versus flow

good and what can be improved. Examining the reason(s) for the variation from the benchmark is more important than the variation itself.

2.4 ESTIMATE POTENTIAL ENERGY SAVINGS AND PERFORMANCE IMPROVEMENT

Virtually any facility can reduce its energy cost, no matter how efficient or how well operated it is. In fact, experience shows that as the plant successfully implements ECMs the staff's energy awareness increases and more opportunities are found for further cost savings. Energy cost reduction becomes an ongoing program and part of the plant's normal operating procedures.

After the data collection and site survey are completed, the next step is to identify the opportunities in the plant for energy cost reduction. Some of the opportunities may require only operational changes with little or minimal investment. Some opportunities may involve upgrading existing process equipment or replacing existing equipment with more energy efficient equipment. A special class of this second category is upgrading aeration systems for higher efficiency and/or better control.

An example of an operational change that requires no investment might be alternating the raw wastewater pumps. Many plants alternate pumps on a weekly basis to even wear and to avoid equipment degradation that may result from operating it too infrequently. The operating procedure may involve starting the standby pump, letting it run for several minutes to stabilize and verify proper operation, and then stopping the lead pump. Inspection of the utility demand profile may show a spike at 3:00 P.M. every Wednesday, which coincides with the operator's pump alternation at the end of a shift. By simply rescheduling the alternation to the beginning of the shift, say 7:00 A.M., before peak usage rates begin and before demand charges are in effect, both consumption and demand cost will be reduced with no investment, and without actually decreasing energy use!

An example of an ECM that requires some investment would be replacement of older motors with premium efficiency motors. In a typical application where the motor runs continuously the cost of energy is more than 95% of the total life cycle cost of the motor—initial motor cost and maintenance are almost insignificant. Motor replacement is generally a very cost-effective ECM. Many electric utilities have incentive programs to offset some of the cost of the new motor, making replacement even more attractive.

Identifying potential ECMs is the first step of developing the recommendations in the assessment report. Resources of manpower, time, and money are needed for implementing most ECMs, and all of these are in short supply in most wastewater treatment plants. A necessary step in preparing the report is to prioritize the ECMs and identify the order in which they should be implemented.

The parameter used to prioritize ECMs may vary depending on the circumstances at a particular facility. For example, there may be severe budget limits that make low capital expense the most important criteria for determining order of implementation.

In this case, the accrued savings from low investment options may be used to finance other ECMs. Another constraint may be time-sensitive incentives from the electric utility. In this case, it makes sense to implement a given ECM before the incentive becomes unavailable.

The most common method for establishing priorities is the relationship between projected savings and the cost of implementation. There are a wide variety of techniques used to quantify this relationship. The two most common are the simple payback calculation and the present worth analysis.

Simple payback (Equation 1.11) identifies the period of time required for the savings from an ECM to offset the investment required to purchase and install it. The simple payback analysis is favored for many energy conservation analyses because it is easy to calculate and its meaning is readily understood. The method is adequate for initial assessments or for short-term projects like most energy cost reduction programs.

One of the factors that must be determined in the initial assessment is the owner's acceptable payback period. Many municipal authorities accept 5 years or more as an acceptable payback period, because the facility has a long life and a reliable revenue stream. Many industrial organizations, on the other hand, will not invest in measures with a payback longer than 2 years. There are many investments competing for scarce capital in a typical industrial environment, and many industries must concentrate on short-term earnings to meet management objectives.

It should be noted that many utility incentive programs have limited the minimum and maximum payback period for an ECM to qualify for the incentive. It's common for the program to disqualify measures with less than 1-year payback, for example. The utilities consider that such an attractive investment should be implemented without encouragement, and that financing assistance should not be required with such a short payback period.

For projects with long-term returns or more complex financing means, many analysts prefer more sophisticated financial analysis methods. Commonly used methods include net present value (NPV), also referred to as net present worth (NPW) analysis, return on investment (ROI), and internal rate of return (IRR). Of these methods, NPW is the most generally used in the municipal wastewater treatment industry (Equations 1.12–1.14). All of these more sophisticated methods take into account the time value of money and the cost of money. The time value of money quantifies the fact that a dollar earned today is more useful than a dollar earned in the future. The cost of money refers to the fact that money may need to be borrowed at interest, or alternatively the money could be invested at interest rather than be spent on ECMs (see Figure 2.4). For an irregular savings stream:

Equation 2.4

$$\mathrm{NPW} = \sum_{1}^{n} \frac{S}{(1+r)^{n}} - \text{Investment}$$

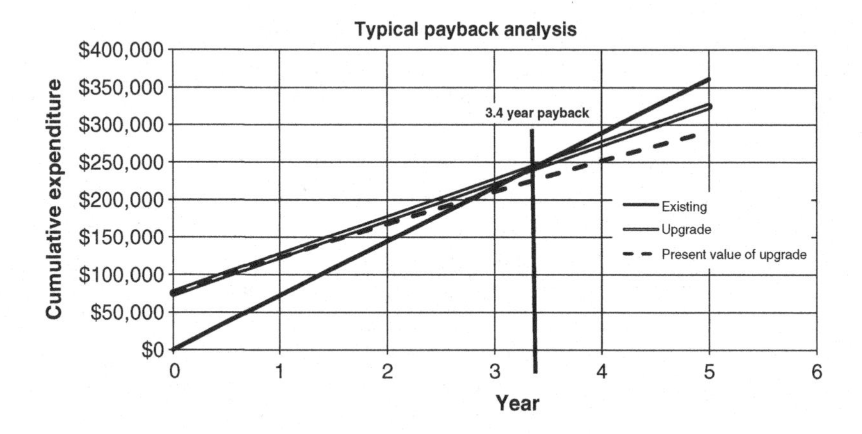

FIGURE 2.4 *Comparison of simple payback and present value*

where

NPW = net present worth, \$, NPW > 0 if the project is financially justified
r = effective annual discount rate, decimal
n = total evaluation period, years
S = net annual savings (energy and maintenance), \$/year

If the savings for each year are identical, which is usually assumed to be the case, then the net present worth simplifies to the form shown in Equation 1.12. For very short time frames the difference between simple payback and NPW is not significant.

Estimating the investment and the potential savings can be straightforward—for example, a simple motor replacement has easily determined cost and readily calculated savings. Aeration system upgrades and aeration controls, on the other hand, require a lot of process knowledge and engineering to calculate the savings.

In the initial assessment, the savings is sometimes based on the composite rate for electric costs:

Equation 2.5

$$S = (\text{existing kW} - \text{projected kW}) \cdot \text{rate}\frac{\$}{\text{kWh}} \cdot 8760\frac{\text{h}}{\text{year}}$$

For most wastewater treatment plants and for aeration controls, a more accurate analysis should be performed. This should take into account the on-peak, off-peak, and demand periods. In some cases, the energy savings will occur primarily at night, when plant loads are low and power cost is reduced. In some cases, the ability to limit the number of blowers operating will reduce the demand cost, and this may be more significant than savings from reduced consumption.

Equation 2.6

$$S = \sum \text{On Peak \$} + \text{Off Peak \$} + \text{Demand \$} + \text{Power Factor \$}$$

Equation 2.7

$$\text{On Peak \$} = \text{On Peak} \cdot \frac{\$}{\text{kWh}} \text{kW}_{\text{saved}} \cdot \frac{60\ \text{h}}{\text{week}} \cdot \frac{52\ \text{weeks}}{\text{year}}$$

Equation 2.8

$$\text{Off Peak \$} = \text{Off Peak} \cdot \frac{\$}{\text{kWh}} \text{kW}_{\text{saved}} \cdot \frac{108\ \text{h}}{\text{week}} \cdot \frac{52\ \text{weeks}}{\text{year}}$$

Equation 2.9

$$\text{Demand \$} = \text{Demand} \cdot \frac{\$}{\text{kW}} \text{kW}_{\text{peak saved}} \cdot \frac{12\ \text{months}}{\text{year}}$$

Equation 2.10

$$\text{Power Factor \$} = \%\text{Penalty} \cdot \left((\text{PF}_{\text{min}} - \text{PF}_{\text{actual}}) \cdot \text{Demand} \frac{\$}{\text{kW}} \text{kW}_{\text{peak}} \cdot \frac{12\ \text{months}}{\text{year}} \right)$$

Note that the above formulas are based on typical time frames and billing practices. The actual rate structure for the facility being assessed should be used for final calculations and project justification.

Experience has shown that automatic DO control can save 25% or more compared to manual control. The controls for most aeration systems can be configured to provide a simple payback of 2–5 years based on energy savings. However, aeration system upgrades and automatic controls generally represent a significant investment. Applying simple rules of thumb to justify this kind of expenditure doesn't represent responsible engineering practice. It is incumbent on the designer to provide a thorough analysis of the process and energy impact of the proposed system before expecting the owner to proceed with the upgrade.

2.5 PREPARE REPORT

After the site survey and analysis are complete, the results must be summarized in an engineering report. This document should identify the baseline data for energy and process, describe the ECMs and prioritize them, provide estimates for costs and savings, and lay out a plan for implementation and verification.

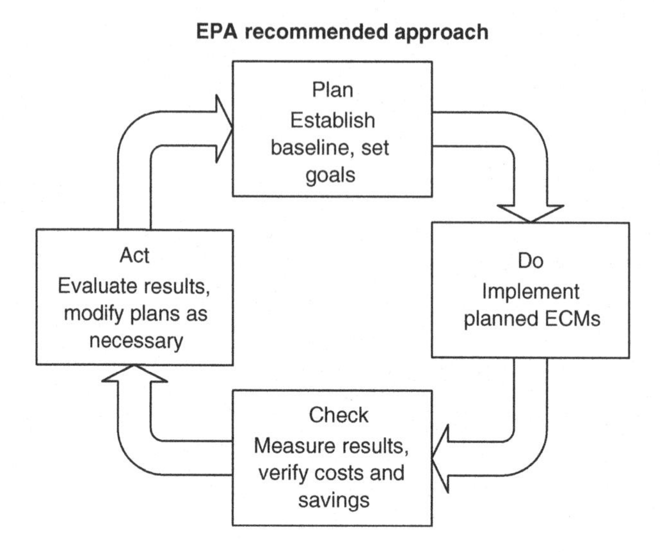

FIGURE 2.5 *EPA approach to ECM implementation*

The engineering report is essentially the road map for energy cost reduction. It is not the final step in the process, but rather the first step in an ongoing program of energy cost reduction. Additional analysis and design may be required before some of the recommendations can be implemented. After implementation some modifications may be required, particularly as process loadings change with time. Testing and verification are critical aspects of a successful, ongoing energy cost reduction program.

The EPA and other organizations recommend a "Plan–Do–Check–Act" approach (see Figure 2.5). The initial assessment is the beginning of the process—part of the "Plan" step. It is the foundation of the following steps. The importance of an accurate assessment and a thorough report cannot be overemphasized. If this portion of the program is not carried out properly, the remaining portions will not be optimized, and may well fail completely to meet the objectives. The report is the documentation of this critical step. It should contain the following sections:

- *Background*: A description of the plant and processes employed.
- *Energy Baseline*: The results of the site survey showing the major energy uses and rate structure.
- *Process Baseline*: The hydraulic and organic loads and effluent requirements—design and actual.
- *Benchmark*: The basis of comparison and the possible range of reductions.
- *Recommended Energy Conservation Measures*: Description of every ECM, with estimated cost, savings, and payback for each.

- *Priorities*: Identify the sequence for ECM implantation.
- *Financing Options*: Indicate potential funding sources such as performance contracting, utility incentives, and O & M budgets.
- *Process Impact*: Indicate the impact of the ECMs on process performance, if any.

Once the report has been accepted by the owner, the next step is to proceed to the detailed design of the proposed ECMs. In the case of aeration and aeration control, this will require detailed design of the system, obtaining cost data for the final design, and accurate calculation of the projected savings. This aspect of the project requires specialized engineering expertise.

EXAMPLE PROBLEMS

Problem 2.1

Refer to the energy bill shown in Figure 2.2. What is the composite electric rate for this facility?

Problem 2.2

The treatment plant with electric rates shown in Figure 2.2 has four 250 hp aeration blowers, but typically operates two at a time. The manufacturer's data shows the average power consumption for each blower is 180 bhp and the motor efficiency is 95%. The blowers are alternated weekly around 9:00 A.M.

(a) What are the potential savings if the blowers are alternated weekly by turning off the lead before starting the lag, compared to starting the lag blower and letting it warm up for 10 min before stopping the lead?

(b) If the operators are concerned about changing the procedure and having reduced air flow during the alternation, what alternatives are available?

Problem 2.3

A facility with the rate structure shown in Figure 2.2 has three existing constant speed positive displacement blowers with 100 hp motors. The blowers were designed for inlet pressure of 14.2 psia, barometric pressure of 14.4 psia, discharge pressure of 9.0 psig, and an air flow of 1800 ICFM. It is estimated that actual average air flow required is 1500 ICFM, and current operating pressure is 8.5 psig. The estimated installed cost of each 100 hp constant torque-rated VFD is $15,000. The estimated installed cost of each turbo blower is $75,000 each, including drive and controls. Assume the efficiency of the existing blowers is 65%, the existing motor efficiency is 94%, and the VFD efficiency is 96%. The estimated efficiency of each turbo blower is 80%, with 98% motor efficiency and 95% VFD efficiency. Provide an initial evaluation to determine if it is more cost-effective to provide VFDs for the existing blowers or to replace them with high-efficiency variable-speed single-stage turbo blowers.

Problem 2.4

For the blower system in Problem 2.3, the operators suggested three alternate ECMs. Estimate the payback and list advantages and disadvantages for each.

(a) Use a V-belt sheave change to reduce the air flow on one or more blowers, with an installed cost of $400 each.

(b) Install only one VFD on the existing blowers.

(c) Install only one new blower.

Problem 2.5

A municipal utility authority operates three activated sludge wastewater treatment plants:

(a) A 0.75 mgd ADF plant with a monthly energy consumption of 180,000 kWh.

(b) A 1.5 mgd ADF plant with a monthly energy consumption of 109,000 kWh.

(c) A 2.5 mgd ADF plant with a monthly energy consumption of 220,000 kWh.

Establish the baseline, typical benchmark, and potential savings for each plant based on Table 2.1. Use the composite power rate from Problem 2.1.

Problem 2.6

Calculate the 5- and 10-year present worth of the turbo blower replacement in Problem 2.3. Assume a 5% discount rate and a 3% inflation rate.

Problem 2.7

A more detailed analysis of the process air demand for Problem 2.3 indicates that the average air demand from 9:00 A.M. to 9:00 P.M. is 115% of the average and the air demand for the rest of the day is 85% of the average. The maximum air demand occurs at noon and is 120% of the average. Use the actual rates shown in Figure 2.2 and recalculate the cost-effectiveness of providing three VFDs for the existing blowers.

Chapter 3

Aeration Processes

There are many variations on the activated sludge (AS) process. All of them share the same fundamental process considerations, and the underlying principles for energy conservation are the same.

Under ideal conditions, the aeration control system simply has to adjust the air supply to the aeration basins to maintain a set DO concentration. On its face this seems like any other feedback control loop. Unfortunately, aeration systems seldom operate at ideal conditions on a consistent basis. In order for the controls to function properly to provide both energy conservation and process performance, it is necessary to understand how the process works, what limits performance, and how to control the system during the inevitable excursions from "normal" operation.

There are other characteristics of the aeration process that make control optimization challenging. The process inherently has long process transport and response times. The response is very nonlinear. Finally, there are wide fluctuations in loads to contend with. Understanding the process is the key to developing and tuning a successful control strategy.

3.1 PROCESS FUNDAMENTALS

The earliest treatment efforts were limited to removing debris and suspended solids from the wastewater (sewage in the then common terminology). The removal consisted principally of mechanical separation by screening and settling of

Aeration Control System Design: A Practical Guide to Energy and Process Optimization, First Edition. Thomas E. Jenkins.

suspended solids in clarifiers. These treatment methods, called primary treatment, proved inadequate for maintaining water quality and were supplemented by secondary treatment. Secondary treatment generally consists of removal of dissolved pollutants by biological means. A variety of technologies are employed, but the dominant secondary treatment method is activated sludge.

Initial development of the activated sludge process began in the early 1900s and widespread implementation dates to the 1940s. Since that time there has been continuing and significant development of the process. There are literally hundreds of variations in operation, and new designs are introduced continually. The modifications address more stringent treatment requirements, improve process and/or energy efficiency, reduce operating or capital expense, or eliminate specific operating problems. Every part of the AS process is included in the modifications: flow patterns, aeration equipment, tank configuration, and control strategy.

Fundamentally, the activated sludge process is quite simple. It mimics in a controlled environment and on an industrial scale the process that takes place naturally in receiving waters in assimilating biological wastes. Microorganisms absorb the waste material and oxygen dissolved in the wastewater. The microorganisms utilize the waste as a food source, and use the oxygen for metabolism. Additional waste is used for microorganism growth and reproduction.

Pollution of the receiving water results when levels of discharged organic constituents exceed the ability of natural processes to assimilate them. Often the first impact of overloading the receiving water is depletion of the oxygen dissolved in the water. This ultimately results in the death of aquatic animals that use the dissolved oxygen to sustain life. The sensitivity to oxygen depletion varies with the species, but the populations of surface waters are all dependent on some level of available oxygen to sustain life.

If the concentration of pollutants in the receiving water increases, the water quality will degrade further. Increases in odors, unsightly appearance, and water borne pathogens are among the consequences.

The pollutants dissolved or suspended in the wastewater and converted during biological treatment are referred to as "substrate." The dominant constituents of the substrate, and the earliest to be addressed by municipal wastewater treatment efforts, are the organic compounds in human waste. The substrate in wastewater does not consist of a single compound, but is a mixture of compounds of varied composition. Some of these are simple compounds, readily and quickly metabolized. Other compounds are quite complex and require an extended time for the microorganisms to achieve removal. The organic compounds are referred to as "carbonaceous," since carbon is the predominant element. These carbonaceous compounds include sugars, fats–oils–greases (FOG), and proteins.

The complexity of the pollutants precludes a simple chemical test to determine their concentration and treatment requirements. In order to quantify the pollutants contained in wastewater, a specific test was developed. The BOD test mimics the action of pollutants in depleting the oxygen in receiving waters. A sample of the wastewater is diluted, seeded with bacteria, aerated to saturation, and held at 20 °C. The depletion of oxygen over a period of time is measured and the result is the biochemical oxygen demand. The usual duration of the test is 5 days, and the resulting value is reported as BOD_5. This measurement reflects the impact of pollutants on the receiving water, and is the basis for process design and treatment permits issued by most governmental authorities.

Another test used to quantify the concentration of compounds is the chemical oxygen demand (COD) test. An oxidizing chemical (potassium dichromate) is added to a wastewater sample to determine the oxygen equivalence of the organic material. The COD test has the advantage of providing results in minutes or hours, rather than in days. Although there is often a correlation between the results of the BOD_5 and COD tests for a given waste, the results are not equal. The BOD_5 test is usually considered the definitive measurement of pollutant concentrations.

Domestic wastewater includes compounds other than carbonaceous ones. These include the nutrients nitrogen and phosphorus, each of which can also degrade the quality of the receiving water. The concentrations of nutrients are generally lower than the concentrations of carbonaceous compounds. However, when they must be removed they have significant impact on both the AS process configuration and the energy required for biological treatment.

The nutrient that receives the most consideration is nitrogen. Nitrogen is contained in many organic compounds, but it is present in the wastewater influent predominantly in the form of ammonia, NH_3. In receiving waters ammonia is converted to nitrate, NO_3, a process referred to as "nitrification." Nitrification increases the oxygen depletion of the receiving water. To minimize this impact, many permits require that nitrification take place as part of the secondary treatment process. This modification of the AS process increases the total oxygen demand of the facility.

Conversion of ammonia to nitrate doesn't eliminate the impact of nitrogen on the receiving waters. "Eutrophication" occurs when excessive nutrient concentrations result in accelerated growth of aquatic plant life, particularly algae. Death and decomposition of the aquatic life further decreases the oxygen levels, as well as creating unsightly conditions and odors.

Nitrate is a primary contributor to eutrophication. Many discharge permits therefore require that the plant "denitrify." In denitrification a zone of depressed oxygen—anoxic—is created. Microorganisms strip the oxygen from the nitrates and uses it to metabolize organic compounds, and the resulting nitrogen gas is released to the atmosphere and removed from the wastewater. Denitrification requires the presence of both nitrates and a carbon food source. The nitrate is usually the result of recirculating nitrified wastewater from the end of the aeration basin to an anoxic zone. The carbon source is most commonly raw wastewater, but additions of other compounds such as ethanol are also used.

Phosphorus also contributes to eutrophication. Chemical/physical methods are often used for phosphorus removal. Some of the variations in the AS process have been developed to provide removal of phosphorus by biological processes. However, because phosphorus removal exerts little influence on oxygen demand, it will not receive attention in this book. Processes for removing nitrogen and phosphorus using microbial activity are referred to as "biological nutrient removal" (BNR). The term

BNR is also used in practice to refer to processes for removing nitrogen only. Biological phosphorus removal also includes creation of an anoxic zone. BNR processes place a process demand on the aeration control systems so that the proper conditions for an anoxic zone are present. This process consideration is more important than energy conservation concerns, and often justifies a more sophisticated control strategy than energy reduction alone would justify.

Activated sludge is a suspended growth process. The contents of the aeration basin are agitated to maintain the microorganisms in suspension. This allows them to absorb the substrate and oxygen from the tank contents.

The floc suspended in the mixed liquor has a complex physical and biological composition. There are multiple species of organisms, acclimated to the specific environment of the aeration basin. Most of these organisms are "facultative," which means they thrive in aerated environments but can also survive in nonaerated environments for some period of time. This is important, since the microbes must be able to survive periods without aeration when they pass through the clarifiers and during power failures.

The floc particles generally have a filamentous organism at their core, with other microbes layered around it. Oxygen and substrate diffuse into the floc from the surface, and the products of metabolism diffuse out into the water. In many processes the diffusion into the floc may be the limiting factor in the process capacity. The nonuniform nature of the floc is sometimes the basis of specific processes that take advantage of an oxygen gradient from the surface to the center of the floc particles.

The combination of wastewater and return activated sludge (RAS) is referred to as mixed liquor, and the concentration of microbes suspended in the mixed liquor is measured as mixed liquor suspended solids (MLSSs). The concentration of MLSS influences the oxygen demand and substrate removal and is a key process control parameter.

There is a progression in treatment through the aeration basin. BOD removal is usually the fastest reaction, although in some industrial wastes this may not be the case. If there is sufficient hydraulic residence time (HRT) in the aeration system and the appropriate population of microbes is present, nitrification will occur. Note that nitrification occurs simultaneously with BOD metabolism, not subsequent to it as is commonly stated. However, because nitrification is a slower process BOD removal is usually complete before full nitrification occurs. BNR is a further refinement, and takes place after nitrification and only when the correct process conditions are created.

Many different tank configurations have been developed as part of developments of the AS process. The design criteria, drawbacks, and benefits of each configuration properly belong in a process design text. Some familiarity with the terminology and physical differences are beneficial in developing the appropriate control strategy, however.

The most common tank design has a length several times the width (see Figure 3.1). This is referred to as a "plug flow" reactor because the wastewater moves through the tank as a coherent "plug" traveling from influent to effluent. As the mixed liquor moves through the tank and the biological process progresses, the

FIGURE 3.1 *Plug flow basin*

concentration of the various pollutants changes and the oxygen demand varies. In some cases the plug flow reactor is arranged in several passes, with the wastewater following a serpentine path through the tank (see Figure 3.2). In plug flow reactors, baffles or intermediate walls may be used to prevent short-circuiting and to create segregated zones for specific processes.

A "complete mix" aeration basin is a round or approximately square tank with the contents essentially homogeneous throughout. There may be higher concentration of pollutants at the point where the wastewater and RAS enter the tank, and lower concentrations at the effluent weir, but overall the variations are not significant. Complete mix tanks are very commonly employed with mechanical aeration.

Another common arrangement with mechanical aeration is the "racetrack" or oxidation ditch (see Figure 3.3). The aerators create a horizontal flow around the tank. Varying levels of constituents and DO are found around the path of the tank. When this arrangement is used with diffused aeration it may be necessary to have mixers to induce horizontal movement of the mixed liquor.

The contact stabilization process separates aeration into two tanks. The stabilization tank aerates the RAS only. Wastewater constituents that were adsorbed and absorbed by the floc are metabolized in the stabilization tank, which is usually two thirds of the total aeration volume. The RAS then flows into the contact tank, where it combines with primary effluent or raw wastewater similar to a conventional aeration basin. Separate aeration control is generally required for each basin.

Extended aeration is not a specific tank arrangement, but refers to a process variant with very long hydraulic and sludge retention times. Significant endogenous

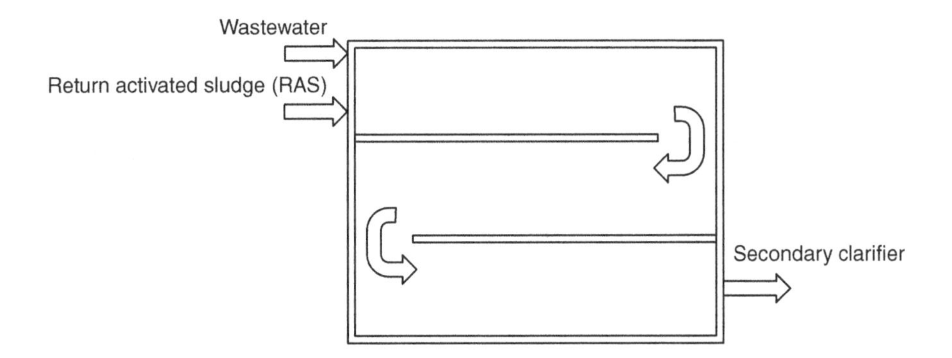

FIGURE 3.2 *Plug flow basin with serpentine flow path*

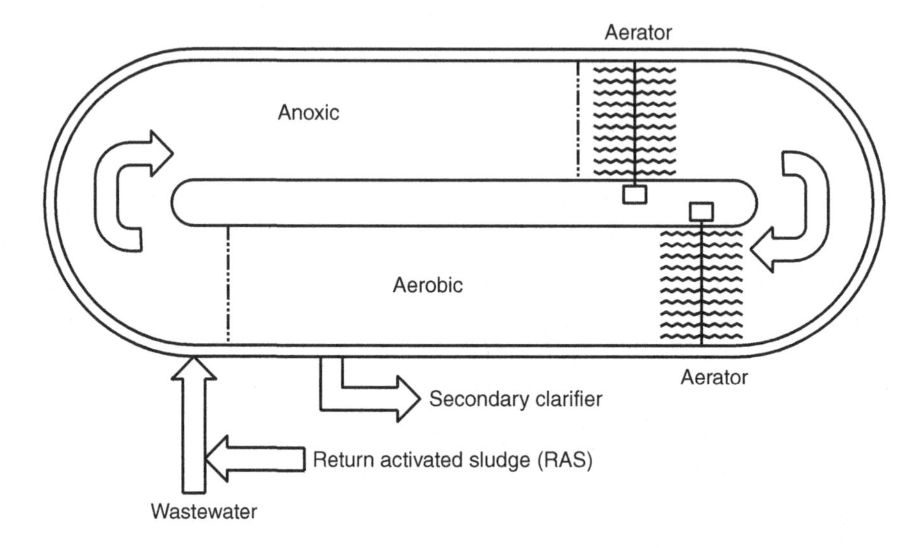

FIGURE 3.3 *Oxidation ditch*

respiration occurs in this process, essentially providing some aerobic digestion of the sludge in the aeration basin.

The sequencing batch reactor (SBR) combines the functions of the aeration basin and secondary clarifier in a single tank. Some type of automatic control is always employed in SBRs. The process operates on a timed basis, with separate cycles of adjustable duration for adding wastewater to the reactor, aeration, and settling (see Figure 3.4). At the end of the settling cycle, the effluent is decanted from the top of the basin, and the sludge that has settled to the bottom of the basin is retained in the process. Usually two or more SBRs operate in parallel to accommodate continuous influent flow. The operating cycles for parallel basins are staggered so the influent

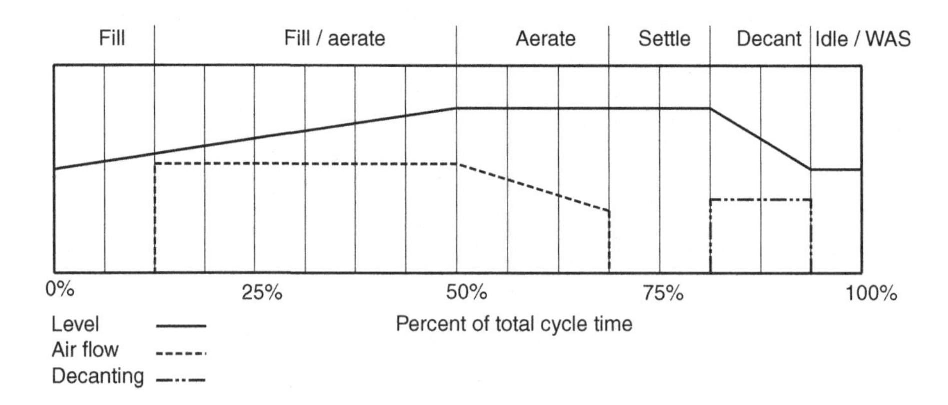

FIGURE 3.4 *Sample SBR cycles*

can flow continuously to one or the other SBR tank. Periodically, excess sludge, called waste activated sludge (WAS), is removed from the basin at the end of the decant cycle. SBRs may combine the BOD and nutrient removal processes in a single basin based on the timing of the various cycles. Some SBRs create a challenge for air flow control because the water level will vary during the aeration cycle and from one cycle to the next as the influent flow rate varies.

Lagoons are usually of earth construction with sloped walls and long hydraulic retention times. They are generally shallower than conventional aeration basins. Lagoons may use mechanical or diffused aeration, but in either case the energy is more likely to be limited by mixing requirements than by process oxygen demand. Many lagoon systems do not have separate clarifiers or digesters, relying on the long retention times to reduce sludge volumes. Because of the long retention time lagoons may not benefit from automatic DO control as much as conventional AS processes.

The aeration system is not always located immediately after primary clarification or the headworks. If the influent contains high strength industrial waste, the aeration basins may be preceded by a "roughing" process to reduce the concentration. This roughing treatment may be a trickling filter or an anaerobic process. The amount of removal achieved by the roughing process must be used in determining the aeration energy requirements.

There are other types of hybrid systems that combine suspended growth and fixed film in an aeration basin. The integrated fixed film activated sludge (IFAS) and moving bed bioreactor (MBBR) processes are similar. They both employ some type of plastic media to provide a growth site for the microorganisms. This provides a higher concentration of bacteria and longer effective solids retention time (SRT) in a given aeration volume, and the attached growth resists solids washout during high hydraulic loading. In the MBBR system, the media is free to move in the aeration basin and in the IFAS system the media may be held in place. These systems may affect control system design by placing restrictions on equipment mounting, particularly DO probes. The solids population may also affect the oxygen uptake rate (OUR) and air flow requirements.

3.1.1 Peripheral Equipment and Processes

The proper operation of the aeration system depends on the performance of processes and equipment upstream and downstream of the aeration basins. These peripheral processes are not part of the aeration control system per se. However, in order to understand and control aeration, it is also necessary to understand how these peripheral processes function and how they affect the aeration system.

The headworks of the plant consists of pumping, solids separation, and sometimes equalization processes.

The solids separation equipment generally includes screens, grit removal, and primary clarification. Screens may be coarse or fine, but both remove large materials from the wastewater stream before they can cause physical damage to pumps and aeration equipment. Rags, for example, are a chronic problem in many plants. They can clog pumps, wrap around mechanical aerator shafts, and obstruct diffusers.

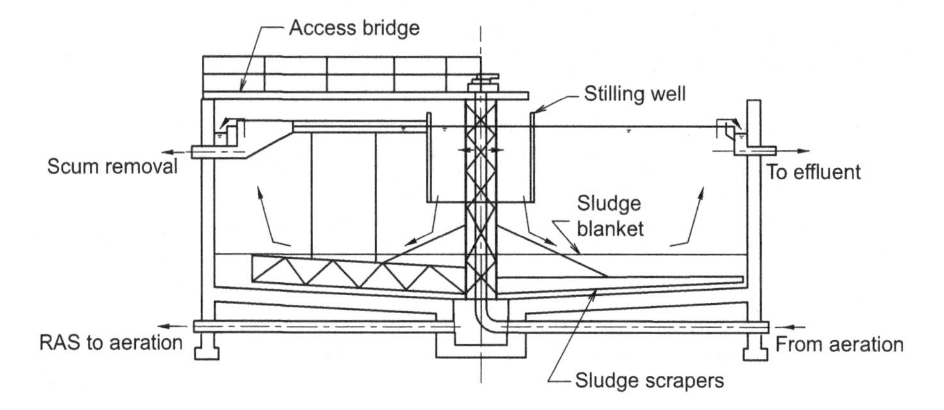

FIGURE 3.5 *Circular clarifier*

Grinders and other equipment may be installed to cut up rags, but the effectiveness of this equipment is often limited.

Grit removal equipment is intended to take nonputrescible solids (solids that do not decompose biologically) out of the wastewater stream. Sand that is washed into sewers with groundwater is an example. In some treatment plants, grit is a major problem. If it isn't removed, it tends to settle to the bottom of aeration basins, and is particularly prone to being deposited in corners and along walls. The accumulated grit can trap organics, which can become septic and produce odors and return BOD to the mixed liquor. In extreme cases, the accumulated grit can interfere with mixing in the aeration basin and may even obstruct diffusers and limit oxygen transfer.

A "clarifier" is a settling tank, where quiescent conditions allow solids to settle to the bottom for removal. The clarified liquid stream then flows out from the top of the clarifier. A clarifier preceding the aeration basins is called a "primary clarifier" and one following the aeration basins is called a "secondary clarifier." They are physically very similar, and both occur in round (see Figure 3.5) and rectangular (see Figure 3.6) configurations. The loading of primary and secondary clarifiers is different because the heavier solids in the raw wastewater settle faster in primary

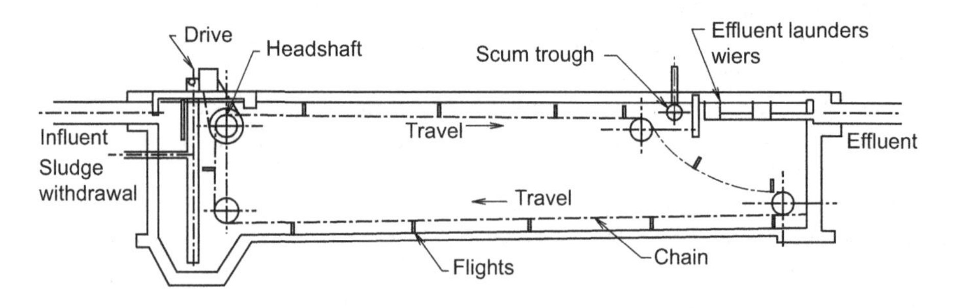

FIGURE 3.6 *Rectangular clarifier*

clarifiers than the lighter solids passing from the aeration basin to the secondary clarifiers.

Primary clarifiers may be omitted in many plants, particularly smaller facilities. Primary clarification, or its lack, affects the aeration system in several ways. Primary clarifiers remove solids—both organic materials and grit. They also remove approximately 25–40% of the BOD in the influent, reducing the organic load to the aeration system. This material doesn't disappear. The settled material, primary sludge, is removed from the clarifier and treated prior to disposal. If the sludge treatment includes aerobic digestion (see below), the plant air requirements are increased to accommodate the additional digester load.

An integral part of the AS process is secondary clarification—the separation of the microorganisms from the wastewater stream prior to discharge to the receiving water. In the secondary clarifier, the settled microorganisms represent a population that is "activated"—adapted to the conditions of the treatment process. The mixture of water and settled microorganisms is referred to as "sludge," and the use of adapted (or activated) microorganisms for treatment gives rise to the term "activated sludge process." The clarifier serves to flocculate the sludge (form the floc particles) and to thicken the sludge (increase the solids concentration) before it is returned to the aeration basins.

Secondary clarifier performance can significantly affect AS process performance. If the sludge doesn't settle properly or the RAS flow is too low there will be insufficient microorganisms returned to the aeration basin. If the food to microorganism (F:M) ratio is not correct, the waste will not be sufficiently metabolized. A common problem with secondary clarifiers is "washout". This occurs when high wastewater flow from a storm event inhibits settling and the microorganisms wash out over the clarifier effluent weir. The result is dilution of the RAS and insufficient microorganism population for proper removal. Sometimes the aeration control system includes logic to prevent solids washout, such as shutting off aeration in a storm event. This allows sludge to settle in the aeration basin which prevents its passing into the clarifier and out of the system.

The operation of the aeration system can also affect clarifier performance. Improper control will affect both mechanical and biological aspects. Both overaeration and underaeration will deleteriously affect the process.

Airlift pumps are sometimes used for sludge pumping and sometimes for influent or effluent pumping, particularly in small facilities. An airlift is a simple, reliable device for low head pumping requirements (see Figure 3.7). There are no moving parts in contact with the pumped fluid. In an airlift pump the blower discharge is directed to the bottom of a vertical pipe. The buoyancy of the air and the reduced density cause bubbles and water to rise and discharge at a higher elevation than the initial water surface. The air pressure required is a function of the submergence of the air entry point, and care should be taken to ensure that the airlift pressure does not exceed the pressure required by the aeration system. The determination of the air flow rate in an airlift pump is empirical in nature. However, as expected the greater the wastewater flow rate and the higher the lift, the higher the air flow rate must be to operate the pump (see Figure 3.8). The percent submergence of the air introduction point is a critical parameter:

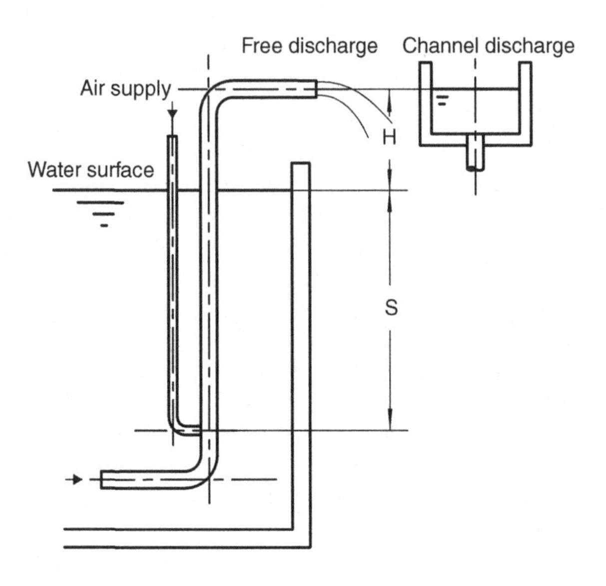

FIGURE 3.7 *Airlift pump schematic*

Equation 3.1

$$\text{Sub} = \frac{S}{S+H} \cdot 100$$

where

- Sub = submergence of airlift pump, %
- S = distance from source water surface to air entry point, ft
- H = distance from source water surface to airlift discharge point, ft

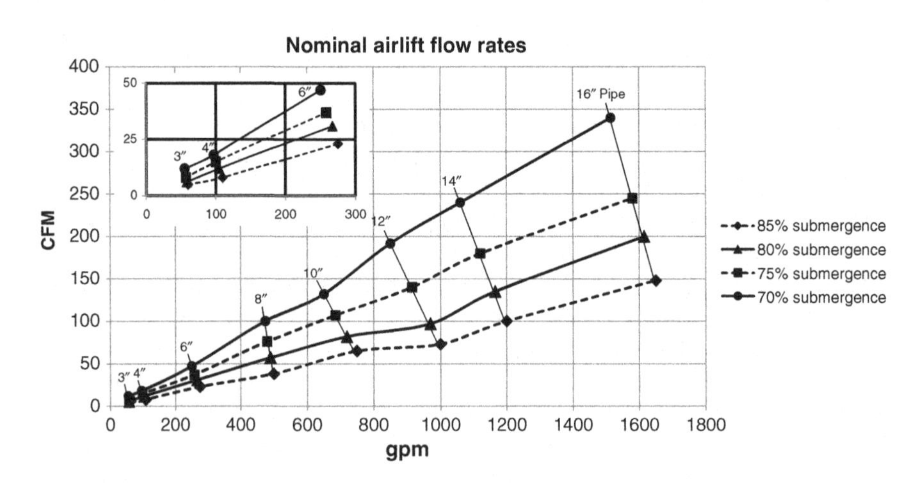

FIGURE 3.8 *Airlift pump air requirements*

If there is significant piping between the air entry point and the water entry point, some pressure drop will occur due to friction. This pressure drop, in feet of head, must be added to the static lift "H" to determine the total dynamic lift in calculating the percent submergence. The submergence is usually kept above 65%.

Open channels are often used for transporting wastewater to and from various process equipment within the treatment plant. Channel aeration is employed to prevent the settling of solids entrained in the wastewater and to prevent septicity and odors. In some cases, hydrogen sulfide (H_2S) entrained in primary effluent is removed by aeration. Channels are usually aerated using coarse bubble diffusers, since mixing and not oxygen transfer is the primary objective. Aeration rates vary, but usually range from 2 to 5 CFM/ft of channel length.

Aerobic digesters are used to treat sludge from secondary clarifiers (WAS) and sludge from primary clarifiers. Aerobic digestion will increase the demand for air and energy, and must be considered in the ECMs for plants where it is employed.

Sludge handling and disposal is a significant part of the operating cost for wastewater treatment plants, and reducing the volume of the sludge minimizes these costs. Digestion processes reduce the volume of sludge by metabolizing the volatile content, resulting in stabilization of the solids, pathogen reduction, and improved dewatering. The term "biosolids" is used to refer to wastewater solids that have been treated and are suitable for beneficial use, such as for fertilizer. There are two general categories of digestion, anaerobic and aerobic. There are many variations in digestion processes, some of which combine aspects of aerobic and anaerobic digestion.

Anaerobic digestion is found in larger facilities, and in addition to sludge stabilization this process can produce digester gas, which is approximately 60% methane. Digester gas can be used as an energy source, and can be a valuable by-product of sludge treatment. Anaerobic digestion requires heating of the digester contents, and some of the digester gas is commonly used for this purpose. Surplus gas is either flared or used to power cogeneration equipment, boilers, or similar energy recovery systems. Supernatant, the liquid periodically drawn off the digester, is very high in BOD, ammonia, and hydrogen sulfide. The supernatant is not discharged from the plant, but is sent to the aeration system for further treatment.

Aerobic digestion equipment is very similar to secondary aeration equipment. A tank receives the sludge from the clarifiers, and mechanical or diffused aeration is used to provide oxygen. Microorganisms in the digester use the oxygen for endogenous respiration (the metabolism of cell tissue) and to metabolize the products of cell breakdown. This reduces the volatile solids in the sludge by 35–50% as well as decreasing the volume. If the digestion includes primary sludge, there is also a BOD content that will be metabolized.

The aerators and diffusers used for aerobic digesters are the same as those used for secondary aeration. In a digester, the contents have much higher solids content than aeration basins, reaching several percent solids. The contents are consequentially denser and more viscous, which will affect aerator performance. This must be accounted for in initial design and in evaluating ECMs. In both cases, the aeration equipment manufacturer should be contacted for specific performance limitations.

Aerobic digestion is limited by mixing requirements and by oxygen requirements. Mixing requirements for digesters with mechanical aeration range from 0.75 to 1.5 hp per thousand cu ft. Diffused air systems require approximately 20–40 CFM/1000ft^3, but this is based on coarse bubble diffusers. Specific mixing recommendations from diffuser manufacturer should be obtained for fine-pore systems.

Oxygen requirements can vary widely from plant to plant and from time to time within a given plant. The reduction of cell tissue requires approximately 2.3 lb O_2/lb of volatile suspended solids (VSS) destroyed. If primary sludge is included in the digestion process it requires an additional 1.6–1.9 lb O_2/lb BOD removed by the primary clarifiers.

The air requirements for aerobic digestion can equal or exceed the air required for secondary aeration. Aeration for digesters is further complicated by the variable water level encountered as the digester is operated through supernatant decanting, sludge withdrawal, and fill cycles. As the water level in the digester changes, the mixing, air flow and power requirements change. In diffused aeration systems, the back pressure is a function of diffuser submergence. The variable level makes control difficult if a single blower system is used to supply aeration and digestion, since the pressure drop across the valves will vary. It may be cost-effective to provide separate blowers for the digesters that will vary the discharge pressure to match the level variations, reducing the energy required.

In the past, fine-pore diffusers were considered unacceptable for digesters because of concerns over fouling of the diffusers. Recent practice has shown that fine-pore diffusers can provide reliable operation in digesters. The superior mixing and oxygen transfer potential of fine-pore diffusers make them the preferred technology for most systems. They should be considered as an ECM for existing aerobic digesters that have either mechanical aeration or coarse bubble diffusers.

Most aerobic digesters operate in a batch mode. Periodically the aeration is stopped and the solids are allowed to settle. The liquid at the top is drawn off and returned to the headworks of the plant. This liquid, referred to as supernatant, is generally high in BOD and ammonia, and may contain hydrogen sulfide. The supernatant can significantly increase the demand for oxygen in the secondary aeration tanks. The timing and rate of supernatant removal must be considered in the analysis of aeration energy requirements. Settled solids are drawn off from the bottom of the tank. The solids may be sent to dewatering equipment or drying beds. Ultimate sludge disposal methods include land application, incineration, and landfill.

Centrifuges and belt presses are common devices used to thicken sludge after digestion. The water drawn from the sludge by these processes has a high BOD content, high ammonia concentration, and generally contains hydrogen sulfide. The washwater and filtrate from belt presses and the centrate from centrifuges are usually returned to the plant headworks or the aeration basins for treatment. The result can be a large and rapid increase in oxygen demand in the aeration process. This not only creates an increase in energy consumption and power demand, but can also create instability in DO control and rapid drops in DO concentration.

Digester supernatant, belt press washwater, and centrate are examples of internal plant flows referred to as sidestreams. They may dramatically increase the organic load to the aeration system. They must be accounted for in determining the requirements of the aeration system. Because sidestreams don't enter or leave the plant, they are not part of the plant permit requirements, and are often poorly monitored by plant staff.

Filters are used in some facilities to provide additional treatment after the secondary clarifiers. Sand filters are the most common technology. Membrane biological reactors (MBRs) are gaining in acceptance and actually replace secondary clarifiers in many applications.

Sand filters do not directly impact the aeration system during filter operation. Most filters are periodically backwashed to remove solids entrained in the filter bed so that hydraulic losses and filter efficiency can be maintained within limits. The backwash water and solids are generally recycled to the primary clarifiers or to the aeration tanks. However, the concentration of pollutants in the backwash is low and does not generally create an appreciable increase in aeration system loadings. Some filters combine air scouring and water backwashing. The air is generally provided by dedicated blowers, but may be drawn from the aeration system header. Because of the short duration of the backwash cycle the impact on blower energy isn't significant, but timing of the backwash may increase peak demand charges. The static head of the sand filters is generally low. Proper sizing of the control valve is critical to maintaining set air flows.

MBRs also need to be flushed to prevent excessive accumulations of solids on the membrane surface. The cleaning process is generally manufacturer specific, and may include hydraulic backwashing combined with continuous and/or intermittent air scouring. In some cases, dedicated blowers are provided in the MBR system for use during scouring and cleaning, although the membrane scour air may be drawn from the aeration system. Note that MBRs do not eliminate the process oxygen demand for metabolizing BOD and converting ammonia to nitrate. The MBR filtration membranes may be mounted in the aeration tank or separately, but in either case an aeration system is required to provide the process oxygen demand.

Postaeration is the last unit process in most treatment plants. Permits usually require that the plant effluent DO concentration be above a minimum level, which is higher than the concentration required in the aeration systems. When an elevation difference exists between the plant discharge and the receiving water, a simple cascade aerator—a series of steps that the water passes over—may be sufficient. In other cases, a small aeration basin may be provided with diffused or mechanical aeration to raise the DO concentration to the desired value. Because the effluent is essentially clean water, there is little demand for oxygen other than that required to raise the DO concentration. For diffused aeration systems, the demand can be determined from the following:

Equation 3.2

$$Q_s = 0.33 \cdot \frac{Q_{ww} \cdot C_s}{\text{OTE} \cdot 1.024^{(T-20)}} \cdot \ln\left(\frac{C_s - C_i}{C_s - C_f}\right)$$

where

Q_s = theoretical air flow for postaeration, SCFM
Q_{ww} = wastewater flow, mgd
C_s = saturation concentration of O_2 at 20 °C, mg/l (9.08 mg/l at sea level)
OTE = oxygen transfer efficiency, decimal
C_i = initial DO concentration, mg/l
C_f = final DO concentration, mg/l

A safety factor of 1.1 is usually applied to the theoretical air flow to accommodate the difference in oxygen transfer between clean water and plant effluent.

3.1.2 BOD Removal

In the United States, the Clean Water Act of 1972 resulted in secondary treatment requirements for most facilities of 30 mg/l BOD_5 and 30 mg/l TSS or 85% removal of BOD. As time has passed additional legislation and regulation has resulted in more stringent discharge requirements for many facilities. Additional wastewater constituents are now included in permits. Nonetheless, removal of BOD is the first requirement for treatment, and successful process performance generally mandates BOD removal.

The BOD_5 test has many limitations, but it remains one of principal regulatory parameters in most permits. Among the problems is the nonspecific nature of the test. It cannot distinguish between rapidly metabolized compounds and more complex ones. More importantly, the test cannot separate the effect of ammonia conversion from the metabolism of carbonaceous compounds. In many facilities, the conversion of ammonia, the nitrogenous demand, artificially elevates the BOD_5 test results. This gives a false indication of permit violation. In these cases, an inhibited BOD test might be used. A chemical that suppresses the nitrifying bacteria is employed in the test, and the result is the carbonaceous BOD (CBOD). It reflects only the oxygen demand of the carbonaceous compounds in the sample.

The BOD test is sometimes modified by filtering the wastewater prior to running the test in order to remove particulates and suspended solids. The result is the value of the soluble BOD (SBOD).

The duration of the BOD_5 test is another drawback in its use. An operator has to wait 5 days for the test to indicate the result of load or operational changes.

This obviously affects the operator's ability to effectively operate the plant in real time. A number of other tests have been developed as operational tools. Generally regulatory agencies do not accept these tests for verifying compliance.

One such test is the chemical oxygen demand. In this test the organic matter in the wastewater is chemically oxidized. The results of this test can be obtained in a few hours, making it more useful for operational purposes than the BOD test. There is generally a correlation between COD and BOD_5, but the relationship may be variable over time, particularly in the case of industrial loads.

Oxygen uptake rate (OUR) is another test that can provide rapid indication of the organic load in the system. The test, typically performed in a device called a respirometer, indicates the amount of oxygen utilized by the process and is generally expressed as mg/(l h). In the conventional OUR test, a sample of mixed liquor is drawn from the aeration basin and the depletion of oxygen is measured over time. The OUR is readily calculated:

Equation 3.3

$$\mathrm{OUR} = \frac{C_1 - C_2}{t_2 - t_1}$$

where

OUR = oxygen uptake rate, mg/(l h)
C_1, C_2 = initial and final DO concentration, mg/l
t_1, t_2 = initial and final time, h

The specific oxygen uptake rate (SOUR) is obtained by dividing the oxygen consumption rate by the mass of total solids.

OUR is not a direct indicator of BOD concentration, although in general increasing organic load results in increasing OUR. The measurement is not able to distinguish between carbonaceous and nitrogenous oxygen demand. It isn't possible to distinguish between a low OUR due to low loading or from low population of active microorganisms. Endogenous respiration can increase the OUR.

Despite these limitations, OUR can be a very useful parameter for controlling the aeration process. The results can be obtained very quickly, and there are off-gas analyzers and respirometers that can be installed at the aeration basin and provide results in real time or near real time. Even more significant from a control standpoint, the OUR can be used to directly calculate the process oxygen demand. If the aeration system's oxygen transfer characteristics are known, OUR can directly control and optimize the aeration energy. For diffused aeration:

Equation 3.4

$$Q_s = \frac{\mathrm{OUR} \cdot V_t}{16700 \cdot \mathrm{AOTE}}$$

where

Q_s = air flow required to meet demand, SCFM
OUR = oxygen uptake rate, mg/(l h)
V_t = aeration tank volume, cu ft
AOTE = actual site oxygen transfer efficiency, decimal

For mechanical aeration:

Equation 3.5

$$P_a = \frac{\text{OUR} \cdot V_t}{16000 \cdot \text{AE}}$$

where

P_a = aerator shaft power required, bhp
AE = actual site aerator efficiency, lb O_2/(hp h)

For design purposes, it is possible to calculate the expected oxygen uptake rate based on the wastewater flow rate and characteristics:

Equation 3.6

$$\text{OUR}_{\text{Req}} = \frac{\Delta\text{BOD} \cdot U_{O_2}}{\text{HRT}}$$

where

OUR_{Req} = OUR required to meet treatment objective, mg/(l h)
ΔBOD = BOD removed from influent to effluent, mg/l
U_{O_2} = utilization of oxygen, lb O_2/lb BOD
HRT = hydraulic retention time, h

Equation 3.7

$$\text{HRT} = \frac{1.795 \cdot 10^{-4} \cdot V_t}{Q_{ww}}$$

where

Q_{ww} = wastewater flow, mgd

Because the carbonaceous constituents in wastewater are complex and variable organic compounds, it isn't possible to develop a universal stoichiometric ratio for the oxygen required for metabolism. The oxygen utilization is variable from waste to waste and from process to process. The typical range is from 1.0 to 1.3 lb O_2/lb BOD.

The variation is due to the differing composition of the waste and the solids retention time. For example, in an extended aeration system the long SRT results in a higher level of endogenous respiration and a corresponding increase in the oxygen utilization. For domestic wastewater and the conventional activated sludge process, a value of 1.1 lb O_2/lb BOD is a reasonable approximation for the oxygen utilization if site-specific data is not available.

The concentration of BOD and other components of the wastewater is generally expressed as parts per million (ppm) by weight. (Note: 1 ppm = 1 mg/l = 1 g/m^3.) It is often convenient to convert flow rate and concentration to mass loading:

Equation 3.8

$$w = C \cdot Q_{\mathrm{ww}} \cdot 8.34$$

where

w = mass loading rate of constituent, lb/day
C = concentration of constituent, ppm
Q_{ww} = wastewater flow, mgd

It is convenient to estimate air flow requirements or mechanical aeration power directly from BOD loading:

Equation 3.9

$$Q_{\mathrm{s}} = \frac{w_{\mathrm{BOD}} \cdot U_{\mathrm{O_2}}}{\mathrm{AOTE} \cdot 24.84}$$

where

Q_{s} = air flow required to meet demand, SCFM
w_{BOD} = mass loading rate of BOD, lb/day
$U_{\mathrm{O_2}}$ = utilization of oxygen, lb O_2/lb BOD
AOTE = actual site oxygen transfer efficiency, decimal

Equation 3.10

$$P_{\mathrm{a}} = \frac{w_{\mathrm{BOD}} \cdot U_{\mathrm{O_2}}}{\mathrm{AE} \cdot 24}$$

where

P_{a} = aerator shaft power required, bhp
AE = actual site aerator efficiency, lb O_2/(hp h)

The principal products of metabolism for BOD removal are H_2O and CO_2. These are diffused through the floc and into the mixed liquor, and some of the CO_2 will pass into the atmosphere.

Many factors affect BOD removal. DO concentration is one. Hydraulic retention time, MLSS, and wastewater temperature are also important in determining BOD removal.

3.1.3 Nitrification

The next progression in treatment is converting ammonia, NH_3, to nitrate, NO_3. The conversion is actually a two-step process. First specialized organisms, principally Nitrosomonas, convert ammonia to nitrite (NO_2). Then a second group of microorganisms, principally Nitrobacter, convert nitrite to nitrate. Because of the kinetics of the conversion, nitrite is quickly converted to nitrate and the nitrite is generally ignored. For aeration control purposes, nitrification may be regarded as simply converting ammonia to nitrate. By accomplishing the conversion in the aeration system instead of in the receiving water, the depletion of oxygen in the receiving water is eliminated.

The concentration of ammonia in the wastewater is measured as ammonia nitrogen. In other words, only the nitrogen component is considered. Typical influent and effluent concentrations of ammonia nitrogen are 40 and 3 ppm, respectively. However as with most process factors, there is considerable variability in these values.

Unlike BOD removal, nitrification has a stoichiometric relationship that governs the oxygen utilization for ammonia converted and the oxygen required. It takes 4.6 lb O_2 to convert 1 lb NH_3 to NO_3. This is the oxygen utilization rate for nitrification. Despite the lower concentration of ammonia relative to BOD in the influent for most plants, this ratio means that nitrification can easily double the total process air requirement:

Equation 3.11

$$Q_s = \frac{w_{NH_3} \cdot 4.6}{\text{AOTE} \cdot 24.84}$$

where

Q_s = air flow required to meet demand, SCFM
w_{NH_3} = mass loading rate of ammonia, lb/day
AOTE = actual site oxygen transfer efficiency, decimal

Equation 3.12

$$P_a = \frac{w_{NH_3} \cdot 4.6}{\text{AE} \cdot 24}$$

where

P_a = aerator shaft power required, bhp
AE = actual site aerator efficiency, lb O_2/(hp h)

Achieving nitrification requires longer retention time than BOD removal and a longer sludge age, often referred to as solids retention time (SRT). Although both BOD removal and nitrification rates are temperature dependent, the nitrification process is influenced by temperature fluctuations much more than BOD removal. Once the population of nitrifying organisms has developed, it is stable, but the initial development of the correct population, particularly in cold water, can be a problem. Therefore many operators will continue to operate a plant to achieve nitrification year-round even though their permit restricts ammonia discharge only in summer months.

Unintentional nitrification is a common occurrence in underloaded systems, particularly in warm weather. Many plants whose permits only require BOD and TSS removal also nitrify as a side effect of long retention times. This results in an increase in air demand since maintaining the required DO concentration requires meeting the air demand for both BOD removal and nitrification. A greater problem that often results from unintentional nitrification (and sometimes from intentional nitrification as well) is floating sludge in the secondary clarifier. A population of denitrifying bacteria develops, which, in the absence of oxygen in the clarifier, strips oxygen from nitrate to support metabolism. That releases nitrogen gas, which can form bubbles that become trapped in the sludge blanket. The bubbles can create enough buoyancy to make the sludge rise to the top of the clarifier. This can cause the TSS in the final effluent to exceed permit levels and can also decrease the concentration of the RAS.

3.1.4 Denitrification

Biological nutrient removal is used to remove excess nitrogen and phosphorus (nutrients) from the wastewater. Removal of both compounds biologically requires alternation of aerobic and anoxic or anaerobic conditions. This creates a compelling process performance need for accurate aeration control because failure to maintain the proper range of DO in the various process areas will result in process failure. In the context of minimizing energy through aeration control, the nitrogen removal process is more important than phosphorus removal.

Many discharge permits require complete removal of nitrogen—particularly when the receiving water contains populations of sensitive marine organisms (such as shellfish) and where there are concerns of nitrate entering the drinking water supply. Biological denitrification occurs after nitrification has converted ammonia to nitrate. In an anoxic environment, where the DO concentration is less than 0.5 mg/l, bacteria use the oxygen from nitrate to metabolize carbonaceous compounds. The carbon compounds are generally obtained by combining influent wastewater with nitrified wastewater recycled from the effluent of the aeration system.

There is a tremendous variety of process configurations used for denitrification. One of the most common is the modified Ludzack–Ettinger (MLE) process, which itself has many variations. In this process, the first section of the aeration basin is devoted to an anoxic zone in which wastewater and RAS are introduced to the basin. This is followed by an aeration zone for BOD removal and nitrification. Mixed liquor

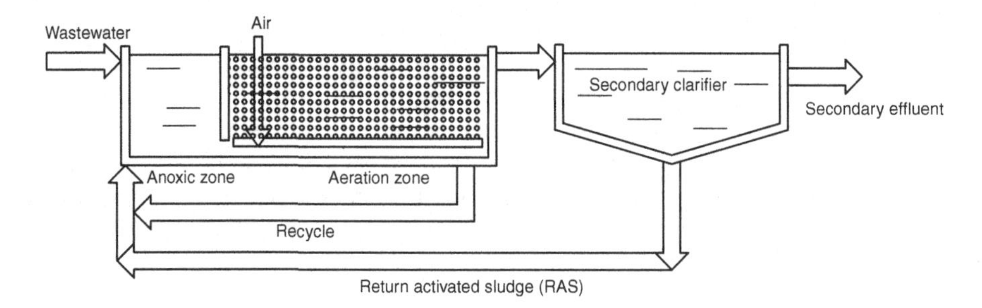

FIGURE 3.9 *MLE process*

from the effluent end of the aeration zone is recycled to the anoxic zone where the nitrates are converted to nitrogen and oxygen using the raw wastewater as the carbon source (see Figure 3.9). SBRs and other systems perform denitrification within a single basin by alternating cycles of aeration and anoxic operation.

There are many other proprietary processes for nitrogen removal, including some that provide simultaneous nitrification and denitrification. These processes rely on the gradients of substrate and dissolved oxygen as they diffuse into the floc. The outer layer provides BOD removal and nitrification. The inner layer of the floc is anoxic because the available oxygen is used by the outer layers. This type of process requires precise control of DO concentration to maintain the proper gradients within the floc.

In BNR systems, the mixed liquor DO concentration is generally maintained at a low level, often near or below 1.0 ppm. This ensures that the anoxic zone operates correctly. Denitrification can be considered an energy conservation measure in and of itself. In the denitrification process, some of the BOD is metabolized using oxygen from nitrate which recovers oxygen previously provided in the aeration system during nitrification. Approximately 25% of the oxygen utilized for nitrification is recovered thus reducing the total air supplied to the aeration system by that amount.

3.2 LOADING VARIATIONS AND THEIR IMPLICATIONS

Many biological processes used in wastewater treatment exhibit a degree of stability and inherently respond to some fluctuations in loads. If the influent to the aeration system was consistent, or even fluctuated within a narrow band, automatic control would be unnecessary. This is far from the case in most plants. Variations in hydraulic and organic loads of 2:1 or 3:1 are common (see Figure 1.4). Without a control system to modulate the air supply to match these load variations, the plant would be overaerated or underaerated for large portions of each day. Although the process may accommodate these fluctuations without treatment failure, they have an unfavorable impact on both energy and process efficiency. There are savings obtainable by matching aeration energy to the process demand because the natural tendency for operators is to match peak demand and then use a little extra air "just in case." Ideally the aeration energy should follow the diurnal loading pattern exactly.

Some impacts of underaeration are obvious. If the mass flow rate of oxygen supplied to the aeration basins is not high enough to meet the biological process demand, the wastes will not be metabolized. This results in high BOD concentrations in the plant effluent. Evidence of inadequate oxygen supply is low DO concentration in the mixed liquor. If the rate at which oxygen is supplied to the basins is lower than the rate it is utilized by the biology, the DO will drop. Conversely, low loads will result in increased DO. There are limits on the fluctuations that the process can tolerate without impacting pollutant removal or microbe activity.

More isn't better—both process and energy efficiency are compromised by overaeration. For both mechanical and diffused aeration, oxygen transfer efficiency decreases exponentially as mixed liquor DO concentration increases (see Chapter 4). Obviously aeration beyond the needs of the process wastes energy.

One of the factors complicating energy conservation in aeration systems is the occurrence of process loading peaks that correspond with peak electrical demand. This isn't particularly surprising because both result from the same pattern of human activity. The impact of the diurnal wastewater flow pattern on energy cost is unfavorable, since the peak air demand almost always occurs during the time of day that has peak energy cost.

The peak process loading can be attenuated by equalization basins. By storing peak flow and pumping it into the aeration system during low loading and off-peak periods, the energy cost can be significantly reduced. This strategy reduces both energy consumption and the peak demand. In many systems, nighttime aeration requirements are dictated by mixing and minimum air flow restrictions. By moving some of the process demand to nighttime, all of the oxygen supplied to the basin can be utilized for process, reducing total energy use as well as cost. There are energy requirements associated with equalization for pumping and aeration to prevent septicity and odors. These must also be included in the system analysis. In general equalization is very cost-effective.

Seasonal variations occur in both process loads and permit requirements. Industrial contributions from food processing may be seasonal or tourism may add to loads during certain times. Many plants have stricter standards for warm weather—particularly for nitrogen. These variations may dictate a high level of flexibility in the process equipment and operations. The well-designed control system must accommodate and enhance that flexibility.

In most climates, seasonal changes also affect the temperature of the plant influent. This has multiple effects. As indicated above, some microbial populations are sensitive to water temperature, and may not develop well in cold water. Almost all biological activities increase with warmer mixed liquor temperature which changes the OUR and removal rates. This may shift demand from one portion of the aeration tank to another, requiring rebalancing of the air in plug flow basins. Counterintuitively, the solubility of oxygen actually decreases in warm water, which decreases the OTE. The result is that air demand generally increases during summer months.

The combined effect of these variations can create challenges for the aeration control system. First of all, the demand for aeration energy and air flow will fluctuate with process changes and a control system must be adapted to the wide range of air demand. The ability to add and remove units is essential to matching aeration energy to process demand. Less obviously, many of these fluctuations are extremely nonlinear in nature. This poses serious challenges for tuning the controls. In some cases, the control algorithm can include temperature compensation and mitigate some of the fluctuations in tuning. In most cases, the expectation should be that seasonal retuning is required. It is very difficult to achieve control stability and reasonable responsiveness with the nonlinearity inherent in the aeration process.

Warm weather generally presents the most operational challenges for aeration systems. Several factors combine to make this the case. First, warm moist air is not as dense and contains less oxygen in a given volume. Second, as mentioned above, warm water has a lower saturation concentration of oxygen. Third, increased biological activity makes the initial process demand for oxygen higher.

The aeration equipment and control systems must be able to accommodate swings in demand from a fraction of the average daily flow to many times the average daily flow. The time frame for load variations spans seconds for some types and months for others. Sidestreams can result in large changes in process demand very quickly. Seasonal water temperature changes occur over weeks or months. The control strategy must be able to provide optimized aeration to meet process needs at minimum energy requirements through all of these fluctuations.

3.3 PROCESS LIMITATIONS AND THEIR IMPACT ON CONTROL SYSTEMS

The variability of the process itself offers a variety of challenges in control system operation. The physical and mechanical limitations of the process equipment also impose further constraints on the operation and may limit optimization opportunities.

Settling of sludge in the aeration basins can occur if insufficient air is supplied to the aeration basins. The minimum air rate required to maintain mixed liquor solids in suspension is the "mixing air" flow rate. The actual value is dependent on the type of diffusers, diffuser arrangement, the MLSS concentration, and the nature of the sludge. A conservative value for tanks with full coverage fine-pore diffusers is 0.12 SCFM/ft^2 of water surface area:

Equation 3.13

$$Q_{\text{mix}} = (l \cdot w) \cdot Q_{\text{req}}$$

where

Q_{mix} = total mixing air flow to basin(s), SCFM
l = tank length, ft
w = tank width, ft
Q_{req} = required mixing air rate, SCFM/ft^2

Different diffuser and tank configurations will have different mixing requirements, and the diffuser supplier should be consulted for recommendations. Note that the typical recommended air flows are quite conservative, and field experience has shown that air flow equal to 2/3 of the "standard" value is usually adequate to keep solids in suspension. Visual observation and testing is recommended during system commissioning to verify the actual requirements.

Systems with mechanical aeration also have minimum mixing requirements. A range of 0.5–1.5 hp/1000 ft^3 of tank volume is typical. Many systems, both diffused and mechanical aeration, employ separate mechanical mixers to maintain solids in suspension. This is often the case where mixing energy requirements exceed the energy needed to meet process oxygen demand. Another strategy when mixing dominates the aeration energy is intermittent aeration. The solids are allowed to settle for a short period and are resuspended when aeration is restored. This is accomplished by combined aeration and mixing cycle alternating with a nonaerated cycle.

Fine-pore diffused aeration systems generally have a minimum air flow per diffuser that must be maintained. Lower air rates may result in fouling of the diffusers by biological growths on the surface exposed to the mixed liquor. Diffusers also have a maximum recommended air rate. This may be dictated by high differential pressures that could cause rupture of membranes. In the case of ceramic or coarse bubble diffusers, the increase in pressure drop through the diffuser could cause discharge pressure to exceed the blower capacity.

Many operators feel that high blower discharge pressure will result in diffuser damage. This is the result of misinterpretation of the physical constraints. Excess pressure drop across membranes may cause damage. However, the pressure on the mixed liquor side of the diffusers is dictated by the static pressure resulting from submergence. The pressure difference across the diffuser is a result of the friction of air passing through the orifices and diffuser pores, which in turn is a function of air flow alone. High blower discharge pressure will only affect the diffusers if it results in high air flow through the diffusers. This is rare because in most cases a high discharge pressure is caused by pressure drop through the control valves upstream of the diffusers.

Nutrients are required for cell growth and metabolism. A minimum ratio of BOD to nitrogen and phosphorus must be maintained or synthesis and removal of the wastewater constituents will be impaired. This is more often a problem in industrial wastes than in plants treating only domestic wastewater.

Improper DO concentrations can limit treatment performance. As indicated above, if sufficient oxygen is not available, the microorganisms can't metabolize BOD or convert ammonia to nitrate. In this eventuality, the process is "oxygen limited." Operation below or above certain threshold concentrations can also cause undesirable organisms to form. Above the oxygen concentration needed to satisfy demand and provide an aerobic floc, increased DO does not accelerate biological activity for BOD removal. The DO concentration to maintain aerobic floc varies with the specific facility, but in general DO above 1.0 ppm is adequate. Some facilities report excellent process performance with DO consistently between 0.5 and 1.0 ppm.

In the case of nitrification, there is an increase in ammonia conversion with increasing DO concentration, but only up to approximately 3 ppm. Furthermore, if there is sufficient hydraulic residence time and adequate microbial population, full nitrification can be achieved within the aeration basin at much lower DO concentration. Since OTE drops with higher DO, it is more energy efficient to operate the process at the lowest DO level that meets the performance objectives.

This is a very important observation! Although DO concentration can limit treatment, it is not a driving force in substrate removal. Maintaining a target dissolved oxygen concentration is an indirect indicator of process performance. Insufficient DO leads to process failure, but the correct DO does not guarantee treatment levels will be met. Adequate treatment can be achieved across a wide range of DO concentrations. More importantly, excessive DO will not improve treatment—*it just wastes energy.*

Low DO concentration will cause additional problems in the biology itself. Sustained operation at DO < 0.5 mg/l will result in the development of excess populations of filamentous organisms. One common problem in filamentous organism is Nocardia. This common filamentous organism can result in blanketing the aeration basin with dark foam that may reach several feet in thickness. Nocardia is a common, but not the only, source for foam on the aeration basins.

Foam from a variety of causes is a chronic problem in many treatment plants. In extreme cases, the foam depth is sufficient to allow it to run over the sides of the tanks and blow about in the wind. This is obviously unsightly and unpleasant. The accumulation on walkways and grating also represents a safety hazard. The foam may pass on from aeration basins to digesters. Foaming in diffused aeration systems is exacerbated by high air flow rates.

Some population of filamentous organisms must be present in order to properly develop floc, since the filaments generally form the core or "backbone" of the floc. However, excess filaments will extend beyond the floc and even bridge floc particles. The result is hindered settling—the sludge won't settle fast enough to be removed in the clarifier. Instead it passes over the weirs and into the effluent. Imagine the difference in velocity when dropping a feather as opposed to a pebble. The poor settling, often referred to as sludge bulking, both increases the TSS of the effluent and decreases the RAS concentration. Beggiatoa is a common microbe that occurs after extended operation at low DO concentrations and causes sludge bulking.

Filamentous populations are more difficult to remove than to create. Various strategies are employed to remove filamentous organisms from RAS and mixed liquor. Some of these are: chlorination, chemicals, changes in sludge wasting, DO increases, and so forth. Because of the difficulty of eliminating filamentous organisms, operators are justifiably adamant about refusing strategies that have a possibility of creating populations of filamentous organisms.

Paradoxically, one of the best mechanisms for preventing filamentous organisms from becoming a problem is preceding the aeration system with an anoxic selector zone. By controlling the residence time and creating the proper conditions, this selector zone interferes with filamentous metabolism and reduces their population.

There are process problems resulting from excess aeration as well. Settling of suspended solids in the secondary clarifier depends on a well-formed floc. The turbulence associated with high levels of aeration can break up the floc particles. This may lead to dilution of the RAS and WAS and excess TSS in the clarifier effluent.

There are filamentous organisms that develop if DO concentrations are high—usually above 6.0 ppm—for extended periods of time. As with organisms caused by low DO, these can extend beyond the floc, hinder settling in the clarifiers, and increase foam formation.

The time dependency of bacterial growth and substrate metabolism and conversion is referred to as "process kinetics." Both of these factors are important. The growth kinetics will determine the amount of sludge produced in the process. The excess sludge yield must be removed as WAS and processed prior to disposal. The kinetics of substrate removal determines the required hydraulic residence time for treatment. As with many biological and chemical processes, the kinetics of treatment can be characterized mathematically by a series of Monod equations, specific to each process. These take the general form of:

Equation 3.14

$$r_{su} = -\frac{\mu_{max} \cdot X \cdot S}{Y \cdot (K_S + S)}$$

where

r_{su} = rate of substrate concentration change, g/(m^3 d)
μ_{max} = maximum bacterial growth rate, g/(g d)
X = biomass concentration, g/m^3
S = growth limiting substrate concentration, g/m^3
Y = yield coefficient, g/g
K_s = half velocity constant, g/m^3

The various coefficients in the Monod equation are empirically determined, and are waste and process specific.

Wastewater reactions are sometimes referred to as involving "zero-order" or "first-order" kinetics. A zero-order reaction proceeds at a constant rate regardless of substrate concentration: $r_{su} = \pm k$. A first-order reaction proceeds with the rate of removal proportional to the remaining substrate concentration, C: $r_{su} = \pm k \cdot C$. The details of reaction kinetics and the determination of the various coefficients are not properly part of the control system design or energy conservation. They are part of the plant process design. However, familiarity with the terms can be helpful. Further, these reaction equations are the basis for software simulations of the process performance.

The simulation of process performance can be a very valuable tool in system design and in analyzing existing operations. In the face of any change to a successful operating methodology, the operators will properly ask what impact the change will have on treatment. This is particularly true if the proposed ECMs involve taking aeration basins or other equipment out of service. Process simulation can provide data on the impact of the change, indicate the sensitivity of the process to the proposed modifications, and provide guidance for other process changes that may impact energy consumption.

It is important to note that process simulation, like any program, is dependent on accurate input data. It should also be remembered that it is based on a mathematical model of a very complex process system. The software may miss the effect of some parameters on performance, while the actual process operation misses nothing. However, process modeling is increasingly used as a design and operational tool, and the benefits for control strategy development and ECM effectiveness should not be ignored. Sensitivity analysis, which identifies the direction as well as the magnitude of the process response to changes in operating parameters, can be particularly useful.

It should be obvious that controlling the aeration process is a complex activity with a level of difficulty far beyond comparatively simple processes such as HVAC control. The interrelated nature of the process equipment must be considered in the system design—particularly in the analysis of performance and projected savings. Attempting to grasp the entire system at once is daunting. Successful implementation requires breaking down the various sections, analyzing them independently, and then pulling the pieces together into a coherent whole.

EXAMPLE PROBLEMS

Problem 3.1

A plant uses airlift pumps for pumping RAS from three clarifiers to the aeration basin influent channel. An operator has proposed replacing them with centrifugal pumps and VFDs to provide better operational control and save energy. The RAS flow is 500 gpm per clarifier and the discharge point is 3 ft above the clarifier water surface with an additional 2 ft of friction head in the piping. The current pumps are 8 in. nominal diameter, and the air entry is 12 ft below the water surface. The variable speed aeration blowers have an average efficiency of 70% and the discharge pressure is 8.0 psig with inlet pressure of 14.5 psia and barometric pressure is 14.7 psia. Assume the new pumps will have an average efficiency of 65%. Both the pump and the blower motors have a nameplate efficiency of 95% and the VFDs are 96% efficient.

(a) Are there energy savings?

(b) If the energy cost is \$0.07/kWh, new pumps cost \$4000 each installed and the VFDs cost \$1000 each installed, what is the payback period?

Problem 3.2

The plant in Problem 3.1 has aeration basins with influent channels 80 ft long total and effluent channels 60 ft long. A small PD blower is proposed to aerate the channels for mixing instead of using air from the aeration tank blowers. The diffuser submergence will be 6 ft. If the blower installed cost is $10,000 and the average efficiency is 65%, will this be cost-effective?

Problem 3.3

A complete mix tank is 40 ft square and has a side water depth of 14 ft. It is equipped with a mechanical aerator, 25 hp with an actual site aeration efficiency of 2.0 lb O_2/(hp·h). The manufacturer recommends a minimum of 0.75 hp/th cu ft for mixing. The average daily flow rate to the tank is 0.5 mgd. Influent BOD is 165 ppm and 90% removal is required. The utilization rate is reported as 1.1 lb O_2/lb BOD removed.

(a) Will the tank be mixing limited or oxygen limited?

(b) If the tank is upgraded with fine-pore diffusers with an actual site OTE of 10% what air rates will be required for mixing and for process performance?

Problem 3.4

In the spring, the mixed liquor temperature at a WWTP increases from 50 to 60 °F. List at least three effects that this may have on the process.

Chapter 4

Mechanical and Diffused Aeration Systems

The first step in the evaluation process is determining the oxygen requirements of the process. The next step is to examine the characteristics of the oxygen transfer equipment.

In the initial system assessment, the energy and oxygen requirements for aeration are identified, and potential ECMs are examined. The assessment should always include an evaluation of the aeration devices, but this is especially true in older facilities. Replacing the existing aeration devices with more efficient devices can provide savings of 50% or more in some cases. A replacement usually comes with high capital cost. It may take months or years to implement therefore the payback needs to be calculated to ensure the replacement is cost-effective. The secondary treatment process is impacted by aeration system replacement so regulatory agency approval or review may be required. If the time required for the upgrade implementation is long, other ECMs should be implemented in the interim—even if they have lower savings or poorer payback. There is no point in forgoing two years of savings from other ECMs while waiting for the aeration system upgrade to be implemented.

Savings calculations for upgrading the aeration equipment cannot be examined in isolation. Because the aerators are such a critical part of both process and energy performance, any analysis must include them. However, the performance of the aeration equipment is also affected by other unit processes. For example, range limitations on the blowers may reduce the potential savings from more efficient diffusers. Oversized valves may make accurate air flow control and DO control impossible. A blower replacement that is cost-effective in an existing system may

Aeration Control System Design: A Practical Guide to Energy and Process Optimization, First Edition. Thomas E. Jenkins.

not provide a reasonable payback at reduced air flows obtained with new diffusers. Conversely, existing blower range limitations may make their replacement essential if the savings from alternate aeration devices are to be obtained.

It is important to understand the type and operating characteristics of the aeration devices employed at a given facility in order to develop the correct control strategy. Many aeration control system designers and programmers skip this step in the design process. They assume that a conventional feedback loop will provide control of the basin DO concentration and the blower air flow rates. This is true if the oxygen demand operates within a fairly narrow range and the process equipment is sized appropriately. This is rarely the case in either municipal or industrial treatment. The aeration equipment performance is very nonlinear in nature, and normal operation often reaches the upper and lower limits of system capacity. If the characteristics of the specific aeration devices employed are not properly accommodated, the control system is likely to be unstable and inadequate.

4.1 OXYGEN TRANSFER BASICS

There are a tremendous variety of commercially available aerators. The fundamental operating principle of all aerators is the same: air and mixed liquor are brought into contact with each other, the oxygen passes through the air/water interface layer then diffuses into solution in the mixed liquor, and then diffuses into the floc and microorganisms within the floc particles. As with any process of this type, the rate and direction of the solution process is a function of the relative proportions of the solute (the substance being dissolved: oxygen) in the original air mixture and the concentration in the solvent (the base substance: mixed liquor, primarily water).

The amount of oxygen dissolved in the water is usually expressed as concentration in milligrams per liter (mg/l). (*Note*: 1 mg/l = 1 ppm = 1 g/m^3):

Equation 4.1

$$C = \frac{m_{\text{solute}}}{V_{\text{solvent}}}$$

where

C = concentration of solution, mg/l
m_{solute} = mass of solute dissolved in the solvent, mg
V_{solvent} = volume of solvent, l

It is possible to express concentration in other terms, such as percent or molarity, but these are not generally used for aeration systems. The concentration may be given as percent saturation, but this is also rarely used and can lead to confusion.

The saturation concentration of oxygen in water is an important consideration in aeration. Saturation concentration is the maximum amount of oxygen that can be dissolved in water. The difference between saturation concentration and actual concentration represents the "driving force" that causes the oxygen to move from the gaseous solution (air) to the liquid solution (mixed liquor). The farther the concentration of oxygen dissolved in the mixed liquor is from saturation, the higher the rate of oxygen dissolution will be (mass transferred per unit time).

The concept of oxygen driving force is extremely important. The savings obtained by DO control are chiefly obtained by improving the oxygen transfer efficiency of the aeration process through operating at the lowest DO concentration which provides proper process performance. It seems intuitive that lower air flow will result in lower DO concentrations, but understanding the mechanism aids in calculating savings. Even more important is realizing that the effect of concentration on oxygen transfer efficiency is extremely nonlinear.

A target DO concentration of 2.0 ppm is commonly used as the basis of comparison for DO control savings and for estimating aerator performance. The conventional wisdom among operators is that this is a "safe" DO level for maintaining process performance. Many regulatory agencies recommend or require that the aeration system be capable of maintaining a DO concentration of 2.0 ppm at worst-case loading. They do not require that the aeration system actually operate at that level at all times. Using 2.0 ppm as a baseline concentration is convenient for demonstrating the effect of driving force on diffused aeration requirements:

Equation 4.2

$$\frac{Q_{\text{as}}}{Q_{2.0}} = \frac{C^*_{\infty\text{f}} - 2.0}{C^*_{\infty\text{f}} - C_{\text{a}}}$$

or

$$C_{\text{a}} = C^*_{\infty\text{f}} - \frac{Q_{2.0} \cdot \left(C^*_{\infty\text{f}} - 2.0\right)}{Q_{\text{as}}}$$

where

Q_{as} = actual required air flow, SCFM
$Q_{2.0}$ = required air flow to maintain 2.0 ppm DO, SCFM
$C^*_{\infty\text{f}}$ = steady-state value of dissolved oxygen saturation concentration at infinite time in field process water, ppm
C_{a} = actual DO concentration, ppm

For mechanical aeration, the oxygen transfer rate (OTR) or aerator efficiency (AE) may be similarly calculated using Equation 4.11. Since aerator power is approximately proportional to the oxygen transfer rate and blower power is approximately proportional to air flow rate, the ratio in Equation 4.2 represents the ratio of required aeration power to actual aeration power when the aeration system is operated at a higher than necessary DO concentration (see Figure 4.1). The impact of driving force is dramatic; it typically requires twice as much power to maintain 6.0 ppm DO as it does to maintain 2.0 ppm.

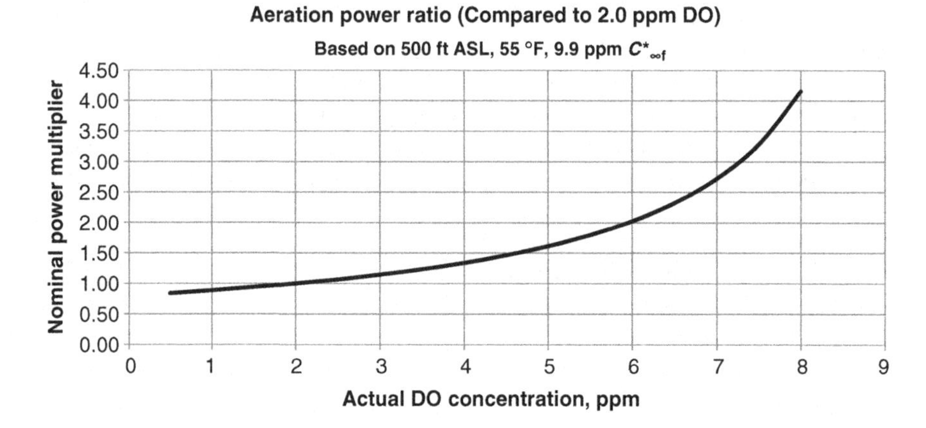

FIGURE 4.1 *Effect of DO concentration on oxygen transfer requirements*

Saturation concentration is not a true constant because the amount of oxygen that will dissolve in a given volume of water varies with the water pressure, the barometric pressure, and temperature. As one would expect, if the barometric pressure decreases, then the solubility of oxygen also decreases (see Table 4.1 and Figure 4.2). If the air is introduced at a submerged point where water and air pressure are higher than atmospheric, then solubility increases. Note that higher solubility not only increases the amount of oxygen that may be dissolved in the mixed liquor, but also increases the rate of the transfer. This is why deeper tanks are often preferred to shallow ones. This is also why the saturation concentration is usually expressed at "infinite time." In practice, this means that sufficient time has elapsed to allow the DO concentration to asymptotically approach the ultimate point of stability. The depth of the aeration basin and barometric pressure also affect mechanical aeration.

The saturation concentration of most gases, including oxygen, decreases with increasing temperature. This is contrary to common expectation, since a solid

TABLE 4.1 Variation of DO Saturation with Barometric Pressure

Barometric Pressure, psia	DO Saturation, ppm				Approx. Elev. FASL
	5 °C (41 °F)	10 °C (50 °F)	20 °C (68 °F)	30 °C (86 °F)	
14.70	12.77	11.29	9.09	7.56	–
14.50	12.60	11.14	8.97	7.45	400
14.00	12.16	10.75	8.65	7.19	1400
13.50	11.72	10.36	8.33	6.92	2400
13.00	11.29	9.97	8.02	6.65	3400
12.50	10.85	9.58	7.70	6.38	4400
12.00	10.41	9.19	7.39	6.11	5400
11.50	9.97	8.81	7.07	5.84	6400
11.00	9.54	8.42	6.75	5.58	7400

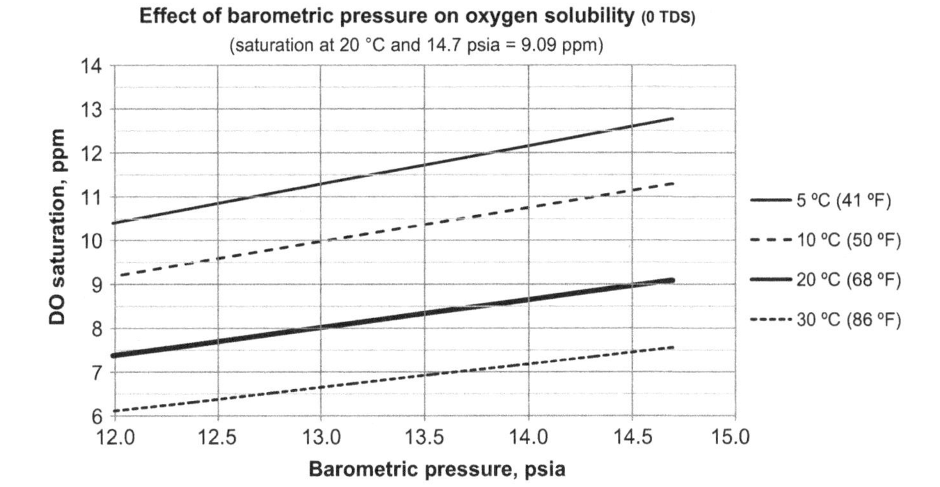

FIGURE 4.2 *Effect of altitude on DO saturation*

dissolved in a liquid usually has greater solubility at higher temperature. The decreased solubility of oxygen at higher temperature is an important factor in aeration system design. Biological activity generally increases with higher mixed liquor temperature thereby increasing the oxygen uptake rate and oxygen demand in the influent end of the aeration basins. Solubility decreases at higher temperature which makes the oxygen transfer more difficult. In defining saturation concentration, the norm is to use 20 °C (68 °F) as the standard condition.

The increase in mixed liquor temperature is generally coincident with increased ambient air temperature, decreasing the capacity of blowers in diffused aeration systems. This explains why warm weather is generally the most problematic period for aeration systems.

Salinity, usually expressed as total dissolved solids in parts per thousand (TDS, ppt), also affects the solubility of oxygen in water. Increased salinity level decreases the saturation concentration of oxygen (see Figure 4.3). In most wastewater, this effect is small because of the low TDS, but salinity at a given site should be verified if possible. It is worth noting that in clean water testing of aerator performance, the chemicals used to deplete the DO concentration during the test may lead to an appreciable concentration of dissolved solids, and the effect of TDS should be included in the efficiency evaluation.

In many cases, adequate accuracy in measuring TDS levels can be obtained by measuring the conductivity of the water and correlating this to TDS concentration. Conductivity has the advantage of being a simple and rapid measurement, but it should be remembered that the resulting TDS value is an approximation. The most accurate measurement of TDS is obtained by evaporating a known volume of water and measuring the mass of the residue.

There are other factors that affect oxygen transfer efficiencies. The oxygen must diffuse through the interface layer between the air and the mixed liquor. For

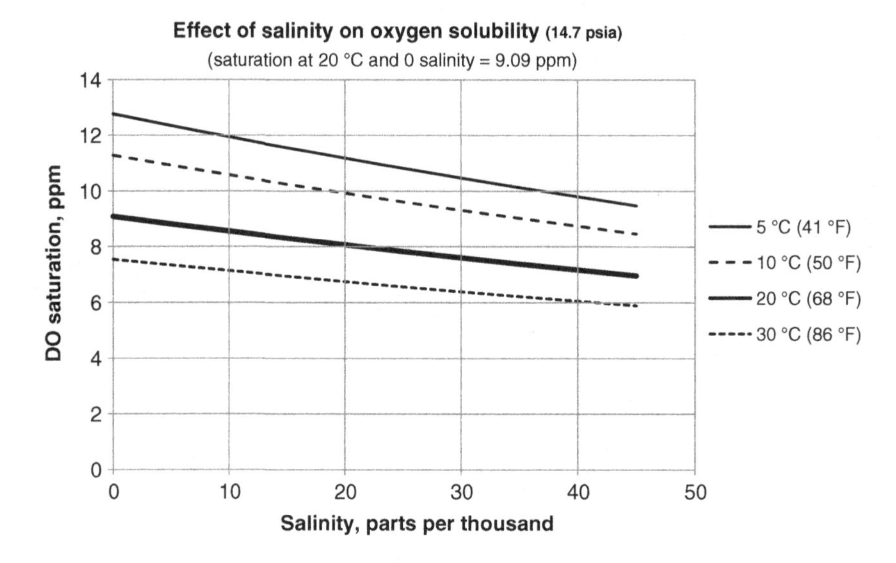

FIGURE 4.3 *Effect of salinity on DO saturation*

mechanical aeration, this may mean diffusing from the atmosphere into droplets of the mixed liquor. In diffused aeration and in many types of mechanical aeration, there is diffusion through the wall of a bubble rising through the wastewater and into the mixed liquor. The diffusion process is affected by surfactants and other compounds that alter the surface tension at the interface layer, by the MLSS concentration, and by other constituents in the mixed liquor.

Aerator effectiveness generally deteriorates with time, particularly for diffused aeration systems. The efficiency of mechanical aerators will be affected by coating of the mechanical components, wear, or physical damage. Diffuser efficiency will decrease with organic and mineral coatings accumulating on the diffuser surface and loss of the elasticity of membrane materials. A fouling factor for diffused aeration is usually applied to the design process to account for these variations. The fouling factor may be ignored for mechanical aeration.

All of the combined effects described above make the oxygen transfer and the power requirements of the aeration devices very site specific. In order to have a rational basis of comparison between aeration devices, the American Society of Civil Engineers has developed a test protocol for determining oxygen transfer characteristics under standardized conditions. These conditions are clean (potable) water, 14.7 psia barometric pressure, 68 °F (20 °C) wastewater temperature, and an initial dissolved oxygen concentration in the water of 0.0 ppm. The procedures are defined in ASCE 2-91, *Measurement of Oxygen Transfer in Clean Water.* Alternate procedures have also been developed for testing oxygen transfer at a specific site and conversion of these the results to standardized conditions, ASCE 18-96, *Standard Guidelines for In-Process Oxygen Transfer Testing*. The most common technique for in-process testing is the offgas method.

For diffused aeration systems, the test results are usually reported as standard oxygen transfer efficiency (SOTE, %), which is the mass of oxygen dissolved in the wastewater as a percentage of the mass of oxygen supplied to the basin. For mechanical aerators, the test results are usually reported as standard oxygen transfer rate (SOTR, lb_m O_2/h), which is the rate of oxygen transfer. The results for either type of aerator may also be reported as standard aeration efficiency (SAE, lb_m O_2/(hp h)), which is the mass of oxygen transferred per unit energy input. If the blower power requirements or efficiency are known and stated, the SOTE data for diffused aeration systems can be converted to SAE:

Equation 4.3

$$\mathrm{SAE_m} = \frac{\mathrm{SOTR_m}}{P_\mathrm{m}}$$

where

$\mathrm{SAE_m}$ = standard aerator efficiency, lb_m O_2/(hp h)
$\mathrm{SOTR_m}$ = aerator oxygen transfer rate, lb_m O_2/h
P_m = aerator power draw, hp

Equation 4.4

$$\mathrm{SOTR_d} = Q_\mathrm{s} \cdot \mathrm{SOTE} \cdot \rho \cdot C_{\mathrm{O_2}} \cdot 60\frac{\mathrm{min}}{\mathrm{h}}$$

where

$\mathrm{SOTR_d}$ = diffuser oxygen transfer rate, lb_m O_2/h
Q_s = air flow rate, standard cubic feet per minute (SCFM)
SOTE = diffuser standard oxygen transfer efficiency, decimal
ρ = density of air at standard conditions, 0.075 lb_m/ft^3
$C_{\mathrm{O_2}}$ = concentration of O_2 in air, decimal, 0.23 (23% by weight)

This simplifies to:

Equation 4.5

$$\mathrm{SOTR_d} = Q_\mathrm{s} \cdot \mathrm{SOTE} \cdot 1.035$$

These equations may be used to calculate actual aerator efficiency and oxygen transfer rates by substituting the values for field OTR and OTE for the values at standard conditions.

There are many formats for the equation to calculate blower power. One convenient format is:

Equation 4.6

$$P_\mathrm{b} = \frac{(Q_\mathrm{s}/60) \cdot \gamma \cdot R \cdot T_\mathrm{i}}{550 \cdot (k-1)/k \cdot \eta_\mathrm{b}} \cdot \left[\left(\frac{p_\mathrm{d}}{p_\mathrm{i}}\right)^{(k-1)/k} - 1\right]$$

where

P_b = blower shaft power, bhp
Q_s = air flow rate, standard cubic feet per minute (SCFM)
γ = specific weight of air at standard conditions, 0.075 lb_f/ft^3
R = gas constant for air, 53.3 ft lb_f/lb_m °R
T_i = blower inlet air temperature, °R (°R = °F + 459.67)
k = ratio of specific heats for air, $C_p/C_v = 1.395$
$(k-1)/k = 0.283$, dimensionless
η_b = blower efficiency, decimal
p_d = blower discharge pressure, psia (psia = psig + barometric pressure)
p_i = blower inlet pressure, psia

Refer to Chapter 5 for additional information on determining blower power requirements.

The US customary system of units sometimes gives rise to confusion in the terms "density," "specific weight," and particularly "pounds." A pound is technically a force, as is weight. A weight of one pound is the force (lb_f) exerted by one pound of mass (lb_m) in a gravitational field of 1 g, 32.17 ft/s^2. Based on Newton's first law, $F = ma$, a constant of proportionality is required. This constant, g_c, is equal to 32.17 lb_m ft/lb_f s^2 but is often given simply as 32.2 ft/s^2. Another unit of mass in the US customary system is the "slug," which is equal to 32.17 lb_m.

The definition of density is mass per unit volume. It should be expressed as lb_m/ft^3 when using US customary units. The symbol used in this text for density is "ρ." In common usage density is intended to reference weight per unit volume, or lb_f/ft^3, although this is technically incorrect. Weight per unit volume is more correctly referred to as "specific weight." In this text, the symbol for specific weight is "γ." In some cases, using weight instead of mass and density instead of specific weight doesn't matter. In other cases, particularly in calculations, it makes a great deal of difference.

Until or unless the metric system completely displaces the US customary units, it is important to define the terms properly and be aware of the correct usage at all times. In the wastewater treatment industry in the United States and Canada, there are some cases where the metric system units are more common. An example is dissolved oxygen concentration, where mg/l is generally used. In other cases, US customary units are most common. An example of this is air pressure, which is usually given in pounds per square inch. The strict definition of "pounds force" is ignored. The units used also vary from location to location. In this text, we will try to use the units most commonly encountered among treatment plant operators in the United States, and clarify when required to avoid confusion.

Combining and simplifying Equations 4.4 and 4.6 yields a more useful formula:

Equation 4.7

$$SAE_d = \frac{2418 \cdot SOTE \cdot \eta_b}{T_i \cdot \left[(p_d/p_i)^{0.283} - 1\right]}$$

where

SAE_d = diffused aeration system standard aerator efficiency, $lb_m\ O_2/(hp\ h)$

It should be noted that the standard aeration efficiency (AE) obtained above is for clean water at defined conditions of wastewater temperature, barometric pressure, and so on. Actual efficiency in field process conditions may be less than 50% of the efficiency under standard conditions. In order to determine oxygen transfer efficiency, aeration efficiency, and oxygen transfer rate under actual field conditions, factors must be introduced to correct for the conditions that affect oxygen solubility:

Equation 4.8

$$OTR_f = \alpha \cdot F \cdot SOTR \cdot \theta^{(T-20)} \cdot \frac{(\tau \cdot \beta \cdot \Omega \cdot C^*_{\infty 20} - C)}{C^*_{\infty 20}} = \alpha \cdot F \cdot SOTR \cdot \theta^{(T-20)} \cdot \frac{C^*_{\infty f} - C}{C^*_{\infty 20}}$$

where

- OTR_f = field oxygen transfer rate for the system operating under process conditions, lb_m/h
- α = correction factor for basin geometry and wastewater characteristics with new aeration devices, dimensionless, typically $\alpha < 1.0$
- F = fouling factor, process water SOTR of a diffuser after a given time in service/SOTR of a new diffuser in the same process water, dimensionless, typically $F \leq 1.0$
- SOTR = standard oxygen transfer rate of new aeration system, lb_m/h
- θ = empirical temperature correction factor, dimensionless, typically $\theta = 1.024$
- τ = temperature correction factor for dissolved oxygen saturation, dimensionless, $\tau = C^*_{sT}/C^*_{s20}$
- β = correction factor for salinity and dissolved solids, = process water C^*_{st}/clean water C^*_{st}, dimensionless, typically $\beta = 0.95$
- Ω = p_b/p_s (for tanks less than 20 ft in depth), dimensionless
- $C^*_{\infty 20}$ = steady-state value of dissolved oxygen saturation at infinite time at 20 °C and a barometric pressure of 14.7 psia, ppm
  - for mechanical aeration the surface saturation concentration at 20 °C is used, and therefore $C^*_{\infty 20} = C^*_{s20}$
  - for diffused aeration $C^*_{\infty 20}$ can be obtained from manufacturer's data
  - for diffused aeration where manufacturer's data is not available $C^*_{\infty 20}$ may be approximated: $C^*_{\infty 20} = 9.09 + 0.1 \times$ ft. submergence
- C = average actual process water dissolved oxygen concentration, ppm
- C^*_{sT} = tabular value of dissolved oxygen surface saturation concentration at actual process water temperature, barometric pressure of 14.7 psia, 100% relative humidity, ppm (see Table 4.1)

C^*_{s20} = value of dissolved oxygen surface saturation concentration at 20 °C, barometric pressure of 14.7 psia, and 100% relative humidity —$C^*_{s20} = 9.09$ ppm

p_b = barometric pressure under field conditions, psia

p_s = standard barometric pressure, psia—14.7 psia

$C^*_{\infty f}$ = $\tau \cdot \beta \cdot \Omega \cdot C^*_{\infty 20}$ = steady-state value of dissolved oxygen saturation concentration at infinite time in field process water, ppm

T = wastewater temperature,

The correction of SOTE to OTE_f (oxygen transfer efficiency) or from SAE to AE_f (field aerator efficiency) can be made using the above equation and substituting SOTE or SAE for SOTR. It is common in the field to substitute the acronyms "AOTE" for OTE_f and "AOR" for SOTR, and these will be used interchangeably.

During the design or evaluation process, the values for the various parameters are assumed to be constant and the values used are estimated based on past experience or manufacturer's recommendations. To do otherwise would make the calculations impractically cumbersome, and in most cases would not result in significant improvements in accuracy. However, it is important for the engineer performing the evaluation to remember that there is in reality a range of values, and this prevents absolute precision in the calculations. Even the most advanced computer modeling is dependent on the accuracy of the characterization of the wastes and treatability and the performance parameters of the aeration system equipment.

Two examples of this variability are the parameters α and blower efficiency. The value of α is the ratio between clean water and process water performance. For diffused aeration a typical assumed value for α is 0.6, and for mechanical aeration a typical assumed value is 0.8. In reality, α varies not only from plant to plant, but by as much as 50% from hour to hour at a given plant as demonstrated by field tests.

Blower efficiency is a similarly variable value. Even for a specific type and model blower, the efficiency is not constant. As with pumps, a centrifugal blower has a best efficiency point (BEP) that is typically close to the design point. Operating at other inlet conditions, air flow rates, and discharge pressures will result in a decrease in blower efficiency. The efficiency of positive displacement (PD) blowers also generally decreases as speed and flow rate decrease.

The variability and uncertainty associated with the calculations should not discourage detailed calculations using all of the information available. Rather, it should induce a realization that the calculations represent a model of the process and energy performance, not the actual performance. If the results of the analysis indicate marginal improvements in energy consumption or a very long payback period on the investment, a common sense based realty check should be applied to verify the justification for the system. The calculations themselves should be gauged to the accuracy of the available data—there is no point in reporting results to 13 decimal places if the original assumptions are only accurate to ±25%.

4.2 TYPES OF AERATORS

Aeration devices fall into two main categories: mechanical aeration and diffused aeration. Within both of these categories, there are a variety of designs and operating characteristics that is almost bewildering. Aeration devices are a very competitive market segment. They represent important and crucial pieces in process performance, energy consumption, and capital equipment cost for an activated sludge wastewater treatment plant. A great deal of ingenuity has been employed in ongoing advancement of the technology. Most designs are proprietary, but broadly similar aeration devices are usually available from multiple manufacturers.

Each category of aerator has advantages and disadvantages. The very earliest aeration systems employed fine-pore diffused air. Over time there have been cycles in the relative popularity of diffused and mechanical aeration. Many factors influence aerator selection, such as:

- Energy consumption
- Simplicity of operation
- Reliability and service life
- Equipment cost
- Maintenance requirements
- Compatibility with existing equipment and tanks
- Site weather conditions
- Land requirements and availability
- Operator experience and preferences

Historically, diffused aeration systems have been the most common. During some periods maintenance requirements have caused fine-pore systems to be displaced by coarse bubble diffusers. Lower energy consumption for mechanical aeration compared to coarse bubble diffusers, combined with process and operating simplicity, led to wide adoption of mechanical aerators in some areas. Further increases in energy cost combined with the development of cost-effective membrane materials for fine-pore diffusers have led to that type of device currently being most common in new applications.

Both mechanical and diffused aeration are used in existing applications, and each has advantages. The trend, however, is to replace mechanical aerators and coarse bubble diffusers with membrane fine-pore diffusers. Increasing energy awareness and expense can be expected to accelerate this trend. Capital expense may keep this replacement from being cost-effective, but in any plant with coarse bubble diffusers or mechanical aeration replacement with more energy efficient aeration should be one of the first ECMs evaluated.

There are common principles in all types of aerators. As indicated above, the physics of oxygen transfer are common to all devices. There are upper limits and lower limits of operation that restrict maximum and minimum oxygen transfer rates and mixing energy. Higher rates of oxygen transfer require higher energy use for all devices.

Two mechanisms are principally responsible for the oxygen transfer for all mechanical aerators. The relative importance of each mechanism is dependent on the specific aerator design. First, the mixed liquor is splashed or thrown into the air as a film or droplets. The surface of the water is in contact with ambient air, and the oxygen can diffuse into the liquid. Second, the action of the aerator impellor displaces the mixed liquor and draws air bubbles into the bulk liquid. The oxygen diffuses from the bubble into the mixed liquor.

Other, comparatively minor, mechanisms for oxygen transfer exist—again with the relative importance dependent on design details. Oxygen can diffuse from the atmosphere into the mixed liquor at the surface of the wastewater, as occurs naturally in lakes, and agitation of the mixed liquor creates greater surface area for this transfer. At the tank walls the velocity of the mixed liquor creates downward movement, which results in some air being carried downward into the wastewater.

The most important aspect for the aeration control system designer is that all types of aerator provide some mechanism for controlling the energy and oxygen transfer rate. This may be directly by controlling the aeration device itself or indirectly by controlling external equipment. Regardless of the mechanism or aeration type, the objective is the same: provide the oxygen requirement of the process at the lowest energy requirement.

4.2.1 Mechanical Aerators

Mechanical aerators fall into two broad categories based on physical configuration (see Figure 4.4). Horizontal shaft aerators are sometimes referred to as "brush aerators" because the construction of some early types resembled bottle brushes. Horizontal aerators are generally used in oxidation ditches (see Chapter 3).

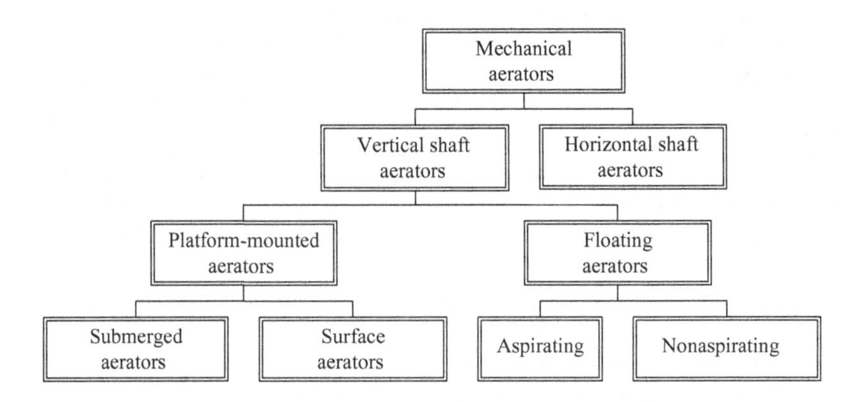

FIGURE 4.4 Types of mechanical aerators

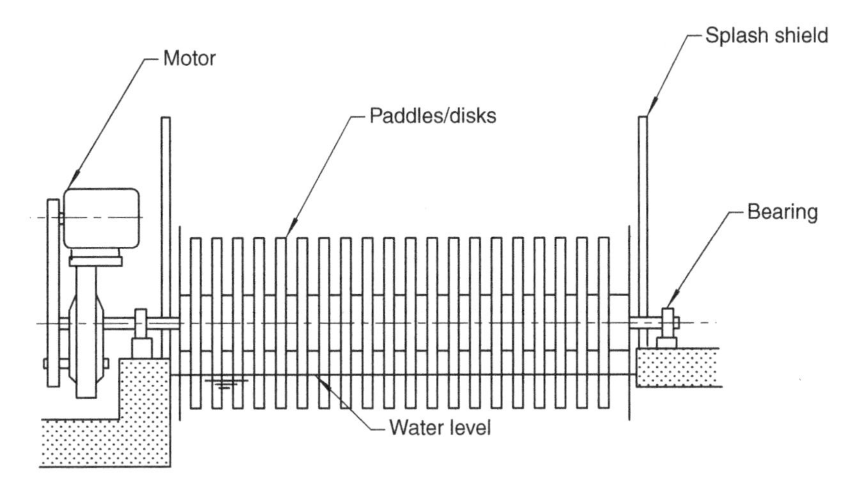

FIGURE 4.5 *Sample horizontal aerator*

Horizontal aerators are mechanically simple (see Figure 4.5). An electric motor is usually used to rotate the aerator shaft through a gear reducer. A long tubular shaft spans the ditch and is supported by ball or roller bearing pillow blocks on each end. Some aerators employ flat blades or paddles to agitate the wastewater, while others employ disks with protrusions or other surface features to provide agitation. In many cases, covers are provided over the aerators to minimize splashing mixed liquor onto handrails and walkways.

The action of the aerator on the mixed liquor provides mixing as well as oxygen transfer. In the typical oxidation ditch, the mixed liquor is propelled around the ditch by the brush aerator and sufficient velocity is maintained to keep the solids in suspension.

Vertical shaft aerators are more varied in construction, application, and performance. Platform-mounted aerators are installed on a fixed platform, generally in the center of a square or round complete mix tank (see Figure 4.6). The platform is supported by columns and an access bridge is provided for maintenance personnel. An electric motor is connected to a gear reducer, and the reducer output shaft is extended to mount the aerator impeller. The exposed impeller is similar to a centrifugal pump impeller, and as it rotates the wastewater is thrown into the air. The outward movement of the mixed liquor and the flow upward into the impeller induces mixing of the tank contents. Some manufacturers install a draft tube below the impeller to enhance mixing. There are also designs that employ a vertical shaft surface aerator at the turn of an oxidation ditch both to provide oxygen transfer and to create circulation of the wastewater.

In some cases, the aerator impeller is mounted under the water surface and air is introduced below the impeller. The impeller mixes the tank contents and disperses the air bubbles. In some designs, the air is supplied by separate blowers, and in other cases, the movement of the wastewater induces air flow, similar to the action of aspirating aerators. This type of aeration system is often referred to as a submerged turbine aerator.

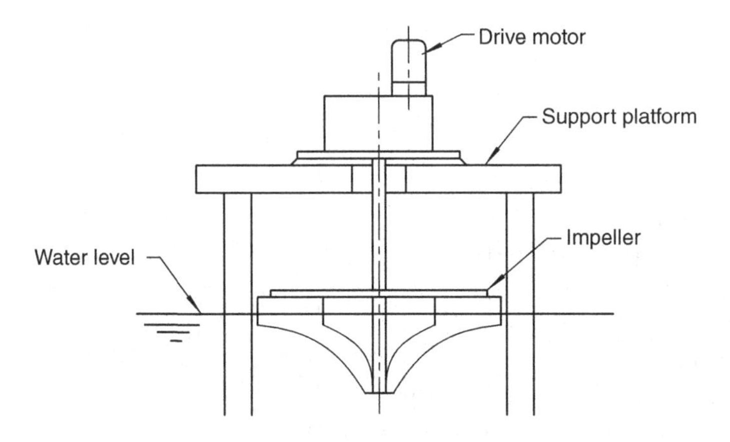

FIGURE 4.6 *Sample vertical shaft aerator*

Many floating aerators have features in common with platform-mounted aerators. They use similar oxygen transfer mechanisms—using an impeller to fling the mixed liquor into the air and mix aeration basin contents. Floating aerators generally operate at higher speeds than platform-mounted units, and are typically lower horsepower. Lagoons often employ an array of floating aerators tethered to the lagoon walls. Floating mixers are very similar, but the design provides circulation of the wastewater without aerating it.

Aspirating mechanical aerators are a variation of floating aerators. The shaft is usually mounted at an angle from vertical. A direct coupled motor rotates a propeller at high speed below the water surface. The shaft of the aerator is hollow, and air is drawn into the eye of the impeller. The air is broken into bubbles, and the mixed liquor velocity creates a plume of bubbles extending down and out from the aerator. In some cases, the aspirating action of the aerator is supplemented by a blower discharging into the aerator shaft. Aspirating aerators may be employed in lagoons, oxidation ditches, and complete mix activated sludge tanks.

These categories are somewhat arbitrary, and many other schemes may be used. The types described are the most common encountered, but the list is by no means exhaustive. The fundamental similarity between all of them is that a mechanical device is used to create or enhance a liquid/air interface and mix the basin contents.

4.2.2 Mechanical Aeration Control Techniques

There are three methods employed for modulating the oxygen transfer rate of mechanical aerators:

- Variable water level
- Variable speed
- Variable operating time

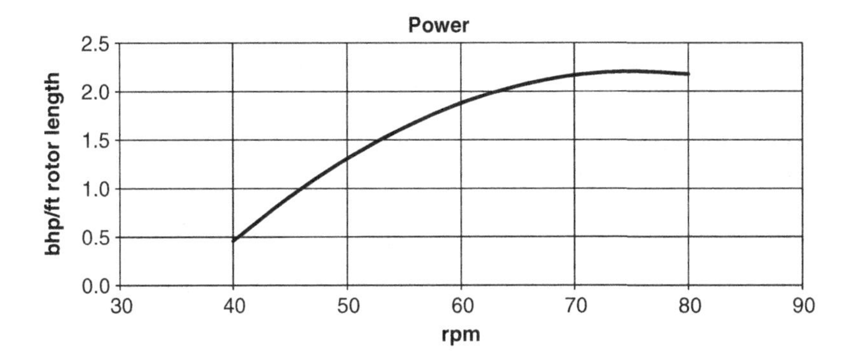

FIGURE 4.7 *Sample relationship of brush aerator power versus speed*

The variable water level is obviously not applicable to floating or submerged aerators, but it is employed with some success for both horizontal aerators and platform-mounted units. An effluent weir or slide gate is adjusted manually or automatically to raise or lower the water surface, usually with an operating range of several inches. As the submergence decreases, the OTR and power draw both decrease as well.

There are some concerns with this technique. The hydraulic profile of the plant must have sufficient head drop between the aeration basin and the secondary clarifier to maintain flow from the aeration basin as its level drops. The mechanical operators for weirs and gates are prone to jamming, particularly if adjustments are infrequent. The changes in OTR and power consumption are nonlinear. This may actually cause the SAE to decrease if the OTR decreases more rapidly than the power consumption. There are successful applications of DO control based on changing water level, but in general it is more feasible to use variable aerator speed for controlling the aeration system.

Varying rotor speed is generally a viable and cost-effective control technique applicable to most mechanical aerators. As the speed of the aerator is reduced, the power drops (see Figure 4.7), as does the OTR (see Figure 4.8). As with level

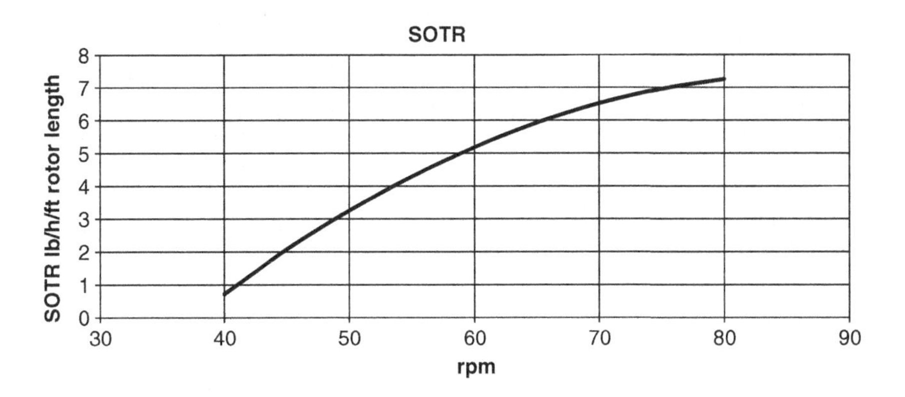

FIGURE 4.8 *Sample relationship of brush aerator oxygen transfer versus speed*

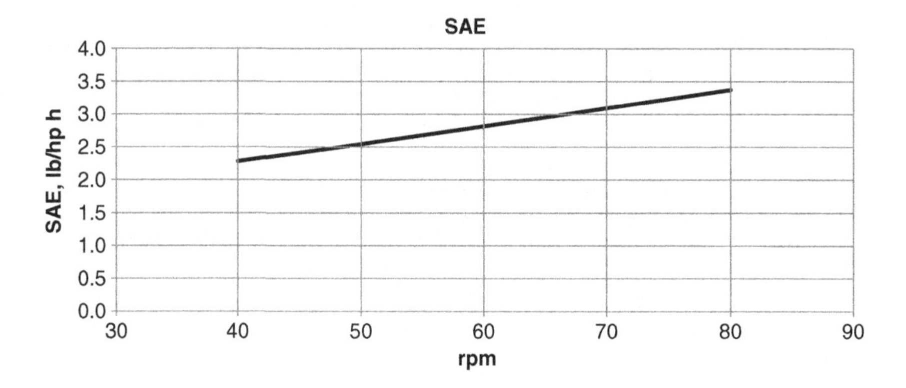

FIGURE 4.9 *Sample relationship of brush aerator aeration efficiency versus speed*

variations, the OTR and power may not be linear functions of aerator speed and the aerator efficiency (see Figure 4.9) may drop at reduced speeds. This will be the result if the OTR decreases faster than the power demand of the aerator. From a process control point of view, as long as the increase in speed increases OTR, the automation will work properly.

The most prevalent technique for varying aerator speed is the variable frequency drive (VFD). These provide continuously adjustable speed control. Some older aerators used two speed motors for coarse DO control.

There are mechanical considerations with variable speed control. In rare instances, the operating speed range may include a critical speed—an unstable operating point where the rotational speed matches the natural frequency of the shaft and creates resonance and uncontrolled vibration. There may also be minimum and maximum operating speeds recommended by the manufacturer because of lubrication concerns or other mechanical considerations.

Mixing is always a concern, and it is important to provide sufficient energy to the aeration system to maintain adequate solids suspension. The required energy varies with tank geometry, aerator design, MLSS concentration, and so on. For mechanical aeration in complete mix activated sludge, a range of 0.5–1.5 hp/1000 ft^3 has been reported as adequate to maintain mixing. For oxidation ditches, the requirement is generally based on maintaining sufficient horizontal wastewater velocity around the ditch to suspend solids. Lagoons present a particular problem for mixing. Often the mixing energy requirement greatly exceeds the energy needed to satisfy the process oxygen demand. In this case, it may be cost-effective to provide separate mixers that can operate at a lower power demand than aerators. That allows separate control of oxygen supply and mixing energy.

Both speed and level variations have an impact on OTR that is design specific. The actual characteristics of the aerator and the manufacturer's performance data should always be referenced during the evaluation and control system design process. Some manufacturers recommend a combination of both methods to allow ranges of OTR in excess of 5:1.

Reversing the rotation direction of some horizontal aerators may be used to change the oxygen transfer rate and aeration efficiency, but this is not a practical technique for automatic DO control. The change in direction is employed manually to respond to long-term process loading changes.

One of the simplest and oldest process control techniques for mechanical aeration is on/off cycling of the aerators. This can be accomplished simply with the high/low "alarm" contacts built into most DO transmitters. When the DO rises above the high setting the aerator is shut off. When it drops below the low setting the aerator is turned on. The wastewater momentum will maintain the solids in suspension for some time after the aerator is stopped, but some settling will occur during the off cycle. If the aerator is restarted within a reasonable time the solids will quickly resuspend without harming process performance. External time delays should be provided to limit off time regardless of DO concentration to prevent excess settling. Additional time delays should also be provided to prevent short cycling the motor — turning it off and on so often that damage from overheating occurs. VFDs and reduced voltage starters can reduce the potential heating problem.

The SAE of mechanical aerators is evaluated using standardized test protocols. Verification of actual field performance is more difficult, but the test methods are identified in ASCE 18-96. For large treatment plants, it may be worthwhile to verify the actual OTR and AE under field conditions. In many cases the equipment specifications for new projects identify a penalty to allow the owner to recoup the difference in energy cost if the aerators don't perform as expected. For a small facility, the cost of the testing may exceed the potential compensation. Judgment should be used in identifying the performance testing requirements.

An example of the physical interactions between the operating parameters of a mechanical aerator will also illustrate the mechanism of the energy reduction achieved by a DO control system. It is important to note that, by definition, at steady state the required oxygen transfer rate (ROTR) needed to maintain process performance must equal the actual oxygen transfer rate (AOTR). The ROTR can be determined from the tank volume and the OUR:

Equation 4.9

$$\mathrm{ROTR} = \frac{\mathrm{OUR} \cdot V_t}{16000}$$

where

ROTR = required oxygen transfer rate, $\mathrm{lb_m/h}$
OUR = oxygen uptake rate, mg/(l h)
V_t = tank volume, $\mathrm{ft^3}$

The AOTR at 2.0 mg/l and various aerator speeds is determined from the manufacturer's data and corrected to field conditions using Equation 4.8. The AOTR at each constant revolutions per minute can be corrected to the AOTR at various mixed liquor DO concentrations:

Equation 4.10

$$AOTR_{act} = AOTR_{2.0} \cdot \frac{C^*_{\infty f} - C_a}{C^*_{\infty f} - 2.0}$$

where

$AOTR_{act}$ = actual oxygen transfer rate at alternate DO, lb_m/h
$AOTR_{2.0}$ = actual oxygen transfer rate at 2.0 mg/l, lb_m/h
$C^*_{\infty f}$ = steady-state value of dissolved oxygen saturation concentration at infinite time in field process water, ppm
C_a = actual DO concentration, ppm

A chart showing this data can be constructed. It shows the variation in AOTR with revolutions per minute and DO concentration (see Figure 4.10).

Assume a mechanical aeration system using four 20 ft long brush aerators with performance characteristics shown in Figure 4.8 is operating at steady state. If the ROTR is 200 lb_m/h and the aerator speed is 60 rpm, the DO concentration will be approximately 3.0 mg/l. This is shown as Point "A." At this speed, the total power requirement for all four aerators is 150 hp. If the process oxygen demand is reduced by 25% to 150 lb_m/h and the speed remains constant, the DO concentration will increase to approximately 4.6 mg/l—shown as Point "B." This is the result of the decreased load and the fixed oxygen transfer rate. The ROTR and AOTR must be equal at the new steady-state operating point. Therefore, the DO concentration rises until the reduced driving force for oxygen transfer reduces the AOTR.

Now assume that the system has a DO control with a VFD for each of the aerators. The control senses the increase in DO and reduces the speed to maintain 3.0 mg/l DO. As the speed drops, the SOTR drops, causing the DO to drop as the process absorbs the oxygen. As the DO drops, the driving force increases. The new steady-state speed

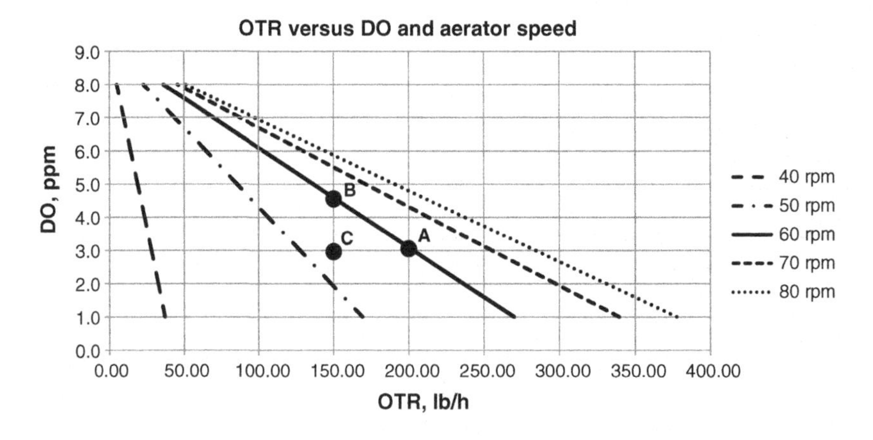

FIGURE 4.10 *Example of brush OTR versus DO concentration*

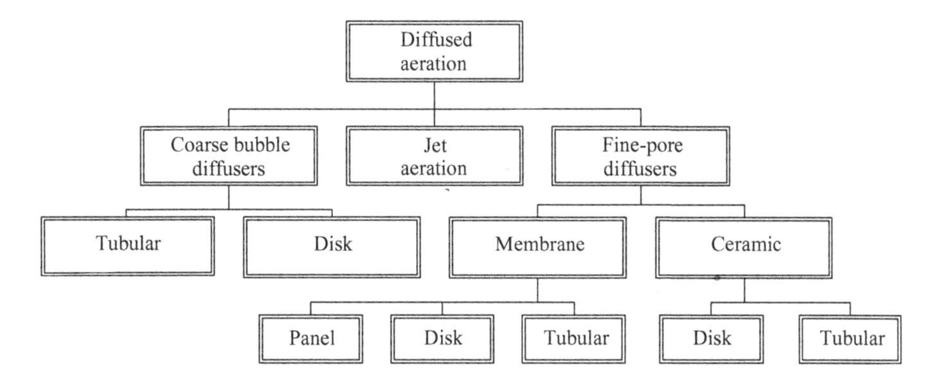

FIGURE 4.11 *Types of diffusers*

is approximately 53 rpm—shown as Point "C." At this speed, the power requirement for the four aerators is approximately 120 hp (see Figure 4.7). Note that a 25% decrease in loading only resulted in a 20% decrease in power, demonstrating the nonlinear nature of DO control.

4.2.3 Diffused Aeration

The variety of diffusers commercially available is at least equal to the variety of mechanical aerators. Each type of diffuser has advantages and disadvantages. As with mechanical aerators, process, cost, and energy considerations are used in their selection. The categorization of diffusers is also somewhat arbitrary, but in general there are two broad types—coarse bubble and fine-pore. Jet aeration is another category, combining diffused aeration and pumping in a single device (see Figure 4.11).

There is no uniform definition as to what constitutes each type of diffuser. Some engineers and operators prefer the term "porous" diffusers to "fine-pore," and the term "nonporous" to "coarse bubble." Some manufacturers promote their products as medium bubble or ultrafine pore. Ultimately, the designations are not as important as the oxygen transfer, maintenance, and energy characteristics of the individual devices. These must be evaluated for each individual treatment facility.

Porous diffusers were used in the early period of development of the activated sludge process. Because energy costs were low and maintenance costs were high, they were eventually displaced by mechanical aerators and coarse bubble diffusers. Energy costs increased and diffuser technology advanced, and the fine-pore diffuser has now become the basis of design for most new facilities. There are still many applications for coarse bubble diffusers, however, and the installed base is very large. Evaluation of both types of diffusers is an important part of determining appropriate ECMs.

The original coarse bubble diffuser consisted of standard pipe with $^1/_4$ to $^1/_2$ in. diameter holes drilled in the side. These exhibited a tendency to clog with biological growth or rags and other debris wrapping around the pipe. Current design coarse bubble diffusers fall into either tubular or disk styles.

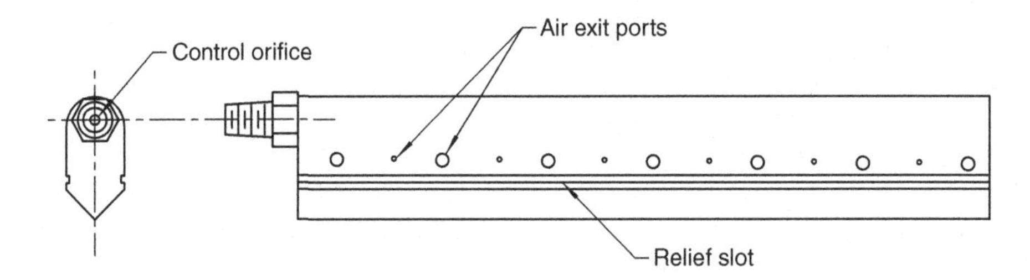

FIGURE 4.12 *Example tubular coarse bubble diffuser*

The most common tubular style is usually 24 in. long and constructed from thin gauge stainless steel (see Figure 4.12). They are often called "wide band" diffusers. They are promoted as "nonclog" by virtue of a long air relief slot that resists closure by biological growth and rags. They are frequently found in existing facilities, and in new applications are used where fouling is a concern, particularly in aerobic digesters. They are also popular in channel aeration where mixing is of greater concern than oxygen transfer. The tubular diffusers are usually arranged along the length of one wall of the aeration basin, and the rising bubbles induce a cross rolling action in the wastewater to maintain mixing. If the diffusers are arrayed along both walls a dual roll action is created.

Disk style coarse bubble diffusers, sometimes referred to as orifice style, have much greater design diversity. Most of them have a check valve feature, which prevents water entry into the air distribution piping when the air flow is stopped. The check valve may be constructed from a rigid ball or disk, but more often an elastomer is configured to seal the diffuser and prevent backflow. Most disk styles have plastic bodies and EPDM (ethylene propylene diene monomer) elastomer, but many other materials are also available. Disk diffusers may be arranged along the tank wall similar to tubular diffusers or arranged in a grid spanning the floor of the aeration basin.

Manufacturers have extensive test data on oxygen transfer characteristics (see Figure 4.13), but if this data is not available an approximate SOTE of 0.75% per foot of submergence may be used for estimates. Field performance, OTE_f, is roughly half of SOTE.

The mixing requirements for coarse bubble diffusers are approximately 15–20 CFM/1000 ft^3 of tank volume. Cross roll arrangements try to maintain a mixing velocity of 2.0 ft/s. The manufacturer's data should be consulted to determine minimum and maximum flow rate and pressure drop characteristics (see Figure 4.14).

Fine-pore diffusers are generally divided into ceramic or membrane style, based on the porous media. Ceramic diffusers use a sintered aluminum oxide to create a tortuous path for the air to pass from the header pipe into the mixed liquor. The bubbles created are much smaller than those generated by coarse bubble diffusers. The result is approximately five times greater surface area to volume ratio. This increases the diffusion of the oxygen and improves the OTR and OTE. Oxygen transfer characteristics vary widely between diffuser designs and basin

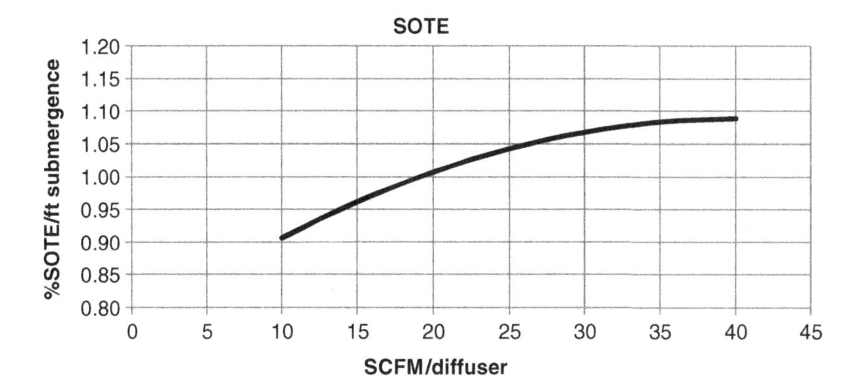

FIGURE 4.13 *Example tubular diffuser SOTE*

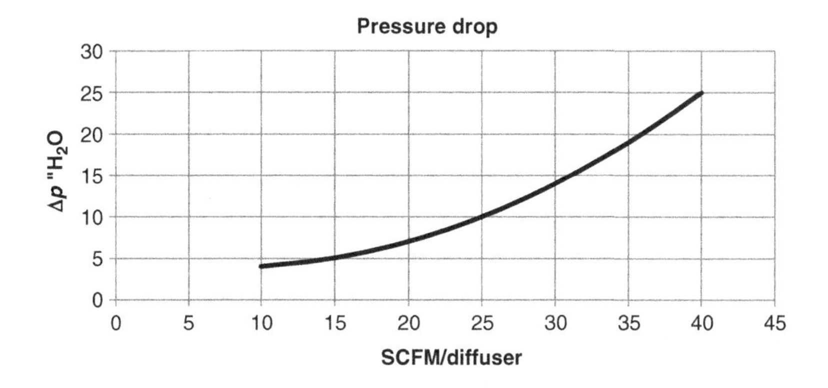

FIGURE 4.14 *Example tubular diffuser pressure drop*

configurations, but for quick approximations a SOTE of 1.5% per foot of submergence may be used.

Membrane diffusers have tiny slits in an elastomeric sheet (see Figure 4.15). These slits create small bubbles as the air passes through. Most membrane diffusers use the elastomer to create a check valve to prevent the back flow of mixed liquor into the air header when air flow is lost. The newest design membrane diffusers provide OTE (see Figure 4.16) equalling or even exceeding ceramic diffusers. Membranes are available in a variety of materials, with many formulations proprietary. EPDM is the most common, but polyurethane and Teflon-coated membranes are available for use in chemically aggressive wastes or where fouling is a problem. Membrane life varies widely and may be limited by either fouling or degradation of the elastomer. In general 5 years or longer is a reasonable projected membrane life. Ceramic diffusers have a longer life, but are considered more prone to fouling, have more labor-intensive cleaning requirements, and are more expensive at initial purchase. The most cost-effective type should be analyzed for each site, but the trend in newer systems is toward membrane style.

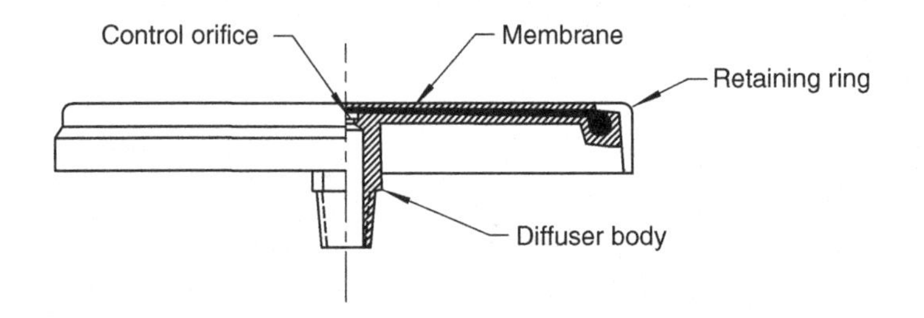

FIGURE 4.15 Example membrane fine-pore diffuser

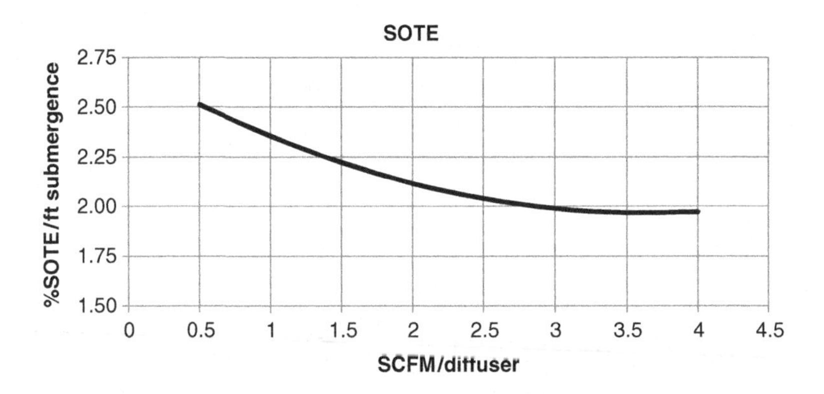

FIGURE 4.16 Example membrane fine-pore SOTE

Historically, the higher content of solids and debris, intermittent operation, and greater potential for biological fouling made most designers reluctant to use fine-pore diffusers in aerobic digesters. The commercialization of membrane style fine-pore diffusers with effective check valve function has changed that. Now there are many successful applications of fine-pore diffusers in aerobic digesters. Digester mixing requirements may be different from those of aeration basins. Digesters are more likely to be mixing limited instead of oxygen transfer limited. Particular attention must be paid to mixing requirements in digesters because of the higher solids concentration.

Replacing mechanical aerator with diffusers in oxidation ditches is more complicated. First, oxidation ditches tend to be shallow, so the replacement diffused aeration system OTE will not match typical performance in deeper tanks. Diffuser performance must be calculated for the specific configuration. Second, most ditches rely on the mechanical aeration device to provide sufficient wastewater velocity around the ditch to maintain mixing. Diffusers cannot induce horizontal movement of the mixed liquor. There are several methods available to maintain mixing in an oxidation ditch when upgrading to diffused aeration:

- Use jet aeration, with the pumps maintaining mixing and the air flow modulated to maintain DO concentration.

- Install mechanical mixers to circulate the mixed liquor around the ditch.
- Provide full coverage aeration, with diffusers installed around the entire ditch floor, including the ends. This effectively creates a plug flow reactor. Care must be taken to avoid short-circuiting between the influent and effluent. Baffles may be needed to create the required hydraulic flow.

In general, higher OTE reduces air flow rates and improves the energy efficiency. Two factors are particularly influential in determining the OTE. One is the flux rate, the volume of air per unit of diffuser surface area, expressed as SCFM/ft^2. The lower the flux rate, in general, the smaller the bubble and the better the OTE obtained. There are limits since below a minimum value the air flow is not distributed evenly across the diffuser surface. This reduces efficiency. The second principal factor in maximizing OTE is the distribution of the diffusers in the tank. The most common arrangement is referred to as a "full coverage grid," where a grid pattern of diffusers is spread across the entire floor of the aeration basin. Each diffuser acts as an airlift pump, creating a vertical column of rising wastewater above each diffusers and a returning downward flow between diffusers. A uniform diffuser distribution creates reduced agitation and lower wastewater velocity, which increases bubble rise time and improves OTE.

A common arrangement in plug flow reactors is tapered aeration. The diffuser density, defined as the number of diffusers per square foot of tank plan area, is higher at the influent end of the tank than the effluent end. This optimizes oxygen transfer, since the oxygen uptake rate is higher at the influent end of the basin (see Figure 4.17). The tapered aeration arrangement minimizes equipment cost and optimizes OTE by employing low flux rates throughout the aeration basin.

The typical arrangement for any diffuser type is several grids of diffusers installed in each aeration tank. The grids are connected by a network of piping with a drop leg running vertically up the tank wall and connecting to a main air header network. The main header connects back to a centralized blower room that supplies low pressure air to the entire aeration system. Older coarse bubble systems utilized articulated swing out drop pipes to permit servicing diffusers without draining tanks. Lagoons

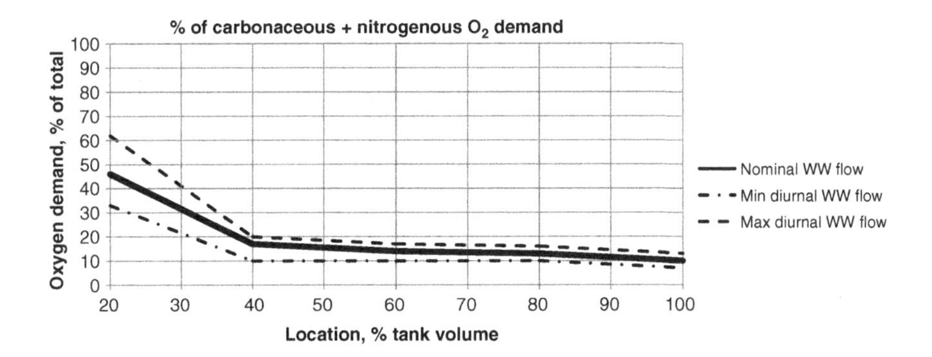

FIGURE 4.17 *Variation of oxygen demand along tank length*

often have floating air distribution pipes running across the water surface with the diffusers hanging below. There are also systems with complete drop-in grids that can be installed without draining the tanks.

There are three typical configurations of fine-pore diffusers. They are tubular, disk, and panel. The different designs balance initial cost and energy requirements. Tubular diffusers have an appearance reminiscent of tubular coarse bubble diffusers. They may be used as direct replacement for coarse bubble diffusers in a cross roll arrangement, but they are more commonly arranged in a full coverage grid. Disk diffusers come in different diameters, but the most common is 9 in. nominal diameter for both ceramic and membrane diffusers. Panel diffusers with elastomer membranes or plate diffusers with ceramic media employ very low flux rates and cover virtually the entire tank bottom. Both designs claim very high OTE that offset higher equipment cost. The panels are typically used in plug flow reactors.

In addition to a minimum air flow rate required to obtain uniform dispersion of the bubbles from the diffuser, a lower air flow limit based on mixing applies to fine-pore diffusers. There is also a maximum air flow per diffuser that is, dependent on configuration. High flow per diffuser also results in high flux rates, which in turn cause lower OTE.

In discussing and comparing diffuser performance, it is common to talk in terms of oxygen transfer efficiency. However, the process performance depends on the oxygen transfer rate. This represents the mass transfer rate of oxygen to the process. Equations 4.9 and 4.10 apply to both mechanical and diffused aeration systems and establish the required oxygen transfer rate of the process. Equations 3.9 and 3.11 can be used to establish the required mass flow rate of air needed to meet the demand for BOD removal and nitrification.

In estimating the impact of DO concentration of OTE and OTR, the reciprocal of the ratio in Equation 4.2 is required:

Equation 4.11

$$\frac{\mathrm{OTR_a}}{\mathrm{OTR_{2.0}}} = \frac{C^*_{\infty \mathrm{f}} - C_\mathrm{a}}{C^*_{\infty \mathrm{f}} - 2.0}$$

The discharge pressure required at the blower for diffused aeration systems is determined by several factors. The predominant factor is the static pressure created by the depth of water above the surface of the diffuser. This can be determined by:

Equation 4.12

$$p_{\mathrm{stat}} = \mathrm{depth} \cdot 0.433$$

where

p_{stat} = static pressure at diffuser, psig
depth = submergence of diffuser surface, ft

Depth is a term that causes a surprising amount of confusion and may be difficult to obtain. The static pressure is based on the distance from the water surface to the air release point of the diffuser. This is the top of fine-pore diffusers and disk-type coarse bubble diffusers. Tubular coarse bubble and fine-pore diffusers use the centerline of the attachment point for determining submergence. The submergence is not usually the dimension provided by operators. Total tank depth—the distance from the top of the tank wall to the high point of the tank bottom—is more readily available. If the freeboard—the distance from the top of the wall to the water surface—is subtracted from total depth, the result is the side water depth (SWD). This is the distance usually given by the operators since it is used in calculating tank volume for hydraulic retention time and other process parameters.

If drawings or diffuser supplier information is available, the distance from the top of the diffuser to the tank bottom may be available, allowing calculation of the submergence. If more accurate information is not available, the height of the diffusers may be assumed to be 9–12 in. above the high point of the tank floor. The conservative approach is to use the lower value, as this will result in a higher estimate of static pressure.

The next most significant contribution to the total system pressure is the friction losses of the air flowing through pipe, fittings, and valves, referenced together as piping losses (see Chapter 6). A typical allowance for piping losses is 1–1.5 psig at maximum design flow. This is usually a conservative estimate, and it is common to observe actual system pressures 0.5–1.0 psig below the design value.

Pressure losses through the diffuser are created by two separate mechanisms. The friction of the air through the diffuser body and air exit passages is one. The other is the pressure drop through the control orifices that may be inserted into the attachment point of the diffuser to the distribution header in the tank. Control orifices are inserted to create a pressure drop high enough to maintain uniform distribution of air flow across all of the diffusers in a grid. Unequal flow distribution can result from pressure drop and velocity head regain through the header. Uneven flow between diffusers is also a consequence of installation errors that result in differences in elevation between individual diffusers.

A common opinion is that replacing coarse bubble with fine-pore diffusers is not feasible because of the higher pressure requirements for fine-pore diffusers. This is, in fact, very rarely the case. Within the normal operating air flow ranges both coarse bubble and fine-pore diffusers have pressure drops of a few inches of water (see Figure 4.18). The differences between the pressure drop of the two types is usually much less than the decrease in piping losses because of lower air flow rates with fine-pore diffusers.

Fouling of fine-pore diffusers can reduce OTE by as much as half of the new diffuser OTE. Fouling is the result of an accumulation of biological growth in a film on the diffuser surface. In some cases, mineral precipitates, such as calcium carbonate, may form on the diffuser surface. The fouling causes the bubbles to coalesce or enlarge as they exit the diffuser, reducing the volume to surface area ratio. Bumping or flexing the diffusers by applying alternating higher and lower than

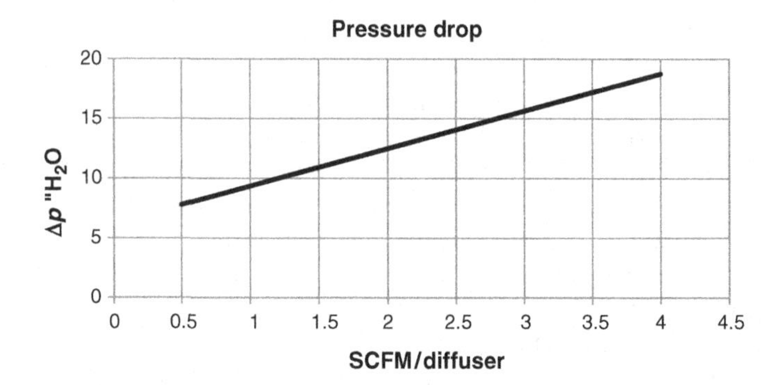

FIGURE 4.18 *Example membrane fine-pore pressure drop*

normal air flows can effectively retard fouling of membrane diffusers. A weekly bumping cycle is often programmed into the aeration controls. Some membrane materials are formulated to resist fouling, and these may be cost-effective in aggressive or industrial wastes.

Eventually mechanical or chemical cleaning is required to restore OTE to a value approaching a new diffuser's. Some suppliers offer automatic chemical cleaning systems that can clean diffusers without draining the aeration basin. The cost-effectiveness of this system needs to be carefully evaluated on a case-by-case basis.

Air side fouling is a common concern with fine-pore diffusers, but its occurrence is actually very uncommon. The phenomenon is perceived as the result of dust or dirt in the air supplied to the diffuser blocking the pores in the diffuser. When it does occur it is usually the result of scale or corrosion products from the distribution piping or from gross entry of mixed liquor solids from a broken distribution pipe or damaged diffuser. When broken pipes or diffusers are repaired, it is a reasonable precaution to examine the other diffusers in the system for contamination.

Concern over air side fouling has often led to the over-specification of blower air filter requirements. The level of filtration required to protect the blower itself is more than adequate to prevent air side fouling from dust in the air stream. Excess filtering adds nothing to diffuser performance or life but does increase blower power requirements.

Jet aeration is a technology that doesn't fit neatly into any other categories. The mixed liquor is pumped through a venturi type restriction. An air stream, supplied by the aspirating effect of the venturi and an external blower, is mixed with the wastewater stream and formed into fine bubbles. The plume of combined air and water is projected horizontally into the tank with the bubbles eventually rising to the surface. Jet aeration allows separate control of oxygen supply by modulating the air stream and mixing energy achieved through the pumped liquid. It is most often found in industrial applications, but is also used in municipal treatment plants.

4.2.4 Diffused Aeration Control Techniques

The control of diffused aeration systems necessarily involves controlling air flow to the diffusers. This is more complex than mechanical aerator control which only requires control of the aerators installed at the tank itself. Control of diffused aeration systems involves control of air flow to each tank or diffuser grid and control of the blower that supplies air to the diffusers. This necessitates a system approach to the aeration control strategy.

The total air supplied to the system is controlled by modulating the blowers. The total blower discharge air flow must be matched to the total process demand. There are a variety of techniques used for blower control (see Chapter 5).

The air piping system is fitted with flow control valves. These are almost exclusively butterfly valves (BFV) because of the high cost of other types in the large diameter piping found in most aeration systems. The most common arrangement has a valve to control each tank and a valve at each drop leg. The valves are used for both isolation and flow modulation. They may be manually or automatically controlled. The purpose of the valves is to balance the air flow between tanks and grids so that the air supply is apportioned according to the relative process demand for air in each zone. These balancing valves should be kept as close to full open as possible to minimize system pressure and blower power.

Loss of air flow to an aeration grid is not an uncommon occurrence. In some cases, the loss is intentional, as with on/off control of air flow for SBRs. In other cases, the loss of flow is the result of equipment failure or power loss. Diffusers with a check valve feature are intended to keep the water from entering the air piping at the bottom of the tank. However, loss of air for extended periods of time will result in water being trapped in the air piping. Tubular coarse bubble diffusers and ceramic fine-pore diffusers admit large quantities of wastewater to the piping. Moisture will also accumulate when warm, moist ambient air is cooled by the water in the tank and condensate forms in the piping. Water trapped in the air piping can restrict or completely block air flow to the diffuser system.

Some mechanism must be provided for removing this moisture. Coarse bubble diffusers simply blow the water out through the diffuser openings. For most fine-pore systems, drainage is accomplished by installing a small (1 in. typical) line in the piping at the bottom of the tank. In some cases, this drain is connected to a coarse bubble diffuser with a check valve feature installed at the diffuser piping invert so that the water can drain out. In many systems, the drain is connected to a ball valve at the tank railing, and manual intervention is required to drain the water.

After extended power outages, the trapped water may restrict air flow so that the blower cannot be started or maintain operation above the safe minimum capacity. A modulating blow-off valve, either on the main header or at each blower, is sometimes used to maintain system pressure for evacuating the trapped water while allowing sufficient air flow for safe blower operation.

Timed operation of diffused aeration is a viable control strategy for diffused aeration systems with check valve diffusers or self-draining headers. This is commonly used in SBRs and systems with time-based nitrification/denitrification processes. It is also a useful technique where minimum blower flow or basin minimum air flows are higher than the flow required for maintaining the desired DO concentration. The air flow to a basin can be shut off on high DO concentration and restored after a lower level is reached or a time delay has elapsed. As is the case with mechanical aeration, settling of solids occurs when the air is shut off, but they are quickly resuspended when air flow is restored. If multiple parallel basins are in operation, the timed operation can be staggered among the several basins provided the blower minimum safe operating flow is maintained. This eliminates the need for frequent starting and stopping of the blowers.

The decision of how many of the air flow control valves to automate is largely based on economics. For small facilities, modulation of the total blower flow rate to the aeration system based on the average or lowest DO measured may provide adequate process control and energy conservation. The distribution of air between the various aeration basins is adjusted manually to achieve approximately equal DO concentrations in each at the time of peak diurnal loading.

The inevitable variations between tanks in hydraulic flow, RAS flow, and air flow will result in some basins operating at a higher DO than others during most of the day. The typical DO variation between tanks is 0.5–1.0 ppm, with the deviation generally being above the system target DO. This indicates that most of the aeration basins are operating with excess air flow for most of the day. When the cost of the energy that can be saved by eliminating this excess flow exceeds the cost of flow transmitters, DO probes, and automatic operation of the air flow control valves, it becomes cost-effective to apply automatic DO control and automatic air flow control to each basin. The breakpoint on this varies with the number of aeration basins, the size of the flow control valves, and the cost of electric power.

As illustrated for tapered aeration, the OUR and air flow demand is not uniform along the length of a plug flow reactor. As much as 50% of the total air demand can take place in the first 20% of the basin, with decreasing demand as the wastewater moves through the process. This distribution proportion may not be uniform throughout the normal diurnal variations, causing some portions of the tank to be overaerated. As with individual tank control, when the incremental energy savings achieved by controlling the DO and air flow in each drop leg separately exceed the additional control system costs, it becomes cost-effective to provide this additional level of control.

A profile of the DO concentration along the length of the aeration basin at several times during the day can be used to determine the OUR distribution and the amount of excess aeration. Another, and more accurate technique, is to use offgas testing along the tank length to determine the demand profile. The offgas measurement is also available for permanent installation at aeration basins, and can determine OUR, OTE, and process air flow demand in real time (see Chapter 7).

Diffused aeration exhibits a response to load changes similar to the mechanical aeration system, except the AOTR is determined by air flow rate instead of revolutions per minute. A similar response chart can be constructed based on loading and diffuser oxygen transfer performance. The ROTR is calculated using

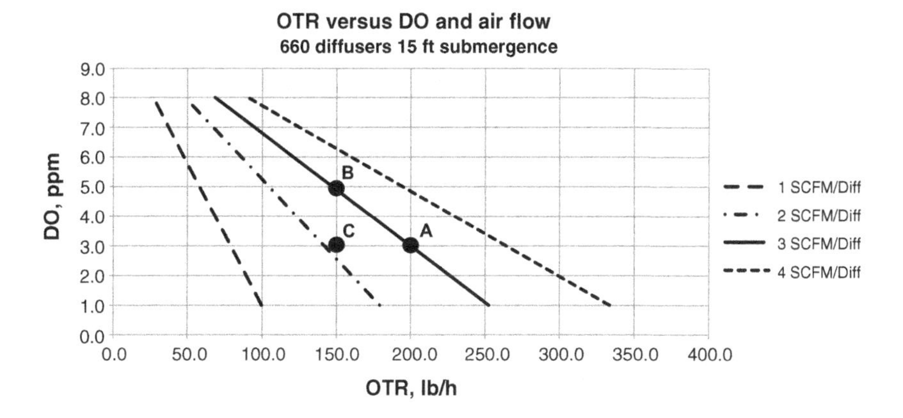

FIGURE 4.19 *Example of membrane fine-pore OTR versus DO concentration*

Equation 4.9. The AOTR is a function of OTE and total air flow, which is the summation of air flow per diffuser. The AOTR at 2.0 mg/l DO concentration can be calculated for various air flows per diffuser using Equation 4.5. The correction to AOTR for other DO concentrations is made using Equation 4.8. The result is a chart showing the relationship between OTR, DO concentration, and air flow (see Figure 4.19).

In the example chart, Point “A” shows the initial steady-state operation at 200 lb_m/h ROTR, 3 SCFM/diffuser, 1980 SCFM total air flow. Assuming a blower efficiency of 70% and typical inlet conditions, the blower power required would be 82 bhp. If the load to this system decreases by 25% to 150 lb_m/h ROTR, the steady state is disturbed. If the air flow rate is maintained, the DO concentration rises and the OTE decreases. This is a result of the DO concentration increasing, causing a corresponding decrease in both oxygen driving force and diffuser OTE. Eventually, a new steady-state equilibrium is reached at 5 mg/l DO, shown as Point “B.”

If an automatic DO control system reduces the flow to the aeration basin to approximately 2.25 SCFM per diffuser, 1485 SCFM total, the DO concentration will drop until the new OTE causes the AOTR to match the ROTR at 3.0 ppm. This is shown as Point “C” on the chart. Using the same blower assumptions as at point “A,” the new blower power will be 62 bhp. As with the mechanical aeration, it is the change in DO concentration and driving force that determines the new equilibrium point.

These charts are also useful in determining the effect of taking aeration basins out of service. The total system OUR is determined by organic load and will remain unchanged. By using the OTR and the target DO concentration, the new aerator operating point can be determined and the impact on energy established. Because of the higher air flow per diffuser with fewer tanks in service, the OTE will decrease. If the existing operating mode is mixing limited, then eliminating overaeration will offset the loss in efficiency—using the chart is one way to establish the new air rate requirement.

Aerobic digester and sludge holding tanks require different control strategies than secondary aeration. Some applications use DO control for digester air flow determination. Other applications, where mixing dominates, use a control based on maintaining air flow proportional to digester level and volume. Timed intermittent air flow is another strategy used for digesters. In other cases, operator preference is to maintain a constant air flow regardless of level or DO concentration. All of these strategies are also applied to equalization basins. Since these are variable level processes, the air flow control can be difficult if the air is drawn from a blower system also supplying aeration basins. This is particularly true if the water level in the digester or equalization basin is lower than the aeration basin, which requires a lot of pressure drop through the digester control valve. Proper sizing of the flow control valve is critical to providing stable flow control across the normal operating range.

It is often convenient to combine Equations 3.9 and 3.11 to simplify calculation of air flow rates, eliminating the need to calculate OUR and HRT:

Equation 4.13

$$Q_s = \frac{0.335 \cdot Q_{ww}}{\text{AOTE}} \cdot (\Delta\text{BOD} \cdot U_{O_2} + \Delta\text{NH}_3 \cdot 4.6)$$

where

Q_s = air flow required to meet demand, SCFM
Q_{ww} = wastewater flow, mgd
ΔBOD = BOD removed from influent to effluent, mg/l
U_{O_2} = utilization of oxygen, lb O_2/lb BOD
ΔNH_3 = ammonia nitrified, mg/l
AOTE = actual site oxygen transfer efficiency, decimal

4.3 SAVINGS DETERMINATIONS

The determination of savings obtained by changing aerator configuration and the savings obtainable by implementing automatic DO control can seem overwhelming. If the calculation of savings is approached incrementally, however, marginal applications can be eliminated quickly. The more detailed analysis can be reserved for projects where the potential savings justify the engineering effort.

One of the tasks accomplished in the initial system assessment is identifying the current power used for the aeration system and comparing the baseline data to the appropriate benchmark. If the current aeration operation is more efficient than standard practice, it would be prudent to explore other ECMs.

The next check is to compare the current aeration devices with recommended best practices. If the current system employs mechanical aeration or coarse bubble diffusers, then conversion to fine-pore diffusers should be examined. This will require working with manufacturers of the existing and proposed equipment to obtain budget estimates for equipment and installation costs and to determine equipment performance data for the site. In many cases, the performance data for the existing equipment can be found in the plant's O&M manuals or in the submittals from the original construction project.

If the existing mechanical aeration equipment is constant speed, adding VFDs is an alternative that should be considered. Even if the AE obtained at reduced speed is not equal to diffused aeration, reduced equipment cost may make this a viable option—particularly if utility incentives for VFDs are available.

The next step is to do a rough estimate of the savings obtainable from implementing DO control. The usual range is 20–40% savings compared to manual control. The savings achieved are a function of the degree of overaeration being experienced and the complexity of the proposed control system. As indicated above, when more control points are added to the system the control precision and energy efficiency improves, but the cost increases. Additional items such as most-open-valve logic and eliminating pressure control also add to the potential energy savings. Often this is possible with no increase in system cost.

Bear in mind that not all control systems are equal. The plant may have an automatic DO control system, but part of the initial assessment is to determine how well it functions. It is common to find controls running in manual because satisfactory tuning couldn't be achieved. Older systems may be based on maintaining constant pressure or may not include efficient blower control. This is where a site visit to see how systems actually function is essential.

The final priority is evaluating blower replacement. This is often the first item on the ECM list because it is an obvious opportunity and the analysis seems straightforward. However, blower upgrades are expensive, improvements beyond 15–20% from one blower technology to another are rare, and the payback may not be as good as other ECMs.

It should be obvious, but it is worth stating that the evaluation should multiply individual savings percentages, not add them. An even better methodology is to avoid percentages altogether and perform a separate savings and payback analysis on each ECM. The savings for each ECM are not always independent, and the installation sequence and interrelationships must be considered. For example, if fine-pore diffusers are installed, the savings from DO control are reduced because the excess DO is achieved with a lower total air flow. If more efficient blowers are installed but are operated at maximum capacity because there is no automatic control, the potential savings will not be realized.

Field experience has shown that theoretical savings calculations are usually optimistic. In some cases, limitations such as mixing or blower turndown may prevent achievement of the potential savings. Operational decisions can have a tremendous impact on the savings achieved. Changes in the number of tanks in operation, DO setpoints, MLSS concentration, and so on will affect the air flow and energy requirements.

In evaluating savings and payback, it is common to use average data. This includes average process loading, average blower and/or aerator efficiency, and the composite power cost. Sometimes this is the only choice; for example, if detailed performance data could not be obtained during the initial assessment or if there is a question about the accuracy of the data obtained. In other cases, the savings predicted on average data are sufficient to indicate a particular ECM is cost-effective.

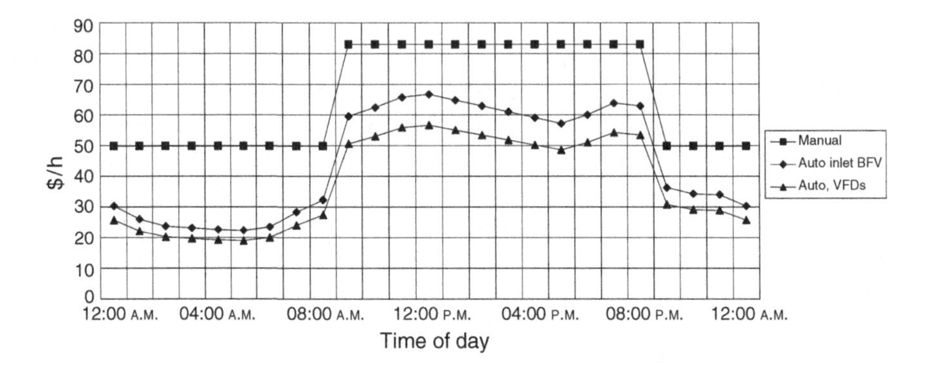

FIGURE 4.20 *Example weekday hourly power cost*

If the energy cost and process data are available, a detailed diurnal study may be used to determine power savings. The basis of comparison is typically manual control at a constant air flow rate to the basins. If a SCADA system is operating, continuous trends or logged data may be available as the basis of comparison. The diurnal analysis is particularly useful if new diffusers are used in conjunction with an upgrade to the DO control system.

The basis of the diurnal analysis is the calculation of hourly aeration system power requirements. The oxygen demand calculation assumes the organic load fluctuates with the hydraulic loading. The power demand is calculated for each hour and is assumed constant over that hour. If side streams or other slug loads are identified, they should be included in the hourly power determination.

Once the typical hourly power variation is determined, the hourly power cost is calculated. The kilowatt used or required is multiplied by the applicable cost per kilowatt-hour. Using a spreadsheet or engineering analysis program makes the calculations manageable and graphing the data simplifies interpretation. For all but the smallest facilities, two sets of calculations are needed: one for weekday costs, when on-peak and off-peak rates apply, and a second for weekend costs when only off-peak rates apply (see Figures 4.20 and 4.21).

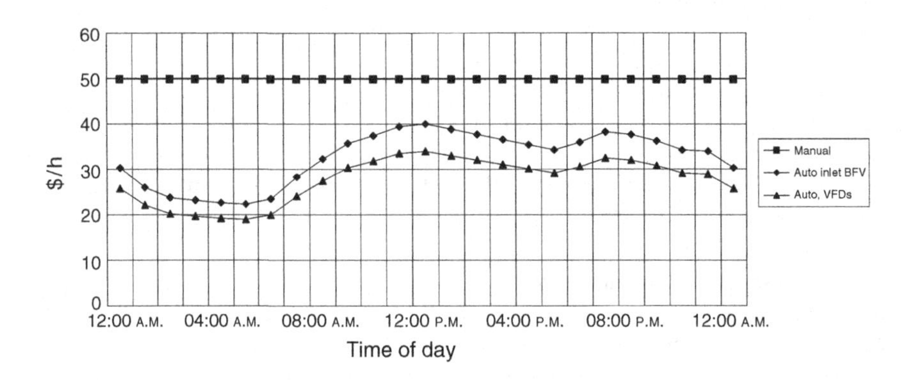

FIGURE 4.21 *Example weekend hourly power cost*

The annual power cost is the summation of the total daily consumption cost, plus any power factor and demand billing (see Equation 2.9). The demand billing may be significant both as a high proportion of the cost and as the basis for many utility rebate programs:

Equation 4.14

$$\frac{\text{cost}}{\text{h}} = \text{kW} \cdot \text{rate}$$

where

cost/h = on-peak or off-peak consumption cost, \$/h
rate = appropriate on-peak or off-peak cost, \$/kWh

Equation 4.15

$$\text{consumption cost} = \left(\sum_{1}^{24} \frac{\text{cost}_{\text{weekday}}}{\text{h}}\right) \cdot 5 \cdot 52 + \left(\sum_{1}^{24} \frac{\text{cost}_{\text{weekend}}}{\text{h}}\right) \cdot 2 \cdot 52$$

Demand charges and power factor penalties, if included in the power rates, are calculated using Equations 2.9 and 2.10:

Equation 4.16

$$\text{annual cost} = \text{consumption cost} + \text{demand charge} + \text{PF charge}$$

The diurnal analysis can be the most accurate method for determining energy cost savings. Plots of hourly cost are very useful in providing insight to the influence of billing structure on energy costs and the importance of the diurnal variations. This is particularly beneficial if flow equalization is being considered. However, it must be remembered that the results are only as accurate as the initial data and assumptions, and that the calculation is only a model of the process.

In most cases, the engineering cost of a full diurnal analysis is not justified by the accuracy of the available data, but the use of rough averages doesn't provide sufficient confidence to proceed with the aeration system, control system, and blower upgrade. A simplified analysis based on typical diurnal flow patterns can provide very accurate results. The following assumptions and input data are used in this method:

- A composite power rate is available based on total costs and total kilowatt-hours, including on-peak, off-peak, demand, and power factor charges (see Chapter 2).
- Aeration system efficiency at average daily flow is available and reflects expected operating conditions, including wastewater temperature and side-stream loads.

TABLE 4.2 Five Point Diurnal Estimate

Hours	% of Time	% ADF
5	20.84	70.00
3	12.50	90.00
2	8.33	100.00
8	33.33	107.50
6	25.00	120.00

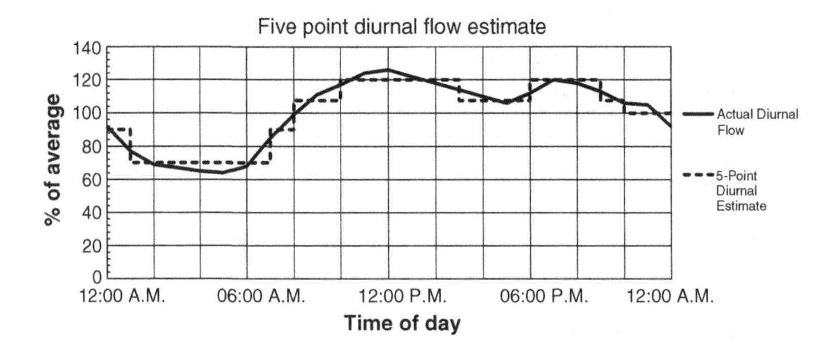

FIGURE 4.22 Five point diurnal analysis

- The oxygen demand is based on average daily flow and average actual organic loads.
- Blower power for diffused aeration systems is calculated based on average annual ambient conditions and average (not full load) blower efficiency.

The diurnal flow pattern for most municipal applications is quite complex. The analysis using the five data points of Table 4.2 properly represent the impact of the diurnal load pattern for most municipal applications. The relative flow rates are combined with the relative hours of operation. These points provide a total process demand (either mechanical aeration or blower power demand) that is quite close to the result obtained with a full 24-hour diurnal analysis (see Table 4.2 and Figure 4.22).

Average daily aeration power is determined by multiplying the aeration system average power required for the process by each %Time and each %ADF in Table 4.2. Multiplying this by 365 provides the total annual operating cost for each option. Deducting this from the current baseline power cost provides the savings:

Equation 4.17

$$\text{annual estimated cost} = 365 \cdot \sum_{1}^{5} \text{kW}_{\text{ave}} \cdot \%\text{time} \cdot \%\text{ADF} \cdot \text{rate}$$

where

kW_{ave} = power required at average daily flow, kW
%time = estimated percent operating at a given demand, from table, decimal
%ADF = estimated proportion of average power, from table, decimal
rate = composite energy cost, $/kWh

The data in Table 4.2 can be simplified for preliminary estimates when on-peak, off-peak, and demand charges apply. If the on-peak power rates run from 9:00 A.M. until 9:00 P.M., which is a common range, the average for each period is:

- On-peak average air flow = 115% of air flow at ADF
- Off-peak average air flow = 85% of air flow at ADF
- Peak air flow for demand charge = 120% of air flow at ADF

Regardless of the aeration type, the accurate determination of current and projected aeration energy cost is a key to identifying the potential savings. The determination of aeration oxygen and power requirements is a critical part of the evaluation and design process. If the aeration system employs mechanical aerators, the next step is to design and evaluate the control system. If diffused aeration is used, then evaluation of the blowers and the piping system follows next.

EXAMPLE PROBLEMS

Problem 4.1

Refer to Problem 2.7. Additional testing has determined that the average aeration system effluent DO concentration is 6.0 ppm at the current operating air rate of 1800 CFM.

(a) Recalculate savings based on reducing air flow to achieve 2.0 ppm DO. Use a mixed liquor temperature of 50 °F.
(b) If the DO control system installed cost is $30,000, what is the revised payback for the VFDs plus controls?
(c) If the blowers have 60% turndown, can they meet the reduced flow requirement?

Problem 4.2

Refer to Problem 3.3. Calculate the standard aeration efficiency for the existing mechanical aerators and the proposed diffused aeration system. The current DO concentration is 3.5 ppm and the mixed liquor temperature is 60 °F. The plant is located at 400 ft above sea level (ASL) and the barometric pressure is 14.5 psia. Assume an average annual air temperature of 55 °F and a blower efficiency of 70%.

Problem 4.3

A treatment plant is permitted for effluent of 10 ppm BOD_5 and 5 ppm ammonia at 1.8 mgd ADF. The plant has five aeration basins, each 15 ft wide × 100 ft long × 18 ft SWD. The diffusers have the characteristics shown in Figure 4.16 and are installed in three equal grids per tank with 100 diffusers in each grid. The diffusers are distributed equally along the length of each aeration basin. The operators indicate that a minimum retention time of 12 hours at a DO concentration of 2.0 ppm is required to meet the treatment requirements. The plant elevation is 100 ft ASL, average annual air temperature is 60 °F, and the average blower efficiency is 70%, inlet losses are 0.1 psi, and discharge pressure equals static pressure plus 1.0 psi. Actual ADF is 0.9 mgd and the primary clarifier effluent is 130 ppm BOD_5 and 45 ppm NH_3. The mixed liquor temperature is 68 °F and the average aeration basin effluent DO concentration is currently much higher than 2.0 ppm. Assume the fouling factor $F = 0.9$, $\alpha = 0.6$, and $\beta = 0.95$. The composite power rate is reported as $0.12.

(a) Are the tanks mixing limited?

(b) If so, how many tanks can be taken out of service?

(c) What energy cost savings can be expected by taking these tanks out of service?

Problem 4.4

A plant is being designed for an ADF of 6.0 mgd and each of four aeration basins is designed to treat 1.5 mgd. The effluent permit requires BOD removal only, with an effluent concentration of 30 mg/l. Primary effluent BOD_5 is 175 mg/l and the design wastewater temperature is 50 °F. The tanks are 30 ft wide × 120 ft long × 16 ft deep. The tanks will have three equal grids of fine-pore diffusers with 100 diffusers per grid. The base design has grids manually controlled with the air flow equally divided between them. The expected AOTE is 11% at an average DO of 2.0 mg/l. It is estimated that the first grid will remove 50% of the total BOD_5, the second grid 30% of the total, and the third grid 20% of the total. The calculated blower power consumption is 34 SCFM per kilowatt. Barometric pressure is 14.5 psia. The installed cost of a DO transmitter, control valve, and air flow transmitter is estimated as $9000 for each grid and the electric power composite rate is $0.15/kWh. For the individual grid control assume a DO profile of 0.75, 1.25, and 2.0 mg/l from influent to effluent grids. Will it be cost-effective to install automatic control on each grid?

Chapter 5

Blowers and Blower Control

The majority of activated sludge systems employ diffused aeration. All of these systems require blowers to provide air under pressure to the process equipment. Blowers provide air to a variety of unit processes:

- Equalization basins
- Grit removal
- Secondary aeration
- Post aeration
- Filter and MBR backwash
- Aerobic digestion and sludge holding tanks
- Channel aeration
- Airlift pumps

Not all treatment plants employ all of these processes, of course, but most treatment facilities have blowers and at least several of these applications. In some facilities, all the low-pressure air requirements are supplied by a single set of blowers, and in others, separate blowers are provided for different processes. Sometimes this is dictated by physical location, changes in the plant over time or by the differences in discharge requirements for each process group. The blower

Aeration Control System Design: A Practical Guide to Energy and Process Optimization,
First Edition. Thomas E. Jenkins.

evaluation for each application requires identification of the processes affected and their impact on blower operation and power.

5.1 COMMON APPLICATION AND SELECTION CONCERNS

There are many similarities between pumps and blowers. These similarities include general physical construction and the types available. Some of the general principles of physics apply to both types of machinery. There are also extremely important differences—many of them a direct result of the blower's operating on a compressible fluid, air. This creates variability and performance complexities that make blower evaluation more complicated than evaluating a pump operating on water, which is an essentially incompressible fluid.

There is no clear or universal definition used to distinguish between fans, blowers, and compressors. All provide the function of moving air under pressure from inlet to discharge. Generally, a machine with a discharge pressure less than 1 psig (28 in. H_2O) is considered a fan. A machine operating above 15 psig discharge pressure is considered a compressor. Machines operating between 1 and 15 psig discharge pressure are considered blowers. This range includes most of the process air systems in wastewater treatment. The term blower will be used exclusively in this chapter since it is the word generally used in the wastewater industry. The blower output is usually identified as "low-pressure air" to differentiate it from HVAC air flow and from "shop air," which is usually in the 80–100 psig pressure range.

An important concept is that a blower is a volumetric device. This applies to all blowers, regardless of type or details of construction. It is common to discuss the blower as producing pressure, but this is technically incorrect. Blowers move air from inlet to discharge. The pressure at the blower discharge is the result of the aeration system's resistance to that air movement. There are limits to the blower's capability to overcome that resistance. A centrifugal blower may be limited by the aerodynamic capability of the impeller and housing. A positive displacement (PD) blower may be limited by the torque of the motor. When the system resistance to flow exceeds the blower's pressure capability, the flow through the blower ceases—usually with unpleasant side effects.

5.1.1 Properties of Air

Air is a mixture of gases. The predominant components, (and those of greatest interest in aeration and blower analysis) are nitrogen, oxygen, and water vapor. Argon and carbon dioxide are present in small quantities and are seldom worth considering in evaluations. The other constituents of air are negligible. The relative concentrations of the various gases may vary with time and location; for example, the air leaving an aeration basin will have a lower concentration of oxygen and a higher concentration of carbon dioxide. The most noticeable and most common variation is the concentration of water vapor. Chemical and physical properties of

TABLE 5.1 Composition of Dry Air

Gas	Percent by Volume	Percent by Weight
Oxygen	20.95	23.20
Nitrogen	78.09	75.47
Carbon dioxide	0.03	0.05
Argon	0.93	1.28

air are usually referenced to either dry air or air with specified moisture content. For the normal composition of dry air, see Table 5.1.

Note that the percentages by weight and by volume differ. Most of the time, aeration systems are concerned with mass flow but the amount of interest in a particular analysis should be identified if there is any potential for misinterpretation.

The units of measurement for air flow can be a source of confusion. The process oxygen demand is based on mass flow rate, for example, lb_m/h. Process control would logically be based on mass flow rate as well. Blowers, on the other hand, are volumetric devices. Blower manufacturers tend to refer to volumetric flow rate, for example, cubic feet per minute (CFM). It is possible to convert from one to the other, but only if the thermodynamic state of the air is known.

Three flow rate units are commonly encountered in blower applications. Actual cubic feet per minute (ACFM) is the volumetric flow rate at the actual air conditions in the flow stream. This may be ambient air, air flowing in an aeration header under pressure, or air flow at any other point of interest. A flow rate identified as ACFM must include the conditions of the air in order to be useful. Inlet CFM (ICFM) is a special case of ACFM, specifically referencing the volumetric air flow at the blower inlet. This usually implies air at ambient temperature and humidity, but at a pressure slightly below atmospheric pressure because of inlet filter losses. When evaluating blower performance, ICFM is a common parameter. The air temperature and pressure should always be identified.

The third unit of measure for flow rate in common use is the most confusing. Standard cubic feet per minute (SCFM) is usually used to specify process flow rates and blower capacity. The unit SCFM is not truly a volumetric flow rate. The specified standard conditions for SCFM define the properties of the air. SCFM is in reality a mass flow rate and should be treated as such.

Like almost everything else in treatment plant operation, the properties of air are variable. Some of these variations are obvious—for example, the impact of temperature on density. Others are more subtle or abstract—for example, the impact of relative humidity on molecular weight. The changes in the property of air with composition and condition have been thoroughly studied and are well documented. The impact of the changes on blower performance are well documented for most types of blowers. ASME performance test codes PTC 9 and PTC 10 identify procedures for evaluating and compensating performance of two common types of blowers, positive displacement and centrifugal blowers.

Because of the variability of air properties, it is convenient to have a standard reference point. In the US wastewater treatment industry, standard conditions are defined as 68 °F, 14.7 psia, and 36% RH. These conditions are the definition of SCFM. At these conditions, the density of air is 0.075 lb_m/ft^3 and the specific weight is 0.075 lb_f/ft^3. By definition, this density represents a specific gravity of 1.0. In other industries, standard conditions may be defined as 60 °F, 14.7 psia, and dry air. Outside the United States, reference conditions for air properties are often referenced as "normalized." The definition varies, but 0 °C, 101.3 kPa, and dry air are the most common reference conditions.

In many calculations, air can be treated as an ideal gas. Boyle's law and Charles' law apply. Boyle's law states that the volume of an ideal gas at constant temperature varies inversely with the ratio of the absolute pressure:

Equation 5.1

$$\frac{V_2}{V_1} = \frac{p_1}{p_2}$$

where

$V_{1,2}$ = volume at condition one and two, length^3
$p_{1,2}$ = absolute pressure at condition one and two, force per length^2

Charles' law states that the volume of an ideal gas at constant pressure varies directly with the ratio of absolute temperature:

Equation 5.2

$$\frac{V_2}{V_1} = \frac{T_2}{T_1}$$

where

$V_{1,2}$ = volume at condition one and two, length^3
$T_{1,2}$ = absolute temperature at condition one and two °K or °R

From these two laws, a relationship for volume, pressure, and temperature can be developed. This is the "perfect gas formula." The general formula is:

Equation 5.3

$$p \cdot V = n \cdot R_0 \cdot T$$

where

p = pressure, lb_f/ft^2
V = volume, ft^3
n = number of moles, dimensionless
R_0 = universal gas constant, 1545 ft lb_f/mol °R
T = absolute temperature, °R where °R = °F + 460

It is usually convenient in aeration control applications to modify this formula specifically for air and for more common units of measure. Note that the gas constant for any gas can be determined by dividing the universal gas constant by the molecular weight of the gas:

Equation 5.4

$$p \cdot V = m \cdot R_{air} \cdot T$$

where

p = pressure, psia
V = volume, cubic feet
m = mass, lb_m
R_{air} = gas constant for air, 0.3704 $ft^3\ lb_f/lb_m\ °R\ in^2$
T = absolute temperature, °R

The ideal gas laws express the impact of temperature and pressure on air volume and represent the largest variations in air volume and flow rate. For many blower calculations, sufficient accuracy can be achieved by including just these two parameters in the corrections and ignoring the effect of relative humidity:

Equation 5.5

$$Q_2 = Q_1 \cdot \frac{p_1}{p_2} \cdot \frac{T_2}{T_1}$$

where

$Q_{1,2}$ = volumetric flow rate at conditions 1 and 2, ft^3/min
$p_{1,2}$ = pressure at conditions 1 and 2, psia
$T_{1,2}$ = temperature at conditions 1 and 2, °R

Note that in these conversions, temperature and pressure ratios should be in an absolute scale. For US customary units, the absolute temperature is expressed as degrees Rankine. Absolute temperature in °R = °F + 459.67 but most calculations use °R = °F + 460 to simplify the arithmetic. The difference at standard conditions is less than 0.1% which is negligible in most calculations. Outside the United States, absolute temperature is often expressed in degrees Kelvin (°K). The precise conversion is °K = °C + 273.16 but this is typically rounded to °K = °C + 273.

Absolute pressure is usually expressed in US customary units as pounds per square inch absolute (psia). Pressure is always measured as the difference between two points. Gauge pressure is the difference in pressure between the inside and outside of a container—typically the difference between ambient air and air inside a pipe. Absolute pressure is the difference in pressure between the point of measurement and a vacuum.

Equation 5.6

$$\text{psia} = \text{psig} + \text{barometric}$$

where

psia = absolute pressure, psia
psig = gauge pressure, psig
barometric = barometric pressure, psia

Barometric pressure at sea level and standard conditions is 14.696 psia—usually taken as 14.7 psia for convenience in calculations. Barometric pressure is not constant but varies with weather conditions and decreases with increasing altitude. The variation in barometric pressure with weather is 0.25 psia or less. Because actual barometric pressure variations are indeterminate and not related directly to flow rate or ambient conditions, this variation is generally ignored in the blower calculations for energy comparisons between systems. For all except the most extreme cases, barometric pressure at a given altitude can be calculated:

Equation 5.7

$$p_{\text{bar}} = 14.7 - \frac{\text{Alt}}{2000}$$

where

p_{bar} = barometric pressure, psia
Alt = altitude, feet above sea level (FASL)

Assuming ideal gas behavior and neglecting the effect of relative humidity, the conversion from ACFM to SCFM can be simplified to:

Equation 5.8

$$Q_{\text{s}} = Q_{\text{a}} \cdot \frac{p_{\text{a}} \cdot 35.92}{T_{\text{a}}}$$

where

Q_{s} = mass flow rate, standard cubic feet per minute
Q_{a} = volumetric flow rate, cubic feet per minute
p_{a} = actual pressure, psia
T_{a} = actual temperature, °R

It is sometimes convenient to have the mass flow rate in pound per minute. SCFM can be converted to pound per minute by multiplying by the density:

Equation 5.9

$$q_{\text{m}} = Q_{\text{s}} \cdot \rho$$

where

q_{m} = mass flow rate, $\mathrm{lb_m/min}$
Q_{s} = mass flow rate, standard cubic feet per minute
ρ = density of air at standard conditions, 0.075 $\mathrm{lb_m}$/standard cubic foot

Compressibility, denoted by "Z," is the departure of a real gas from the behavior of an ideal gas. At the temperatures and pressures encountered in aeration systems, Z is essentially unity, and air can treated as an ideal gas without appreciable error.

It is important to verify both the actual air conditions at the point of interest in an aeration system and the standard conditions used as the basis for the process calculations. This is particularly true in comparing instrument readings. It is very common to have flow meters in aeration piping that measure ACFM at the blower discharge pressure and temperature while other flow meters measure SCFM. These readings are obviously going to disagree. Unless the units and conditions are clearly stated confusion will result. Note that if a velocity-based meter, such as a Pitot tube or orifice plate, is used to measure ACFM the measurement can be corrected to SCFM. This requires first measuring pressure and temperature and then using Equation 5.8 for the conversion.

5.1.2 Effect of Humidity

Humidity fluctuations have an impact on the molecular weight of air and therefore on blower performance. The molecular weight of water (18.016) is lower than the molecular weight of dry air (28.97). Higher water vapor content (higher relative humidity) therefore reduces the molecular weight of air. Dalton's law is the basis for humidity corrections. It states that the total pressure of ideal gases is equal to the sum of the partial pressures of the individual gases in the mixture:

Equation 5.10

$$p_{\mathrm{total}} = p_{\mathrm{a}} + p_{\mathrm{b}} + p_{\mathrm{c}} + \cdots$$

where

$p_{\mathrm{a,b,\ldots}}$ = partial pressure of gases a, b, and so on, psia

For aeration blower applications, the water content of air is generally specified as percentage relative humidity (%RH). The vapor pressure of water (p_{v}) is the partial pressure of water vapor in any gas mixture, including air, which is saturated with water. The saturation vapor pressure of water (p_{sat}) is a function of temperature (Table 5.2). It can be obtained from steam tables or a psychrometric chart. The relative humidity data may be obtained directly by hygrometers. In other cases, the relative humidity isn't measured directly, but rather the dry bulb and wet bulb temperatures are measured using a sling psychrometer. This device has two thermometers. One thermometer has the bulb directly exposed to the air, measuring

TABLE 5.2 Saturation Water Vapor Pressure

°F	p_{sat}, psia
32	0.0886
40	0.1217
60	0.2561
68	0.3390
80	0.5068
100	0.9492
120	1.6927

the dry bulb temperature. The other thermometer has the bulb wrapped in a moist cloth so that the effect of evaporative cooling is measured with the result being the wet bulb temperature. A psychrometric chart is used to obtain relative humidity and other pertinent data from these two temperatures.

Reading a psychrometric chart can be intimidating at first glance, principally because of the amount of information presented in most charts. For most blower applications, the only information of interest is:

- Dry bulb temperature, °F
- Wet bulb temperature, °F
- Relative humidity, %
- Vapor pressure, psia

The locations of the critical readings are shown in Figure 5.1. Psychrometric charts are widely available online and from other sources. Once the actual vapor pressure p_v is known, the saturation vapor pressure p_{sat} can be calculated using:

Equation 5.11

$$\%RH = \frac{p_v}{p_{sat}} \cdot 100$$

where

$\%RH$ = relative humidity, %
p_v = actual vapor pressure, psia
p_{sat} = saturation vapor pressure at actual dry bulb temperature, psia

When the relative humidity and vapor pressure are known, the correction from actual to standard conditions can be more precise. This is most commonly applied in converting specified process requirement in SCFM to the corresponding ICFM at various inlet conditions. This conversion is typically necessary to interpret the blower manufacturer's data and determine power consumption at site conditions.

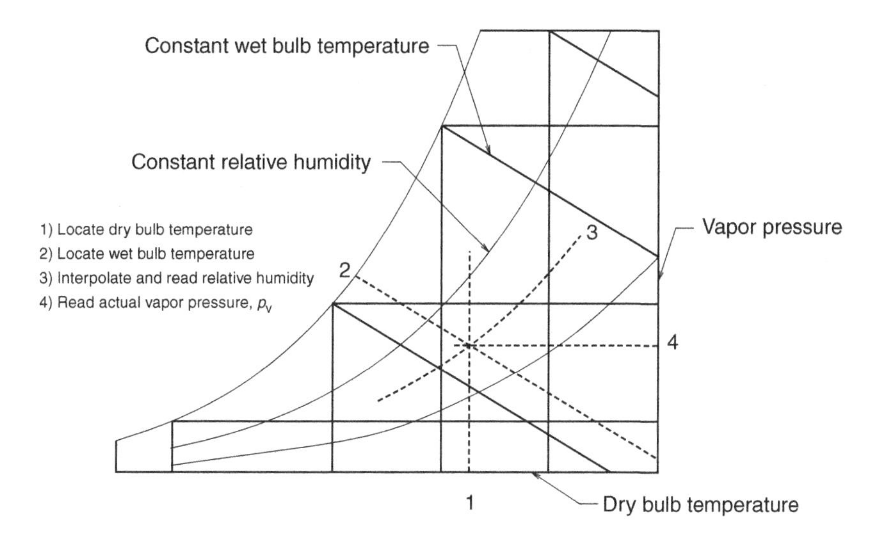

FIGURE 5.1 *Reading a psychrometric chart*

Equation 5.12

$$Q_i = Q_s \cdot \frac{14.58}{p_b - (\mathrm{RH} \cdot p_{sat})} \cdot \frac{T_i}{528} \cdot \frac{p_b}{p_i}$$

where

Q_i = inlet volumetric flow rate, ICFM
Q_s = mass flow rate, SCFM
p_b = barometric pressure, psia
RH = relative humidity, decimal
p_{sat} = saturation vapor pressure at actual dry bulb temperature, psia
T_i = actual inlet air temperature, °R
p_i = actual inlet pressure, psia

Note that the saturation vapor pressure of water at 68 °F is 0.339 psia, at 36% relative humidity the actual vapor pressure is 0.12 psia, and 14.7 − 0.12 = 14.58.

When blower manufacturers calculate performance, relative humidity is included. In design specifications, the blower design point should also include relative humidity. For process calculations and preliminary energy savings estimates, the loss of accuracy caused by omitting relative humidity is less significant than the errors introduced by the process performance assumptions. The effect of relative humidity may be safely neglected in these calculations. However, it is important to be aware of the full correction procedures so that the elimination of superfluous factors will be made consciously.

In general, conversion between standard conditions outside the United States and the 68 °F, 14.7 psia, and 36% RH standard can be made on a volumetric basis, using

Equation 5.5 or Equation 5.12. One minor difference should be noted, however, and included if significant. Outside the United States, many calculations use dry air (0% RH) as the basis of air properties. At 36% RH, the proportion of oxygen is slightly lower than it is in dry air since 0.8% by weight of the air is water vapor. Since process demand is based on oxygen demand, precise calculation methodology requires recognition of the difference. To convert from normalized cubic meters per hour to SCFM, it is necessary to multiply by a factor to compensate for this difference in oxygen content:

Equation 5.13

$$Q_s = Q_n \cdot 0.6386$$

where

Q_s = mass flow rate, SCFM, which will provide equal oxygen to the process
Q_n = normalized mass flow rate, N m^3/h, 0 °C, 101.3 kPa, 0% RH

Selecting inlet conditions for calculating performance and energy savings can be challenging. Many different methodologies are used for selecting evaluation conditions. They range from overly simplistic to incredibly complex and cumbersome. Each method has its advantages and disadvantages. It should be remembered that any analysis is a mathematical model of the process. Inherently it involves assumptions and estimates.

One extreme of analysis methodology is a level of complexity that exceeds the accuracy of the initial assumptions and creates a costly and unwieldy analysis. The opposite is a single-point analysis that ignores the effects of variable loads, fluctuating blower output, and control technique. These variations can have a significant and measurable impact on energy efficiency and process performance. The selected analysis methodology should be a reasonable compromise between these two extremes.

A commonly applied technique uses a tabular approach—typically using five conditions with different flow rates, temperatures, relative humidity, and duty cycle for each. A common example of this procedure uses values of maximum design flow, 100 °F, 90% RH, and maximum design discharge temperature for 40% of the annual operating time. This procedure also uses minimum design flow at −5 °F, 45% RH, and 15% operating time. Other examples have variations from worst-case summer design conditions of 105 °F and 90% RH to minimum anticipated winter conditions of 0 °F and 10% RH. In every case, the evaluation was done at maximum design discharge pressure.

The problem with this procedure is that it is quite arbitrary and usually has little resemblance to actual process fluctuations. As seen in Chapter 3, the most significant load fluctuations occur on a diurnal pattern, not seasonally. The simultaneous occurrence of maximum process load and maximum ambient temperature is possible, but this would represent an insignificant percentage of annual operating hours. The complexity of this approach is not justified by a corresponding increase in accuracy in the results.

The opposite extreme is represented by performing the energy analysis based on the design conditions for the blowers—worst-case inlet temperature, relative humidity, and discharge pressure. This approach is not likely to provide satisfactory predictions of energy savings for blower and control upgrades. First, the operation at simultaneous worst-case conditions is an extremely rare occurrence. Second, variation in blower and control system energy requirements across the flow range represents the most significant variable in power demand. Finally, evaluation at design pressure negates the influence of control strategies specifically designed to minimize discharge pressure.

A recommended procedure uses the flow variations and duty cycles identified in Chapter 4. If the control system operation is based on maintaining constant discharge pressure, the maximum pressure anticipated in normal operation should be used. This is typically 0.5 psig below maximum design pressure.

Keeping the analysis effort reasonable and commensurate with the anticipated accuracy generally requires making the evaluation at a single set of inlet conditions. Establishing the inlet temperature and humidity to use is somewhat arbitrary, but can have a significant effect on the results. Some blower and control configurations exhibit the lowest energy demand at high inlet temperature, others provide better performance at low inlet temperature. The recommendation is to use the location's average annual inlet temperature and relative humidity for the evaluation.

This data is available from a variety of sources, including the National Oceanic and Atmospheric Administration (NOAA) and the American Society of Heating, Refrigeration, and Air Conditioning Engineers (ASHRAE). In the case of ASHRAE tables, the average of the 97.5% winter and 2.5% summer temperature provide an approximation of the annual average temperature. The relative humidity corresponding to the 5% dry bulb and coincident wet bulb temperatures provides a conservative but reasonable relative humidity value for use in the calculations.

5.1.3 Pressure Effects

The discharge pressure used for the evaluation is also critical to the comparison of control alternatives and different blower designs. For example, throttled centrifugal blowers show little variation in power consumption versus discharge pressure, while the correlation is directly linear with positive displacement blowers. If actual discharge pressure and flow data can be obtained during the initial system evaluation, this should be used. Otherwise a reasonable but conservative evaluation can be made based on the static pressure and the design discharge pressure. This evaluation incorporates the variation with flow rate that is expected with control strategies that optimize pressure.

The first step is to establish the static pressure associated with the diffuser submergence using Equation 4.12. This is the minimum discharge pressure at which air flow will occur.

The actual discharge pressure includes friction losses from air moving through the various air distribution components and the diffusers. Pressure drop is proportional to the square of the air flow rate. The pressure drop can be calculated from the piping layout when a high level of accuracy is required. For preliminary calculations, reasonable and conservative results can be achieved based on the design flow and design discharge pressure for the aeration system. The first step is to determine the constant of proportionality for the friction losses:

Equation 5.14

$$k_f = \frac{(p_{des} - p_{stat})}{Q_{des}^2}$$

where

k_f = constant of proportionality for friction losses, psi/SCFM^2
p_{des} = total system pressure at design flow, psig
p_{stat} = static pressure, psig
Q_{des} = design system air flow rate, SCFM

This will allow the development of a system curve showing the relationship between flow and pressure:

Equation 5.15

$$p_{tot} = p_{stat} + k_f \cdot Q^2$$

where

p_{tot} = total system pressure, psig

Systems that operate at constant pressure are a special case of the system curve. The system curve is a horizontal line, k_f is equal to zero, and a fixed value is added to the static pressure to accommodate friction losses. This is generally the case when a control system is employed to maintain the pressure. In this type of operation, the blowers are modulated to maintain flow while operating against a changing system resistance.

By applying Equation 5.15 to each of the five flow rates used to characterize the diurnal flow variations, a reasonable discharge pressure will be available for the blower power analysis. This will reflect pressure variations and their impact on energy for control strategies that optimize discharge pressure. As indicated above, control strategies based on maintaining constant pressure must be analyzed based on the actual or anticipated pressure setpoint regardless of flow.

The system curve is a very useful tool in analyzing and developing control strategies. Comparing the system curve to field observation can identify over-design or excess capacity in the blowers. The system curve can also help identify the available operating range of the blowers.

The system curve is necessary for determining both the operating flow and the power demand of the blower system. By superimposing a system curve on a blower characteristic curve, the actual operating point at a specific condition can be identified (see Figure 5.2). In the case of centrifugal blowers, the intersection of

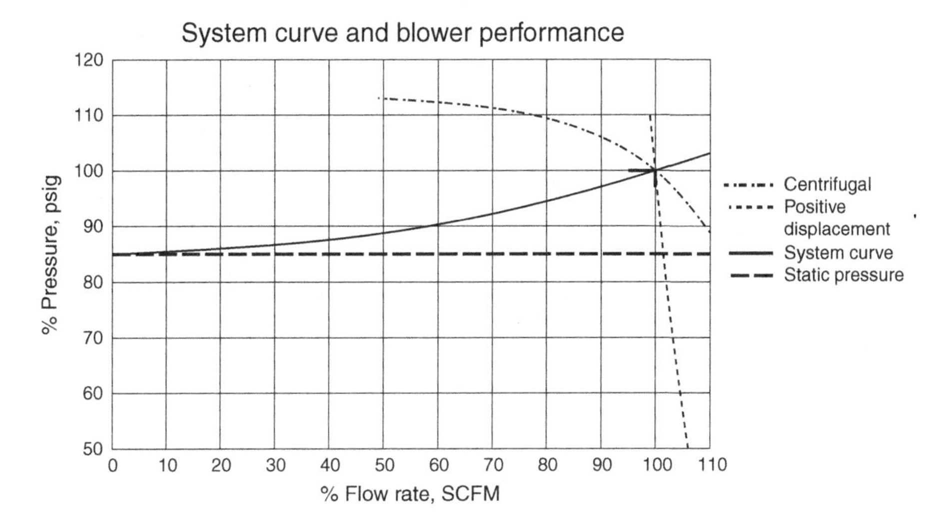

FIGURE 5.2 *Example system curve*

the system curve and the blower characteristic curve establishes the flow rate. The purpose of blower modulation is to shift the blower's characteristic curve so the intersection point coincides with the desired flow rate. In the case of positive displacement blowers, the flow rate is established directly by modulating the blower but the resulting discharge pressure determines the blower power draw.

Filtering the air prior to the blower inlet doesn't have a direct effect on control strategy, although many control systems monitor filter pressure drop for alarms. Inlet filtering does have an impact on power requirements. Many systems have filters with dust removal ratings far in excess of the level required for protecting the blowers. In early implementations of fine-pore diffusers there was concern over air-side fouling (plugging of the diffuser media by dust and dirt). This led to the specification of filters with very stringent contaminant removal ratings and corresponding high-pressure drop across the filter. Experience has shown that air-side fouling is virtually nonexistent in practice. The rare occurrences have resulted from scaling and contaminants in the distribution piping downstream of the blowers. Blower inlet filters should follow the manufacturer's recommendation for equipment protection which usually means capture of 95% of particles 10 μm or larger. The allowance for pressure drop for blower performance calculations for this filter rating is approximately 6 in. water column (0.20–0.25 psi). This allows for an increase in filter pressure drop as plugging of the filter media occurs with accumulation of dirt.

5.1.4 Common Performance Characteristics

The power consumption of any blower is a function of flow, pressure, and efficiency. Air flow rate is primarily a function of process demand for oxygen. Pressure is primarily a function of the physical configuration of the diffusers and air distribution system. Blower efficiency is primarily a function of the mechanical design of the blower.

The control system performance is dictated and limited by these factors. The control system designer is often involved in the evaluation of the alternate ECMs and the selection of the aeration equipment. The influence of the control system designer extends well beyond this, however. The control strategy is capable of optimizing the performance of the blowers and the entire aeration system: matching flow to process demand, minimizing discharge pressure by eliminating wasteful throttling, and controlling the blowers to optimize efficiency. In order to maximize the aeration system performance, it is necessary for the control system and control strategy to accommodate the similarities and differences among the various types of blowers. The design should also integrate the blower control with the rest of the aeration system.

The thermodynamics of the compression process is common to all blowers—regardless of design. The mechanical differences show up in the efficiency of the blower and in the response of the blower to variations in flow, pressure, and control technique.

As previously discussed, the term "efficiency" must be used with care. It appears often in discussions of power consumption and is a term encountered in blower energy consumption calculations. Provided the term efficiency is clearly defined and used appropriately, it may provide insight into the operation and control of aeration blowers. Incorrectly applied, it can confuse and obscure.

There are a variety of efficiencies associated with blowers. Mechanical efficiency is generally applied in reference to losses external to the compression process itself—bearing friction, lube system power requirements, and so on. Volumetric efficiency is generally used with positive displacement blowers—referring to the difference between the theoretical and actual volume moved by the blower.

Compression efficiency for the blower applies to the thermodynamic efficiency of the compression process itself. It is defined as the ratio of power contained in the compressed gas stream to the actual power input to the blower. The compression losses are generally much higher than the mechanical losses. The compression efficiency of an ideal gas occurring without any heat transfer is the adiabatic efficiency, also referred to as the isentropic or reversible efficiency.

An important dimensionless parameter in blower power calculations is k—the ratio of the specific heat at constant pressure, C_{p}, to the specific heat at constant volume, C_{v}. For air, the heat capacities vary with temperature, pressure, and relative humidity. When precise calculations are required, for example when computing power for performance guarantees, the variations must be included. For estimating the effectiveness of an ECM or calculating savings, the variations are not as significant as many of the other assumptions required and the ratio is assumed to be a constant:

Equation 5.16

$$k = \frac{C_{\mathrm{p}}}{C_{\mathrm{v}}} = \frac{C_{\mathrm{p}}}{C_{\mathrm{p}} - R_{\mathrm{air}}} \approx 1.395$$

where

k = ratio of specific heats for dry air, dimensionless
C_p = specific heat at constant pressure, for dry air at 68 °F ≈ 0.240 BTU/lb °R
C_v = specific heat at constant volume, for dry air at 68 °F ≈ 0.172 BTU/lb °R
R_{air} = gas constant for dry air, ≈0.0686 BTU/lb °R

Equation 5.17

$$\frac{k-1}{k} \approx 0.283$$

Another useful parameter in blower calculations is the pressure ratio, also called the compression ratio:

Equation 5.18

$$r_p = \frac{p_d}{p_i}$$

where

r_p = pressure ratio, dimensionless
p_d = discharge pressure, psia
p_i = inlet pressure, psia

A related term that often causes confusion is "compressor head," or "blower head," expressed in units of ft lb/lb. The compressor head is similar to pump head in that both identify the work done by the system. In pumps, head is often used interchangeably with discharge pressure since pumps work with incompressible fluids. Pump head can be measured directly with a pressure gauge. Compressor head—because of the compressible nature of the air—cannot be directly measured. It must be calculated from other parameters:

Equation 5.19

$$H_{air} = \frac{R_{air} \cdot T_i \cdot \left[r_p^{(k-1/k)} - 1 \right]}{(k-1)/k}$$

where

H_{air} = adiabatic head for air, ft lb_f/lb_m
R_{air} = gas constant for air, 53.34 ft lb_f/lb_m °R
T_i = inlet air temperature, °R

For convenience the adiabatic factor is defined as:

Equation 5.20

$$X = r_{\mathrm{p}}^{(k-1)/k} - 1$$

Applying the mass flow rate and simplifying generates an equation for calculating the gas power:

Equation 5.21

$$P_{\mathrm{gas}_{\mathrm{ad}}} = \frac{q_{\mathrm{m}} \cdot T_{\mathrm{i}} \cdot X}{175.1}$$

where

$P_{\mathrm{gas}_{\mathrm{ad}}}$ = adiabatic isentropic power in the gas stream, hp
q_{m} = mass flow rate, lb/min
T_{i} = inlet air temperature, °R
X = adiabatic factor, dimensionless

Another common form used to calculate gas power is:

Equation 5.22

$$P_{\mathrm{gas}_{\mathrm{ad}}} = q_{\mathrm{m}} \cdot C_{\mathrm{p}} \cdot \Delta T$$

where

C_{p} = specific heat, 5.66 hp min/(lb °R)
ΔT = difference between inlet and discharge temperature, °R

By definition, the efficiency can be obtained from the gas power and measured power:

Equation 5.23

$$\eta = \frac{P_{\mathrm{gas}_{\mathrm{ad}}}}{P_{\mathrm{meas}}} \cdot 100$$

where

η = efficiency, %
P_{meas} = measured power, hp

As previously stated, the efficiency of blowers is widely variable. Every type, model, and manufacturer will have differing efficiencies. For a given blower the efficiency will vary across the operating range and depends on the control method used. The wire-to-air efficiency values shown in Table 5.3 may be used for preliminary approximations, but they should not be relied on for evaluating alternate

TABLE 5.3 Approximate Wire-to-Air Efficiency of Various Blowers at Mid-range Flow (for Preliminary Estimating Use Only)

Blower Type	Approximate Efficiency, %
Lobe type PD	60
Screw PD	65
Multistage centrifugal, throttled	65
Multistage centrifugal, VFD	70
Single stage centrifugal	70
Turbo blower	70

blower types or final determination of energy consumption. In all cases, manufacturer's data should be used for detailed design and final analysis. It should be noted that the efficiencies in Table 5.3 are intended to represent performance at the mid-range flow rate, not at design point or BEP.

Because the actual compression process deviates from ideal behavior, the polytropic efficiency is commonly used by blower manufacturers. The polytropic exponent is determined from test data and substituted for the adiabatic ratio of specific heats, k. Either the adiabatic or the polytropic efficiency may be used with sufficient accuracy so long as the value is applied consistently in determining efficiency from test data and in calculating power using the efficiency. In general, using adiabatic power is more convenient than using polytropic power.

Several different forms of the equation for evaluating blower power consumption are available. The form chosen depends only on the data available and the preferences of the designer. One convenient form is developed by simplifying Equation 4.6:

Equation 5.24

$$P_{\mathrm{wa}} = \frac{Q_{\mathrm{s}} \cdot T_{\mathrm{i}}}{\eta_{\mathrm{wa}} \cdot 3131.6} \cdot X$$

where

P_{wa} = wire to air power, kW
Q_{s} = mass flow rate of air, SCFM
η_{wa} = wire to air system efficiency, decimal

Another form is convenient if volumetric flow rate and pressures are known:

Equation 5.25

$$P_{\mathrm{wa}} = \frac{Q_{\mathrm{i}}}{\eta_{\mathrm{wa}} \cdot 86.9} \cdot \left[\left(p_{\mathrm{i}}^{0.717} \cdot p_{\mathrm{d}}^{0.283}\right) - p_{\mathrm{i}}\right]$$

where

P_{wa} = wire to air power, kW

Q_i = volumetric flow rate of air, ICFM
$p_{i,d}$ = inlet and discharge pressure, psia

The value used for measured power and the various components included in the measurement must be verified when comparing alternate blower and control systems. Many testing standards—for example, ASME PTC 9 and PTC 10—are based on bhp at the blower shaft. This is not particularly useful for current technology. Most blowers are systems with the blower, motor, and flow control system provided as a complete package. The evaluation of power cost in the field must include all of these items evaluated at multiple points in the operating range. This is particularly important for dynamic compressors with variable speed control. Additional peripheral components such as filters and silencers affect performance and should be included in the package evaluated for energy consumption.

The system approach to energy evaluation has led to the development of the concept of "wire-to-air" power. This approach is embodied in newer test standards such as ASME PTC 13. The test methodology measures mass air flow rate delivered to the process at the blower discharge and the power consumed by a complete system. The package tested must include all items required to provide a serviceable blower system including: controls, variable speed drives, motors, filters, and related auxiliary items. The power measurement in this methodology is the total kilowatts required for this system which provides much better data for determining life cycle cost and payback.

In comparing power requirements for alternate systems, it has been customary to use the blower manufacturer's characteristic curves for the bare blower. A better approach is to provide the data for wire-to-air electrical power versus mass air flow rate at several pressures within the anticipated operating range and corrected to site conditions (see Figure 5.3). This data can be calculated or obtained from the system suppliers. The data presented in this format allows comparisons of various alternate designs—including the differences related to discharge pressure fluctuations. This method of comparison eliminates confusion over efficiency by evaluating power demand directly.

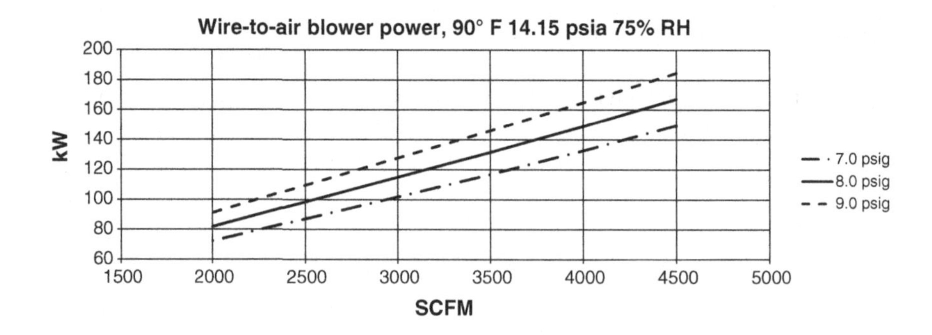

FIGURE 5.3 *Example mass flow rate versus power consumption*

It is often useful to determine the blower discharge air temperature. This can affect the life of gaskets and seals in valves and the pressure rating of piping and diffusers—particularly when plastic and elastomers are used. A rough approximation is an increase of 10 °F over inlet temperature for every 1 psig pressure rise through the blower. A more accurate determination can be made if the blower efficiency is known:

Equation 5.26

$$T_d = T_i + \frac{T_i \cdot X}{\eta}$$

where

T_d = discharge temperature, °R

A portion of the heat of compression through the blower is rejected into the blower room or blower enclosure. The heat created by motor and variable frequency drive (VFD) inefficiency is also rejected into the room. It is good design practice to provide forced air ventilation for the blower room to keep inside temperature at a reasonable level. It is not usually practical to provide air conditioning for the entire blower system. If climatic extremes require cooling for control panels and VFDs, it is generally more economical to mount them in a separate room and cool that space.

The heat rejected by the motors and VFDs is easily calculated from their efficiency:

Equation 5.27

$$H_d = 2544 \cdot P_{mot} \cdot (1 - \eta_{mot} \cdot \eta_{VFD})$$

where

H_d = heat rejected by drive system, including motor and VFD, BTU/h
P_{mot} = motor power, hp
$\eta_{mot,VFD}$ = motor and VFD efficiency, decimal

The heat rejected by the blower itself is a function of the temperature of the blower case and its effective area. The blower case is usually not a smooth shape but has numerous ribs and convolutions. The effective area may be estimated by multiplying the surface area of the approximate cylinder or prism by a factor—typically 1.25. The heat rejected by the blower may be estimated by:

Equation 5.28

$$H_b = 2.4 \cdot 1.25 \cdot A_b \cdot (T_d - T_a)$$

where

H_b = heat rejected by the blower case, BTU/h

A_b = nominal surface area of the blower case, ft^2
$T_{d,a}$ = discharge and ambient air temperature, °F

The air flow used for ventilation is supplied by exhaust fans and drawn into the blower room or blower enclosure through louvers. The louvers may be motor operated or gravity dampers that open when the ventilating fans are turned on. The quantity of air required to maintain a reasonable temperature 10–20 °F above ambient temperature can be calculated based on the heat rejected:

Equation 5.29

$$Q_{fan} = \frac{H_d + H_b}{1.08 \cdot (T_r - T_o)}$$

where

Q_{fan} = required ventilating fan air flow rate, CFM
$T_{o,r}$ = outside and room air temperatures, °F

Blower turndown is a very significant parameter in determining energy use—in many cases, turndown is more important than power consumption at design flow rates. Whether for a new plant or replacing old blowers as part of an ECM, the ability to match the blower flow to the actual process demand is critical. If the minimum safe operating flow rate of a blower is greater than the minimum process demand, it is impossible to match blower capacity to process demand. The result is wasted power—regardless of how "efficient" a given blower system might be.

Minimum and maximum flow rates from the blower system are critical to optimizing process performance. Maximum flow rate available is generally limited by motor power capacity. Minimum flow rate is generally established by mechanical limitations of the blower. In the case of dynamic blowers, this minimum limit is generally the "surge" flow. For positive displacement blowers, the lower limit is typically set by thermal constraints of the blower. Turndown may be expressed as the ratio of the minimum flow to the maximum flow the blower system can provide and is usually stated as a percentage (see Equation 2.3).

The average blower system can provide 50% turndown but there is tremendous variation in this parameter. The blower type and design are obvious factors. The control method can also have a dramatic effect on turndown. Even the control system tuning has a bearing since instability or lack of control precision can prevent utilizing the full theoretical blower operating range.

Most regulatory agencies require redundancy in blower system design so that the worst-case process air demand can be met with the largest unit out of service. This means that a blower system must include at least one standby blower.

Some designers try to minimize equipment cost by using two blowers with each sized to meet 100% of the worst-case air flow requirement. If the selected blowers can provide 50% turndown, it is unlikely that they can operate at the reduced flow rate the process will require during low flow periods. This is particularly true during the early years of a plant's life when the loading is lower than the design load. If the worst-case design blower air flow is taken as 100% capacity, another common arrangement is three blowers with each sized to provide 50% of design capacity. This arrangement is also typically unable to provide enough turndown during low loading periods.

As shown previously, it is not uncommon to experience an 8:1 ratio between worst-case design oxygen demand and minimum demand. This is particularly true at night during the early years of a plant's operation. It is difficult to provide that much range but the blower system should provide a ratio of maximum to minimum flow of at least 5:1.

There are several ways of accomplishing this. One is to install four blowers with each sized at 33% of design flow. A system that provides even more flexibility is to provide four blowers—two sized at 50% of design capacity and two sized at 25% of design capacity. This provides a turndown to 12.5% of design capacity, an 8:1 ratio. As process demand approaches 100% of design capacity, it also allows operating the blowers close to their best efficiency point thereby minimizing power use throughout the plant's life.

Using several smaller blowers may appear to increase installation and equipment cost. However, because each blower and the associated piping and electrical service is less expensive than for a large unit, the difference is not as significant as often thought. Even more importantly, the initial equipment and installation cost is a small fraction of the total energy and life cycle cost for the aeration system.

Analysis of blower performance can be a complex and confusing task. In general, the manufacturer's assistance should be obtained if possible. Familiarity with blower performance is valuable in coordinating controls with the blower suppliers and in comparing performance between alternate designs. It isn't necessary for the aeration control system designer to be an expert in the selection and analysis of blowers. However, it is essential to be familiar with the general operating principles and performance characteristics of the various types of blowers. An understanding of blowers is needed to judge performance under various conditions. It allows the designer to determine if various suppliers are providing directly comparable data.

In developing the aeration control design, it is important to understand the construction, limitations, and response to various control strategies of the various types of blowers commonly encountered in aeration systems (see Figure 5.4).

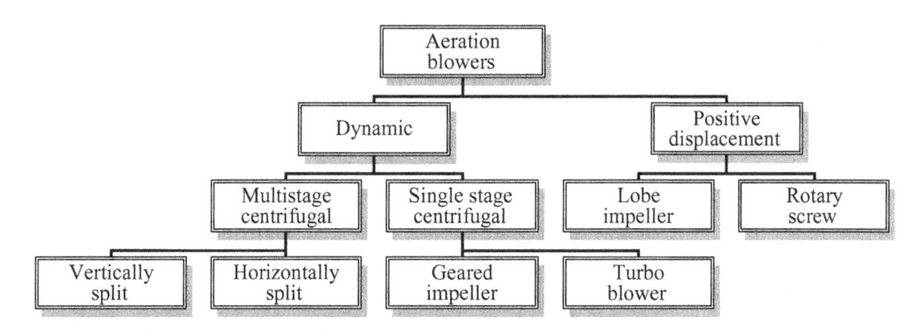

FIGURE 5.4 *Blower categories*

5.2 POSITIVE DISPLACEMENT BLOWERS AND CONTROL CHARACTERISTICS

PD blowers may be found in all but the largest wastewater treatment plants. They are most common in smaller facilities, but are available up to several hundred horsepower. Earlier PD blowers were generally considered less efficient than dynamic blowers, but recent advances in PD blower design and control systems have resulted in tremendous reductions in their energy demand.

5.2.1 Types and Characteristics

The term "positive displacement" refers to the volume swept by the rotating components of the blower in each rotation, forcing this volume to move from inlet to discharge. The volume is the blower's displacement, generally referenced as cubic feet per revolution. The movement of the volume through the blower is theoretically positive—regardless of resistance that volume will be moved through the blower every rotation.

There are two types of PD blowers widely applied to aeration systems, lobe type and screw type. Both designs share important characteristics:

- Volumetric flow rate directly related to blower rotational speed
- Mass flow rate at a given speed proportional to inlet density
- Discharge pressure inherently matches system back pressure
- Constant motor torque required at constant discharge pressure
- Some pulsation occurs in the discharge air flow

5.2.2 Lobe Type PD Blowers

The earliest PD blowers were of the lobe type. These are sometimes referred to as "Roots type blowers" because they were invented and commercialized by the Roots brothers. That is properly a trade name for the company started by the Roots brothers which still manufactures blowers.

The lobe type blower has two counter-rotating shafts, generally with an identical two- or three-lobed impeller on each shaft (see Figure 5.5). The two shafts are connected by gears that maintain the proper clearance between the impellers and transfer torque from the driven shaft to the other shaft. As the shafts rotate, the impellers sweep a fixed volume of air from the inlet to the discharge side of the blower.

The clearance between the two impellers and between the impellers and the outside and end plates of the blower housing creates a small leakage path for air back to the inlet side of the blower. This leakage represents "slip," the difference between the theoretical and actual air volume moved in each revolution. The ratio of actual to theoretical delivery is also referred to as the volumetric efficiency of the blower.

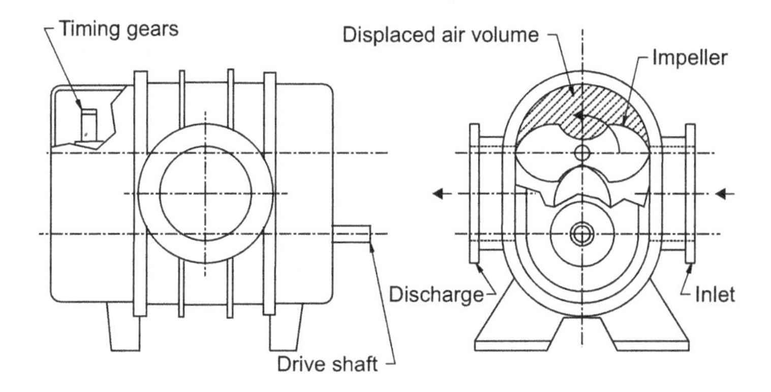

FIGURE 5.5 *Simplified diagram of lobe type PD blower*

In establishing the volumetric flow rate through the blower, the slip must be accounted for. In essence, the blower must move the volume of air that leaks past the seals twice. In the United States, this is accounted for by adding "slip rpm" to the theoretical speed needed to deliver the required volumetric flow rate. The slip rpm is the additional rotation that must be used to compensate for the air flow lost to internal leakage:

Equation 5.30

$$N_a = \frac{Q_i}{\text{CFR}} + N_s$$

where

N_a = actual speed required to provide flow rate, rpm
Q_i = volumetric flow rate, ICFM
CFR = blower displacement, cubic feet per revolution
N_s = actual blower slip, rpm

An alternate formula based on volumetric efficiency may be used:

Equation 5.31

$$N_a = \frac{Q_i}{\text{CFR} \cdot \eta_v}$$

where

η_v = volumetric efficiency, decimal

Slip is a function of the pressure difference between the inlet and discharge side of the blower and the air density. The flow back through the blower impellers approximates air flow through an orifice. The equation to correct slip is therefore similar:

Equation 5.32

$$N_s = N_1 \cdot \sqrt{(p_d - p_i) \cdot \frac{14.7}{p_i} \cdot \frac{T_i}{528}}$$

where

N_s = actual slip at operating conditions, rpm
N_1 = slip at 1 psig differential pressure, rpm
$P_{i,d}$ = pressure at blower inlet and discharge, psia
T_i = blower inlet air temperature, °R

Note that the flow rate in a PD blower is volumetric. The required mass flow rate needed to meet the process demand must be corrected for temperature and pressure at the blower inlet using the formulas above. This must include pressure drop through inlet filters, silencers, and piping.

The pressure acting on the face of the impellers causes a torque reaction on the shafts directly proportional to the pressure difference between inlet and outlet. This is the source of the constant torque characteristic of PD blowers which has an impact on motor and control selection—particularly in the sizing of VFDs. For any rotating machine, power is a function of speed and torque:

Equation 5.33

$$P = \frac{T \cdot N}{5252}$$

where

P = power required, hp
T = torque, ft lb
N = rotating speed of the machine, rpm

In addition to the power required to move the air, essentially the gas power, the total power required includes mechanical losses for bearings, gears, and seals. This is referred to as the "friction power" in the United States and as mechanical efficiency in other locations. The frictional losses are proportional to blower speed:

Equation 5.34

$$\mathrm{FHP_a} = \mathrm{FHP_{nom}} \cdot \frac{N_a}{1000}$$

where

$\mathrm{FHP_a}$ = actual friction power loss, hp
$\mathrm{FHP_{nom}}$ = nominal friction power, hp/1000 rpm

The total power required is the sum of gas and friction horsepower:

Equation 5.35

$$P_{\mathrm{PD}} = [0.0044 \cdot N_{\mathrm{a}} \cdot \mathrm{CFR} \cdot (p_{\mathrm{d}} - p_{\mathrm{i}})] + \mathrm{FHP}_{\mathrm{a}}$$

where

P_{PD} = positive displacement blower power required, hp

The factor 0.0044 in Equation 5.35 is the theoretical value. There can be some variation in this parameter between lobe type blower designs, and specific information should be obtained from the manufacturer if possible.

Lobe type PD blowers are often connected to the motor by a v-belt drive. This has the advantage in constant speed operation of matching the operating speed to the anticipated flow rate. The disadvantage is a slight power loss from belt slippage. This is typically 3–5% of the blower shaft power. Poorly adjusted belts will have a higher power loss.

Lobe type blower efficiency varies with the operating speed and discharge pressure. This may not be significant in some applications since this type of blower can have 60% turndown. The efficiency is therefore a percentage of a smaller power requirement. It is recommended that the SCFM versus kilowatt be determined for the appropriate flow and pressure range. Using this value for power comparisons will provide a more accurate analysis than efficiency comparisons.

The discharge temperature of a PD blower is a critical parameter. If the temperature difference between inlet and discharge is excessive, the result will be warping of the blower end plates and mechanical binding and failure. High discharge temperature will also result in loss of lubrication and bearing or seal failures. The discharge air temperature most often becomes a problem at low speeds and flows. The internal losses remain essentially constant and there is less air flow to remove the resulting heat. The discharge temperature for PD blowers may be estimated from the following:

Equation 5.36

$$T_{\mathrm{d}} = T_{\mathrm{i}} + \frac{T_{\mathrm{i}} \cdot P_{\mathrm{PD}}}{0.01542 \cdot p_{\mathrm{i}} \cdot Q_{\mathrm{i}}}$$

where

$T_{\mathrm{d,i}}$ = inlet and theoretical discharge air temperature, °R

The actual discharge temperature may be lower than the theoretical value because of heat transfer through the blower case. If the blower efficiency is known, the discharge temperature can also be calculated from Equation 5.26.

In addition to maximum discharge air temperature and differential temperature, several other operational limits to PD blower operation should be accommodated in the control strategy. High differential pressure puts strain on the shaft bearings and can cause deflection of the lobes and shafts leading to mechanical failure. Many PD blowers employ splash lubrication for gears and bearings. Therefore low speeds may cause loss of lubrication and mechanical failure. Larger PD blowers may have an integral lube pump and forced lubrication which minimizes this concern. High speeds can also cause mechanical damage from vibration or bearing failure.

Because of the pulsating air flow, lobe type blowers have traditionally generated high levels of noise. The three lobe designs are not inherently quieter than two lobe blowers but the higher frequency noise produced is easier to attenuate and has led to their increased adoption in new systems. Regardless of design, lobe type blowers should be fitted with silencers on both the inlet and the discharge piping of the blowers. These silencers act as mufflers to decrease noise but add pressure drop on the inlet and discharge air stream. The typical loss is 0.2 psi for each silencer.

Positive displacement blowers must always be equipped with a pressure relief valve on the discharge piping. This is necessary to prevent a blockage in the air distribution piping from creating high pressure on the blower discharge. The relief valve will open which vents the air flow to atmosphere and avoids catastrophic failure of the blower or piping.

Older systems consisted of factory assembled or field erected open packages with silencers, relief valve, blower, motor, and belt drive on a common fabricated frame. Occasionally, a field erected sound enclosure was added to reduce noise in the blower room. Currently, it is common practice to provide factory assembled blower packages with all of the components preassembled in a factory tested sound enclosure.

5.2.3 Screw Blowers

Screw blowers are old technology but their application to wastewater treatment is recent. The technology offers the benefits of excellent turndown and low energy consumption. They share many characteristics of lobe type PD blowers including operational simplicity. However, the mechanical design of screw blowers is considerably more complex than lobe types and the determination of energy requirements is less straightforward.

Like the lobe type blower, the screw blower has two counter-rotating shafts connected by gearing at one end. However, the rotors (impellers) on each shaft are helical—one helix with a male profile and one with a female profile (see Figure 5.6). The rotors each have a different number of helixes and rotate at different speeds. The air enters the female helix at one end; as the shafts rotate, the male helix moves the air axially along the shaft, squeezing and compressing it as the chamber is made progressively smaller. At the discharge end of the rotors, the air is expelled into the discharge port of the blower (see Figure 5.7). Clearances are maintained between the rotors and the casing. No lubricant enters the air stream.

Each manufacturer has proprietary variations in the design of the blower case and the rotors. This makes generalized performance calculation procedures a challenge. The geometry of the two rotors determines the displacement of the blower. They also establish a nominal pressure ratio. Operation is not limited to this pressure—the blower discharge for a screw blower will match the pressure required to move the discharge air

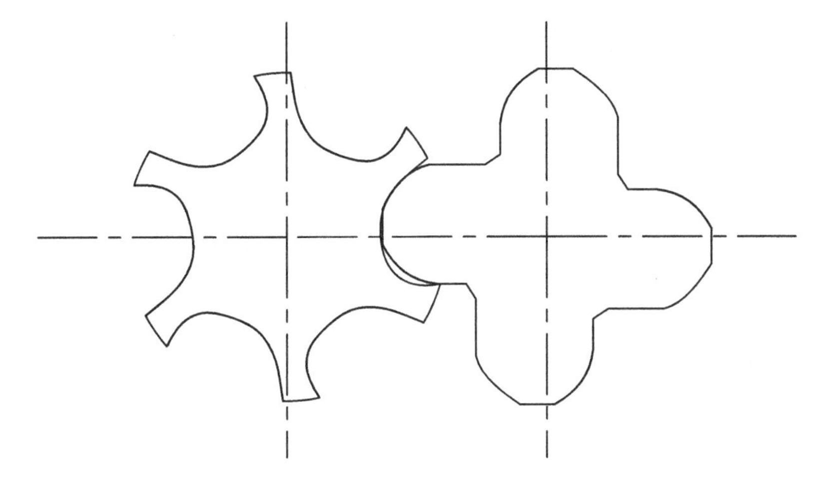

FIGURE 5.6 Example screw blower rotor profiles

flow through the aeration system. Therefore, screw blowers must be fitted with a pressure relief valve. There is some decrease in efficiency if the discharge pressure is either above or below the nominal pressure ratio (see Figure 5.8).

The volumetric flow rate of a screw blower is a function of speed. There is internal leakage between the rotors and casing and between the rotors themselves. This leakage is proportional to both speed and pressure differential. It is a function of the geometry of the two helixes, their length, and their diameter. For screw blowers, it is more common to refer to the volumetric efficiency rather than slip:

Equation 5.37

$$\eta_{\text{v}} = \frac{Q_{\text{i}}}{\text{CFR} \cdot N_{\text{a}}}$$

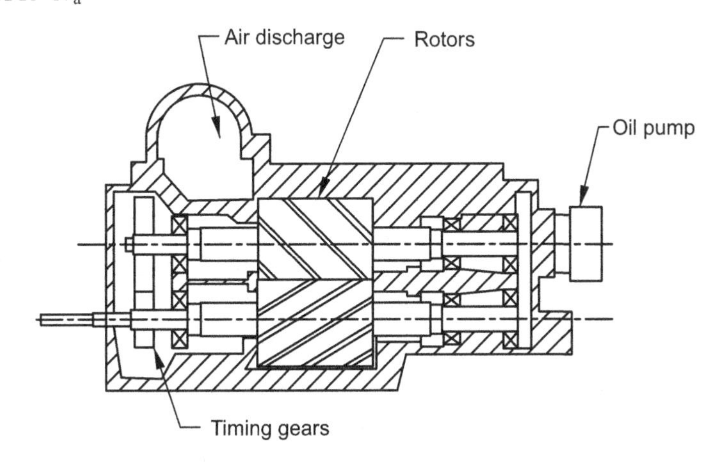

FIGURE 5.7 Simplified diagram of screw blower

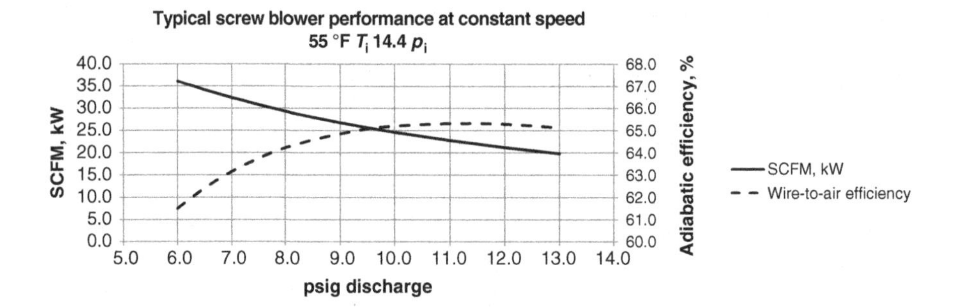

FIGURE 5.8 *Example variation of screw blower performance with discharge pressure*

where

η_v = volumetric efficiency, decimal
Q_i = volumetric flow rate delivered to process, ICFM
CFR = blower displacement, cubic feet per revolution
N_a = actual speed, rpm

The volumetric efficiency is not constant (see Figure 5.9). It varies with speed and pressure. The inlet flow rate must be corrected to SCFM to determine mass flow rate and match blower output to process demand. As with lobe type PD blowers, there is also some leakage to atmosphere through seals and so on. This is usually negligible and may be ignored in calculations.

Many screw blowers have the motor directly coupled to the blower shaft. They may include additional internal gearing between the input shaft and the impeller shafts to achieve the required operational speed. In this case, the input shaft speed will match the motor speed. Other designs use a belt drive to connect the input shaft to the motor. In this case, the belt ratio must be included in calculations and the flow rate is based on the blower input shaft rpm—which will differ from the motor speed.

Determining the power required by the screw blower is more complex than for a lobe type PD blower. In addition to losses from the internal slip, there is power lost to

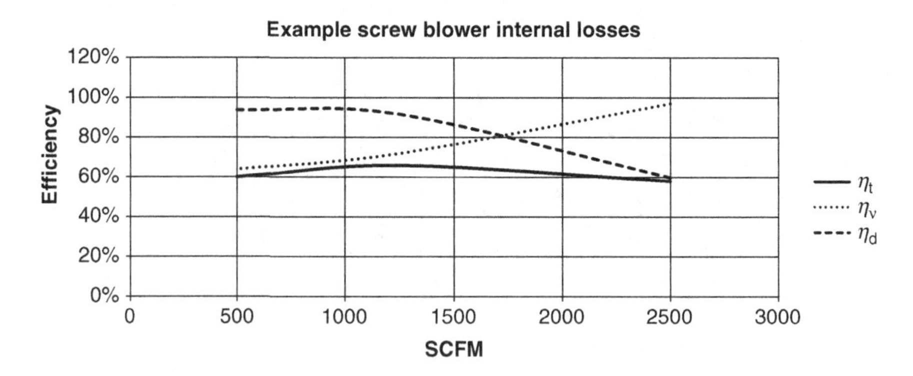

FIGURE 5.9 *Example variation of screw blower performance with speed*

the friction in bearings, seals, and gears. There are also dynamic flow losses (η_d) from friction in the air stream as it passes through the internal passages of the blower. Losses in both of these categories increase with blower speed and flow rate. The dynamic efficiency includes these frictional losses. It is a complex function of rotor geometry, speed, and inlet pressure.

The screw blower air stream does not have the high level of pressure pulsation found in lobe blowers. However, they are usually equipped with inlet and discharge silencers and with noise abatement enclosures. These are typically provided in a factory assembled sound enclosure.

The total blower efficiency is the product of the volumetric efficiency and dynamic efficiency. The best efficiency point is quite often at mid-range for the blower operation. In determining the total wire-to-air power, the factors that must be considered include:

- Thermodynamic losses inherent in any compression process
- Losses from over or under compression if the operating pressure ratio deviates from the nominal pressure ratio
- Pressure drop through filters and silencers
- Volumetric losses from slip (internal leakage)
- Friction losses for bearings, seals, gears, and belt drives
- Dynamic losses from internal air stream friction
- Motor losses (essentially I^2R losses)
- Losses from VFDs or other variable speed drives if provided

Precise power determination for screw blowers is best obtained from the manufacturer. For preliminary calculations where only design point data is available, it may be necessary to back calculate overall efficiency using Equation 5.22 or Equation 5.23 and assuming a constant efficiency. If the actual operation is close to the design point conditions, reasonable accuracy can be obtained. For final design calculations the operating points should be submitted to the manufacturer for a more accurate power determination.

5.2.4 Control and Equipment Protection Considerations

Positive displacement blowers must never be controlled by inlet throttling. This creates significant increase in power requirements by increasing the pressure rise across the blower, but has an insignificant effect on flow rate.

The most practical control method for any type of PD blower is variable speed. With current technology that generally means variable frequency drive control of motor speed. If process performance is critical—for example, if high DO is inhibiting denitrification—it's possible to reduce flow to the process. This may be achieved by dumping some of it over a blow-off valve and venting to atmosphere, however this does not provide any energy savings.

Older systems sometimes employed two-speed motors to provide a high and low flow rate. This limits the control flexibility and energy savings. Some systems also use an internal combustion engine (often fueled from digester gas) to drive the blower. This can provide a limited range of speed variation but comes with high equipment cost and maintenance requirements. Vibration and resonance between the engine and blower has been a problem in some of these installations. Given the current state of technology and economics of equipment, a VFD will be the most cost-effective option. If digester gas is available, it is generally more economical to use it for a motor-generator or microturbine to provide electric power into the plant or the distribution grid and then use electric motors and VFDs for the blowers.

Some existing systems use multiple identical blowers with different sets of sheaves in the belt drive system to achieve different blower speeds and flow rates. This arrangement is most effective as a manual control option to change the operating blower on a seasonal or other basis. It lacks the flexibility to produce step-less modulation of air flow for accurate process control and optimum energy consumption.

Sheave changes should be examined for belt-driven PD blowers even if a VFD is being added to modulate air flow. The belt ratio for constant speed operation is usually selected to provide a specified design flow rate. This is usually not the maximum flow rate of the blower. By using different sheaves it is possible to obtain more flow at a higher efficiency point using the existing blower. The motor rating must be checked to make sure that it is not going to be overloaded at the higher speed. If the operating pressure is lower than the specified design pressure, which is typically the case, this should not be a problem.

The blower turndown is usually limited by either blower discharge air temperature or motor heating. As the blower speed decreases, the flow of cooling air through the motor falls off—particularly for totally enclosed fan-cooled (TEFC) motors. The lower limit on speed is usually 50% of nominal unless special conditions exist. Motor temperature and blower discharge monitoring may allow fine tuning these lower limits to actual field conditions.

A common concern is retrofitting VFDs to existing motors. This concern results from both motor heating and the possibility of harmonics or dV/dt insulation stress. If the existing motor has a 1.15 service factor and Class F or better insulation and the grounding system is good, the existing motors should work in conjunction with a VFD without problems or reduced life.

In selecting a VFD for use with a PD blower, the constant torque nature of the load should be considered. This may require oversizing the VFD compared to the nominal power.

Individual blower flow measurement is not usually necessary with PD blowers. Because the flow rate is proportional to speed, calculating the ICFM can be readily incorporated into the programming using the VFD speed signal as the input. There will be some error because of variable slip and so on, but this is readily accommodated by the DO control feedback loops. To calculate and display SCFM requires compensation for inlet pressure and temperature.

Inlet pressure and temperature measurement is useful for blower protection as well. The inlet filter loss can be monitored to detect a dirty filter. Inlet temperature can be used in conjunction with discharge temperature to calculate differential temperature—an important protection parameter. Discharge air temperature should be monitored and used to stop the blower at high temperature.

Vibration is generally monitored on only the largest PD blowers. Usually a case mounted accelerometer is used. Bearing temperature is another protection parameter incorporated into the controls on large PD blowers. Lube oil temperature and lube oil pressure may be monitored on large units for additional protection.

Pressure as a control parameter for the blower itself is not the optimum approach, but it is used. The only control output is blower speed which in turn controls blower flow. Older control strategies may try to maintain constant discharge pressure. In this case, a cascaded control loop is used with the output of the pressure control used as the input to the flow control loop. If the pressure setpoint is higher than the necessary system requirement, this wastes power in direct proportion to the excess pressure. A more efficient approach is to use direct flow control or some other most-open-valve (MOV) strategy.

5.3 DYNAMIC BLOWERS

Dynamic blowers, generally called centrifugal blowers or sometimes turbocompressors, are mechanically simple machines but have more complex performance and operational considerations. They are the dominant type of blower for aeration systems in large treatment plants, however they can also be found in all but the smallest facilities. For many decades the designs were quite stable, but in recent years the technology has rapidly evolved as energy costs have escalated and control capabilities expanded.

Historically, PD blowers have been considered as providing constant flow at variable pressure. Centrifugal blower performance was more often viewed as variable flow at nearly constant pressure. New developments have made both views inaccurate. Just as variable speed has expanded the capabilities of PD blowers, new technology—including variable speed—has greatly expanded the capabilities and efficiency of dynamic blowers.

Dynamic blowers are essentially kinetic energy machines. Air enters near the center of an impeller that looks very much like a centrifugal pump impeller. The vanes on the rotating impeller fling the air outward. This creates high velocities and imparts kinetic energy to the air. The air leaves the impeller and passes through a diffuser section in the blower case where it slows. Some of the kinetic energy—velocity pressure—is converted to potential energy—static pressure. The air stream passes into a volute or scroll at the outside diameter of the case. There additional energy conversion from kinetic to potential energy takes place and the flow is collected into the discharge pipe connection.

The aerodynamics of the impellers and casings are complex, and will not be addressed in this text except in the most general terms. Instead the discussion will concentrate on performance, power calculations, application considerations, and, of course, control strategies.

In addition to centrifugal blowers, there are axial flow compressors that fit the category of dynamic blowers. However, they are rarely applied in wastewater treatment applications and will not be discussed in this text.

5.3.1 Types and Characteristics

There are three general types of centrifugal blowers:

- Geared single stage centrifugals
- Multistage centrifugals
- Turbo blowers

The turbo blower is a very recent development. Historically, only the first two types were applied to wastewater treatment aeration. In consequence, the "geared single stage" blowers are often referred to as simply "single stage" to differentiate them from "multistage" centrifugal blowers. The newer turbo blowers are technically single stage centrifugals but in the wastewater treatment industry they are seldom referred to in that manner.

The three types differ in the manner used to create discharge pressures high enough to move air into the process. Because of the low density of air, it is more of a challenge to generate enough kinetic energy to create pressure with a blower than it is with a pump. Single stage and turbo blowers create the required pressure by spinning a single impeller very rapidly thereby creating a high peripheral speed at the outside of the impeller. Multistage centrifugals operate several impellers in series with each one creating a portion of the total pressure increase required.

The dynamic blower performance is usually presented in a characteristic curve, often called the blower curve. The characteristic curve usually consists of two plots: a plot of pressure versus flow and one of power versus flow. The flow is often given in ICFM rather than SCFM. In some cases, head is substituted for gauge pressure. It is important to note that the blower curve identifies the performance capabilities of the blower. A specific operating point can only be determined by superimposing the system curve on the blower curve. The intersection of the system curve with the blower curve showing pressure versus flow capability identifies the operating air flow rate. The power consumption at this flow can be read from the curve showing power versus flow. Although it is not the customary format, a curve set showing mass flow versus power consumption at various pressures (as in Figure 5.3) can be very useful.

Regardless of the data format, it must be remembered that the blower curve is only valid at a specific set of inlet conditions of temperature, humidity, and absolute pressure. The curves are also only valid at the specific operating speed indicated. Kinetic energy or the air moving through the blower is a function of mass and velocity^2—whatever affects either of these parameters will affect performance and power.

The performance of a centrifugal blower is more dependent on air density than a PD blower's. Centrifugal blowers are volumetric machines. The inlet density affects the relationship between mass flow and volumetric flow. The density also affects the pressure capability of the blower with discharge pressure at a given flow decreasing as density decreases. A dynamic blower's performance is also affected by relative humidity. This changes the density of the air and the molecular weight. The operating speed also has an effect on a dynamic blower's performance that is more complex and dramatic than the impact of speed on a PD blower.

Because power is a function of flow and pressure, it is affected by the variations in either inlet conditions or speed. However, the effect is not consistent and will vary depending on the blower control method. For example, an inlet-throttled blower will have its highest power draw at low inlet temperature. A variable speed controlled blower will have its highest power draw at high inlet temperature. This fact indicates that the energy analysis must include the control methodology in order to provide an accurate comparison between system designs.

The varying inlet parameters and speed actually affect blower head (see Equation 5.19) rather than changing discharge gauge pressure. Since a blower's head cannot be measured directly and process requirements and operator custom reference discharge gauge pressure, it is more convenient to present results in terms of discharge pressure. The precise calculation methods are shown in the ASME performance test codes and are used by blower manufacturers in projecting site performance. The manufacturer's data should be obtained whenever possible.

The most accurate calculation methods identified in the performance test code and other sources are too cumbersome for routine use in energy evaluations. Fortunately, reasonable approximations can be achieved by using simplified methods based on density changes from temperature and pressure (ignoring humidity) and the affinity laws for speed changes.

The effect of temperature and pressure on discharge gauge pressure is approximately equal to their effect on density. It is possible to compensate for the affect of changing inlet conditions on a blower's performance, neglecting relative humidity, using:

Equation 5.38

$$p_{\mathrm{da}} = p_{\mathrm{c}} \cdot \frac{T_{\mathrm{ic}}}{T_{\mathrm{ia}}} \cdot \frac{p_{\mathrm{ia}}}{p_{\mathrm{ic}}}$$

where

p_{da} = actual pressure increase from inlet to discharge, psig
p_{c} = discharge pressure from curve, psig
$T_{\mathrm{ic,ia}}$ = inlet temperature for curve and actual, °R
$p_{\mathrm{ic,ia}}$ = inlet pressure for curve and actual, psia

Equation 5.39

$$P_{\mathrm{a}} = P_{\mathrm{c}} \cdot \frac{T_{\mathrm{ic}}}{T_{\mathrm{ia}}} \cdot \frac{p_{\mathrm{ia}}}{p_{\mathrm{ic}}}$$

where

P_a = actual blower power, hp
P_c = blower power from curve, hp

These calculations do not include energy losses in motors, VFDs, and so on. The actual inlet pressure should include losses through piping and filters. Note that the density ratios are calculated in terms of absolute pressure and temperature while the discharge pressure is expressed as gauge pressure. The designer should be aware of the flow units in the curve—ICFM or SCFM—and make the necessary corrections to accommodate process requirements.

Historically, inlet throttling was the most common form of controlling centrifugal blower flow rate—particularly for multistage blowers. The control device, typically a manually or automatically operated butterfly valve, is simple and inexpensive. Throttling is used to control flow with two mechanisms. First, although we are generally concerned with the discharge gauge pressure, the centrifugal blower actually functions by producing a differential pressure from inlet to discharge. Inlet throttling creates a pressure drop at the blower inlet so the discharge gauge pressure is reduced.

The second effect of throttling is a reduction in inlet density caused by the pressure drop across the valve. This density reduction results in reduced blower discharge pressure, as indicated in Equation 5.38. Since the centrifugal blower is essentially a volumetric device, the effect of throttling and density changes must be reflected by correcting the ICFM to SCFM.

An interesting aspect of using inlet throttling for control is that the effect of most-open-valve (MOV) control and similar pressure minimization strategies is reduced. Since the pressure ratio is essentially fixed at a given volumetric flow rate, shifting the pressure ratio from discharge to inlet has only a minimal effect on the blower power required.

The simplified method for determining the effect of inlet throttling on blower performance is to calculate the pressure drop across the valve at several flow rates, then use Equations 5.38 to calculate the result. The pressure drop across the inlet valve is not constant—for a given valve position it varies with Q^2. The pressure drop across the valve must be subtracted from the pressure rise across the blower to estimate the new discharge pressure. The approximate power consumption may be calculated using Equation 5.39. Plotting the new performance curve against the old performance and the system curve will provide the change in flow rate and power consumption (see Figure 5.10).

Varying the blower speed is a more efficient technique for flow control of any type of dynamic blower. Inlet throttling maintains the pressure ratio of the blower, dissipating the power used to create the pressure difference by parasitic throttling. Variable speed control changes the pressure ratio of the blower by changing the peripheral speed of the impeller and reducing the kinetic energy imparted to the air stream. Rather than dissipate the energy of air compression through throttling, variable speed control reduces the power demand by reducing the pressure ratio and

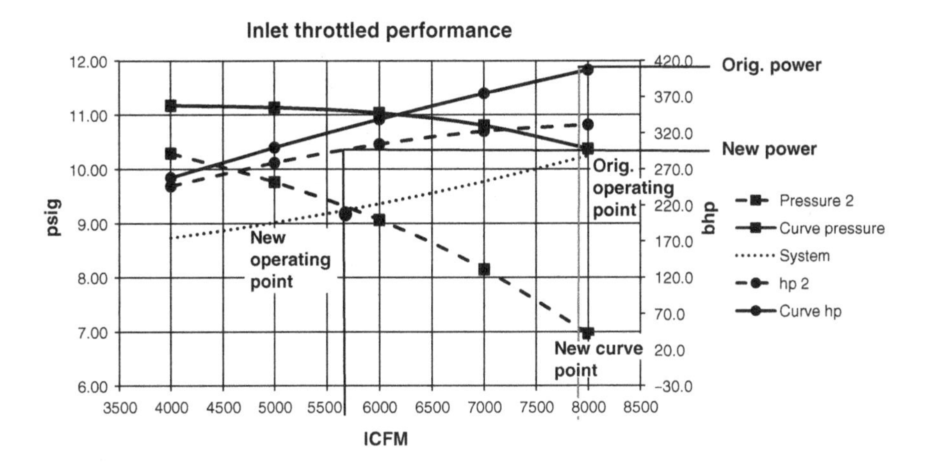

FIGURE 5.10 Effect of inlet throttling on dynamic blowers

energy used in compression. Variable speed control can reduce the power of a blower by 10–20% compared to other control techniques.

The theoretically correct calculation methods for predicting the effect of variable speed operation are no less demanding than those for inlet condition changes. Again, reasonable accuracy can be achieved by applying affinity laws—similar to those used for variable speed pumping. These are often referred to as the "fan laws":

Equation 5.40

$$Q_a = Q_c \cdot \frac{N_a}{N_c}$$

where

$Q_{a,c}$ = actual and curve volumetric flow rate, ICFM
$N_{a,c}$ = actual and curve rotational speed, rpm

The pressure relationship of the affinity laws is often stated as the discharge pressure being proportional to the square of the speed ratio. This relationship is essentially correct for pumps—which move an incompressible fluid—and for fans, which have a very low-pressure ratio. It may be used for estimating the effect of small changes in blower speed. However, a more accurate expression of the relationship is that blower head (Equation 5.19) is proportional to the square of the speed ratio. In practice, the adiabatic factor (Equation 5.20) is used instead of head:

Equation 5.41

$$X_a = X_c \cdot \left(\frac{N_a}{N_c}\right)^2$$

where

$X_{a,c}$ = actual and curve adiabatic factor, dimensionless

Once the new adiabatic factor is determined, it can be solved for the pressure ratio using Equation 5.20 and for the discharge pressure using Equation 5.18.

Once the speed has been determined, the power required at the new speed and flow rate can be calculated:

Equation 5.42

$$P_a = P_c \cdot \left(\frac{N_a}{N_c}\right)^3$$

where

$P_{a,c}$ = actual and curve blower power, hp

The procedure for calculating variable speed performance begins with picking a flow point on the manufacturer's curve and reading the associated values for pressure and power. An actual speed is selected and the new flow calculated. Then the new pressure and power at that new flow are calculated from the speed ratio and the pressure and power at the original flow (see Figure 5.11). Note that the operating flow will still be the intersection point of the system curve and the new blower curve. This new operating flow will typically be much lower than the original flow times the speed ratio since the reduced pressure will move the intersection point further to the left.

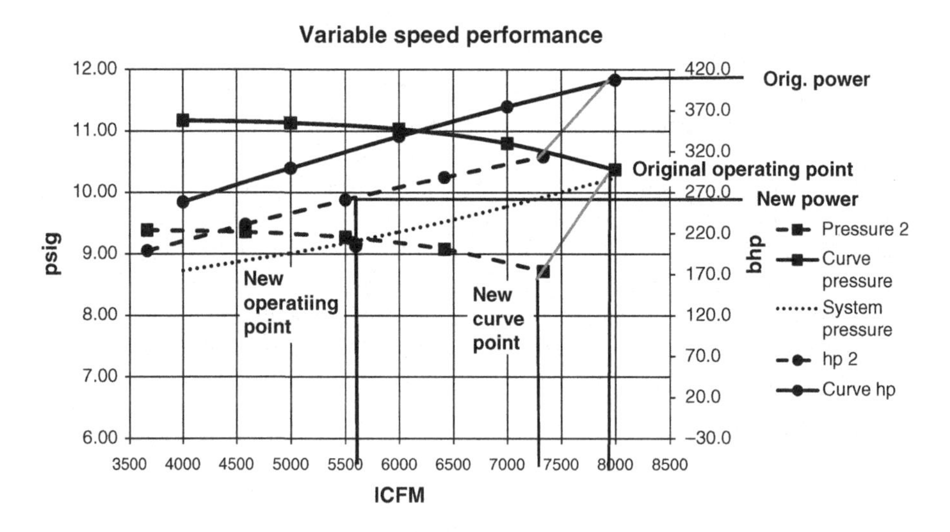

FIGURE 5.11 Effect of variable speed on dynamic blowers

When it is necessary to establish power requirements for variable speed operation, it is most convenient to create performance curves at various speeds using the method described above. The operating speed needed to develop a specific flow and pressure can be taken from the graph as the point where the performance curve intersects the system curve and then the power can be calculated.

There is a tendency for designers to select a blower with the best efficiency point (BEP) as close to the design point as possible. Most existing blowers were selected in this manner. In actual service blowers operate most of the time at flows lower than design and at inlet conditions more favorable than worst case design conditions. To optimize the energy efficiency of the blower system, pick a blower with the BEP to left of design point (i.e., at a lower flow than design capacity). During normal operation the blower will modulate from the right to the left side of the BEP (from flows higher to lower than flow at BEP). If the BEP is in the middle of this range the average efficiency will be better. If the BEP is at the maximum flow capacity then modulation to lower flows will result in steadily decreasing efficiency.

Common control methods for single stage dynamic blowers are variable inlet guide vanes (IGVs) and variable diffuser vanes (VDVs). These are generally employed on large blowers. Horizontally split multistage centrifugal blowers may also be controlled with inlet guide vanes. There are some turbo blower designs that incorporate VDV control into the design while some single stage systems have combined IGV or VDV with variable speed control.

Inlet guide vanes operate by causing the air at the blower inlet to spin—referred to as "pre-rotation." This reduces the velocity difference between the impeller and the air stream thereby reducing the kinetic energy gain. At reduced IGV openings, in addition to pre-rotation, there is an inlet throttling effect that creates a pressure drop. The effect of IGVs on the blower performance curve is a shift to the left and downward. The slope of the curve also increases.

Variable diffuser vanes are installed in the blower casing in the diffuser section between the impeller outside diameter and the volute. They affect the conversion of kinetic to potential energy—the conversion of velocity pressure to static pressure. Changing the angle of the diffusers has the effect of moving the blower curve to the right or left.

Large geared single stage blowers may have both types of vanes. By coordinating the movement of the vanes, it is possible to maintain the blower at or near the BEP across a wide range of flows and discharge pressures. Dual vanes also provide significant improvement in turndown. The design of the vanes and coordination of vane movement is proprietary to the blower manufacturer. To determine the performance of the blower throughout the operating range usually requires interpolation of the performance curves but compensation for changing inlet conditions can be done using Equations 5.38 and 5.39.

There is a tremendous variety in impeller construction and geometry. Open impellers have the blades or vanes exposed on one side. Closed impellers have a shroud or disk covering both sides of the vane. Radial impellers have essentially straight vanes radial from the shaft. They provide a fairly flat performance curve and are typically used in conjunction with inlet throttling control. Backward curved impellers have a more complex geometry with the blades turning away from the direction of rotation at the outside diameter. These impellers generally have a steeper curve and are preferred for variable speed operation.

Surge is a complex phenomenon that only occurs at low flow conditions in centrifugal blowers. The term refers to a pulsating reversal of flow in the blower. If the discharge pressure from the system resistance to flow causes a pressure exceeding the capability of the blower then air flow at the blower discharge ceases. Since the system contains air under pressure, air will flow from the piping system backwards into the blower. If the blower discharge is fitted with a check valve, it will slam shut and create a disturbingly loud noise. The pressure at the blower discharge drops and the blower resumes pushing air into the system. If the pressure again exceeds the blower capability, another cycle of flow reversal will occur.

This pulsating flow from surge can cause physical damage to the blower bearings and impellers. The loss of air flow can also result in high temperature in the blower with thermal expansion of the impeller resulting in interference with the casing and catastrophic failure. The potential for surge damage varies with the blower design and characteristics. Small multistage centrifugals can operate in surge for many minutes before damage occurs. Geared single stage blowers are more sensitive to surge. Turbo blowers are very sensitive to surge and can experience catastrophic failure in a few seconds.

The surge "point" is actually a moving target. The occurrence of surge is to some extent dependent on the system piping and resistance to flow. The surge point is a function of the relationship of flow to pressure in the blower's design. High inlet temperature and lower air density lower the blower discharge pressure curve and surge point. Reduced operating speed has the same effect. Most manufacturers will provide a surge point on a curve plotted at a specific set of conditions and a surge limit line (SLL) that defines the surge point at variable inlet conditions, speed, or guide vane position.

Rise to surge is a parameter used to identify the blower's sensitivity to increase in discharge pressure and the potential for surge. Rise to surge is the difference between the pressure at the blower design point and the maximum pressure or pressure at the surge point. The greater the rise to surge and the steeper the blower curve, the more stable the control operation will be. A high rise to surge is particularly important with variable speed operation. The general recommendation is a minimum of 1.25 psi rise to surge for variable speed operation with a steadily increasing pressure as flow is decreased. In general, this characteristic requires backward curved rather than radial impellers.

5.3.2 Multistage Centrifugal Blowers

Multistage centrifugal blowers have several impellers mounted on a common blower shaft (see Figure 5.12). The blower case is constructed to direct air flow discharged from one impeller to the eye of the next impeller in series. As the air moves through successive impellers, one per stage, the pressure is increased. After the final stage the air is directed by the volute to the blower discharge. In general, higher discharge pressures require more stages.

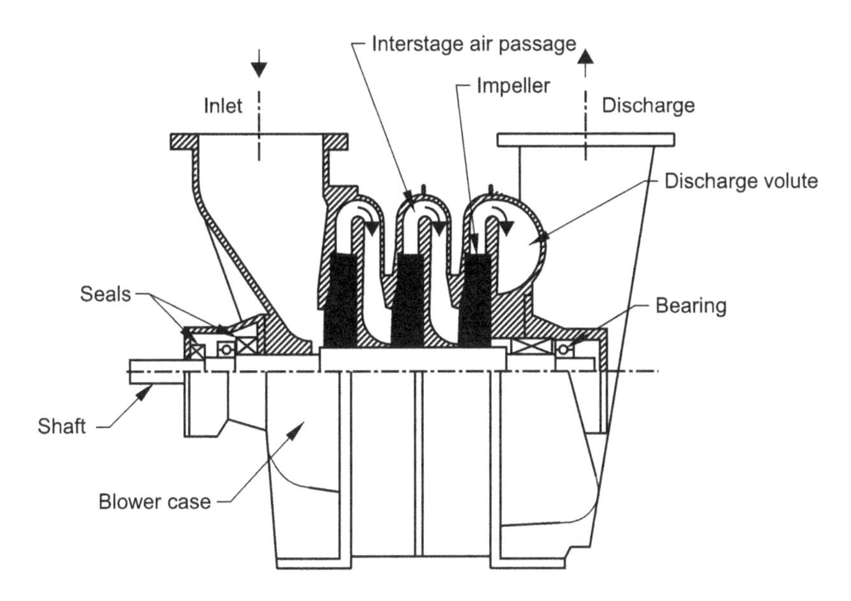

FIGURE 5.12 *Simplified diagram of multistage centrifugal blower*

The casing of the multistage blower is almost always cast iron. Impellers are usually aluminum, but alloy steel and stainless steel may be used. Multistage blowers generally use ball or roller bearings for shaft support. These may be oil or grease lubricated. Shaft seals are typically noncontact labyrinth style, but carbon ring face seals are also available.

The blower is almost always direct-coupled to a two-pole motor. In the United States, this results in a nominal operating speed of 3600 rpm. Because Europe and other countries have 50 Hz power instead of the 60 Hz used in the United States, the nominal speed is only 3000 rpm. This results in lower discharge pressure. Multistage blowers are not common in locations with 50 Hz power. VFDs may be used to provide higher frequency at the output rather than the input frequency, making application of multistage units possible in these areas.

The blower case can be either vertically split or horizontally split. The vertically split design is more common. Historically, this type of blower is the most common in midsize applications—between 100 and 750 hp. This is the result of low cost, robust construction, simple flow modulation, and good efficiency. Turbo blowers have recently displaced the multistage blower in the lower horsepower range. However, combining VFD control with multistage blowers can result in excellent energy performance at lower capital cost. The total life cycle cost for both options should be carefully evaluated during the design process to establish the most cost-effective system.

Horizontally split multistage blowers are generally found only in very large systems—those exceeding 1000 hp. This type of blower has aerodynamically superior passages in the casing and larger diameter impellers. In aeration applications only two or three stages are normally needed to achieve required discharge pressures. The horizontally split multistage blower has pressure-lubricated journal bearings. Control of blower flow rate is usually provided by inlet guide vanes. This provides better efficiency than throttling control.

5.3.3 Geared Single Stage Centrifugal Blowers

These blowers offer very high flow rates and excellent efficiency. Flow ranges exceed the capabilities of other styles of blowers. They are generally found in large treatment plants but are available in a wide range of capacities. Outside the United States the single stage blower is found in most midsize treatment plants because the design eliminates the performance disadvantage associated with lower speed motors and multistage centrifugal blowers.

Single stage blowers achieve high discharge pressure by rotating the impeller at very high speeds—several thousand rpm. The impellers may be aluminum, alloy steel, or stainless steel. Pressure-lubricated journal bearings are used throughout for both radial and thrust loads. The casing is cast iron for wastewater aeration applications, and may be horizontally or vertically split.

In order to achieve the high impeller speeds which greatly exceed the capability of standard motors, a speed increasing gear box is provided between the motor and the impeller shaft. The gear box may be integral with the blower (see Figure 5.13) or it may be a separate unit direct-coupled to the blower shaft. Because of the ability to

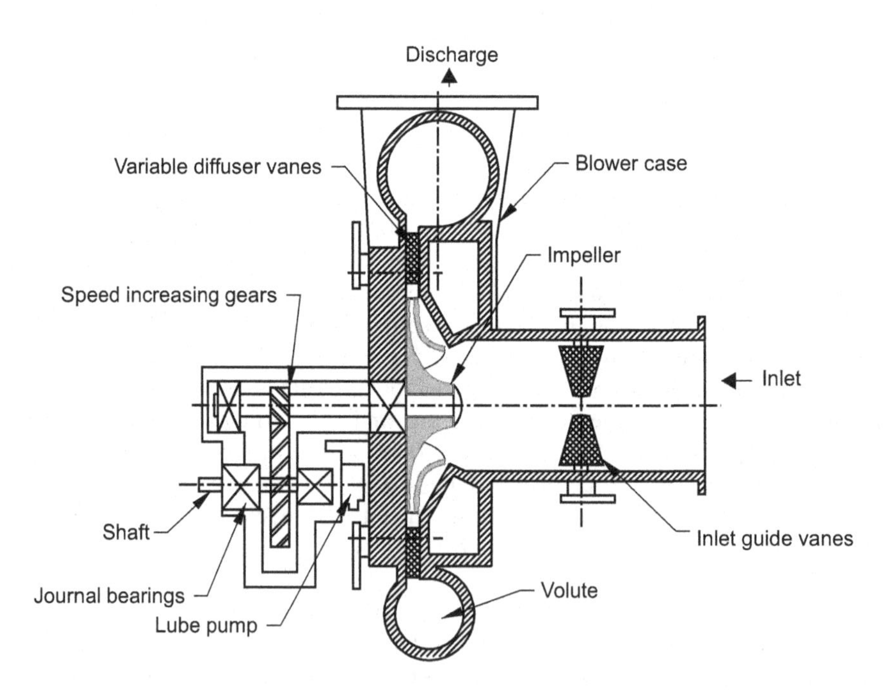

FIGURE 5.13 *Simplified diagram of geared single stage centrifugal blower*

select the gear ratio to meet the specific needs of the application, the single stage blower performance is not dependent on nominal motor speed.

The single stage blower is more complex mechanically than other types and is correspondingly more expensive. In the past, it has also been the most efficient type of blower. This has offset the initial equipment cost.

The high efficiency of these blowers is partially due to the inherent efficiency of the single stage design. It eliminates dynamic losses from air movement between stages. Part of the efficiency is due to the ability to create complex and highly efficient impeller geometry. Part of the efficiency is due to the control techniques used for modulating blower air flow.

The single stage blower typically uses IGV and/or VDV for capacity control. Blowers employing both IGV and VDV are referred to as "dual vane" machines. Modulating air flow with IGV or VDV control is much more efficient than the inlet throttling control that had historically been the only alternative. Furthermore, the ability to manipulate dual vane machines to maintain operation near the BEP provides high efficiency across a wide range of operating conditions. The comparatively low energy requirement of this design is particularly beneficial in the high horsepower ratings.

Single stage blowers are usually very high horsepower machines. The electric motors used to power them are typically "medium voltage." The most common electrical supply systems are 2400 VAC or 4160 VAC. This inhibited the application of variable frequency drives to this type of blower since medium voltage VFDs have been extremely expensive in the past. However, recent technology has made VFDs very cost-effective in the range of power found in wastewater treatment. Variable speed is even more energy efficient than IGV control. The long equipment life and high power rating of single stage blowers makes this an attractive option. This is particularly true for ECMs to existing facilities where the single stage blowers can be retained but efficiency across the operating range improved by retrofitting VFD control.

Many installations of high horsepower single stage blowers have impellers custom designed to meet specified design conditions. The blower design point for both flow and pressure exceeds the actual operating requirement in many of these facilities. A cost-effective ECM is replacing the original impeller with one that is more closely matched to actual operating experience. This can be combined with an upgrade to VFD control. The result is a dramatic decrease in aeration energy use without compromising process performance.

When VFDs are applied to single stage blowers, two aspects of the mechanical package must be verified. First, the critical speeds for vibration must be checked to ensure that the normal operating speed range will not include resonant frequencies. Second, the pressure lubrication system must be checked to make sure that proper oil flow and pressure will be maintained as operating speed is reduced. Neither of these are commonly issues but good design practice requires examination of both factors.

Because of the greater sensitivity to surge and the high equipment cost, single stage blower installations are virtually always equipped with blow-off valves and sophisticated surge prevention and protection systems.

5.3.4 Turbo Blowers

The turbo blower has rapidly been adopted as the design of choice for many systems. This relatively new technology has become very common in both new and retrofit aeration applications. The turbo blower offers low power requirements, low maintenance, and compact packages with integrated blower and controls. They are generally limited to ratings below 400 hp but the available range is expanding. As always, careful analysis is required to ensure that this is the most cost-effective and appropriate technology.

The turbo blower is a single stage unit. Instead of gearing to achieve high impeller speeds, the impeller is mounted directly on the motor shaft. Several features are characteristic of the turbo blower:

- Motor and blower impeller an integral unit
- High efficiency very high speed permanent magnet synchronous motors
- Integral VFD providing frequency of several hundred Hz to provide high motor speed
- High-technology lubrication-free bearings
- Compact package construction with blower, controls, and accessories factory assembled in a sound attenuating enclosure (see Figure 5.14)

There are, of course, many variations between and within the offerings of the various manufacturers.

The most common bearing technology is the air foil bearing. This is functionally similar to the journal bearing used in geared single stage blowers. In the most

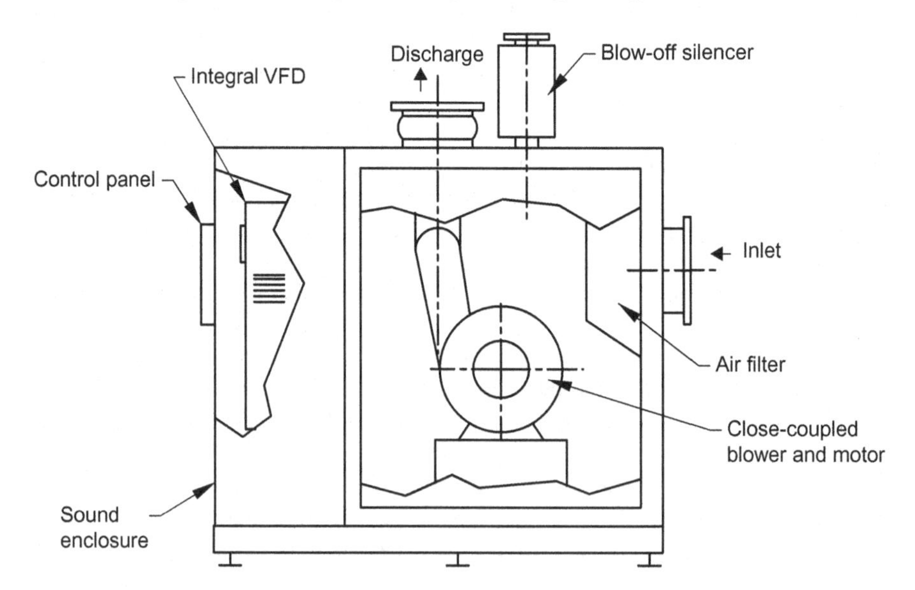

FIGURE 5.14 *Simplified diagram of turbo blower package*

common turbo blower configuration, however, the shaft floats on a film of air instead of the oil film of a conventional journal bearing. Because of the extremely high operating speed, air forms a hydrodynamic film between the shaft and the bearing. The shaft floats on the resulting cushion of air—all but eliminating friction and wear. Special alloys and geometries are employed to prevent damage to the bearings during the very short-duration start cycle.

A variation of the turbo design is the magnetic bearing. This is most often found in the higher horsepower ranges. A magnetic field is created around the shaft—essentially floating it in a manner similar to a maglev train. The result is also virtually friction-free shaft support with no maintenance.

The package's VFD and motor are very efficient but they do generate heat. Many turbo packages use air for cooling their enclosures. In some cases, this is a separate flow stream with forced ventilation and the cooling air is exhausted from the enclosure. In other designs, the process air flow is used for removing heat and the air is ducted to the blower inlet. This does have the effect of raising the inlet air temperature, which affects blower performance, but because of the high volume of process air the impact is generally not significant. Other systems, particularly higher horsepower units, use water cooling for the motor and drive. In most of these systems, the blower cooling is a closed loop system with the cooling water recirculated between the blower package and an external heat exchanger. The external heat exchanger may be air-to-water or water-to-water.

Most turbo blowers include inlet filters as part of the package. Many use louvered enclosures and draw the process air directly from the blower room. This can create uncomfortable conditions for the operators in cold climates. Many turbo packages offer the option of piped inlet air connections.

The very high-frequency output from the VFDs exceeds the capabilities of commercially available drives. The characteristics of the synchronous motors used in most turbo packages also require special drive considerations. Consequently, both the motor and VFD for turbo packages are proprietary designs rather than commercially available motors and VFDs.

The blower/motor arrangements also vary for the turbo blowers. Many use a single impeller and single-ended motor shaft. Others incorporate a double-ended design with an impeller mounted on each side of the motor and the two blower units operating in parallel. Still other units achieve high flow rates in a compact package by mounting what are essentially two complete blower/motor assemblies in a single package with the two blowers operating in parallel.

5.3.5 Control and Protection Considerations

Avoiding surge is one of the primary objectives of any dynamic blower control system. The best technique is to prevent surge from occurring by modulating the blower control and system restriction to prevent operating at pressures and flows that will result in surge. A surge control line (SCL) is usually used as the lower limit of safe operation. The SCL is usually constructed by adding a surge margin to the SLL. The control system must take corrective action if blower flow reaches the SCL.

A number of protection features may be used to prevent operation below the SCL. The control element (inlet butterfly valve, IGV/VDV, or VFD) is modulated to increase the flow rate and move operation back to safe limits. Many blowers are equipped with a blow-off valve. If modulation fails to relieve the impending surge condition, the blow-off valve is opened. This vents some or all of the blower flow to atmosphere thereby reducing system restriction and increasing flow rate. Some blow-offs are modulating. This results in safe system pressure and blower flow while maintaining high enough pressure to create air flow into the aeration system. In other cases, the blow-off is opened fully when an impending surge is detected and it dumps all flow to atmosphere.

If the surge prevention system cannot eliminate the low flow, the blower must be stopped before damage results. The time delays allowed before safety shutdown occurs must be coordinated with the blower manufacturer's requirements.

Starting a dynamic blower can prevent challenges to the control system designer. This is particularly true when starting a second or third blower discharging into an aeration system operating under pressure. The blower must pass through the surge point during acceleration and the starting technique must ensure that this does not result in damage. Throttled multistage blowers are sometimes started, with the inlet valve closed to reduce inrush current. After the blower is up to speed, the inlet valve is opened—the blower passes quickly through surge—and normal operation is achieved. Other manufacturers recommend opening the inlet throttling valve to a minimum position during starting so that the blower quickly passes through surge without pulsation.

Variable speed blowers generally accelerate through the surge area fast enough to prevent surge pulsations.

If the blower system is equipped with a blow-off valve, it is generally opened prior to starting the blower. This permits acceleration with low back pressure and prevents surge. After the blower has reached operating speed and a safe air flow, the blow-off valve is closed.

An alternate to the blow-off valve is a bypass valve. Rather than venting blower flow to atmosphere, the bypass directs it back to the blower inlet in a closed loop. This technique is used principally in applications compressing gases which may not be discharged to atmosphere. The blower generates heat during the compression process so venting discharge flow back to the inlet results in increasing inlet air temperature. Extended operation in this mode could result in an inlet temperature higher than safe limits. Therefore, aeration blowers should be fitted with blow-off valves instead of bypass valves.

The primary blower control parameter is air flow rate. There are a variety of devices and systems for measuring air flow but any dynamic blower control should include air flow measurement.

Pressure is a secondary measurement parameter. Even systems that use automatic control to maintain constant blower discharge pressure will require a second control loop to modulate blower flow to increase or decrease discharge pressure.

There are a wide variety of other parameters for monitoring equipment health and providing safety shutdown. Large single stage blowers may have complex systems with dozens of sensors on each blower. Small multistage blowers may operate successfully with only a simple surge shutdown based on motor current. Among the parameters monitored and used in blower protection systems are:

- Bearing temperature
- Bearing and/or blower case vibration
- Discharge pressure
- Inlet (suction) pressure
- Inlet temperature
- Discharge temperature
- Lube system pressure
- Lube system temperature
- Motor winding temperature
- Motor load

Turbo blowers generally incorporate a complete blower control and protection system as part of the package. These systems generally have communications capability for data reporting and control interface with aeration control systems and plant SCADA packages. Most will also accept run/stop commands from external dry contacts and air flow setpoints from analog inputs. The controls also provide dry contacts for run confirmation and fault and provide an analog signal for actual air flow.

The process load establishes the oxygen demand of the aeration system and the diffuser efficiency determines the air flow rate needed to supply the oxygen to the process. The blower system provides this air flow rate to the process. Tying these two "ends" of the process equipment together is the piping system—an important but often neglected component of the aeration system.

EXAMPLE PROBLEMS

Problem 5.1

The specifications for a blower system include the following data points:

SCFM	°F	%RH
3500	40	20
4000	60	36
4500	80	60
5000	100	80

The site barometric pressure is given as 14.2 psia. Convert the flow rate to ACFM and lb/h of air while

(a) ignoring relative humidity,

(b) including relative humidity, and

(c) calculate the percentage error for each point if humidity is ignored.

Problem 5.2

A lobe type PD blower is connected to a 1750 rpm 100 hp motor by a belt drive with a 1.7:1 speed increasing ratio and 3% slip. Motor efficiency is 94%. The blower has a displacement of 0.662 CFR, a 1 psi slip of 108 rpm, a maximum rated speed of 3400 rpm, and friction hp of 1.24 bhp/1000 rpm. The actual discharge pressure varies between 7.0 and 8.0 psig. Worst-case ambient conditions are 100 °F and barometric pressure is 14.4 psia.

(a) What is the current flow rate in SCFM?

(b) If the belt ratio is changed, what is the maximum flow in SCFM that can be obtained from the blower?

(c) A proposed ECM includes installing a VFD on the blower. Construct a chart for kW versus SCFM at 7.0 and 8.0 psig with the new belt ratio and variable speed.

Problem 5.3

A plant has four geared single stage blower's with inlet guide vane control. Plant SCADA trends show a pattern generally conforming to Table 4.2 with an average air flow of 28,500 SCFM and occasional excursions during storm events to 56,000 SCFM (see Figure 5.15). The blower performance characteristics and ambient

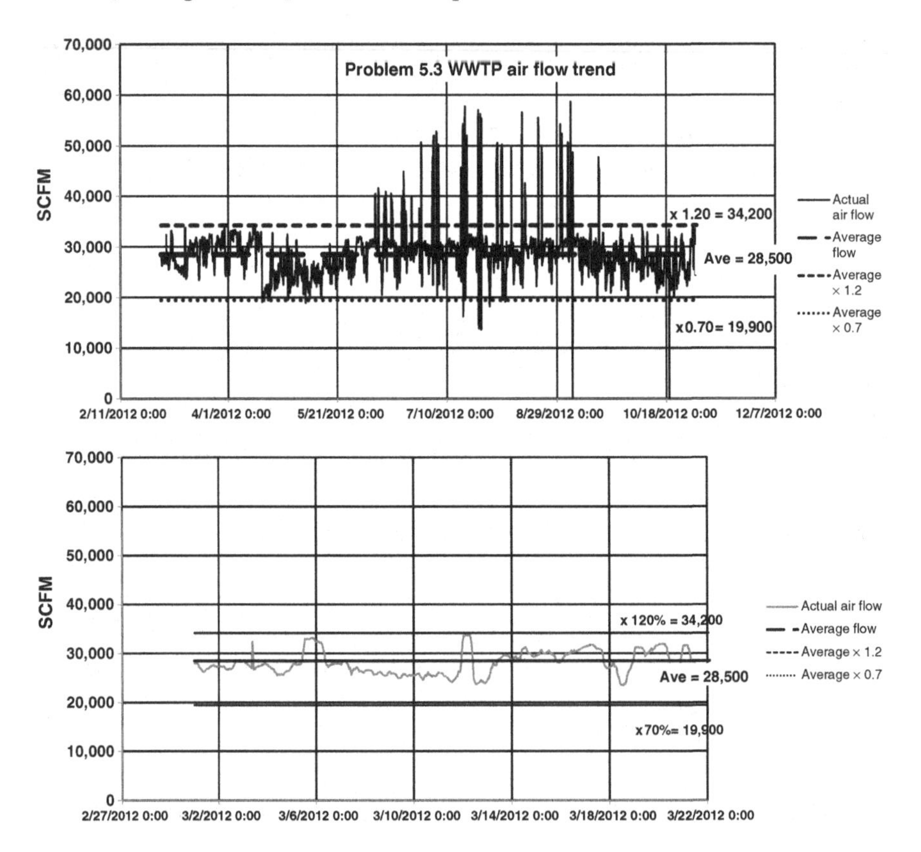

FIGURE 5.15 Problem 5.3 plant airflow trend

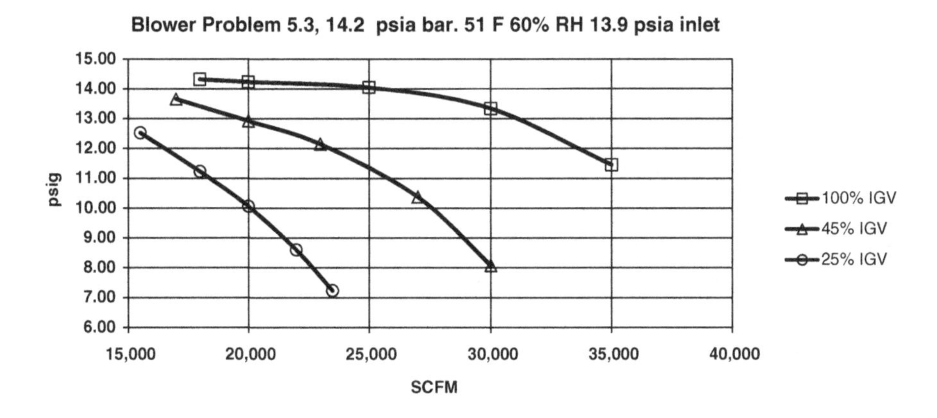

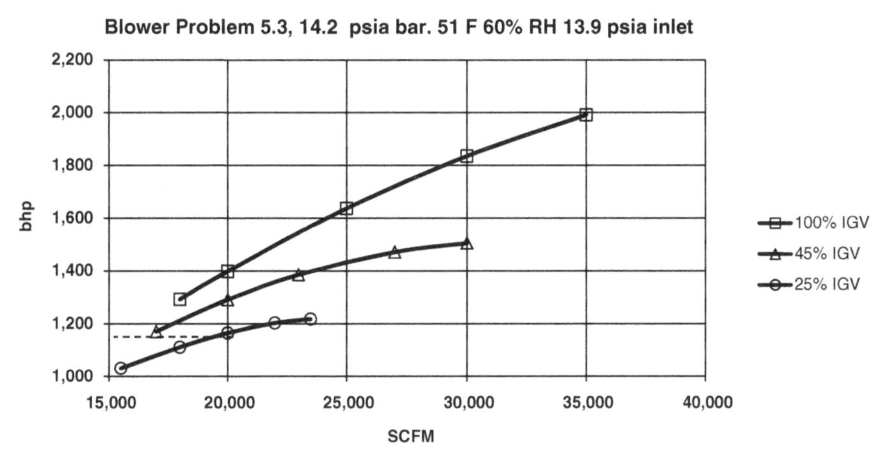

FIGURE 5.16 *Problem 5.3 blower performance curves*

conditions are shown in Figure 5.16 for 51 °F, which is the annual average temperature at the site. The motors are 2000 hp, 4160 VAC 3 Phase power, and have an average efficiency of 96%. Current controls maintain a constant discharge pressure of 10.3 psig. It is proposed that the guide vane control be replaced with 97% efficient VFD control. The energy cost is $0.06 /kWh on and off peak plus a demand charge of $5/kW for the highest demand within a 12-month period (ratchet charge). Diffuser submergence is 19.5 ft.

(a) If each VFD has an installed cost of $300,000, including control system upgrades, will the system be cost-effective?

(b) Adding a synchronous transfer system to permit controlling two blowers from one VFD adds $80,000 to the installed cost of a VFD for each blower. What are the pros and cons of this system?

(c) How is the payback and system selection affected if a utility rebate of $50/hp is available for each VFD?

Problem 5.4

A 75 hp screw blower with an inlet filter pressure drop of 0.2 psi and a discharge pressure of 9.0 psig has an efficiency of 70%. The blower case is 24″×12″×9″. The motor efficiency is 97% and the VFD efficiency is 97%. The package has a temperature rating of 40 °C (104 °F). The ambient air temperature is 90 °F and the altitude is 800 ft. above sea level.

(a) What is the discharge temperature of the process air flow?

(b) How much ventilating air will be required to keep the enclosure temperature within safe limits?

Chapter 6

Piping Systems

The aeration system piping simply connects the blowers to the process, conducting the air to the aeration basins and the diffusers. This is simple and straightforward. The piping system doesn't directly consume power and is for the most part a passive part of the plant's process equipment. Piping is generally ignored in the initial evaluation of baseline energy use. The piping is given little attention in the evaluation of ECMs and the design of aeration control systems.

This lack of attention is a mistake. The layout of the piping will often dictate the quantity, size, and locations of monitoring and control instrumentation. This has a direct bearing on the cost of the control system and on the payback from energy savings. Pressure drop in the piping adds to the energy requirement of the blowers, but contributes nothing to process performance.

Poorly designed piping directly affects energy requirements in other ways. Undersized piping obviously creates excessive pressure drop. This is, however, rarely a problem. The opposite extreme, oversized piping, is a problem encountered frequently. Valves and air flow meters are commonly sized to match the pipe in which they are installed. If this piping is oversized, the flow meters won't provide accurate data at the reduced flows commonly encountered. Oversized valves make the air flow difficult to control. The lack of precision in air flow control will have a direct effect on the ability of the system to control DO and air flow to match changing process demands.

Inspection of the existing piping system is a critical part of the initial assessment. Accurate evaluation of the sizes, locations, and air flow ranges should be made for

Aeration Control System Design: A Practical Guide to Energy and Process Optimization,
First Edition. Thomas E. Jenkins.

each segment of the system. The evaluation should start at the blower discharge and extend all the way to the diffusers. In an upgrade to an existing treatment facility, budgetary constraints will usually prohibit replacement of the piping system. Determination of necessary modifications to accommodate proposed ECMs can and should be included in the design process.

Piping condition is also an important part of the assessment of existing aeration systems. It is surprising to find air leaks that go uncorrected for years but this is often the case. Walking the route of buried piping during or immediately after a rain can be quite instructive. It is very common to observe air bubbling up through the sod or out of cracks in pavement and sidewalks—the result of corroded pipe or failed joints in the buried air piping. Watching these leaks is watching dollar bills blow into the air and disappear!

Piping systems are usually outside of the experience and comfort zone of control system designers. It is important that they grasp the fundamentals, however, in order to achieve optimized energy consumption and control stability. An amazing number of problems in commissioning and unsatisfactory control system performance can be traced directly to poor design of the aeration piping and improper control valve sizing. Understanding the basics of piping and accommodating the requirements in the control system design will enhance the performance of the control system.

6.1 DESIGN CONSIDERATIONS

There are three major considerations in piping design:

- Layout
- Size
- Material

6.1.1 Layout

The basic layout of the air distribution piping is determined by the relative positions of the blowers and the aeration basins. Some systems have a single blower dedicated to a single tank. This arrangement is most commonly found in SBRs and small package plants. In this configuration the piping is minimal, control valves are not required to balance the air between multiple tanks, and sizing and design are not as critical as in larger systems.

A more common arrangement is multiple blowers supplying air to multiple aeration basins. The blowers are installed in a blower room and discharge into a common main header. The main header connects the blower room to the aeration basin and has branches for each aeration basin. Each basin branch has multiple drop legs in each tank. Each drop leg provides air to multiple horizontal pipes in the bottom of the tank, supplying air to the diffuser grid (see Figure 6.1).

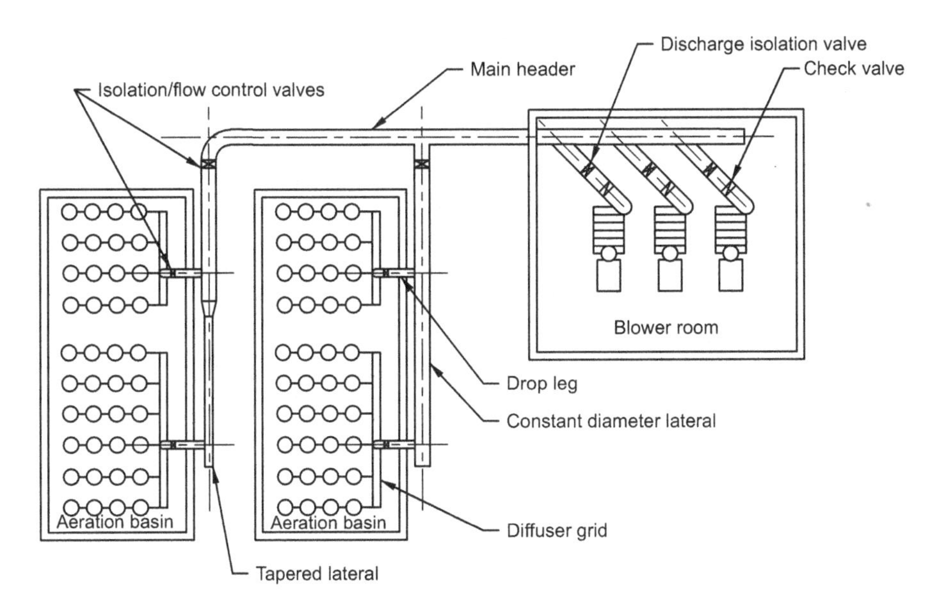

FIGURE 6.1 *Piping layout example*

Isolation valves, typically manually operated butterfly valves (BFV), are used to segregate various parts of the aeration system. The isolation valves may also be used for flow control. In large systems the valves are generally equipped with electric motor operators to allow the control system to modulate air flow as well as to provide operator convenience.

Each blower should have an isolation valve on the discharge side to allow removal of the blower and piping for maintenance. If the blowers share a common inlet filter and inlet header there should be isolation valves for each blower inlet as well. Each lateral branch off the main header should have an isolation valve to allow taking a complete basin out of service for maintenance. Every drop leg should have an isolation valve. This valve is also used for either manual or automatic control of the air flow to each grid.

The main header is often run underground between the blower building and aeration basin—particularly if the pipe must cross roads or drives. This should be avoided whenever possible. It is just as cost-effective to run the pipe up and over roadways which allows inspection, maintenance, and repair of leaks. Buried pipe is usually forgotten and rarely maintained.

In the blower room, connections to the main header should be at 45° if the air travels along the header in one direction to feed all basins. Experience has shown this arrangement greatly simplifies starting multiple blowers, even though theory indicates that a 90° connection should perform just as satisfactorily. If the main blower room header feeds aeration basins from both ends then 90° connections are preferred. The main header should be generously sized to minimize pressure drops and pressure differences between multiple blowers. The main header velocity should be approximately half the normal velocity recommendation for distribution piping shown in Table 6.3.

The pipe run should be as straight as possible with a minimum of bends and fittings in order to minimize pressure drop. Elbows should be long radius style where possible. Mitered 45° and 90° elbows should be avoided.

The main header usually has multiple takeoffs along its length to supply air to a lateral for each basin. Each lateral in turn has multiple takeoffs—one for each drop leg. There are two variations in the handling of the pipe size at each takeoff point. In some cases, the diameter of the header or lateral is reduced at each takeoff and tapered so that a nearly constant velocity is maintained as air is drawn off. The usual justification for this is lower pipe cost. The alternate arrangement is to maintain a constant diameter along the entire length of the header or lateral. The justification for this is to reduce the pressure drop.

The basis for reducing diameter or maintaining constant diameter should actually be assisting the air distribution along the header by taking advantage of pressure changes along the length. There is an expectation that the takeoffs closest to the blower will naturally receive the most air and those furthest from the blowers receive the least. This expectation is based on observation of the pressure drop due to friction along the header. In fact, in a constant diameter header the takeoff furthest from the blowers will tend to get the most air. In a header with the diameter tapered along the length the takeoff closest to the blowers will tend to get the most air.

This is the result of "velocity head" on the pressure distribution in the header. The velocity head is explained by Bernoulli's law. Bernoulli's law defines the energy balance between static pressure (potential energy), velocity pressure (kinetic energy), and pressure loss due to friction (energy lost by conversion to heat in a moving fluid). In its basic form for incompressible fluids like water, Bernoulli's law is usually represented as:

Equation 6.1

$$\frac{p_1}{\gamma} + \frac{V_1^2}{2 \cdot g_c} + z_1 = \frac{p_2}{\gamma} + \frac{V_2^2}{2 \cdot g_c} + z_2$$

where

$p_{1,2}$ = static pressure or head, lb_f/ft^2
γ = specific weight of fluid, lb_f/ft^3
$V_{1,2}$ = velocity of fluid, ft/s
g_c = acceleration of gravity, 32.17 ft/s^2
$z_{1,2}$ = height of liquid column, ft.

This form of the equation assumes no work is done by the fluid stream, there is no heat transfer, and there is no change in internal energy of the fluid.

In the case of a moving stream of air in the real world, it is necessary to include pressure drop in the equation. For a gas, such as air, static pressure from the height of a column of air is negligible and may be ignored. The normal pressure units are pounds force per square inch. Accommodating these factors, Bernoulli's equation may be expressed as:

Equation 6.2

$$p_1 + \frac{\rho \cdot V_1^2}{3.335 \times 10^7} = p_2 + \frac{\rho \cdot V_2^2}{3.335 \times 10^7} + \Delta p_f$$

where

$p_{1,2}$ = static pressure, psi
ρ = density at air flow conditions, lb_m/ft^3
$V_{1,2}$ = air velocity, ft/min
Δp_f = pressure drop due to friction, psi

Density and specific weight vary with temperature, pressure, and relative humidity. For piping calculations, the effect of humidity can be ignored. Compared to the other assumptions involved in these calculations, the impact of humidity is insignificant:

Equation 6.3

$$\rho_a = \frac{2.692 \cdot p_a}{T_a}$$

where

ρ_a = density at air flow conditions, lb_m/ft^3
p_a = actual pressure, psia
T_a = actual temperature, °R

The term $V^2/2g$ in the various forms of the Bernoulli's law equation define the dynamic pressure of the moving fluid, more commonly known as the velocity head:

Equation 6.4

$$p_d = \frac{\rho \cdot V^2}{3.335 \times 10^7}$$

where

p_d = velocity head or dynamic pressure, psi

Bernoulli's equation is the basis of many types of flow monitoring instruments such as orifice plates and venturis (see Chapter 7).

An equally important concept is conservation of mass and its consequence, the equation of continuity. For a given air flow through pipes of changing diameter the velocity changes if the pipe diameter changes:

Equation 6.5

$$Q_a = A_1 \cdot V_1 = A_2 \cdot V_2$$

where

Q_a = volumetric flow rate at actual conditions, ACFM (ft^3/min)
$A_{1,2}$ = cross-sectional area of pipe, ft^2
$V_{1,2}$ = velocity, ft/min

If the diameter of the pipe is constant and the flow changes as a result of a takeoff from the header to a diffuser grid drop leg, the equation of continuity tells us that the velocity must increase:

Equation 6.6

$$V_2 = \frac{Q_2}{Q_1} \cdot V_1$$

If the velocity decreases then so does the dynamic pressure. Since the total pressure must stay constant, the static pressure increases by an amount equal to the decrease in dynamic pressure—a phenomenon known as "velocity head regain." The increased static pressure in the pipe causes an increase in flow rate through subsequent drop legs and diffusers. This accounts for the tendency of the aeration grid furthest from the blowers to have the highest flow rate if the header diameter is constant.

Conversely, if the diameter changes to maintain a more or less constant velocity then there is no velocity head regain. In this case, the pressure drop due to friction causes the grid furthest from the blower to exhibit lower static pressure. The natural result is a tendency for the furthest drop leg to receive the least air. If the air flow at this drop leg must be higher, then the control valves in the rest of the system must be throttled to create a higher system pressure and offset the pressure drop due to friction.

The practical application of this in piping system layout is to minimize the throttling required at the various valves to control air flow. In a plug flow reactor, the diffuser grid closest to the wastewater influent generally requires the highest air flow. This is needed to meet the demands of the high oxygen uptake rate at the influent end of the basin. If the influent end of the tank is furthest from the blowers then a constant diameter air pipe is recommended to take advantage of the velocity head regain. If the influent is at the end of the tank closest to the blowers then the air pipe should be tapered so the tendency of the air flow to be highest closer to the blower room assists in the air distribution. Phrased differently, if the air flow direction is opposite the hydraulic flow direction the piping should be constant diameter. If the air flow and hydraulic flow are in the same direction, the pipe diameter should be tapered. If the air velocity is kept low, it may be possible to ignore the velocity head effect but this should only be done after calculations to determine the magnitude of the velocity head effects.

TABLE 6.1 Nominal Pipe Support Spacing, Air Service, feet

Nominal Size (in.)	Schedule 40 Steel	Schedule 5S	Schedule 80 PVC (140 °F)
4	17	26	4.5
6	21	32	5
8	24	33	5.5
10	26	36	6
12	30	40	6.5
14	32	41	7
16	35	43	7.5

Condensation can sometimes be a problem in air mains, although to a much lesser extent than in submerged diffuser grids. The air distribution piping should include drain or purge valves to allow draining of condensate. The piping should also have a slight pitch to the drains. Some systems use eccentric reducers at size transitions to maintain the invert (bottom of the pipe) at a constant elevation to enable proper drainage.

Detailed piping support design is beyond the scope of this text but the control system designer should be aware that support systems will influence and restrict the pipe layout. Both pipe diameter and material influence the spacing and type of supports (see Table 6.1). The supports must accommodate thermal expansion (see Equation 6.7) and therefore electrical connections to valves and transmitters must be made with flex conduit. Control valves may represent significant piping loads and steps should be taken to verify that the pipe and support systems can accommodate the weight of valves and operators. It is particularly easy to overlook this precaution when new electric operators are added to existing valves in control retrofits.

Thermal expansion can be a very significant factor in air piping systems. The change in pipe temperature from being off-line in winter to high blower discharge air temperature during the summer can approach 300 °F in some climates. The change in length is proportional to the length of the pipe run. Many plants have very long pipe runs. Air mains and laterals extend along the entire length of an aeration basin, so the total change may be quite large. The total thermal growth is a function of the material of the pipe and is greater for nonmetallic pipe. Change in length may be calculated from:

Equation 6.7

$$\Delta L = L_0 \cdot \alpha \cdot \Delta T$$

where

ΔL = change in length, in.
L_0 = initial length, in.
α = linear coefficient of thermal expansion, in./in. °F
ΔT = change in temperature, °F

TABLE 6.2 Typical Linear Coefficients of Thermal Expansion and Young's Modulus (Properties will vary with Temperature and Composition)

Material	α, in./in. °F	E, psi	Notes
Carbon steel	6.5×10^{-6}	30×10^{6}	
Stainless steel	9.9×10^{-6}	28×10^{6}	
Cast (ductile) iron	5.9×10^{-6}	13×10^{6}	
Aluminum	12.8×10^{-6}	10×10^{6}	
PVC	28.0×10^{-6}	0.4×10^{6}	
HDPE	90.0×10^{-6}	0.1×10^{6}	(High-density polyethylene)
FRP	14.0×10^{-6}	1.0×10^{6}	(Fiberglass reinforced plastic)

If the pipe is restrained to prevent movement then stress will be induced in the pipe in the axial direction. On decreasing temperature the stress will be tensile, and on increasing temperature the stress will be compressive (Table 6.2). The stress can be calculated as:

Equation 6.8

$$\sigma = E \cdot \alpha \cdot \Delta T$$

where

σ = stress in pipe, psi
E = Young's modulus (modulus of elasticity), psi

The pipe supports are usually designed to hold one end of the pipe run in a fixed position and allow expansion in one direction. The support that provides this function is commonly referred to as a pipe anchor. The adjacent pipe supports are designed to allow the pipe to move through or over them during expansion and contraction. Intermediate locations that require a fixed position—such as drop legs to diffuser grids—are equipped with expansion joints designed to accommodate axial and lateral movement of the two ends of the joint. For low-pressure aeration piping these joints are usually reinforced elastomer sleeves, often corrugated, with flanged connection to the adjacent piping. The elastomer should be selected to provide protection from ultraviolet (UV) damage due to sunlight and should be rated for the highest anticipated air temperature.

Many expansion joints are equipped with restraint rods. They are provided to prevent axial thrust loads from pulling the joint apart. The restraints (or thrust rods) should not restrict the normal axial movement of the expansion joint. It is, unfortunately, common to see the thrust rods tightened to the point that they have drawn the expansion joint ends inwards so that normal movement is not possible. The manufacturer's recommendations and procedures should be consulted and followed during installation to prevent this.

An important consideration in both piping layout and sizing is the requirements of instrumentation and controls. A common problem with obtaining accurate flow meter readings is the lack of sufficient straight pipe. A minimum length of ten diameters is required to establish a uniform velocity profile across the pipe.

The air velocity in any pipe is not constant across the diameter of the pipe. The air velocity is zero at the pipe wall and in a long, straight pipe the velocity will be at the maximum value at the center of the pipe. This variation is referred to as the "velocity profile." For low flows or small pipes, the velocity profile is parabolic—a condition referred to as laminar flow. In laminar flow, the air flow is essentially axial with the pipe. For the velocities and sizes typically encountered in aeration systems, the air movement is full of turbulence—a swirling of the air. The overall movement is in the direction of air flow. In this condition, known as fully developed turbulent flow, the velocity profile is more nearly uniform across the pipe, but the velocity at the center is higher than close to the pipe walls. The centerline velocity is typically 122% of the average velocity.

Most flow meters require fully developed uniform air flow for accurate measurement. As air passes through an elbow, it is pushed to the outside radius as it makes the turn which creates a higher velocity on that side of the adjacent pipe. Passing through a butterfly valve causes the air to move to the outside of the pipe as it moves around the disk and stem. These or similar obstructions immediately prior to the meter create an uneven velocity profile that causes the flow meter reading to be incorrect. This is particularly true for single point measurements. In general either flow straightening vanes or a minimum length of straight pipe is required to ensure accurate flow measurements (see Chapter 7).

The "Y-wall" tank configuration is common in older facilities, particularly in plants with plug flow aeration basins (see Figure 6.2). It is compact and

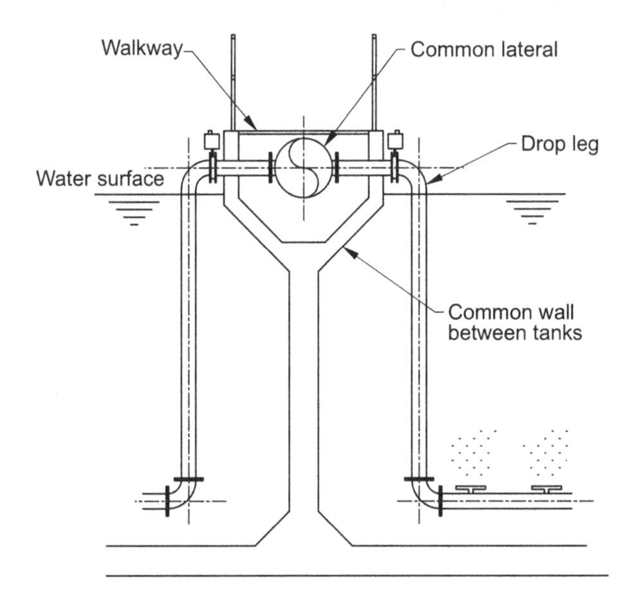

FIGURE 6.2 *Typical "Y-wall" configuration*

economical. This configuration causes problems if air flow measurement is required for each basin or each drop leg. The close proximity of the takeoff tee, control valve, and elbow make it virtually impossible to measure the flow in each drop leg because of the distorted velocity profile. Because two tanks share the takeoff point from a common lateral, it is impossible to determine the air flow to each tank. Theoretically each tank should receive equal flow, but that assumption may be incorrect. Unequal process demand in each tank may not require equal flow to each in any case. Some modification to the piping is required if accurate flow control and DO control are desired.

If basin air flow control is required but not control of each drop leg, there are two options:

- Run a second lateral parallel to the existing one and provide flow measurement and flow control valves on each
- Monitor each drop leg individually and totalize the flow for each basin.

Because of space and cost, it may actually be less expensive to monitor and control the air flow in each drop leg rather than install a second lateral or remove the existing lateral and replace it with two new ones. The total air flow to the basin can be controlled by totalizing and controlling the flow to each drop leg. Separate DO control may be provided at each drop leg or the flow to each may be controlled proportional to the number of diffusers or tank volume served by each drop leg (Chapter 9).

The typical Y-wall drop leg configuration does not lend itself to accurate flow measurement or monitoring but it may be modified to obtain accurate measurement and control. There are three options available (see Figures 6.3 and 6.4):

- Install a flow meter below the water level.
- Use a horizontal or vertical goose neck to provide the necessary straight piping and reconnect to the existing drop leg.
- Use a horizontal takeoff of the required length and connect to a new drop leg asymmetrical to the diffuser grid.

The first method utilizes existing piping and only requires adding a half coupling or similar fitting to allow installation of the primary sensing element. This technique is most often employed in conjunction with averaging pitot tubes but is also used with thermal mass flow meters. The drop leg generally has more than enough straight pipe to meet measurement needs. There is often concern about installing the sensing element in an inaccessible location below water level but service needs are usually virtually nonexistent—particularly in the case of a pitot tube.

The goose neck, either horizontal or vertical, is most convenient and cost-effective when the existing diffuser grid and drop leg are being maintained. This system allows all or most of the existing piping to be retained as is and short lengths of small diameter pipe are installed at existing connections. Note that the vertical

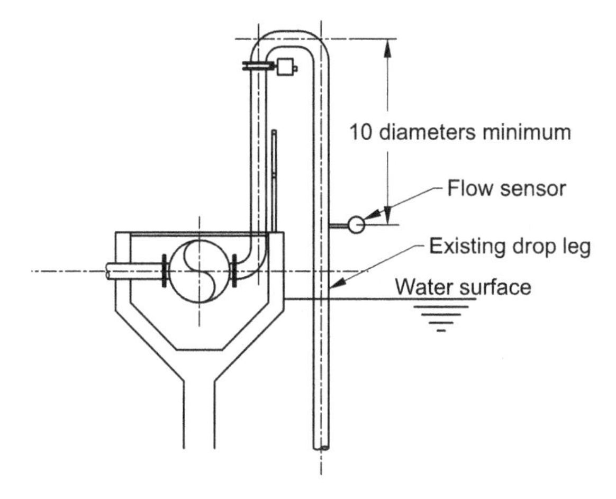

FIGURE 6.3 Vertical goose neck

goose neck requires less new pipe since the drop leg provides more than enough straight pipe downstream of the flow element.

If the existing diffusers are being replaced, then a horizontal pipe from the existing takeoff point to a new drop leg minimizes piping modifications and cost. The only precaution is that the horizontal distribution piping in the bottom of the tank must be sized to ensure even flow distribution to all diffusers. This is not usually a problem.

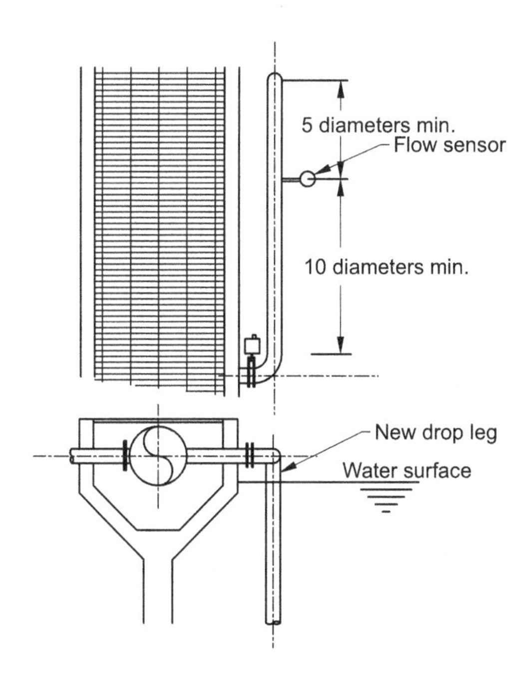

FIGURE 6.4 Horizontal takeoff

6.1.2 Pipe Size

There are two principal parameters that determine pipe size. Both are ultimately linked to economics. The first parameter is the initial installed cost of the pipe, fittings, and supports. Since installation labor is greatly influenced by diameter, the size of the pipe is the dominant factor in initial equipment cost and capital investment. Obviously, the larger the pipe, the greater the initial cost.

The second parameter is the frictional pressure drop through the piping system. This is directly related to operating cost since higher pressure drop adds directly to the discharge pressure required at the blowers and therefore the power consumption and energy cost. As one would expect, the larger the pipe diameter is, the lower the pressure drop and operating costs are. The relationship is very nonlinear, however, with pressure drop decreasing as the fifth power of diameter (Equation 6.11).

Experience has shown that selecting the pipe diameter to create an air velocity within a specified range is a good method for initially establishing an appropriate pipe diameter. Normal ranges of velocity are shown in Table 6.3. The sizing is usually based on keeping the velocity within the range at the maximum anticipated air flow rate. The velocity at minimum anticipated air flow is usually checked, although this is not a particularly critical parameter. The minimum velocity in air mains is of interest principally in verifying that it is within the acceptable range for flow measurement devices and control valves.

In addition to pressure drop considerations, the velocity of the air in the piping system is directly related to the noise produced by the air flow. If velocities are high, a surprising level of noise results from the air moving through pipe, fittings, and valves. This noise is radiated to the surroundings, and can be particularly troublesome with thin wall piping. Limiting air velocity will reduce the noise generation.

For a given flow rate and velocity the required cross-sectional area of the pipe can be readily determined:

Equation 6.9

$$A = \frac{Q_a}{V}$$

where

A = cross-sectional area of pipe, ft^2
Q_a = volumetric flow rate at actual conditions, ACFM (ft^3/min)
V = design velocity, ft/min

TABLE 6.3 Typical Distribution Piping Air Velocities

Nominal Pipe Diameter, in.	Design Velocity, ft/min
1–3	1200–1800
4–10	1800–3000
12–24	2700–4000
30–60	3800–6500

Or, more conveniently:

Equation 6.10

$$d = 24 \cdot \sqrt{\frac{Q_a}{\pi \cdot V}}$$

where

d = diameter of pipe, in.

The calculated diameter must be adjusted to the nearest inside diameter of commercially available pipe (Table 6.4).

Specifying pipe and identifying dimensions for piping can be very confusing. For example, there is nothing in a 4 inch steel pipe that actually measures 4 inches. Rather, pipe sizes are always given as "nominal" diameter. In general, the outside diameters for a given nominal pipe size are equal so that fittings will be compatible but there are many exceptions to this. It is always necessary to verify the actual pipe dimensions with supplier's data for the specific pipe under consideration. Note that the pipe material will influence dimensions—even for pipe with the same schedule or rating.

In addition to the pipe diameter, the pipe material and rating or class must be identified. For steel and stainless steel pipe the rating is given as a schedule number. Schedule 40 is standard pipe—commonly used for water and other services.

TABLE 6.4 Typical Steel and Stainless Steel Pipe Dimensions

Size, in.	Rating	ID, in.	OD, in.	Inside Area, ft^2
3	Sch. 10	3.260	3.500	0.05796
	Sch. 40	3.068	3.500	0.05134
4	Sch. 10	4.260	4.500	0.09898
	Sch. 40	4.026	4.500	0.08840
6	Sch. 10	6.357	6.625	0.22041
	Sch. 40	6.065	6.625	0.20063
8	Sch. 10	8.329	8.625	0.37837
	Sch. 40	7.981	8.625	0.34741
10	Sch. 10	10.420	10.750	0.59219
	Sch. 40	10.020	10.750	0.54760
12	Sch. 10	12.406	12.750	0.83944
	Sch. 40	11.938	12.750	0.77730
14	Sch. 10	13.624	14.000	1.01236
	Sch. 40	13.124	14.000	0.93942
16	Sch. 10	15.624	16.000	1.33141
	Sch. 40	15.000	16.000	1.22718
18	Sch. 10	17.624	18.000	1.69409
	Sch. 40	16.876	18.000	1.55334

Schedule 80 is heavier-walled pipe intended for higher pressure service and as a rule is only found in low-pressure aeration systems using nonmetallic pipe. Schedule 10 and the lighter Schedule 5 pipe are often used in aeration systems. The most common piping for main distribution headers and laterals is Schedule 10 or 5 stainless steel pipe. Although stainless steel is more expensive than carbon steel, the thin wall, lower fabrication cost, and elimination of painting or other corrosion resistance measures offset the cost of stainless steel.

Threaded pipe connections and fittings are standardized. Threaded fittings are not generally employed except for Schedule 40 or heavier steel pipe, and for Schedule 80 or heavier nonmetallic pipe. Threaded connections are rarely used for pipe larger than 3 in. nominal diameter.

Instrument connections for flow, pressure, and temperature penetrate the pipe wall. The instrument usually requires a threaded connection, usually between $^1/_4$ and 1 in. nominal. In heavy walled pipe, such as ductile iron or Schedule 40 steel, it is adequate to drill and tap the aeration pipe wall directly. In thinner walled steel and stainless steel pipe a hole is drilled through the pipe wall and a half coupling or special fitting with the required pipe thread is welded to the pipe OD. Clamp-on saddle fittings are also used for instrument connections on pipes of all material and wall thickness.

Pipe threads are tapered to create an interference fit between the male and female threads. Standard pipe threads may have a small clearance between the root of the female thread and the crest of the male thread thereby providing a leakage path. It is standard practice to use a sealant, such as pipe dope or Teflon® tape, on the threads to prevent air leaks at the connection.

Flanged connections are typical for pipe and fittings used in aeration systems. The typical flange connection is the so-called ANSI 150 lb flange. The dimensions are standardized which allows connection of piping and fittings of various materials and manufacturers to be joined.

6.1.3 Pipe Material

The selection of piping system material is a function of economic and functional considerations, with the functional considerations influencing the economics.

The primary functional constraints are:

- Pressure rating
- Thermal rating
- Corrosion resistance

The factors influencing cost are:

- Material cost
- Installation and joining requirements (including labor)
- Support requirements
- Maintenance requirements

There are two broad categories of piping material: metal pipe and nonmetallic (plastic) pipe. Both are used in aeration piping with many systems combining both types.

The metals commonly used in aeration piping are carbon steel, ductile iron, and stainless steel. Carbon steel is usually found in interior locations, such as blower room headers and blower inlet and discharge piping. Schedule 40 pipe, painted for corrosion resistance, is commonly used for these applications. Carbon steel pipe is also available in Schedule 10, and this type is frequently hot dip galvanized (coated with zinc) for better corrosion resistance. Galvanized pipe is also used for air mains and laterals.

Cast iron—and more commonly ductile iron—is much less susceptible to corrosion than carbon steel. That is why it has been used for buried pipe and sewer lines for many years. Older aeration systems often used cast iron pipe for aeration because it was readily available, corrosion resistant, and in common usage for process piping. Ductile iron pipe and fittings have much heavier walls and higher pressure ratings than needed for aeration systems. Ductile iron pipe is not generally the most economical material for distribution piping in new construction.

In most new facilities and upgraded aeration systems, stainless steel has become the material of choice. Stainless steel includes alloy metals, principally chrome and nickel, which impart corrosion resistance. Stainless steel is more expensive per pound than carbon steel or ductile iron but the higher cost per pound is offset by using thin wall pipe—either Schedule 10 or Schedule 5. Thin wall pipe would be unsatisfactory when fabricated from carbon steel since rust could easily penetrate the thin pipe walls and create leaks. The corrosion resistance of stainless steel eliminates this concern.

There are many varieties of stainless steel available. A common reference is "18-8" stainless steel—referring to the percentages of chrome and nickel and identifying austenitic stainless steel. However, it is better practice to refer to the various alloys by the American Society for Testing and Materials (ASTM) designations. The most common material for aeration piping is ASTM-304L, although ASTM-316L is also used. The ASTM-316L has greater corrosion resistance and a higher cost. Since ASTM-304L provides more than adequate corrosion resistance for most aeration applications there is little justification for the more expensive grades in aeration service.

Joints between sections of pipe and fittings are typically made with welded or flanged connections. If welded connections are employed, it is imperative to use "L" grades of stainless steel. The "L" designation indicates low carbon. These grades are formulated to prevent corrosion in the heat-affected zones immediately adjacent to welds. In normal grades of stainless steel, the heat of welding causes the chrome and nickel to form carbides in the heat-affected zone, preventing them from imparting the needed corrosion resistance. The low carbon grades reduce the carbide formation and reduce the corrosion potential at welds.

Another measure employed to ensure the corrosion resistance of stainless steel is passivation. The stainless steel fabrication is dipped in a bath, generally nitric acid. This removes any carbon steel particles that may be on the surface to prevent them from initiating corrosion. The passivation also creates or enhances the impervious oxide layer that imparts the corrosion resistance to stainless steel.

Stainless steel pipe is used for air mains and laterals, and is usually extended to a point below the water surface in drop legs. This is due to the high air temperature from the blower discharge. Metal piping can easily withstand these temperatures but many nonmetallic pipe materials cannot. Stainless steel pipe is used to the point where cooling of the pipe and air stream brings the mean pipe wall temperature to a safe level and nonmetallic (plastic) pipe may be safely used.

There are a wide variety of nonmetallic pipe and fitting materials available. Aeration systems generally employ one of the following:

- PVC—polyvinyl chloride
- CPVC—chlorinated polyvinyl chloride
- HDPE—high density polyethylene
- FRP—fiberglass reinforced plastic (commonly referred to as "fiberglass")

All plastic pipe materials have excellent corrosion resistance, smooth interior walls for low friction losses, and are light weight. They are limited in service temperature, and therefore should not be used on the blower discharge in locations where the heat of compression results in air temperature above the recommended service temperature. Because of the low pressure involved in aeration systems, pressure rating is not an issue. Note, however, that nonmetallic pipe should not be used in high-pressure air or gas applications such as shop air. Manufacturer's data sheets should be consulted for application restrictions.

The most common material for diffuser grids and submerged aeration piping is PVC. For most systems Schedule 40 is selected. If threaded connections are required Schedule 80 PVC should be used, since the threads reduce the mechanical strength of Schedule 40 below an acceptable level. PVC pipe is commonly joined to fittings using solvent cement which is similar to adhesive bonding. The joint area should always be prepared with primer before applying solvent cement to provide maximum joint integrity. Flanges are available to connect pipe sections, valves, and pipe of other materials. The use of "Van Stone" flanges, which have an adjustable ring, are recommended because they eliminate concerns about bolt-hole alignment. The standard temperature rating of PVC pipe is 140 °F.

For slightly higher temperature rating, CPVC pipe is recommended. It has a maximum normal service temperature of 200 °F. CPVC fittings joints, and flanges are similar to PVC fittings. CPVC and PVC fittings and pipe should never be joined to each other with solvent welding.

PVC and CPVC pipe dimensions match the steel pipe dimensions for corresponding schedules and sizes. This includes pipe flanges, which have bolt-hole patterns matching standard 150 lb steel and cast iron flanges.

Fiberglass pipe has excellent corrosion resistance, mechanical strength, and stiffness. It is generally manufactured by winding the resin impregnated fiberglass over a mandrel and curing. There are two resins in common use—polyester and epoxy. Polyester is the least expensive and most common. Fittings and flanges are available in both materials. Fiberglass is not commonly found in aeration systems because of the high cost of pipe and fittings. The cost is aggravated by joining techniques which usually require tapering the pipe ends prior to adhesive bonding with epoxy adhesive.

Many lagoon systems use HDPE pipe for laterals spanning the lagoon width. HDPE pipe is flexible and light, making it the material of choice for floating laterals. Most lagoons are shallower than conventional aeration basins, which reduces blower discharge pressure and air temperature. HDPE pipe is usually rated for 140 °F, but the manufacturer's data should be consulted for all applications. Joining is usually accomplished by heat fusion, with the two ends of the pipe heated and then forced together by specially designed devices.

Aeration piping installed outdoors must withstand exposure to low temperatures and potentially damaging UV light in sunlight. This is not a problem for metal pipe but is a consideration for plastics. Protection from UV is generally incorporated into the plastic during manufacturing. For example, HDPE pipe generally contains carbon black to inhibit UV effects. Low temperatures can also result in brittleness and make the piping system prone to mechanical damage.

Aeration grids are submerged under water during normal service, which provides protection from UV and cold temperature. However, tanks are often taken out of operation to permit service or eliminate wasted energy from excess process capacity. It is common to keep 1–2 feet of clear water over the diffusers and aeration grids to maintain UV protection. If the climate can cause the water to freeze then there is the possibility of physical damage to the diffusers and the grid pipes. Some operators maintain a minimal air flow through the grid during cold weather to prevent this. In all cases, the diffuser manufacturer should be consulted for the proper preventive actions.

In addition to the pipe and fittings, there are a variety of elastomers used in piping systems. Elastomers are used in expansion joints, flange gaskets, O-rings and other sealing members, and valve seats and valve stem seals. Over time, the elastomers can degrade from heat, UV exposure, and ozone in the air stream. The degradation is accelerated by elevated temperature. There is also a minimum service temperature for most elastomers, below which they become brittle and loose elasticity. Commonly used elastomers such as natural rubber, SBR (styrene butadiene), Buna-N, and neoprene are generally unsuitable for use in aeration systems. Aeration applications require higher temperature ratings. The most commonly used elastomers are EPDM, Viton® Fluoroelastomer, and silicone rubber compounds (Table 6.5).

TABLE 6.5 Typical Temperature Ratings of Common Elastomers

Elastomer	Minimum Temperature, °F	Maximum Temperature, °F
Red rubber	−20	+160
SBR	−20	+170
Neoprene	−20	+170
Buna-N	−20	+170
EPDM	−30	+250
Viton	−20	+400
Silicone rubber	−70	+500

Flange gaskets may be full face—with the gasket ID and OD matching those of the flange. Full face gaskets have bolt holes. The gaskets may be ring gaskets—which are cut so the gasket OD fits inside the joint bolts. For the pressures in aeration systems, ring gaskets are adequate. It usually is not necessary to provide a gasket for flange joints on each side of a butterfly valve. Most butterfly valves include a raised surface on the seat that functions as a gasket.

Thermal insulation is often applied to air piping—particularly within the blower room where piping close to floor level exposes operators to the full discharge air temperature from the blowers. The pipe surface will often exceed 200 °F. Personnel comfort and safety may require insulation below 8 feet above grade. Thermal insulation will also reduce noise from air movement and may be continued on all piping within the blower room.

6.2 PRESSURE DROP

One of the most frequently encountered concerns in piping system design is the frictional pressure drop that air flow creates in the piping between the blower discharge and the diffusers. The pressure drop is identified as Δp_f in Equation 6.2. Once a pipe size has been established using the nominal velocity, the actual pressure drop should be calculated to verify that the blower discharge pressure capability will be adequate.

There are a variety of methods available for calculating the pressure drop in air flow. The classic approach includes calculating the Reynolds Number and using the Darcy–Weisbach equation in conjunction with a Moody diagram. However, a more convenient method includes direct calculation of the pressure drop from known parameters. For clean steel pipe the pressure drop can be calculated from:

Equation 6.11

$$\Delta p_f = 0.07 \cdot \frac{Q_S^{1.85}}{d^5 \cdot p_m} \cdot \frac{T}{528} \cdot \frac{L_e}{100}$$

where

Δp_f = pressure drop due to friction, psi
Q_S = air flow rate, SCFM
d = actual pipe inside diameter, in.
p_m = mean system pressure, psia
T = air temperature, °R
L_e = equivalent length of pipe and fittings, ft.

Equation 6.12

$$p_m = p_{initial} - \frac{\Delta p_f}{2} = p_{final} + \frac{\Delta p_f}{2}$$

where

p_{initial} = system pressure at the beginning of the pipe section, psia
p_{final} = required system pressure at the end of the pipe section, psia

The pressure drop using this equation can be solved by iteration: entering an assumed value of p_m, solving for Δp_f, and repeating until the change in Δp_f is negligible. The required level of accuracy determines the number of iterations. There are many other methods for calculating pressure drop, some of which don't require iteration. Many of these methods require more effort and additional steps which make them less convenient. This is particularly true if software is available for the iterative solution of Equation 6.11.

The "equivalent length" in Equation 6.11 is used to account for the pressure drop of air flowing through the fittings—elbows, tees, and so on—in the piping system. The pressure drops through valves and fittings may be several times as high as the losses through the piping itself. The simplest way to include the restriction caused by fittings is to use the length of straight pipe that will result in the same pressure drop as the fitting. The equivalent length is given as the ratio of length to diameter (*L*/*d*) and converted to length (Table 6.6):

Equation 6.13

$$L_e = \text{Ratio} \cdot \frac{d}{12}$$

where

L_e = equivalent length of fitting, ft.
d = nominal pipe diameter, in.
Ratio = *L*/*D* from Table 6.6

The total equivalent length for a segment of the piping system is the sum of the actual pipe lengths and all of the equivalent lengths of pipe for the fittings in that segment. Each segment analyzed should be a single diameter. Pressure drops for each segment in series should be totaled for a given run of piping.

Most piping systems consist of a network of segments. This usually includes similar branches in parallel providing air to individual aeration basins. Typically the

TABLE 6.6 Equivalent Length of Pipe Fittings

Description	Ratio, *L*/*D*
90° standard elbow	30
45° elbow	16
Tee, flow through run	20
Tee, flow through branch	60
BFV, 100% open	20
Transition	20

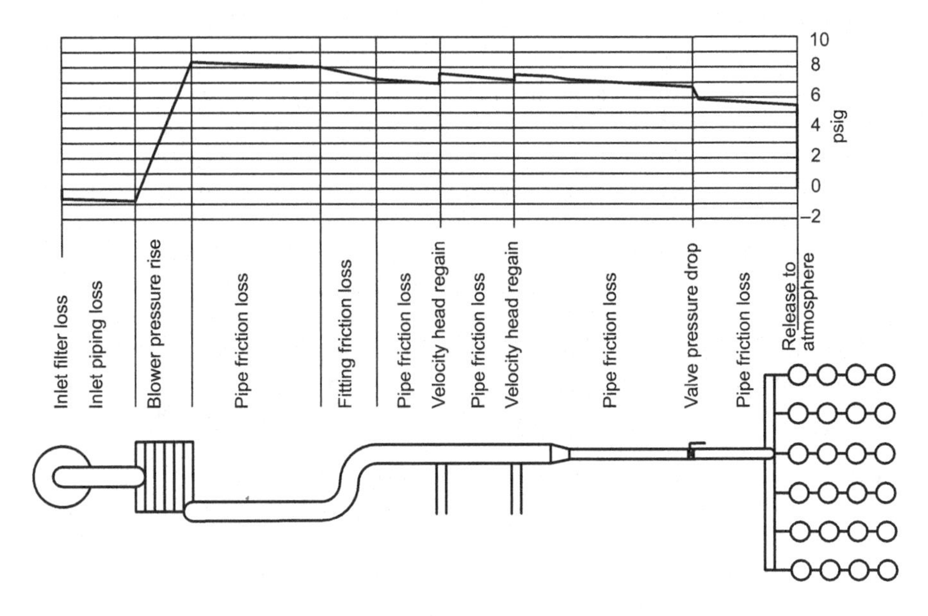

FIGURE 6.5 *System pressure gradient*

total pressure drop is calculated for the longest run of pipe—or for the most complex segments—using the highest anticipated flow rate for each segment. The total pressure drop is added to the pressure drop for the diffusers and the static pressure to determine the worst-case discharge pressure for the blowers (see Figure 6.5).

It is useful to develop a system curve for the aeration piping system. This can be calculated from the pressure drop at a single flow rate using Equation 5.14 and then calculating the total pressure at various flow rates using Equation 5.15.

Parallel piping branches behave similarly to parallel resistors in an electrical system. The total pressure drop between the junction point of the sections and the discharge point will be identical. The total flow will equal the sum of the flow through each parallel segment. Unlike electrical resistors, the flow through each branch will be proportional to the square root of the resistances to flow. For segments in parallel, the proportion of flow passing through each can be calculated from:

Equation 6.14

$$Q_1 = Q_2 \cdot \sqrt{\frac{k_{f2}}{k_{f1}}} = \frac{Q_{total}}{1 + \sqrt{k_{f1}/k_{f2}}}$$

where

$Q_{1,2,total}$ = air flow rate through segment 1, 2 and total flow, CFM
$k_{f1,f2}$ = constant of proportionality for friction losses in each segment, psi/CFM^2 (from Equation 5.14)

For two segments in series, the flow through each will, by definition, be equal.

It may be useful to calculate the total constant of proportionality for segments in series or parallel to use in determining the system curve. For segments in series:

Equation 6.15

$$k_{\text{f total}} = k_{\text{f1}} + k_{\text{f2}}$$

For segments in parallel:

Equation 6.16

$$k_{\text{f total}} = \frac{1}{\left(1/\sqrt{k_1} + 1/\sqrt{k_2}\right)^2}$$

It is occasionally necessary to determine the pressure drop resulting from air flow through an orifice. One example of this is evaluating the control orifices used in diffusers to compensate for level differences. The pressure drop for air flow through an orifice is a function of the diameter and flow rate:

Equation 6.17

$$\Delta p_{\text{o}} = \frac{\rho \cdot Q_{\text{a}}^2}{992.2 \cdot C_{\text{o}}^2 \cdot d^4}$$

where

Δp_{o} = pressure drop through orifice, psi
ρ = density at air flow conditions, $\text{lb}_{\text{m}}/\text{ft}^3$
Q_{a} = air flow rate, ACFM
C_{o} = orifice coefficient, dimensionless,
= 0.60 for sharp edged orifice
= 0.80 for knife edge beveled orifice
d = orifice diameter, in.

Designers strive for accuracy in calculations and system modeling but precise determinations of pipe friction are not possible. The level of accuracy achievable in pressure drop calculations is generally no better than ±10%. This is partly due to the complicated physics involved, but even more to the many assumptions implicit in the calculations. Pipe roughness, fluid viscosity, and air temperature are just a few of the parameters that are both subject to error in the initial assumed values and subject to change throughout the life of the system. The best practice is to make assumptions based on worst-case conditions and then provide sufficient adjustability in the system to accommodate variations.

6.3 CONTROL VALVE SELECTION

Control valves are employed to isolate a segment by shutting off air flow completely. They are also used to modulate the flow to a pipe segment by throttling the air flow. Throttling is essentially the creation of a variable restriction in the air line, resulting in a pressure drop across the control valve. This is a parasitic pressure drop and increases the energy requirement of the blowers. Minimizing the amount of throttling should be one objective of the aeration control strategy. However, throttling may be an economical method for controlling centrifugal blower flow rate. Further, throttling is the only practical way to control the distribution of air between multiple points in the process.

Two types of valves are normally used for throttling aeration systems. For small diameters, generally 3 inch nominal or less, ball valves or plug valves are used. These are economical and provide a reasonable flow control range. They are available with threaded connections—the most common joining method in small diameter air pipes.

For larger control valves the choice is generally BFVs. They are economical, available in a wide variety of materials, and provide a selection of form factors. Generally butterfly valves perform well in both shutoff and throttling service. Specialized control valves, such as diamond port knife gate valves, are encountered in aeration control system. Higher cost makes their use uncommon.

The three principal metal components in control valves are the body, disk (or ball) and the stem. Cast iron is the material of choice for bodies and disks. It is corrosion resistant and durable, with a service life measurable in decades. Some designers are tempted to use stainless steel bodied valves with thin wall stainless steel piping. While this has a certain intellectual appeal, there are no practical advantages to justify the high cost of stainless steel valves. The life of a cast iron valve body will exceed the useful life of the aeration system.

Valve bodies for flanged connections are generally available in wafer style or lug style. Wafer style bodies are generally less expensive and more convenient to install. They are more common in air flow control service than lug style. Wafer style valves are designed to be trapped between two flanges, and usually have a few alignment holes for the flange bolts to pass through. The remaining flange bolts pass across the outside diameter of the valve body.

Lug style valves have threaded holes for all flange bolt holes. Threaded rod is installed in the tapped lugs and nuts are used to trap the valve between the flanges on each side. This construction has the advantage of allowing dead end service. That means the butterfly valve is held tightly to a flange on one side if the flange and piping on the other side is removed. It should be noted that exposure to sunlight for extended periods will cause degradation of the seats and seals. For long-term service a blind flange should be installed to protect the valve elastomers.

The disk of butterfly valves and the balls and plugs for smaller valves are available in a variety of materials. These control members are only exposed to the working fluid—low-pressure air. Therefore, the manufacturer's standard materials of construction are generally acceptable. For butterfly valves the disks may be bronze, cast iron, or ductile iron. For some manufacturers the standard disk is uncoated, and for other suppliers plated, epoxy-coated, or vinyl-coated disks are standard.

Undercut disks are recommended for air flow control service. In this modification of standard valves the outside diameter of the disk is machined. This reduces the breakaway torque required to move the valve away from the fully closed position. For valves in low-pressure air service, the breakaway torque is the highest torque requirement. Use of the undercut disk may reduce the size and cost of the motor operators (Chapter 8). Undercut disks have lower bubble-tight seal pressures than full disks, but are rated well above the pressure requirements of most aeration systems.

The disk of a butterfly valve will protrude past the faces of the valve body when the valve is opened. There is normally clearance between the disk and the inside diameter of adjacent pipe and fittings, so disk movement is not impeded. However, there are infrequent occasions where the inside diameter of the adjacent system component is less than the outside diameter of the disk. This creates interference and prevents full disk movement. This occurrence is most common when an expansion joint rated for high pressure is installed on one side of the valve. These joints may have an inside diameter smaller than the normal pipe and fittings. The possibility of interference should be examined in this arrangement. If necessary a short pipe spool should be placed between the valve and the expansion joint.

PVC and other nonmetallic ball and butterfly valves are available, but the air temperatures in most systems are above the safe operating range for these materials. There are also high-performance butterfly valves available. These were originally designed for pulp and paper or chemical process service. These have corrosion-resistant metal construction, such as stainless steel, and usually have PTFE (Teflon) or stainless steel seats. They don't provide functional or service life advantages to offset their higher cost and are not recommended for aeration systems.

The valve stem operates the disk and is exposed to atmospheric corrosion as well as the process fluid. Stainless steel is recommended for stems. Pitting or rusting from corrosion will affect the structural strength of the stem and can cause premature failure of the stem seals.

The most critical selection in most aeration control valve applications is related to seat and seal material. Elastomers are used to seal the disk for shutoff service, to seal the stem to prevent leakage to the outside, and to seal between the valve body and adjacent pipe flanges. The temperature rating must be compatible with the process air temperature. The ratings in Table 6.5 are generally applicable to valve materials as well however, as always, the specific manufacturer's data must be verified. Long-term exposure of elastomers to excess temperature will cause the material to harden and loose its resilience. This increases operating torque, which may lead to motor operator failure. The loss of resilience also prevents proper sealing for shutoff and isolation.

There is a category of butterfly valves used for throttling air flow that does not include resilient seats. These valves have what is referred to as a "swing-through" disk. They are also called "blast gates" or occasionally "dampers." The disk is manufactured with a diameter slightly smaller than the metal inside diameter of the valve body. The small clearance makes this type of butterfly valve unsuitable for shutoff service, but they are useful where only throttling for control of air flow is needed. An example application includes inlet throttling of blowers with individual intake filters. Another application is installing an inexpensive swing-through valves in an upgrade to a system that already has resilient seated valves for shutoff.

Many applications in wastewater treatment plants specify American Water Works Association ("AWWA") butterfly valves. The AWWA developed the C504 standard to ensure adequate service life for buried valves in high-pressure water lines. The AWWA valve is generally longer length and employs heavier construction and more expensive materials compared to standard butterfly valves. For low-pressure air service, the features of AWWA valves do not provide longer service life and the additional expense is unjustified.

The actual effect of opening and closing a specific air flow control valve must be examined in relationship to the rest of the blower and piping system and the operating characteristics. For example, throttling a valve between the blower and the common discharge header will increase the blower pressure. If the blower is a centrifugal type, the air flow will be decreased and power demand may drop. If the blower is a positive displacement type, the air flow will be unchanged and the power consumption will increase. Throttling the blower discharge won't affect the relative proportions of air flow delivered to multiple aeration basins.

If several basins are in parallel off a common air system, throttling the flow control valve to one will usually affect the air flow to all basins—as well as the blowers. The air flow through the valve being closed will decrease because of the increased pressure loss through the valve. The affect on the air flow to the other basins will depend on the nature of system supplying air to the basins:

- If the blower is a manually operated PD, the other basins' air flow rate will increase by the same amount the first basin's air flow decreases. The blower discharge pressure will increase because the total system restriction increases.
- If the blower is a manually controlled centrifugal blower, the air flow to the other basins will increase. However, the increased restriction and higher back pressure will slightly decrease the total blower air flow. Therefore the increase will not exactly equal the decrease in the first basin's air flow.
- If the system pressure is essentially constant, there will be little change in the second basin's air flow rate. This will occur in large systems or if an unusually effective automatic control system maintains constant pressure. Note that in order for the pressure to remain constant with the increase in total system resistance, the blower air flow rate must decrease.
- If the blower control system is coordinated to maintain system air flow in response to demand, the blower discharge air flow rate will be decreased to match the drop in air flow from the throttled basin valve. This will also maintain the original system pressure.

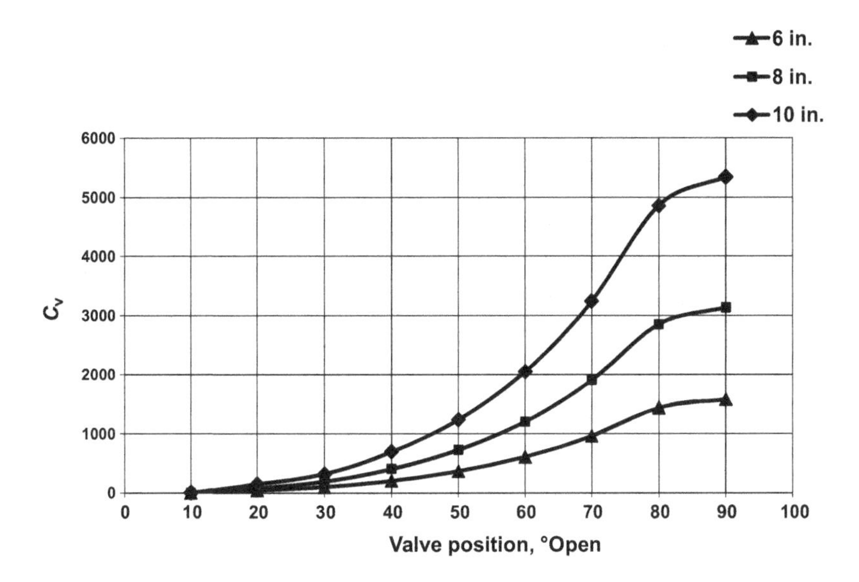

FIGURE 6.6 *Typical butterfly valve C_v*

These effects are greater in small systems. The impact of a single valve movement is less in large systems with many takeoff and control points.

One of the problems in control logic development is to understand the effect of basin throttling changes on the entire system and incorporate logic that will eliminate—or at least minimize—these effects. This involves coordinating all of the basin air flow control valves and the blower control into a coherent strategy. Like so many other aspects of aeration control, the effect of valve position on air flow is nonlinear. This complicates the control algorithms.

The pressure drop characteristic of a valve is expressed in terms of "C_v". This is established by the manufacturer using test data for a specific valve (see Figure 6.6). By definition, C_v is the flow rate of water, in gpm, that will pass through the valve with a 1 psi pressure difference across the valve. For air flow calculations, the pressure drop also includes the effects of temperature and pressure because they impact the air density. The pressure drop is calculated from:

Equation 6.18

$$\Delta p_v = \left(\frac{Q_s}{22.66 \cdot C_v}\right)^2 \cdot \frac{SG \cdot T_u}{p_u}$$

where

Δp_v = pressure drop across the valve, psi
Q_s = air flow rate, SCFM
C_v = valve flow coefficient from manufacturer's data, dimensionless

SG = specific gravity, dimensionless, =1.0 for air
T_u = upstream absolute air temperature, °R
p_u = upstream absolute air pressure, psia

In most aeration systems, the valves operate in a narrow range of upstream pressure. Since the downstream pressure is primarily the result of static pressure from diffuser submergence, the pressure drop through the valve also operates in a narrow range.

In many cases, the pressure difference through the valve is known, and it is necessary to determine the flow through the valve:

Equation 6.19

$$Q_s = 22.66 \cdot C_v \cdot \sqrt{\frac{p_u \cdot \Delta p_v}{SG \cdot T_u}}$$

In other cases, it is necessary to determine the valve position corresponding to a known pressure drop and flow. The C_v is calculated, and then the valve position with the corresponding C_v is determined from the manufacturer's data:

Equation 6.20

$$C_v = \frac{Q_s}{22.66} \cdot \sqrt{\frac{SG \cdot T_u}{p_u \cdot \Delta p_v}}$$

Because of the nonlinearity of control, the air flow rate through a butterfly valve is very sensitive to movement when the valve is nearly closed. A very small change in position will result in a dramatic change in flow rate. Conversely, when a butterfly valve is in a full open position, a change in position will result in very little change in flow rate. The normal range of effective valve movement is between approximately 20 and 70% open. Control below that percentage tends to be erratic and above that control tends to be ineffective (see Figure 6.7). It's recommended that the control system limit basin butterfly valve travel to this range—with the obvious exception that 0% is necessary for shutoff and isolation.

A commonly encountered piping problem is sizing flow control valves to match the diameter of oversized pipe. This is particularly a problem when an existing aeration system and piping is retrofitted with more efficient diffusers. The original pipes were most likely sized conservatively for worst-case maximum air flow rates based on the original diffusers. When the system has the operating air flow rates reduced by the more efficient diffusers, the flow rates through the control valves may become quite low. This tends to push the operating valve travel below the minimum allowable range, causing erratic control.

The valve sizing should always be checked against anticipated flow ranges for both new systems and upgrades. If necessary, a section of pipe should be replaced with smaller diameter pipe and valve to ensure adequate control. While this may

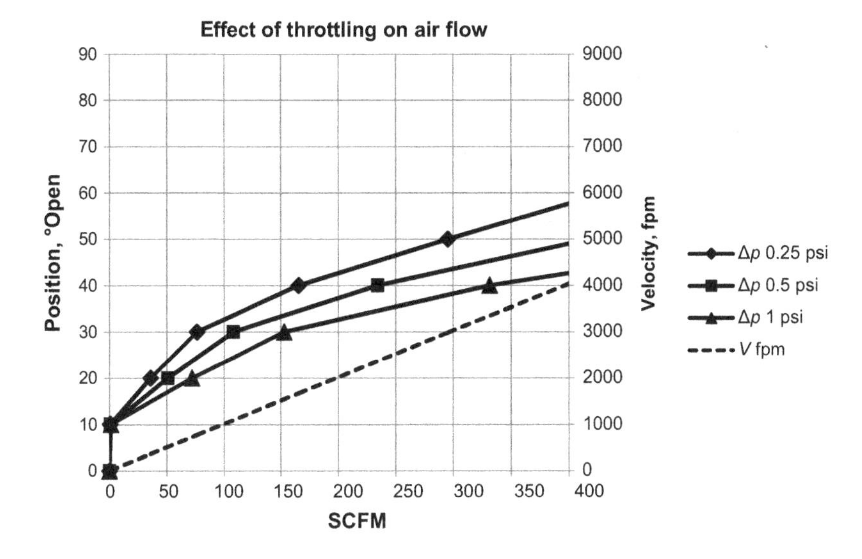

FIGURE 6.7 *Four inch (4 in.) butterfly valve throttling performance*

increase the initial cost of the control system, it may also prevent the control being inoperable and abandoned because stable control cannot be achieved.

It is obvious that the details of piping system design are complex and numerous. It isn't necessary for the control system designer to be aware of all of these details. A qualified specialist should be enlisted for final design. However, the piping system can have a tremendous impact on the success of the aeration control system. It is essential that the system designer be aware of some of the special considerations. It is also important to have sufficient familiarity with piping systems to contribute to the design process and be able to discuss the concerns of the piping system with the specialists.

EXAMPLE PROBLEMS

Problem 6.1

Refer to the piping layout in Figure 6.8. The plant is at an altitude of 1000 ft ASL. Ignore friction losses and assume equal air flow to all diffuser drop legs. Calculate the velocity head and the static pressure at points "A" and "E" if two blowers are operating at rated capacity. Determine if the effect is significant.

Problem 6.2

An orifice is to be installed in fine-pore diffusers to compensate for errors in the installed height of a diffuser. This will be done by introducing $\frac{1}{2}$ in. of additional

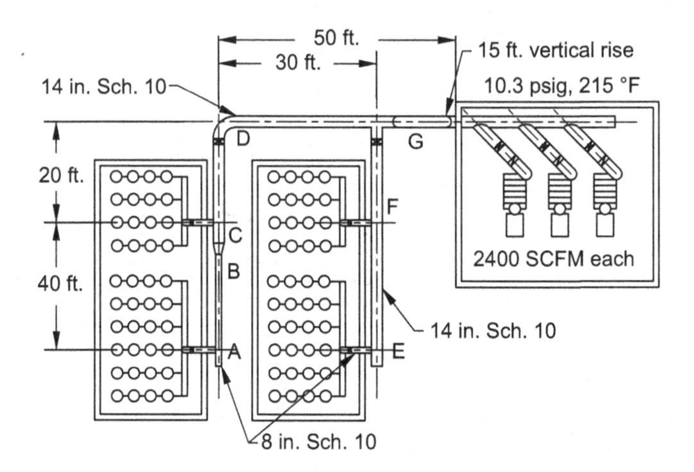

FIGURE 6.8 *Problem 6.1 piping layout*

pressure drop at minimum flow. The nominal depth is 18 ft. The flow rate through each diffuser will range from 1.0 to 4.0 SCFM and the nominal air temperature at the diffuser face is 50 °F.

(a) What is the recommended orifice diameter?

(b) What is the total pressure drop at minimum and maximum air flow rates if the pressure drop through the membrane itself is shown in Figure 4.18?

Problem 6.3

Refer to Problem 6.1. Calculate the pressure drop from friction between the blower room and points "A" and "E". Assume the blowers are at maximum capacity, all valves are full open and the air flow is split equally among all four grids.

Problem 6.4

The existing 8 inch butterfly valves for the diffuser drop legs in Figure 6.8 are being considered for replacement. The control valve is limited to travel between 20° and 70° open. The C_v for a 6, 8, and 10 in. nominal valve are shown in Figure 6.6 and are tabulated below:

°Open	6 in. C_v	8 in. C_v	10 in. C_v
20	45	89	151
30	95	188	320
40	205	408	694
50	366	727	1237
60	605	1202	2047
70	958	1903	3240

The upstream temperature is 215 °F and the upstream pressure is 10.1 psig. The flow will vary between 600 and 2400 SCFM. Static pressure is 7.8 psig and the pressure drop for the diffusers and grid piping is shown in the table below:

SCFM	Δp, psi
600	0.50
1200	0.77
1800	1.03
2400	1.30

Calculate the valve positions for each of the four flow rates and make a recommendation for the valve size.

Problem 6.5

The system in Figure 6.8 has the drop leg valves controlled manually. The air flow rate at point "D" is 2400 SCFM. The grid connected to point "C" has 400 diffusers with an estimated pressure drop through the butterfly valve, drop leg, grid piping, and diffusers of 0.75 psi at 900 SCFM. The grid connected to point "A" has 600 diffusers with an estimated pressure drop (including the above plus the piping) between "C" and "A" of 0.90 psig at 1200 SCFM. The static pressure is 7.8 psig. Ignoring velocity head regain, how much air flow will go to each grid if the butterfly valves are not used for throttling?

Problem 6.6

The 60 ft long pipe section from "D" to "A" in Figure 6.8 is subject to a worst-case temperature change of 225 °F between being out of operation in winter and maximum discharge air temperature in the summer. The pipe is anchored at point "D."

(a) How much movement will be expected at point "A" if the pipe is stainless steel?
(b) How much movement will be expected at point "A" if the pipe is fiberglass?

Chapter 7

Instrumentation

You can't control what you can't measure.

Regardless of the control logic employed, the first step in any automatic control procedure is measurement of the process variables of concern. Only then can the system implement control actions intended to prevent problems with the process or deviations of the controlled variable.

Instrumentation is essentially information conveyance. The process variable must be measured, the information about the measured parameter converted to a transmittable signal, and the signal received by a controller or display device. This device converts the information into a format useable by the operator or by the controller.

That description of the function of instrumentation seems quite straightforward. However, the combinations of measured variables, the sensors used, and the methods of information conveyance are nearly infinite. Some instruments find broad use across many industries. Others are very process specific and find use only in wastewater treatment.

Further, the instrumentation field is very dynamic. New technologies for measurement, transmission, and display of process data are constantly being introduced. It is important, therefore, to understand the underlying principles of instrumentation so the control system can optimize the application of existing technology and employ new technology as it becomes commercially available.

Aeration Control System Design: A Practical Guide to Energy and Process Optimization, First Edition. Thomas E. Jenkins.

There are only seven fundamental units of measurement:

- Mass, kg
- Length, m
- Time, s
- Current, A
- Temperature, °K
- Amount of a substance, mol
- Luminous intensity, cd

All other units of measure, whether metric (SI, International System of Units) or US customary units, are derived from these fundamental measurements. The US customary units are generally defined in relation to the SI base units.

This text primarily uses the US customary system of units as a convenience—the units used in the text are those most commonly encountered in the wastewater treatment field in the United States. Note, however, that the US customary units are far from "pure." Motors are rated in horsepower, for example, but the electricity is paid for in kilowatt-hours. In fact, common usage in most countries (including many of those nominally "all metric") is a hodgepodge of customary and SI units. Local preference dictates the units of measurement.

The eclectic use of units of measurement may be offensive to purists, but in point of fact it makes little difference pragmatically. The units that are most comfortable for the operators—the ones that most clearly and quickly convey information to them about the way the process is performing—are always the "right" units.

In formulas and calculations the units of measurement need to be consistent. They must also correspond to the units on which the formula is based. It is recommended practice to confirm the accuracy of any unit conversions by writing the units into the formula and canceling them algebraically to verify the units of the final result. Where the formula has been simplified for convenience, the intended units should be indicated. This caution is particularly pertinent to control system programming. The logic very often has only the required arithmetic operations and units of measurement are not indicated. Verification of units may be tedious or time consuming, but failure to verify consistency and compatibility can be disastrous!

Ultimately, the operator and the control system convert the measurement into one of three values: the measured variable is "OK," it is "too high," or it is "too low." For both the automatic control system and the operator, this information determines whether or not corrective action is required and in what direction the corrective action should move the measured variable.

7.1 COMMON CHARACTERISTICS AND ELECTRICAL DESIGN CONSIDERATIONS

The terms sensor, transducer, and transmitter are often, and incorrectly, used interchangeably. The distinctions between the three are not always precise and definitive, but in general the following apply:

- A sensor detects a physical quantity or characteristic.
- A transducer converts a physical quantity or characteristic to an electrical signal, usually a low energy level signal.
- A transmitter takes an electrical signal, usually a low energy level signal from a transducer, and amplifies or modifies it to a more standardized signal, often for transmission over long distances.

Based on these definitions, transducers contain sensors and transmitters contain transducers.

Sensors and transducers by themselves are seldom used in wastewater automation systems. Transducer output is usually measured in millivolts or microamperes. The physical distances involved, the existence of electrical interference from magnetic fields (electromagnetic interference, EMI), and the nature of controller input construction require that a transmitter of some type be used for virtually all process measurements. There are three types of sensor or transmitter output signals used in most control systems.

The first type of output is the "discrete" signal, so-called because it contains two discrete states—on or off. In the past this type of signal was called "digital" because it represented either on (1) or off (0) states. In today's control environment this causes confusion with digital communications systems and the term discrete is preferred. A pressure switch would be one example of a discrete output.

Discrete outputs can be divided into mechanical contacts and solid-state contacts. Mechanical contacts have the advantage of switching any voltage up to the contact rating. The rating is based on the voltage that can be switched without excessive arcing between the contacts, known as the breakdown voltage. Mechanical contacts are also rated on current, which is based on the maximum amperage that the contacts can handle without thermal problems. A common contact designation is NEMA A600. This specifies that the switching device is rated for up to 600 V and 10 A continuous current. NEMA A600 has separate ratings for make (closing) of 7200 VA and break (opening) of 720 VA. Mechanical contacts often have lower ratings for DC because of the greater tendency for arcing. Arcs between the contact points erode the contact surface and will eventually result in failure by either high contact resistance or "welding" the contacts closed. Diodes or resistor/capacitor circuits are often used in parallel with mechanical contacts to reduce arcing.

Mechanical contacts are further designated by the number of poles and contact arrangements. A very common arrangement is the single pole double throw (SPDT) or form "C" contact. This arrangement has one common connection to two sets of contacts. One set is normally open (NO) and one is normally closed (NC). The contacts change status when the switch is activated. The "normal" state refers to the switch or contact when it is not installed and not powered (see Figure 7.1).

Solid-state contacts have the advantage of being able to switch very rapidly and repeatedly without damage. They have the disadvantages of being restricted to a specific range of voltages, lower current limits, and in some cases of allowing slight current leakage through the contact in the off state. Transistor outputs or switches are

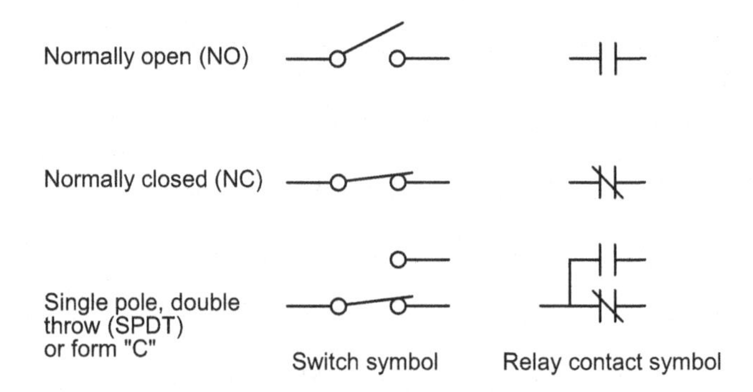

FIGURE 7.1 *Contact arrangements*

usually limited to 24 VDC. They are typically categorized as sinking or sourcing, although some combine sinking and sourcing capability. Sinking transistors (NPN type) are wired to switch current on the negative side of the circuit—after the load and sourcing transistors (PNP type) are wired to switch on the positive side of the circuit—before the load (see Figure 7.2). Many designers prefer sourcing switches, as this removes power from the load when the switch is "open." Triacs are used to switch AC and are typically limited to 120 VAC. Solid-state switches often have a minimum load requirement, and will not consistently operate if the load created by the switched device is below this value. In this case it may be necessary to add a resistor to the circuit.

Note that the solid-state DC inputs of many PLCs and controllers are also designated as sinking and sourcing. In a sourcing input the current flow is from

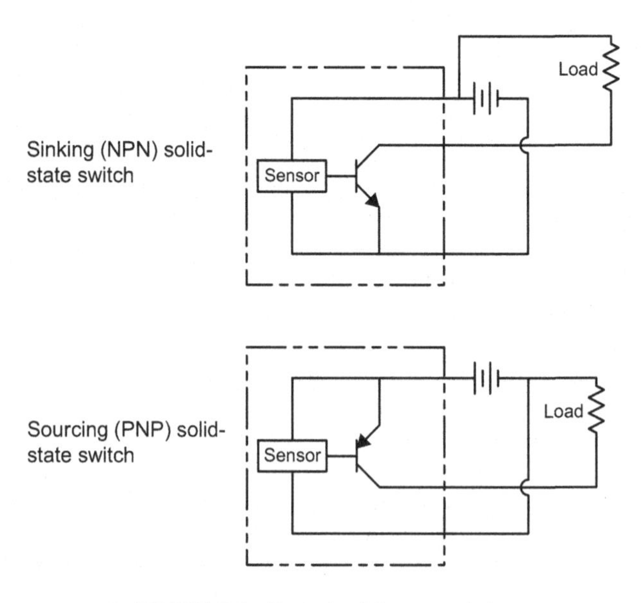

FIGURE 7.2 *Typical solid-state switches*

the power supply, through the input, through the switch, and back to the power supply. In a sinking input the current flow is from the power supply, through the switch, through the input, and then back to the power supply. The sinking type input is often preferred because it does not have voltage at either terminal in the off state, simplifying diagnostics.

Pulse outputs are a variation of discrete signal—where the frequency of the on/off pulse is proportional to a measured process variable. Two examples are using a proximity switch to pick up the rotational speed of a shaft and the output of a flow meter, where each pulse indicates that a specific volume of fluid has passed through the meter. Pulse outputs are often used when a totalizer is used to track daily flow volumes. A high-speed counter input is available with many controllers to convert the pulse frequency to analog process data.

The second type of transmitter output is the analog signal. This is the type of output most commonly employed for measuring process variables. The output of the analog transmitter varies the signal magnitude in proportion to the value of the measured variable. A flow transmitter is one example of an analog transmitter.

A variety of outputs are available, including 0–5 VDC, 0–10 VDC, ±10 VDC, 0–20 mA, and 4–20 mA. In the United States 4–20 mA current output is the most common, but in other areas 0–10 VDC is more frequently encountered. Pneumatic outputs with a range 3–15 psig are still used for some control systems, but they are increasingly being displaced by electrical signals.

Voltage output have a minimum load rating, which is based on preventing too much current passing through the transmitter. A common minimum load is 10 kΩ (Ohm). Current output transmitters generally have a maximum load specified, typically from 500 to 1000 Ω. This represents the maximum total resistance in the output loop that the transmitter can drive the signal through. In many cases, the current transmitter output is a function of the supply voltage.

A 4–20 mA current signal has the advantage over voltage outputs of greater noise immunity from EMI and RFI (radio frequency interference) from adjacent power wiring, motors, and switches. This makes current signals more suitable for long-distance transmission. The 4 mA value is transmitted at the low end of the process measurement range and 20 mA is transmitted at the high end of the range. The 4 mA is a 20% offset and represents a "live zero" in the signal. This provides the control system with the ability to distinguish between a zero process measurement at 4 mA and a failed transmitter or broken wire at 0 mA.

Analog transmitters are further categorized as two-wire, three-wire and four-wire (Figure 7.3). These categories represent the way the transmitter is powered. Two-wire transmitters have 4–20 mA outputs. They are generally powered with 24 VDC wired into the signal loop in series with the transmitter signal, which is why they are often referred to as "loop powered." This arrangement simplifies the wiring by eliminating separate power connections. The energy required to operate the transmitter electronics is derived from an internal resistor or diode, which creates a voltage drop to provide internal power. The two-wire transmitter has a minimum voltage requirement in order to create sufficient voltage for both the internal electronics and the output loop.

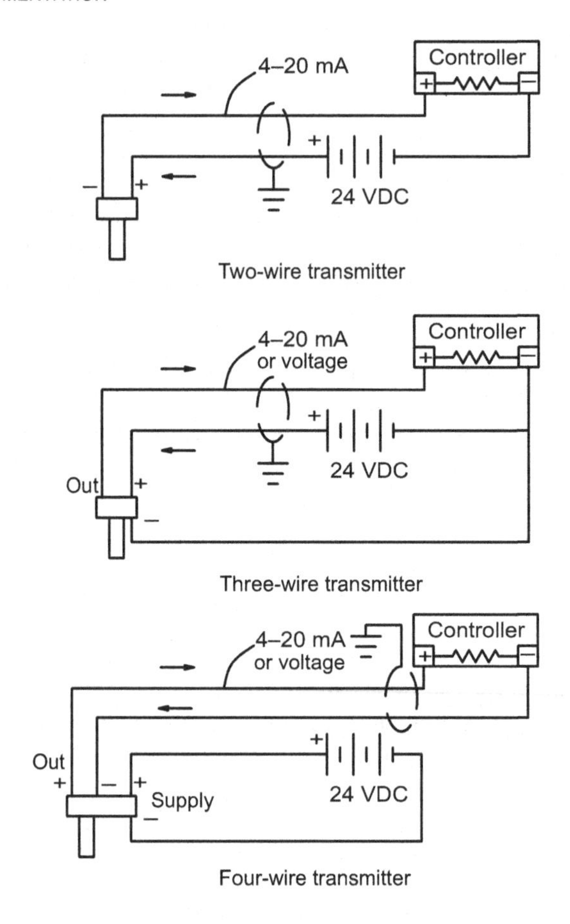

FIGURE 7.3 *Transmitter wiring*

Particular attention must be paid to loads and power requirements when a loop powered transmitter is used in conjunction with a loop powered indicator. The loop powered indicators generally have an internal diode to create a voltage drop that provides power for the indicator electronics. This voltage is generally 3 VDC or more. At 4 mA loop current, this creates a load on the transmitter output circuit equivalent to 750 Ω—which is at or near the maximum load for many transmitters. As a result the transmitter output may shut off at low process values.

For many process instruments the power available from the signal loop isn't sufficient to operate the instrument. Most analytic instruments, for example, fall into this category. In a four-wire transmitter there are two terminals for the output signal and two terminals for the external power supply. The external power is typically 24 VDC or 120 VAC. Three-wire transmitters are actually just a special case of DC powered four-wire transmitters, with one terminal for the positive side of the power supply, one terminal for the transmitter output, and one terminal common to the power and output signal negative connection.

It is often necessary to connect multiple devices to the output of a single transmitter. A flow meter output, for example, may be connected to a chart recorder, a digital indicator, and a controller or PLC input. If the analog signal is a voltage then the devices may be connected in parallel so they may all read a single transmitter output. If the transmitter output is current then the devices may be connected in series.

The third type of output is the digital communications signal. The information on the measured process variable is sent over a communications network. The network's communications medium, protocol, and topology can be simple or quite complex (Chapter 10). Digital communication systems can be more problematic to design and commission, but they are able to provide much more information about both the process and the transmitters themselves. The use of communications networks may also simplify the installation wiring by eliminating the need to wire the transmitter outputs individually back to the control system. A single communications link can connect all transmitters. The power for the transmitters can often be tapped into existing convenience outlet or lighting power at the basins. The more traditional analog and discrete transmitter outputs are gradually being displaced by digital communications.

One type of communications output, HART (Highway Addressable Remote Transducer), includes both digital communications and analog output signals on a single pair of wires. The analog signal is a 4–20 mA current output, which can be read by any conventional display or controller. Superimposed on the DC analog current signal is 1200 Baud digital data communications link using an open, standardized protocol. Because of their high frequency, the communications bits are readily filtered out and ignored by analog devices. Special controller communications modules or handheld HART communicators are used to read the extensive digital data transmitted by the processor. HART instruments are more expensive than conventional transmitters but they can be conveniently calibrated and rescaled using the communications capabilities.

Most process measurements are more complex than the seven basic measurement parameters and use units derived from these basic ones. For example, volumetric flow rate is a combination of length^3 and time. Pressure is a combination of force and length^2. Force itself is a unit derived from mass, length, and time using Newton's First Law. From a process control viewpoint this doesn't really matter—the key factor is that a process variable can be measured so the information can be used in manipulating the process to maintain desirable operating conditions.

Some sensors work more or less directly based on the measured parameter. For example, the millivolt output of a thermocouple is a direct function of the temperature. Other measurements must rely on indirect measurement techniques. They usually employ a law of physics or measurement of a secondary effect of the measured parameter on the sensor. One example would be measuring flow rate with an orifice. The actual measurement is the difference in pressure across the orifice and Bernoulli's law is used to calculate the flow rate from this measurement. Another example would be a resistance temperature detector (RTD), with the actual measurement being the resistance. The resistance changes as a function of temperature and is in turn measured by means of current and voltage.

Four terms important in defining transmitter performance are "accuracy, "resolution," "precision" and "repeatability." Each can be significant in control system performance and should be addressed separately.

Accuracy is the agreement between the transmitter output and the actual process variable. Several characteristics, such as deviation, linearity, deadband, hysteresis, and offset are included in the term. Accuracy is usually expressed as a percentage of full scale (FS) or of reading. For example, if a pressure transmitter has a specified accuracy of 0.5% of full scale and the full scale output is 15 psig then the possible error at an actual system pressure of 7.5 psig would be ±0.075 and the transmitter output could be between 7.42 and 7.58 psig. If the accuracy is defined as 0.5% of reading, the error at 7.5 psig would be ±0.038 and the output could vary between 7.46 and 7.54 psig.

Resolution is often confused with accuracy. Resolution references the smallest change in the measurement that can be transmitted or displayed. A pressure gauge that has a dial marked in 0.1 psig increments has better resolution than one marked in 1 psig increments. However, if the gauge with higher resolution is calibrated with a 0.5 psig offset from true zero it will not be as accurate as a lower resolution gauge calibrated correctly. The incorrectly calibrated gauge will merely provide an incorrect reading with more digits.

Considerations of resolution are particularly subject to misinterpretation when the transmitter output is connected to a PLC input. It is frequently assumed that a 16 bit input will be more "accurate" than a 12 bit input, for example. The difference may have very little practical effect, however. The 12 bit input will have a resolution of 1 part in 4096 ($2^{12} = 4096$), or 0.02% and the 16 bit input will have a resolution of 1 part in 65,536 ($2^{16} = 65{,}536$), or 0.001%. Very few process transmitters can provide accuracy better than ±0.1% so the additional resolution will not improve the control system performance. The error in reading exceeds the resolution. Additional digits in the PLC logic won't provide additional exactness in the process control performance or monitoring.

Precision indicates the agreement between outputs or readings of the same value of the process variable at successive readings with the process variable unchanged. It indicates the degree of stability of the instrument or transmitter.

Repeatability, like precision, refers to the ability of the measurement system to provide the same reading for identical values of the measured process variable when taken at different times. Repeatability may be a function of correct installation as well as the transmitter itself.

In most process control applications, good repeatability is more important than high accuracy. An operator may recognize that the process performs well when the value measured for a certain process parameter is at a specific reading. It isn't important if the "real" process value is greatly different from the display. If the process performs well at that reading, it is only important that the same set of process conditions produces that same measured value. On the other hand, if the measurement is not repeatable the operator's manipulation of the process to achieve that value won't result in consistent process performance.

Proper wiring is critical to instrument performance—particularly for low voltage analog and communications signals. Because of the low current in most analog loops, the conductor size is more likely to be dictated by mechanical strength than by voltage drop. Suppressing EMI- and RFI-induced voltage and signal noise is more of a problem. Lighting ballasts, motors, contactors, and adjacent wiring can all induce noise and error into instrumentation circuits. To minimize the impact, the following measures should be included in wiring system design:

- Use shielded twisted-pair wiring for analog signals.
- Ground all shields at only one end.
- Do not run analog or communications wiring in the same conduit as AC power and control wiring.
- Make all field terminations and splices with terminal blocks.
  - If necessary, crimp connectors may be used to splice wiring.
  - Never use wire nuts to splice analog signal or communications wiring.
- Verify the type of communications wiring required by the device manufacturer.

A common problem in field wiring is having reversed polarity at the controller input card. This is true for both analog and discrete I/O. The result can range from failure to function to physical destruction of the I/O card. Most, but not all, I/O have some level of over voltage, over current, and reverse polarity protection. Checking these ratings should be part of the selection process.

Ground loops are created when two or more devices in a current signal loop are tied to ground. If both a transmitter and a controller have the negative signal terminal grounded, and the two grounds are at slightly different voltage levels, some of the current will pass through the ground path and cause erroneous readings. A common situation is using a power supply with a grounded negative, a chart recorder with a grounded negative, and PLC input with a grounded negative. The first device in the loop after the transmitter may read correctly, but the signal will pass from that device directly back to the power supply through the ground loop, and not pass through the second device at all (Figure 7.4).

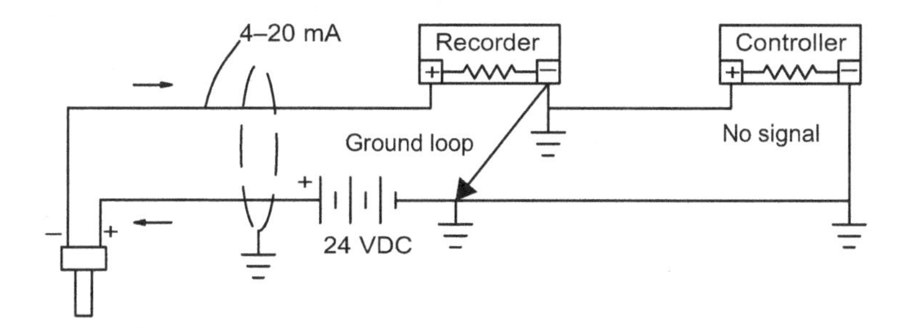

FIGURE 7.4 *Ground loop*

Signal isolators are used to eliminate ground loops. They are available in two types—transformer and optical. Most new designs use optical isolation, which creates a complete electrical separation between the input and output signals, and often between the two signals and the power supplies. In addition to eliminating ground loops, a signal isolator can also provide a protective barrier between the field instrument and the controller or other panel mounted components. This reduces the possibility of damage to the panel from surges, lightning strikes, and similar electrical disturbances.

There are a variety of related devices that combine isolation with other functions. Signal splitters provide two outputs that mimic the input value. They are useful if a single transmitter analog output signal is required at two instruments with internal grounds or at two widely separated locations. Signal conditioners are extremely useful in converting one type of analog signal into another. Examples include the following:

- 4–20 mA to 0–5 VDC or 0–10 VDC
- 0–5 VDC, 0–10 VDC, or 0–20 mA to 4–20 mA
- Thermocouple or RTD to 4–20 mA
- 0–5 A AC to 4–20 mA

Dropping resistors, which are simply standard resistors with close tolerances, are sometimes used to convert current signals to voltage. A common application is converting a 4–20 mA signal for use with a PLC having 0–10 VDC or 0–5 VDC analog inputs. By placing a 250 Ω resistor across the input terminals, a 1–5 VDC signal is created. The program logic must include the proper scaling to accommodate the 20% offset in converting the signal to engineering units.

The range of a transmitter represents the upper and lower values that the instrument can measure and transmit. The lower limit usually corresponds to the minimum output signal, and the upper limit, therefore, usually corresponds to the maximum output signal. The span is the total value between the upper and lower limits of the range—the algebraic difference between the two. The term full scale is usually used in reference to the maximum measured value or transmitter output value but may be referring to the span—particularly if the transmitter is zero based. If a transmitter's minimum measured value is not zero engineering units then the difference is said to be the offset. This can be best understood by comparing two temperature transmitters:

Transmitter	Range	Output	Span	Full Scale	Offset
A	0–300 °F	4–20 mA	300 °F	300 °F	0 °F
B	−50 to 250 °F	0–10 VDC	300 °F	250 °F	−50 °F

The transmitter output for any measured variable can be readily calculated (see Figure 7.5):

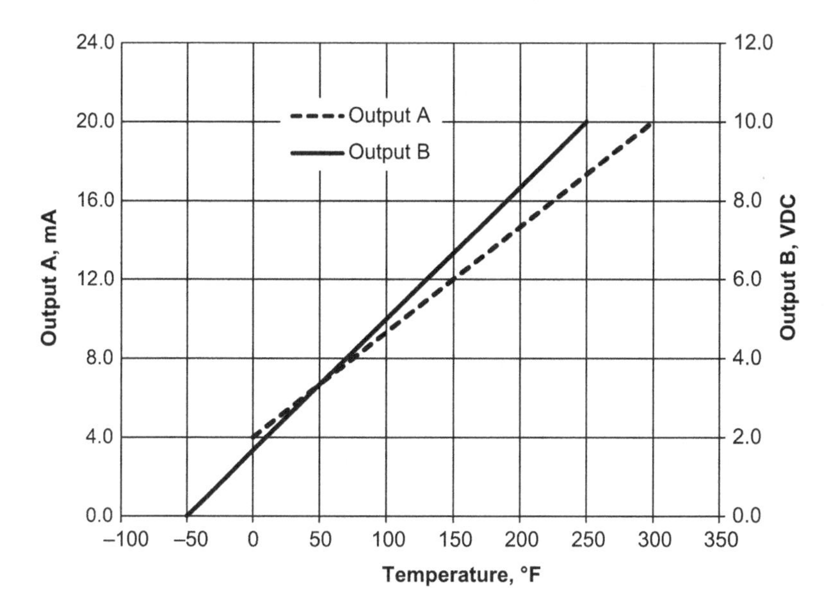

FIGURE 7.5 *Example transmitter output*

Equation 7.1

$$S_{\text{Out}} = \left[\frac{\text{PV} - \text{EU}_{\text{Min}}}{\text{EU}_{\text{Span}}} \cdot (S_{\text{Max}} - S_{\text{Min}})\right] + S_{\text{Min}}$$

Equation 7.2

$$\text{EU}_{\text{Span}} = \text{EU}_{\text{Max}} - \text{EU}_{\text{Min}}$$

where

$S_{\text{Out, Max, Min}}$ = Transmitter actual output signal, maximum signal, and minimum signal, signal electrical units
PV = Measured process variable, engineering units
$\text{EU}_{\text{Max, Min}}$ = Maximum and minimum engineering units measured
EU_{Span} = Engineering unit span

Initial calibration of transmitters is usually performed at the factory. Factory calibration should be performed with test instruments traceable to the National Institute of Standards & Technology (NIST) to ensure accuracy. Periodic recalibration, typically performed in the field, is necessary to maintain consistent accuracy. As a minimum, zero calibration should be performed regularly. A zero process measurement is usually easy to create. A correct output at zero doesn't ensure calibration is correct, but an incorrect reading at zero means measurements at other points in the range are incorrect. A full calibration usually includes creating the process measurement with special calibration tools and verifying the output signal at 0, 25, 50, 75, and 100% of the measurement range. Some instruments are not conveniently field calibrated and must be removed from service and sent to the factory or a facility with specialized equipment for recalibration.

Many transmitters include a local display of the measured process parameter, typically with a liquid crystal display (LCD) digital indicator integral with the transmitter enclosure.

Field instruments are increasingly categorized as "smart" transmitters. There is no universal definition of what constitutes a smart transmitter and the meaning is often whatever a manufacturer's marketing department wants it to be. In general, however, a smart transmitter will have most or all of the following:

- Local digital display
- A microprocessor and solid-state memory
- The ability to change ranges and output signals using a communicator, keypad, or pushbuttons
- Calibration capabilities using a communicator, keypad, or pushbuttons
- Digital communications capability—HART or other

Although most process transmitters have some characteristics in common, there are a tremendous variety of instruments available. Many measurements are more or less standard and are found in almost all process control applications and are used in most industries. Other measurements are specific to the wastewater treatment field and are rarely encountered outside of it. The aeration control system designer must be familiar with the general characteristics of the most commonly used types of instrumentation to ensure proper application and use.

7.2 PRESSURE

Pressure measurements have a wide variety of uses in monitoring, controlling, and protecting process equipment. In addition to the obvious use for measuring the force per unit area of a fluid in a pipe or vessel, pressure measurement is also an integral part of many flow and level measurement devices.

Pressure is always measured as the difference in pressure between two points. This is the source of one kind of classification of pressure instrumentation:

- Gauge pressure, for example, psig, is the difference between the pressure inside a pipe or vessel and the atmospheric pressure around it.
- Absolute pressure is the difference between the pressure in the fluid being sensed and a vacuum.

- Differential pressure measurement requires two sensing ports and measures the difference between these two points, for example, upstream and downstream of an orifice.
- Suction pressure is essentially the same as gauge pressure, but implies that the pressure inside the vessel is below atmospheric, such as in the suction piping of a pump or blower.
- Compound pressure measurement is also a special case of gauge pressure and implies that the measurement may be above or below atmospheric.

Pressure measurement relies on the pressure creating a force on some type of moving member, with the movement resisted by some type of opposing force. In the simple U-tube manometer, the moving column is opposed by the weight of the fluid on the other side of the manometer. Other sensing elements may be a moving piston or diaphragm with the opposing force created by a spring. The moving member may operate a variable resistor or trip limit switches. Traditional pressure gauges and switches use this method, with the diaphragm finding application primarily in low-pressure or low differential pressure applications.

In more sophisticated designs, the movement may be stretching or deforming a metal disk with inherent opposition to the forced deformation. Strain gauges or other sensing elements can turn the deformation into an electrical signal.

As with most process measurements, some applications require only local indication and others require switching for status indication that the pressure is above or below the acceptable value. These are adequate for simple equipment protection and alarming.

Pressure transmitters with an analog output signal are the most common instrumentation used in aeration control applications. The variety of mechanical and electrical configurations is tremendous. The traditional NEMA 4X round aluminum enclosure is used when local indication is required and is the style available for most smart pressure transmitters. Because of their familiarity and larger size, some designers assume that they are more accurate than the cartridge style. This is not always the case. The cartridge style transmitters can be very accurate. Their small size makes them more convenient for installations where local indication is not needed or impractical. The instrument specifications should be used to determine accuracy and suitability. The minimum specification points used in transmitter selection should include the following:

- Type of measurement—gauge, absolute, differential, and so on
- Range available and calibrated range
- Accuracy—including temperature effects, linearity, hysteresis, repeatability, and so on
- Output signal type, power required, and electrical connections
- Type of local display (if required) and field calibration and scaling capabilities
- Enclosure rating—NEMA 4X, explosion proof, and so on—as required

- Proof pressure and/or burst pressure, which define the physical limits
- Materials of construction—significant for corrosion resistance
- Temperature limits for the process fluid and the electronics/enclosure
- Process and electrical connections (NPT, male/female, conduit or pigtails, etc.)

The various sensing technologies developed each have advantages in reliability, accuracy, cost, and so on. The various technologies include strain gauge, capacitive, piezoelectric, and others. For the aeration control system designer, the details of measurement technique are less significant than the supplier's reputation for quality, service, and reliability. It is false economy to use cheap instrumentation that fails to reliably provide process information of sufficient accuracy. It is also poor practice to overspecify the instrumentation to achieve a level of performance beyond the requirements of the control system.

Most pressure transmitters are installed with a male or female pipe thread. The most common arrangement is a male thread on the transmitter screwed into a half coupling, tapped hole, or a saddle tee on the pipe. If the fluid is a liquid then the transmitter should be installed at approximately the centerline of the pipe. If this is not feasible then the transmitter range should be adjusted to compensate for the head difference between the tap location and the transmitter.

Some air lines have substantial pressure pulsations. An example is the discharge piping on a positive displacement blower. In order to protect the transmitter from damage caused by the pulsations and to provide more stable output readings, gauge snubbers should be installed between the piping and the transmitter connection. Snubbers have a small orifice or sintered metal passage that damps the vibration but does not reduce the mean pressure reading.

In some cases, the transmitter must also be separated from corrosive fluids. Isolation diaphragms provide a fluid-filled chamber and diaphragm that prevents direct contact between the process fluid and the transmitter. This device is most often applied in chemical feed applications, not in air system monitoring.

Block and bleed valves are recommended for most installations. They provide a means to isolate the transmitter connection from the process piping for service and calibration and a method for venting the connection to bleed condensate. In the case of differential pressure transmitters—particularly those used for flow indication—a three-valve manifold should be used. This permits isolation from the process line connections, frequently referred to as impulse tubes, and also provides a convenient means for connecting the two ports for zero calibration (see Figure 7.6).

Differential pressure transmitters are often employed to measure the suction in inlet piping for blowers caused by filters and piping. The high side of the transmitter is open to atmosphere and the low side is connected to the piping. (This concept is apparently counterintuitive to most contractors, as experience shows that the connections are frequently reversed.) In order to prevent insects and dust from obstructing the high side port, plugs with screens or filters should be used to block contaminants while allowing free air movement.

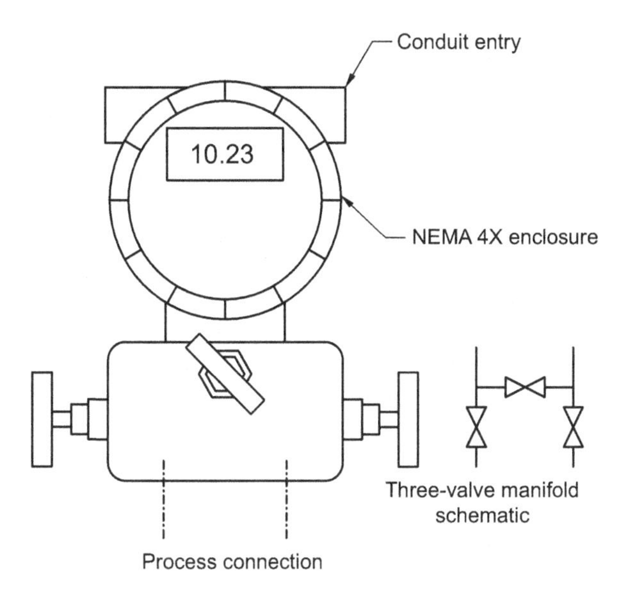

FIGURE 7.6 *Typical differential pressure transmitter*

The heat of compression and high blower discharge air temperature require consideration in mounting pressure transmitters for this service. Temperature ratings should accommodate anticipated ranges for both the fluid (sensing) side and the electronics enclosure. In some cases, a gauge siphon, also called a standoff or pigtail, (curled pipe nipple) may be used as a precaution to provide distance and a temperature gradient between the fluid and the transmitter. Installing the transmitter above hot piping should be avoided.

7.3 TEMPERATURE

Temperature measurement, like pressure, is one of the fundamental process measurements and is widely applied in most industries. Consequently there is a great variety of instruments used—ranging from simple liquid-filled thermometers to noncontact infrared thermography. In aeration control the applications may include blower bearing monitoring, measuring mixed liquor temperature, and providing air temperature for both monitoring and use in correcting volumetric air flow rates to mass air flow. Most measurements are in °F or °C. The calculations used in aeration control typically require that the measured temperature be converted to an absolute scale in program logic.

Temperature instrumentation includes simple indication, high/low switches, and smart temperature transmitters with analog and communications capabilities. Unlike pressure measurement, however, most temperature measurements require direct contact with the fluid or surface to obtain accurate measurements.

The primary sensing element for temperature measurement uses a thermally induced change in physical or electrical characteristics to indirectly identify changes in temperature. For example, the simple liquid-filled thermometer uses the calibrated thermal expansion of the liquid as an indication of the bulb temperature.

Glass thermometers are not sufficiently robust to survive in the rigors of most treatment plant installations. However, a very similar concept is used in many temperature switches. A liquid-filled bulb is inserted into the process fluid and is connected by a capillary tube to the switch. A bellows or diaphragm moves as the fluid volume changes and provides temperature indication and trips limit switches. Temperature switches and indicators may also rely on bimetallic springs that utilize the difference in thermal expansion of the two metals to induce a mechanical movement. The typical home furnace thermostat commonly uses this mechanism.

The most common temperature transmitters rely on the change in electrical characteristics of a primary sensing element. The sensing elements include thermistors, thermocouples (T/C), and RTDs.

Thermistors are usually limited in temperature range. They are typically used in wastewater treatment for protecting motors from overheating. A special type of thermistor, the positive temperature coefficient (PTC) thermistor, exhibits a sharp change in resistance at a fixed temperature established by its material properties. The PTC thermistors are embedded in the motor windings and wired to a special sensing relay that detects the resistance change and opens the control circuit to the motor control.

To many process engineers, temperature measurement is synonymous with the T/C. This is not an accurate viewpoint. In most aeration control applications thermocouples are not the best selection for the primary temperature sensor.

A T/C is most useful when measuring a wide range of temperatures or very high temperatures. The T/C is constructed by joining two dissimilar metals at the sensing point. The dissimilar metals generate a very slight voltage, with the value of voltage increasing with temperature—a phenomenon known as the Seebeck effect. The voltage variation, typically only a few millivolts, is also dependent on the two metals used. Over the years, many different types of T/Cs have been developed for measuring different ranges of temperature. The most common is the iron-constantan Type J, but there are also Types K, T, and E in frequent use, plus other more uncommon constructions.

Thermocouples require a high level of care in installation and wiring practices. Because of the very low signal level, they are very susceptible to EMI- and RFI-induced errors. Furthermore, special wire and terminal blocks are required to eliminate potential errors from additional junctions of dissimilar metals. For example, copper wire and the iron T/C lead will produce its own millivolt signal if there is a temperature gradient between the T/C and the splice or between the splice and the transmitter or display. T/Cs are not as stable or linear as other sensor types. This nonlinearity may result in inaccuracies. Some of these disadvantages can be overcome by placing a signal conditioner or transmitter in close proximity to the T/C. However, overall there is very little need to use a T/C for aeration system temperature measurements.

Most aeration control applications use RTDs as the primary sensing element. In the temperature ranges generally encountered they are linear, stable, accurate, and mechanically robust. They are comparatively resistant to EMI and RFI and can be connected to other devices using standard copper instrumentation wire. A common application is winding RTDs into the motor stator coils for motor temperature protection.

The RTD senses temperature change using the predictable change in resistance with changing temperature. The most common RTD has a platinum (Pt) resistance element, using platinum wire wound about a ceramic or glass core and hermetically sealed. The RTD typically has a thermal coefficient of resistance (α) of 0.00385 Ω/(Ω °C), corresponding to IEC 6075 and ASTM E-1137 standards. This is occasionally referred to as the European value of α. The sensor is usually made to have a resistance of 100 Ω at 0 °C (32 °F), leading to the common designation of "100 Ω Platinum RTD" in specifications and the literature.

RTD transmitters are widely used to convert the RTD signal into a voltage or 4–20 mA output. The transmitters may be mounted integrally with the RTD probe. A common arrangement is a so-called "hockey puck" two-wire transmitter mounted in a head attached to the probe.

Because of the stability of the RTD signal, it is possible to run the wiring between the RTD and the controller for considerable distances (up to several hundred feet of wire in some cases) without using a signal conditioner or transmitter. Standard shielded copper wire may be used. Most PLCs and controllers have inputs available that will read the RTD signal directly and provide a temperature reading. This is the simplest and most economical approach unless there are only one or two RTDs in the system. In that case it may be more economical to use a signal conditioner to provide an analog signal compatible with other instrumentation.

There are three types of RTD wiring arrangements. The different wiring does not change the function or characteristics of the RTD itself. They have been developed to minimize the measurement error of the RTDs.

The simplest arrangement is the two-wire RTD. This style is the least expensive, both for the RTD itself and the wiring and installation cost, and is most commonly used for monitoring motor windings and blower or motor bearings. The measurement of resistance for two-wire RTDs includes the resistance of the connecting leads andwiring in the measurement. This results in an error in the temperature measurement—with the measured temperature being higher than the actual temperature.

In order to eliminate the error introduced by the wiring, a three-wire RTD is used. A second wire is added to one side of the sensor. Measuring the resistance between this pair of wires allows the controller or PLC input card to cancel out the wire resistance which increases the accuracy of the measurement. The three-wire RTD is the most common configuration for process temperature measurements. The four-wire RTD adds a second wire to the other side of the RTD, further increasing accuracy.

There is no standard color code for RTD wiring. The sensor itself is polarity independent so the connections for two-wire RTDs are independent of the connection. For three-wire RTDs, both leads on one end of the sensor are the same color and the other lead will be a different color. The two leads from one end may be connected in either terminal at the input device. Four-wire RTDs have the leads on each end in the same color, with different colors on each end. If the controller input is configured for two-wire RTDs, the duplicate wires on three- and four-wire sensors may be left unterminated. If the input is configured for more leads than the RTD provides, jumpers may be used between the appropriate lead and the additional terminals. This will result in a slight inaccuracy but in most applications the error won't be significant.

Temperature instruments are point measurements and the sensing element must be in contact with the surface or fluid. Measuring fluid flows in piping requires insertion of the temperature sensor through the pipe wall. The tip of the instrument should be inserted 2 or 3 inches into the flow stream. The probe may be inserted through a compression fitting for low-pressure systems. This permits removal of the probe without twisting the connecting flex conduit or wiring. A nonmetallic ferule or sealing O-ring should be used to prevent crimping of the probe sheath. Many probes are fitted with pipe threads to allow them to be screwed into female threads in the pipe wall or half couplings.

Thermowells are hollow tubes designed to accept the temperature probe and separate it from the process fluid. This is useful when the process fluid is very corrosive and when it is necessary to remove the temperature probe without allowing leakage. Thermal contact must be maintained between the end of the probe and the thermowell. This may be accomplished by using spring-loaded tip sensitive probes or by partially filling the thermowell cavity with grease or a heat conductive fluid. A similar case is encountered when the temperature of a bearing must be measured. In this case, the cavity in the housing may be filled with grease to ensure good thermal conduction from the bearing to the temperature probe (see Figure 7.7).

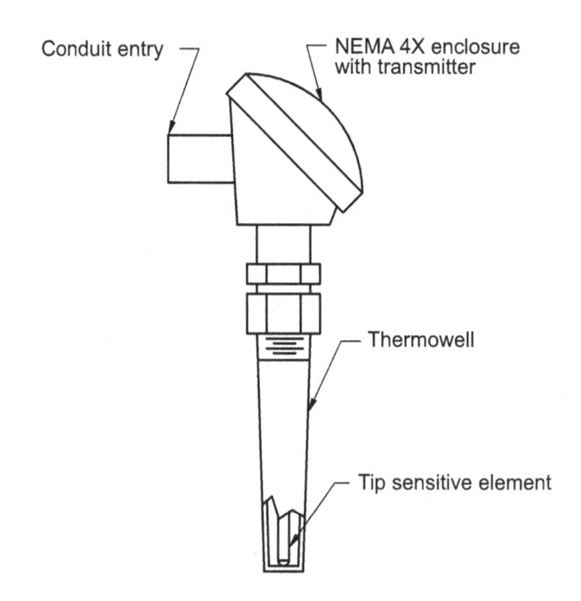

FIGURE 7.7 *Typical temperature transmitter*

If access to the process pipe is limited or the designer is faced with a retrofit into a single tapped hole in the pipe, it is possible to use a tee and connect both a temperature probe and a pressure probe to a single connection. The temperature probe passes through the run of the tee into the fluid and the pressure probe is connected to the branch. Clearance must be maintained between the temperature probe and the pipe and tee inside diameter so the pressure may be transmitted to the pressure transmitter. This arrangement, of course, cannot be used if the temperature probe has a thermowell.

7.4 FLOW

Flow measurement employs a combination of several fundamental units. The flow may be measured as volumetric or as mass flow. Volumetric flow rate is measured in units of length cubed per unit of time. Mass flow rate is measured in mass (or weight) per unit of time. If sufficient information is available about the process fluid and the configuration of the flow conduit, it is possible to convert from one to the other.

There are positive displacement flow meters constructed similarly to positive displacement blowers. These provide very accurate directly measured volumetric flow data and are employed in custody transfer (selling fluids) applications. These meters are generally too small and expensive for wastewater applications.

Many flow meters, for both liquid and air flow, measure velocity. The volumetric flow rate is calculated from the continuity equations and the pipe diameter (see Equation 6.5). Since the density of liquids varies only slightly in most wastewater applications, the volumetric flow rate can be directly converted to mass flow rate. The temperature and pressure of gases must be measured to convert the volumetric flow rate to mass flow rate (see Equation 5.8).

Some flow instrumentation employs point measurement—the velocity is determined at a point and the average velocity across the pipe or duct is assumed from this point value. Other devices measure the velocity across the diameter or even across the entire flow stream. There are two general categories:

- Differential pressure based, which rely on Bernoulli's law to determine velocity (see Equation 6.1), including pitot tubes, venturis, and orifices
- All other types—including Coriolis, thermal mass flow meters, and magmeters

Open channel flow measurement for liquids is occasionally required as part of the control system. Weirs and Parshall flumes are the typical devices used for open channel flow measurement. Both employ level measurement to determine the flow rate (see below).

Weirs are also used for controlling level and flow in tanks, for example, at the effluent of aeration basins. They may be straight-edged or v-notch. A saw-tooth weir is a group of v-notch weirs in parallel. The flow over a long straight-edged weir may be calculated from the water height using the Francis formula:

Equation 7.3

$$Q_w = 1496 \cdot l \cdot h^{1.5}$$

where

Q_w = flow rate of water, gpm
l = length of weir, ft
h = height of water over the weir, ft

In the case of a v-notch weir:

Equation 7.4

$$Q_w = 239 \cdot C \cdot h^2 \cdot \sqrt{64.4 \cdot h} \cdot \tan\frac{\alpha}{2}$$

where

C = discharge coefficient, typically $C = 0.58$, dimensionless
α = included angle of v-notch, °

Parshall flumes are used in open channels for measuring water and wastewater flow rates. They use Bernouli's law to calculate the flow based on changes in water level as it passes through a specially configured variable cross section. Flumes are typically fabricated from fiberglass. There are many modifications and varieties offered by various manufacturers and flow versus head characteristics may vary. In general, the flow through a Parshall flume may be estimated from the empirical formula:

Equation 7.5

$$Q_w = 1795 \cdot w \cdot h_a^n$$

where

Q_w = flow rate of water, gpm
w = width of flume throat, ft
h_a = depth of water at flume inlet, ft
n = empirical coefficient ($n = 1.55$ for 1 ft wide flume; $n = 1.522 \cdot w^{0.026}$ for wider flumes), dimensionless

Flumes and weirs are ancient and well-developed technologies and the literature contains extensive information regarding the restrictions and limitations of their application. The above formulas are applicable to most installations, but it is recommended that the literature and manufacturer's data be consulted for extremely precise requirements.

Water flow rate instrumentation for full pipes is usually divided into clean water and water containing contaminants. Clean water flow, either potable water or secondary effluent, may be measured with orifices, venturis, and so on. Because of the potential for clogging the taps for impulse tubes, differential pressure-based methods are not used with wastewater or other contaminated flow.

Magnetic flow meters or magmeters are very commonly applied on raw wastewater and sludge lines. The magmeter uses a conductive fluid passing through a magnetic field to create an electrical current—in essence using the fluid flow to create a generator. Magmeters are usually installed in-line, but there are insertion models available that greatly reduce cost on large pipes.

Ultrasonic flow meters for measurement in pipes are broadly divided into transit time and reflective or Doppler shift types. They are usually supplied as insertion types, but there are clamp-on units available. The clamp-on types should be applied with caution since lined pipe may significantly reduce accuracy.

Most flow measurements in aeration control applications involve low-pressure air for blower and aeration basin control. Differential pressure measurement is very common for this service but newer technology is increasingly accepted for many systems.

The differential pressure devices commonly encountered include orifice plates, venturis, flow tubes, and pitot tubes. Pitot tubes are usually the least expensive of these elements. Orifice plates are also relatively inexpensive and can provide excellent accuracy but have the disadvantage of a larger "permanent" pressure loss than other types. All of these primary flow elements have the characteristic of providing direct measurement of velocity. Volumetric flow rate is calculated using the pipe or meter cross-sectional area.

The velocity and flow rate for these primary sensing elements are proportional to the square root of the differential pressure. Many differential pressure transmitters have the ability to extract the square root in the transmitter and provide an output signal proportional to velocity. With proper scaling, this signal can also be linearly proportional to volumetric flow rate. In this case the differential pressure transmitter acts as a flow transmitter. However, for many transmitters the square root extraction becomes inaccurate at very low flow and low differential pressures. This limits the turndown on the flow signal. It is very common to have the transmitter go to a zero flow output below 10% of the maximum range. The selection of the flow transmitter must consider the upper and lower ranges of anticipated differential pressure to ensure reliable operation throughout the full anticipated operating flow range.

Flow meter measurement devices and the calculation and installation procedures are identified in ASME PTC-19.5. The general equation for determining mass flow rate for orifices, flow tubes (nozzles), and venturis is provided by the ASME code:

Equation 7.6

$$q_{\mathrm{m}} = 300 \cdot \frac{\pi}{4} \cdot d^2 \cdot C \cdot \varepsilon \cdot \sqrt{\frac{2 \cdot \rho \cdot \Delta p_{\mathrm{m}} \cdot g_{\mathrm{c}}}{1 - \beta^4}}$$

where

q_m = mass flow rate, lb_m/h
d = bore or throat diameter of the flow element, in
C = discharge coefficient of the flow element, dimensionless
$C \approx 0.6$ for orifice plates
ε = expansion factor for compressible fluids, dimensionless
$\varepsilon \approx 1$ at typical air flow conditions
ρ = density of fluid at flow conditions, lb_m/ft^3
Δp_m = differential pressure of flow measurement element, lb_f/ft^2 (psi)
g_c = 32.174 lb_m ft/lb_f s^2
β = ratio of bore or throat to pipe diameters, dimensionless

Equation 7.7

$$\beta = \frac{d}{D}$$

where

d = diameter of bore or throat of the flow element, in
D = inside diameter of the pipe, in

All flow meters present some obstruction in the flow of the fluid and consequently there is some permanent pressure drop in the fluid. This drop may be negligible for many flow meters or it may be large enough to affect total system pressure requirements. The permanent pressure drop through orifice plate flow meters is the larger than that produced by most other types of flow meters. It can be estimated:

Equation 7.8

$$\Delta p_f = \Delta p_m \cdot \left(1 - \beta^2\right)$$

where

Δp_f = permanent friction pressure drop, psi

The discharge coefficient C and β are specific to each type of flow element and the specific geometry of each meter. If accurate flow measurement is required, the manufacturer of the meter should be consulted for the specific parameters and the equation for calculating flow from differential pressure. The generalized equations should only be used if the manufacturer's specific data is not available. The location of the taps for measuring static pressure and differential pressure also affect the flow rate calculations.

Venturis are often applied where accurate measurement is required and the pressure drop through an orifice plate is unacceptable. Another common primary element used for air flow measurement is the flow tube, which has many characteristics in common with the venturi but requires much less length of pipe for installation. There are proprietary versions of the flow tube available that even further reduce straight pipe length requirements compared to the typical flow tube.

It is common to find the differential pressure at a specific flow rate stamped on the flow element or flow transmitter nameplate. This information can be useful in replacing the transmitter or calculating the flow from the differential pressure in a PLC or controller. For air flow the rated differential pressure is specific to the design pressure and temperature and some error will result if the air stream is at any other conditions. For aeration basin control this may not be significant—as stated above repeatability is more important than the actual measurement value. Certain control algorithms can accommodate the errors in flow rate (Chapter 9). Assuming equal pressure and temperature, the flow as a function of differential pressure can be calculated:

Equation 7.9

$$Q_2 = \cdot Q_1 \cdot \sqrt{\frac{\Delta p_{\mathrm{m2}}}{\Delta p_{\mathrm{m1}}}}$$

where

$\Delta p_{\mathrm{m1,m2}}$ = differential pressure at flow 1 and 2, force per area
$Q_{1,2}$ = flow rate at differential pressure 1 and 2, mass or length^3

Pitot tubes, also referred to as pitot-static and pitot-Prandtl tubes, are point flow measurement elements. To determine the flow rate from any point flow measurement, the relationship between the average flow velocity and the velocity at the measurement point must be known or assumed and the cross-sectional area of the pipe or duct must be known:

Equation 7.10

$$V = \sqrt{\frac{3.336 \times 10^7 \cdot \Delta p_{\mathrm{p}}}{\gamma}}$$

where

V = velocity of air stream, ft/min
Δp_{p} = differential pressure at pitot tube, psi
$\Delta p_{\mathrm{p}} = p_{\mathrm{total}} - p_{\mathrm{velocity}}$
γ = specific weight of air at flow conditions, $\mathrm{lb_f/ft^3}$

Pitot tubes are generally mounted at the pipe centerline. For fully turbulent uniform profiles, the center velocity is 1.22 times the average velocity. Note that elbows, tees, valves, and similar fittings can severely distort the velocity profile. A number of proprietary designs for averaging pitot tubes are available and find wide application in air flow measurement. The averaging pitot tube consists of a small diameter tube spanning the pipe with multiple upstream ports spaced along the length (to sample total pressure at equal area increments) and a downstream port (to measure static pressure). These can improve measurement accuracy over a standard pitot tube. However, they are effective in one plane only and cannot correct for gross velocity profile distortions. The manufacturer should be consulted for specific calculation data and application guidelines.

A useful technique applied with pitot tubes and other point velocity measurements is taking the velocity profile of the pipe to determine the distortion in the velocity profile and then compensating for it. In some cases, a simple profile—in one plane or two planes at right angles—is used to determine the point of average velocity. To achieve more precise measurements, often used when field calibrating another flow meter, take multiple readings across the pipe in increments that create equal areas for each reading. The summation of the flow rates then reflects the total flow rate in the pipe.

In many cases, the averaging techniques are only accurate at a single set of flows and readings. If the velocity profile is not repeatable and proportional at all flow rates encountered, the results will not be useful at conditions other than those exactly matching the conditions at the time of measurement.

A relatively new air flow measurement instrument, at least compared to orifices and pitot tubes, is the thermal mass flow meter. This is also referred to as a calorimetric flow meter. Like the pitot tube, this is generally a point measurement. There are averaging versions available—somewhat similar in construction to averaging pitot tubes. These may improve accuracy in unfavorable pipe configurations and may also simplify installation.

Thermal mass flow meters use a sensor consisting of two RTDs with a small gap between them. The sensor is inserted into the air flow stream. One RTD is heated by passing a small current through it and the temperature difference between them is monitored. Air passes over the RTDs and cools the heated element. The degree of heat transfer is a function of both the velocity and the density of the air stream since the amount of heat removed depends on the mass of the air passing over the heated RTD. Some units maintain a constant current to the heated element and monitor the temperature difference. Other designs maintain a constant temperature difference between the two RTDs and measure the current required to maintain that difference. In either case, an empirical formula is used to convert the measurement into an output proportional to velocity.

The base output of the thermal mass flow meter is typically standard feet per minute. By calibrating the meter with the inside diameter or cross-sectional area of the pipe, the mass flow rate in SCFM or other units can be transmitted and displayed directly—without the requirement for temperature and pressure corrections. This is useful in aeration basin air flow control since meeting the oxygen demand of the process is a function of the mass flow rate.

The direct reading of mass flow rate is one reason that this technology is rapidly being applied to aeration control. Others include the following:

- Very high turndown, up to 100:1 maximum to minimum reading
- Simple installation, usually inserted directly through a compression fitting
- Competitive costs compared to a full set of differential pressure, gauge pressure, and temperature instrumentation for mass flow measurement with other devices
- Smart transmitter and communications capabilities included in most designs

There are also disadvantages, of course. Differential pressure transmitters can be field calibrated using simple handheld pressure simulator but field calibration across the flow range is not possible for most thermal mass flow meter designs. Some designs allow zero calibration in the field, but a full calibration requires removal and return to the factory. Although most thermal mass transmitters provide a rapid response to varying flows, some designs require several seconds to stabilize after a flow change—making these models unsuitable for blower control.

As with RTD probes, the compression fitting used for thermal mass flow meter insertion should not use metal ferrules. These will crimp to the probe and make correction for installation errors or relocation of the probe impossible. Many probes are installed through an insertion valve assembly that allows the probe to be removed without leaking the process fluid. The probe is passed through a full bored ball valve. During removal the probe end is slid past the valve, which is then closed. A safety tether is usually fastened to the probe to prevent blowout from the air pressure during removal.

It is nearly impossible to overestimate the importance of proper installation for all flow measuring instruments. Experience has shown that the most common problem preventing accurate flow measurement is lack of sufficient straight pipe for establishing a uniform velocity profile across the pipe diameter. A minimum length of uniform straight pipe equal to 10 diameters should be upstream of the flow meter and straight pipe equal in length to at least 5 diameters should be downstream of the flow meter (see Figure 7.8). These distances should be increased if a butterfly valve (BFV) is upstream of the flow meter.

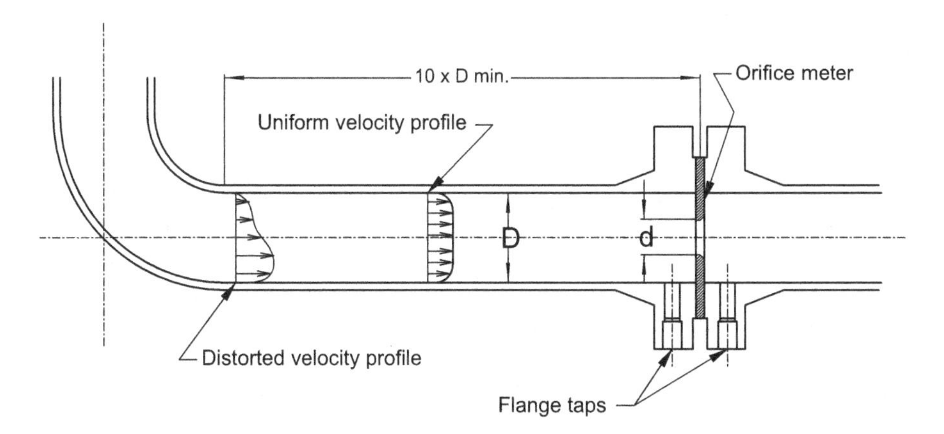

FIGURE 7.8 *Example orifice meter installation*

BFVs upstream of flow meters are particularly problematic since they shift the air velocity in constantly variable patterns as the valve disk moves. Field observations have shown that a single point measurement at pipe centerline may actually display reduced flow as the valve opens and the velocity profile shifts. Valves should be placed downstream of the flow meter if at all possible.

There are flow straighteners and flow conditioning devices that mitigate straight pipe requirements to some extent. The designs include tube bundles, perforated plates, grids, and various proprietary designs. These can be helpful but they are not a panacea. The straighteners themselves require some length and they do not reduce the straight pipe requirement to zero. Straighteners also introduce some frictional pressure drop—the amount varies greatly with design.

As indicated in Chapter 6, a common situation is the occurrence of very short lengths of pipe between takeoffs from the aeration basin lateral and the drop legs to the diffuser. One solution to this is to install the flow element below the water surface in the long drop pipe. An averaging pitot tube is an excellent choice for this. Installation is simplified since the probe is installed in a half coupling or similar threaded adapter on the drop leg. Any leakage will be air out of the drop leg rather than water in. The electronics are all installed externally above the tank surface. The only connections to the submerged element are flexible impulse tubes connecting the pitot tube to the transmitter. If mass flow rate is required, the temperature and pressure transmitters can be installed above the water line. Since the distance is short only negligible error is introduced.

Another technique is available for this situation when there is only one set of takeoffs from the lateral and the flow or differential pressure transmitter is connected to a controller. The air flow in the lateral may be measured ahead of the first takeoffs and between subsequent takeoffs. The controller can subtract the air flow measurement downstream of the takeoff from the measurement upstream of the takeoff to determine the flow through each drop leg. Some minor error may be introduced because two instruments are required, but this is negligible compared to the error introduced by inadequate piping arrangements.

This section discusses only the flow measurement devices commonly encountered in aeration systems. There are many additional technologies for flow measurement. Some are only suitable for flow rates much lower than those commonly encountered in aeration systems. Others are not cost-effective, particularly in large pipe sizes.

7.5 ANALYTIC INSTRUMENTS

The category of analytic instruments is very broad and covers measurement of a variety of chemical and physical properties. Many of these instruments are industry specific and find little application outside of the wastewater treatment industry. Most of these devices are complex—involving multiple parameters and sophisticated algorithms for determining the process variable.

Some analytic instruments involve automated offline titration or chemical addition to a mixed liquor sample. Some of these instruments operate in real time; others require considerable time for the chemical reactions and measurements. They are not suitable for use in automated control strategies.

Titration is a method of analysis where a solution of known concentration (the titrant) is used to determine the unknown concentration of a solution (the analyte)—typically mixed liquor. The titrant is added in known quantities to the mixed liquor and a physical or chemical response in the analyte is measured to determine the measured parameter. This technique usually involves sample pumps withdrawing a known quantity of mixed liquor to the measurement system and other pumps adding the titrant. These systems may be maintenance intensive, as the mixed liquor sampling system is subject to clogging from particulates and biological growth in the sample lines.

7.5.1 Dissolved Oxygen

Dissolved oxygen (DO) concentration is the fundamental parameter for most aeration control systems. DO is the basis for confirming process oxygen demand is being met. Adjustment of the air flow rate to the process has a direct correspondence with increasing or decreasing DO. Reliable and repeatable DO measurement is critical to proper performance of the aeration control system.

DO transmitters incorporate a probe directly immersed in the aeration basin mixed liquor. The probe is usually connected to a separate transmitter enclosure. Some transmitters can accept inputs from multiple probes. Because the solubility of oxygen is a function of water temperature, the probe measures mixed liquor temperature as well as the level of oxygen. Both the transmitter and the probe are usually mounted to the handrail on the basin walkway. Systems that clamp to the handrail are preferred to allow adjustment of the probe location. It is recommended that sufficient extra signal cable, 25–50 feet, be added in the probe to transmitter connection to allow relocation of the probe as needed.

The probe is normally installed at the end of a pipe that extends out into the basin. It is important to avoid installing the probe in stagnant areas that do not reflect the bulk mixed liquor. This can be a problem in lagoons with sloping sidewalls. In order to get the probe extended into the aerated zone, it may be necessary to provide a platform, a retrievable floating mount or ball, or a very long support for the mounting pipe.

Some manufacturers use a design employing a floating ball to keep the tip of the probe at the desired submergence in the mixed liquor. This design is beneficial for digesters and SBRs where the water level may fluctuate. For conventional aeration systems, the probe is installed at the end of a pipe and submerged in the mixed liquor. A submergence of 1–2 ft is used. The best results are obtained when the pipe is at an angle of 5–15° from horizontal to prevent entraining bubbles on the probe tip. This also provides some flushing action on the probe tip from the movement of the mixed liquor (see Figure 7.9).

The location of the DO probe in the tank is important. Some trial and error is often involved in achieving the appropriate feedback to the control system.

The probe's location in complete mix basins is not critical so long as the probe is not adjacent to raw wastewater or RAS influent points. In plug flow basins the

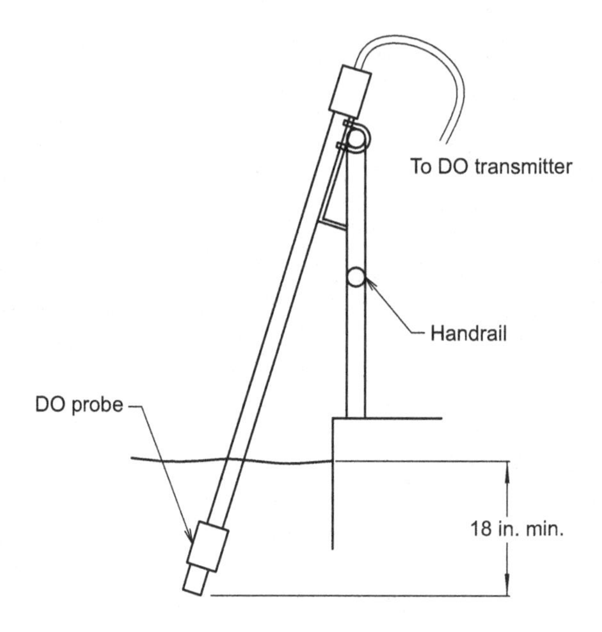

FIGURE 7.9 *Example DO probe installation*

selection of the optimum location along the length of the basin depends on the process requirements and the complexity of the system.

In large facilities where the control system provides independent DO control of multiple zones and drop legs, the probe should be located in the approximate center of the diffuser grid. If grids are controlled individually, the target DO is usually varied along the length of the basin. The setpoint at the influent end is generally low, as it may require inordinate amounts of air to increase the DO where the oxygen uptake rate is very high and the impact of slug loads and sidestreams directly impact the DO concentration. The effluent DO setpoint may be higher for BOD removal and nitrification systems, but in denitrification systems it should be low to avoid high DO in the recycle flow.

If there is one DO probe in a plug flow basin and the air flow to the entire tank is controlled at one point, the location is more subjective. Experience shows that the residence time in the basin usually exceeds the time required for completion of the biological reaction. The recommended initial location is approximately $\frac{3}{4}$ of the distance from the influent to the effluent end of the tank. At this point the process has progressed to the point where the desired DO concentration can confirm that the process is not oxygen limited. There is sufficient residence time remaining so that if the measured DO is low, increased air flow can raise it before the exit from the aeration basin.

The location may require adjustment during the commissioning process. A traverse of the tank at stable process and air flow conditions with a handheld DO probe—to establish the DO profile—is extremely helpful. The test should be performed at a time close to the peak diurnal flow. The point where the DO begins to increase is generally the best location. Installing the probe immediately after the tank influent should be avoided because slug loads and/or low DO from anoxic zones upstream will cause sudden increase in oxygen demand. This will result in dramatic changes in air flow. As the slug moves downstream, the impact is attenuated and leads to more stable control—without decreasing process performance.

The installation of DO probes in oxidation ditches with brush aerators is a similar challenge. If the probe is too close to the aerators, the DO will be too high. If the probe is installed in an anoxic zone, it will never measure a DO high enough to enable modulating the aerators. As with the plug flow reactor, a traverse of the ditch to obtain a DO profile will provide guidance in establishing the proper probe location.

Early DO probes were not reliable and had high maintenance requirements. In recent decades the technology has evolved to provide consistent performance with a reasonable level of maintenance. The transmitters generally fall into the smart transmitter category with calibration, communications, and display capabilities. Most DO transmitters will provide scalable analog outputs for both DO concentration and mixed liquor temperature.

There are two categories of DO probes in current use. The oldest technology uses an electrochemical reaction to measure the mixed liquor DO concentration. In this design a permeable membrane allows oxygen to pass through and diffuse into an electrolytic solution. The electrolyte is in contact with dissimilar metal electrodes. A chemical reaction, similar to that in a battery, crates a millivolt level signal. The transmitter linearizes this signal, compensates for temperature, and converts it to a DO concentration in ppm (mg/l). This type of probe is generally applied where measurement of very low DO concentrations is needed.

The electrode and electrolyte solution are slowly consumed in the reaction. Their replacement along with membrane replacement is part of the required routine maintenance. This is typically needed at one year intervals, but the frequency may vary widely depending on DO concentrations and waste characteristics.

Some inconvenience during system commissioning may be experienced with the electrochemical probes because a "seasoning" interval is usually required for the membranes. Immediately after immersion the rate of oxygen diffusion through the membrane is not stable. A period of approximately 24 hours is usually required for the membrane to achieve normal operation.

A more recent development is the optical DO sensor. This technology is gradually replacing electrochemical units because the maintenance, although not necessarily less frequent, is simpler. The optical probes use a transparent disk coated with a special fluorescent or luminescent coating that is sensitive to oxygen. A light source in the probe causes the compound to absorb and then emit light. The light emission is affected by the oxygen concentration and sensors in the probe measure the change in emission. Various manufacturers use different proprietary compounds and measurement techniques, but all fundamentally operate in this manner. Maintenance for these probes consists of replacing the photosensitive disk—typically on an annual basis. As with membrane type electrochemical probes, however, the frequency of maintenance and calibration are very site specific.

Any item immersed in mixed liquor is subject to fouling from biological growth and mineral depositions such as calcium carbonate. For DO sensors, regardless of type, the principal source of fouling on the probe is a biological film growing on the face of the probe. This film impedes the diffusion of oxygen to the sensor face and will cause depressed readings. In most municipal wastes the film reaches a thickness that substantially interferes with accuracy after one or two weeks of immersion. The development of the film is waste dependent and the frequency for cleaning must be adjusted based on site experience.

For most probes of both types, the cleaning is performed by pulling the probe from the mixed liquor and wiping the film away with a shop rag or towel. As with any instrumentation, the manufacturer's literature should be consulted for more specific information.

The interval between manual DO probe cleanings can be extended by using self-cleaning systems. It is important to note that manual cleaning and routine maintenance are not eliminated with automatic self-cleaning, but only reduced. Small facilities with less than five or six DO probes may not achieve sufficient benefit from the self-cleaning feature to offset the higher equipment cost. A wide variety of designs have been tested and many have been found unsatisfactory. Designs that use moving brushes, grindstones, or other mechanical components immersed in the wastewater have been particularly failure-prone. Field experience shows that periodically blowing a stream of compressed air across the sensitive portion of the DO probe is the most effective technique. The air stream removes the biological film and extends manual cleaning intervals.

The DO signal in mixed liquor exhibits measurable oscillations around the mean DO concentration. Many transmitters have adjustable output signal filtering that will dampen these signal fluctuations. It is also possible to use a digital filtering algorithm in the controller program to dampen these fluctuations.

All analytic instrumentation is more prone to drifting off calibration than less sophisticated instrumentation such as pressure transmitters. DO transmitters are no exception. It is common to have DO transmitters require calibration checks at monthly intervals—recalibration every two or three months should be expected. The calibration check is usually performed by comparing the transmitter's indicated DO concentration against a handheld DO probe. Complete agreement is unlikely and unnecessary. If the two readings are within 0.5 ppm of each other, recalibration is not indicated. Using the Winkler method or other laboratory-based DO measurements as the basis of comparison is not recommended since the DO concentration will degrade in the time required to get the sample from the basin to the laboratory.

An instrument that is easily recalibrated is more likely to receive the necessary maintenance and is also more likely to be satisfactory to the plant operations staff. There are three types of calibration in common usage for DO transmitters. Many models are capable of calibration using all three procedures:

- *Air Calibration*: In this method the probe is exposed to ambient air and the calibration is initiated. Since the oxygen concentration of air is known, the transmitter self-adjusts the internal algorithm to compensate for differences in output and the known correct signal value.

- *Solution Calibration*: Two samples of potable water are prepared. One has the DO concentration suppressed to zero using sodium sulfite dissolved in a sample container. The probe is immersed in the sample, allowed time to stabilize, and the zero calibrated. A second sample of potable water that has reached oxygen saturation by agitation or spraying is used to calibrate the maximum output.
- *Calibration to a Known Reading*: This method is preferred by many operators, who have implicit (although often unjustified) faith in the accuracy of the handheld DO probe used in their laboratory. In this method, the transmitter output is set to match the known DO concentration and the internal algorithm compensation is performed by the transmitter similar to the air calibration method.

Because DO concentration is the base process variable for most aeration control systems, it is critical to provide reliable DO instrumentation. Most manufacturers will provide short-term test instruments so that the designer and plant staff can select the instrumentation that performs best in the specific process conditions at that plant. Testing two or three units to determine operator preference is strongly recommended. This ensures that the installed instrumentation will meet the requirements of the aeration control system and the operators.

7.5.2 Offgas Analysis

One of the difficulties in the development of aeration control strategies is the lack of direct measurement technology for two key parameters affecting air requirements: organic loading in the process and oxygen transfer capability of the diffusers. There is no BOD or OUR probe that provides direct indication of the oxygen demand of the aeration tank. Even if one did exist, it could not be used directly for process control without knowing the OTE or OTR of the aeration system (see Equation 3.4).

There are instruments available that can provide indirect information on the process demand. These are generally offline instruments based on some modification of respirometry. They are not typically real-time *in situ* measurement techniques.

The basic principle of most respirometry devices consists of withdrawing a fixed volume of mixed liquor and using aeration or a chemical to increase the oxygen concentration in the sample. The sample is then allowed to proceed with the biological reaction. The rate of change of DO is monitored and from this the OUR or COD of the mixed liquor is calculated. After a sufficient number of samples and comparison to lab BOD_5 measurements taken at the same time, the BOD_5 can be inferred from the respirometry measurement with some degree of confidence. Other instruments use optical instrumentation calibrated to determine concentrations of specific compounds or elements.

Similar optical techniques are used to determine nutrient concentrations. The compounds of interest are ammonia (NH_3), nitrate (NO_3), and phosphorus. These

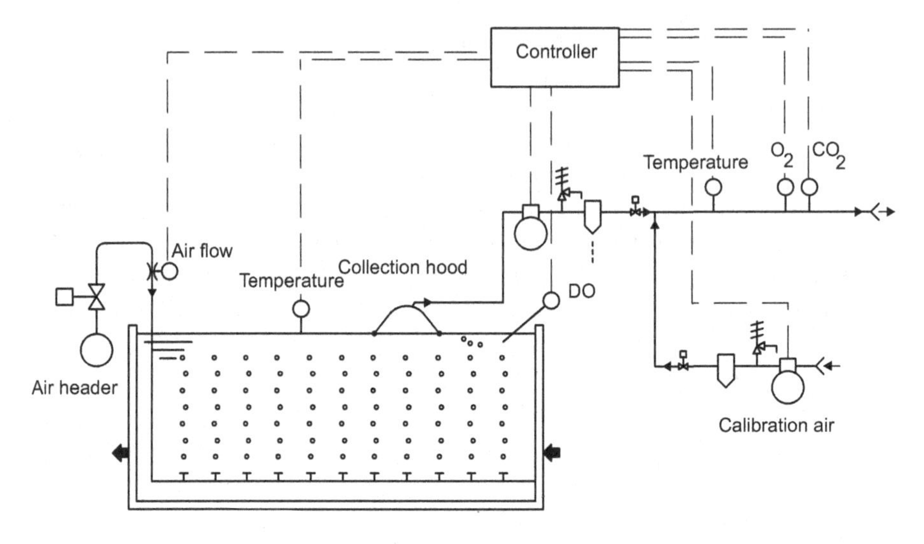

FIGURE 7.10 *Offgas analysis*

measurements are generally applied to samples pumped from the aeration basin. Other instruments use titration—sometimes combined with colorimetric techniques—to determine nutrient concentrations.

These measurements are useful for detecting toxicity in the wastewater and for general process monitoring. However, knowing the loading cannot be used to directly control the air supply to the process. Without knowing the OTE, the air flow required to meet the process demand cannot be determined.

There is technology available for determining these factors indirectly in real-time for diffused aeration systems. The American Society of Civil Engineers standard ASCE-18-96 includes procedures for the offgas analysis methodology for determining the oxygen transfer efficiency of diffused aeration systems in process conditions (see Figure 7.10). The application of this technique to aeration control is actually an instrumentation system rather than a single instrument—much the way determining mass flow rate using differential pressure, temperature, and pressure requires multiple instruments and an analytic unit.

A collection hood is floated or installed at the surface of the actively aerated area of the aeration basin. The hood captures bubbles that rise from the diffusers to the surface under the hood. The captured air is continuously drawn to a transmitter that measures the oxygen concentration in the offgas air. By comparing the concentration of oxygen in the sample to the known concentration of oxygen in ambient air, the actual oxygen transfer efficiency (AOTE) of the diffusers is determined:

Equation 7.11

$$\text{AOTE} = \frac{\text{O}_{2_\text{i}} - \text{O}_{2_\text{o}}}{\text{O}_{2_\text{i}}} \cdot 100$$

where

AOTE = oxygen transfer efficiency, %
$O_{2_{i,o}}$ = concentration of oxygen into aeration (ambient) and out of aeration, %

The system includes measuring the mass air flow to the aeration basins using the appropriate instrumentation in the air supply piping to the basin or zone being controlled. Once the OTE and the mass air flow rate, SCFM, to the aeration system is known the actual oxygen transfer rate (AOTR) can be calculated:

Equation 7.12

$$\text{AOTR} = Q_s \cdot 60 \cdot \frac{\text{AOTE}}{100} \cdot \rho \cdot C_{O_2}$$

where

AOTR = actual oxygen transfer rate, lb_m/h
Q_s = mass flow rate, SCFM
AOTE = oxygen transfer efficiency, %
ρ = density of air at standard conditions, 0.075 lb_m/ft^3
C_{O_2} = concentration of O_2 in air, decimal, 0.23 (23% by weight)

The OUR, which indicates the total process demand for oxygen is calculated (see Chapter 4):

Equation 7.13

$$\text{OUR} = \frac{\text{AOTR} \cdot 16{,}019}{V_t}$$

where

OUR = oxygen uptake rate, mg/(l h)
V_t = tank or control zone volume, ft^3

As with other analytic instrumentation, the above calculations are internal to the unit. Output signals, digital data communications, or local indication is provided to allow the operator or control systems to directly access the values of OUR, AOTE, AOTR, αF, and the process air demand in SCFM. Other values, such as DO concentration and actual air flow, may also be accessed at the offgas controller.

The steady-state OUR calculated includes oxygen required for metabolizing BOD and converting ammonia to nitrogen. In addition to the steady-state demand, the total air flow required at the basin or control zone may be dynamically adjusted to include nonsteady-state OUR resulting from changing loads. Any change in air flow needed to move the current DO to the desired concentration can be calculated for control purposes. This in turn allows calculation of the total air flow rate needed to meet process control objectives by using Equation 3.4. The offgas system may be used to monitor the current state and operating conditions of the process and the actual demand of the process. Unlike many process instruments, it will also provide a direct air flow control setpoint to the aeration control system using feedforward control.

The offgas system provides real-time information on the process demand. It can be used in conjunction with feedforward control algorithms to replace or supplement conventional feedback control (Chapter 9).

7.5.3 pH and ORP

Measurement of pH is used to indicate the relative acidity and alkalinity of a solution. A pH of 2 indicates a very acidic state, 7 indicates a neutral state, and 14 indicates an extremely alkaline state. The operation of the aeration control system is not concerned directly with pH. Although the pH may decrease during nitrification, the aeration control does not usually include control devices or logic specifically intended to control pH. Some systems do include pH probes as a convenient process input monitored by plant staff.

Most pH meters have a glass and metal electrode. The electrode is generally immersed in the mixed liquor. Various constructions and materials are employed for electrodes. As pH changes so does the conductivity of the solution and this is measured by the instrument.

The same instrument technology is also used to measure oxidation–reduction potential (ORP or Redox). The value of ORP, indicated in ±mV, is sometimes used as part of the aeration control strategy. The value of ORP changes as the process moves from nitrification to denitrification. The value indicating the shift is site and waste specific. Some process control strategies use this shift to control the beginning and end of intermittent aeration cycles or to shift the DO setpoint.

Experience indicates that *in situ* pH and ORP measurements are not as robust as DO transmitters or offgas monitoring instruments. The exposed electrodes of pH and ORP probes are subject to fouling and physical damage in the hostile environment of the mixed liquor. The calibration of the instrument shifts considerably and precise calibration typically involves preparation of special solutions. The maintenance often proves time-consuming for the operators and is rarely justified by an improvement in process or control system performance.

7.6 MOTOR MONITORING AND ELECTRICAL MEASUREMENTS

Heat is the principal enemy of electric motors. Most motor failures are the result of cumulative thermal damage to the insulation of the motor windings. Motor monitoring is principally concerned with detecting excessive temperature in the motor or preventing damage by detecting high currents from overloads that will cause excess temperature.

Many large motors (>100 hp) employ RTDs embedded in the motor windings to measure winding temperature and shut down the motor. The temperature level

TABLE 7.1 Typical Motor Temperature Limits

Insulation Class	Rated Rise at 40 °C (104 °F)	Maximum Temperature
B	80 °C	120 °C (248 °F)
F	105 °C	145 °C (293 °F)
H	125 °C	165 °C (329 °F)

permissible is a function of the class of insulation used. Class F is the most common (Table 7.1). Note that motors are rated at 3300 feet above sea level. At higher altitudes the decreased heat transfer capabilities may necessitate derating the motor. Using RTDs is recommended for all but the smallest motors if they are controlled by a VFD.

Winding RTDs are typically supplied in a two-wire configuration, but many control inputs are designed for three-wire RTDs. A third wire may be connected at the motor terminal box and run back to the controller for lead resistance compensation. Alternatively, two terminals at the controller may be jumpered—although this will result in a slight error in temperature readings.

Current monitoring is the most common method of monitoring motor load and this is adequate for many applications. Since motor heating is a function of I^2R losses in the motor, overcurrent protection is sufficient for most constant speed applications. Because of the output harmonics produced by VFDs (Chapter 8), current monitoring may not reflect the total temperature rise in VFD controlled motors.

Many motors also include provisions for protection of the motor bearings by measuring bearing temperature and/or vibration.

The current draw for most motors exceeds the level that can be input directly to controllers and current transmitters. Most constant speed motor current monitoring applications involve a current transformer (C/T). The output of most C/Ts is 5 A AC. They are usually identified by the ratio of input current to 5 A. For example, a C/T with a 200:5 ratio will provide an output of 5 A when the monitored current is 200 A.

The allowable burden, or output load, for small C/Ts may be inadequate to drive the output through long leads between the C/T and the transmitter or ammeter. It is common practice to use a larger C/T and pass the motor leads through the C/T window several times so that the same current is measured multiple times. If a motor lead is passed through a 200:5 ratio C/T four times, it will provide a 5 A output when the motor draws 50 A.

NEVER LEAVE THE OUTPUT TERMINALS OF A C/T OPEN. This will result in very high and dangerous voltages across the two terminals and presents a severe safety hazard. If the C/T must be disconnected, a jumper or shunt switch should be used to short the two terminals.

Many C/Ts and current transmitters are available in split construction. This allows them to be opened and installed around the motor leads without disconnecting the motor.

Because the output current of VFDs is not a true sinusoidal waveform, the output of many C/Ts and current transmitters is not accurate for variable speed installations. True RMS (root mean square) transmitters are available, but even these may not be accurate. In VFD applications that require monitoring the motor current, it is recommended that the VFDs own current or motor power output be used. Most VFDs provide this signal as a standard analog output from the drive.

Multistage centrifugal blowers are often provided with a "calibrated ammeter" that indicates blower flow rate as well as motor current. These ammeters are based on the relationship between blower power requirements and flow rate (Chapter 5). Usually measuring current on only one leg of the three-phase power is considered sufficient.

The calibrated ammeter is subject to error when used as an air flow meter because of variability in voltage, air density, and performance calculations. The calibrated ammeter may be used for blower control in constant speed applications—provided the critical points, such as surge, are verified by field testing. However, they should never be used as the control feedback signal for VFD applications because the accuracy is inadequate for proper control and protection.

Although amperage is an adequate indication of motor output power for most circumstances, true motor power monitoring requires a more sophisticated system. Motor power factor and other considerations dictate measuring the current on all three phases. Potential transformers are also required to measure the voltage on all three phases. The power transmitter will monitor both magnitude and phase shift and include power factor in the calculation of power. It should be noted that this measurement provides input kilowatt to the motor, not shaft output horsepower. Motor shaft power must be calculated by multiplying the input electrical power by the motor efficiency (Equation 1.7).

Motor protection relays (MPR) are often used to provide protection and monitoring for large motors. The MPR will monitor motor power as indicated above. They also provide inputs for monitoring motor winding and bearing temperature and vibration, with interlocks available for other external devices. Protection functions may also include limiting the frequency of motor starts, detecting phase imbalance, and so on. MPRs are smart transmitters with most having communications capability to allow remote reading of the alarms and monitored parameters.

7.7 MISCELLANEOUS

Level measurement and control is occasionally required as part of the aeration control process. SBRs, equalization basins, sludge holding tanks, and aerobic digesters are variable level processes. Varying the air supply as a function of level is one method used to optimize the air supplied to these processes.

The most common level sensing device is probably the float switch. One or more floats with an internal switch are tethered to a pipe or tank wall. The float lifts and an internal switch, often a mercury switch, changes state to indicate the level. Float switches are available in a variety of materials and contact configurations.

Conductive metal rods are the basis of another common level switch—although these are more often employed in clean water than in wastewater. As the water level rises it contacts the end of the rod and completes the circuit between the sensing rod and a ground rod. A special relay senses completion of the circuit and trips contacts to signal the level.

Bubblers are one of the older level measurement technologies and they are still widely employed due to their simplicity and reliability. In a bubbler, compressed air flows through a tube with the end submerged in the process fluid. The constant flow of air helps keep the submerged tube free of biological growths that foul other submerged instrumentation. Mitering the end of the tube aides in minimizing fouling. Measuring the air pressure at the top of the tube the submergence and the level. Bubblers are also employed with stilling wells to measure the depth of flow in Parshall flumes and thereby indicate the wastewater flow rate.

Pressure is often measured to indicate water level since the pressure is proportional to the height of water above the transmitter (Equation 4.12). In order to minimize the impact of fouling from biological growth in the sensing port, many pressure transmitters used in level sensing have a sealed fluid-filled chamber with a flexible membrane separating the sensing port from the wastewater. This system has proven to be simple, reliable, and inexpensive.

Ultrasonic level transmitters are often used because they do not require contact with the wastewater and are, therefore, immune to fouling. A transponder emits a high-frequency sonic pulse and the time required for the return echo is proportional to the distance traveled and the level. Care must be exercised in mounting the transponder—particularly in tanks with large level changes. Echoes from tank walls or other obstructions can cause erroneous readings. In outdoor installations the possibility of wind causing the echo to miss the receiver is potentially a problem. Foam on the surface of some process tanks can also cause errors if the surface of the foam blanket is picked up instead of the actual water surface.

There are other technologies such as radar and flexible tapes that are used in general industries. However, these have not found wide application in the wastewater treatment industry.

Clarifier blanket depth is used to sense the level of the interface between the settled sludge in a secondary clarifier and the water above it. Devices that measure this typically rely on an increase in turbidity between the sludge and the clarified effluent. Many of these units involve a moveable probe that is winched up and down through the clarifier. Others have rigid probes with a series of optical detectors to sense the turbidity and identify the top of the settled sludge. The rigid units may be mounted on a swinging bracket to allow sludge and scum scrapers to pass. Blanket depth measurement is occasionally used in aeration control if the RAS flow is being controlled.

Suspended solids meters, measuring the sludge density or concentration, may also be part of RAS control. Specialized instrumentation is used to sense the sludge's solids concentration. Some instruments use ultrasonic sensors for density detection, some use light and turbidity, and some use nuclear detectors. Sensors may be mounted in the sludge pumping line or located in the clarifiers, depending on manufacturer.

Vibration in process equipment can indicate the occurrence of an actual or impending failure. A common application for aeration control is monitoring blower and blower motor bearings to detect bearing wear or imbalance in the rotating assembly. Vibration switches are still used—particularly for small blowers—but current practice is to use a vibration transmitter connected to a controller with the controller providing warning and shutdown alarms.

Installation practice for vibration monitoring varies depending on blower design and size. PD blowers and small centrifugal blowers may mount a single transmitter on the blower case. It is more common to install the transmitters or sensors directly on each blower bearing housing. Motor bearings on large units may also be monitored.

Vibration is measured in either as displacement in mils (0.001 in.) or as velocity in inches per second (ips). For journal bearings, displacement is preferred. With case-mounted units and for ball or roller bearings (so called antifriction bearings), velocity is the preferred vibration measurement. If the speed of rotation is known, it is possible to convert from one unit to the other:

Equation 7.14

$$d = \frac{19100 \cdot v}{f}$$

Equation 7.15

$$a = \frac{f \cdot v}{3687}$$

Equation 7.16

$$a = \left(\frac{f}{8383}\right)^2 \cdot d$$

where

d = peak to peak displacement, mils (1 mil = 0.001 in.)
v = peak velocity, in./s
f = frequency, cycles/min
a = peak acceleration, gs ($1g = 386.087$ in./s^2)

Trip points for warning and equipment shutdown are dependent on bearing type as well. A maximum displacement of 3 or 4 mil is common for journal bearings and shutdown in the area of 0.6 ips is common for ball bearings. The safe vibration level is very dependent on equipment design and rotating speed, so the manufacturer's data should always be consulted.

The vibration sensor type required is dependent on the bearing type as well. A journal bearing has a shaft rotating within the bearing, with a film of oil separating the shaft from the bearing. Vibration in journal bearings is measured by a proximity probe passing through a hole in the bearing. The probe measures the distance between its tip and the shaft. It is common to have two probes at 90° in each bearing. They may be used with sensors detecting rotational position for very sophisticated diagnostics. More often, the probe output is connected to a transmitter that converts the output to a 4–20 mA signal proportional to displacement. The transmitter output, in turn, is used to trip protective alarms in a PLC or other controller.

Ball and roller bearings are monitored with accelerometers. These generally have a weight that excites a piezoelectric crystal. They are installed on the case or bearing housing. The accelerometers are usually integral with a loop powered transmitter providing an output proportional to velocity.

In many systems, critical status or position indication is provided by simple limit switches. These devices open and close contacts connected to the controller's discrete inputs.

Valve position may be indicated by limit switches, often called microswitches. These are generally operated by a cam integral with a valve's motor operator. Switch packs are also available to indicate the open and closed position of manually operated valves. These switch packs are often suitable for field mounting in retrofit systems.

For some applications noncontact proximity switches are more suitable. There are a variety of technologies used for proximity switches. Magnetic and capacitive proximity switches sense the presence of a target—most often metal—and change the status of solid-state sinking or sourcing contacts. Note that the metal target for magnetic proximity switches must be ferrous and that some grades of stainless steel are nonmagnetic.

Optical (photoelectric) switches are available in through-beam and retroreflective types. The through-beam design has separate emitter and receiver, with the receiver contacts changing state when the beam is interrupted. Retroreflective switches have the emitter and receiver in a single unit. They rely on the light reflecting from the sensed surface—often a special reflector or reflective tape.

A common application for proximity switches is monitoring rotational speed since mechanical switches are unable to operate at the necessary frequency. Magnetic proximity switches are used to pick up the position of shaft keys or coupling bolts. Photoelectric switches pick up reflective tape attached to the shaft. For either type of sensing, the frequency response of both the proximity switches and the controller input must be checked to verify the operating range encompasses the anticipated speeds.

Operator commands, as well as manual control, are provided by hand switches and pushbuttons. While these are not, strictly speaking, sensors or transmitters, they do provide operator input to the control system and are part of the instrumentation. Selector switches may be momentary with a spring return to the default position or maintained, with a detent to keep the switch in the position set by the operator.

Some confusion exists regarding the relative advantages of "30 mm" versus "22 mm" designations for manual switches and pilot lights. The designations refer to the diameter of the hole punched in control panels for installation. These diameters correspond to the outside diameter of $\frac{3}{4}$ and $\frac{1}{2}$ inch conduit. The 30 mm design is older and was once thought to be more rugged and reliable than the 22 mm devices. This is no longer true. Switch selection should depend on the electrical rating of the contacts. Identical ratings are available in both types.

Potentiometers, frequently called "pots," are used to provide manual analog inputs to a control system. A potentiometer is a variable resistor with three leads. The center lead is connected to a wiper that changes position relative to the two leads at the ends of the resistor. As the potentiometer is moved, the proportion of resistance changes as well. Some PLC and controller inputs can take potentiometer signals directly. However, it is more common to use some type of signal conditioning to convert the potentiometer signal to voltage or 4–20 mA. All control system potentiometers selected should be linear. So-called "audio" pots have a nonlinear output and these will cause difficulties in scaling the control signal.

Regardless of the simplicity or complexity of the instrumentation, it exists to collect information about the process and the process machinery. This information is ultimately used by the operator or the control system to make decisions about changes needed to improve the process. These decisions are put into effect by the control elements.

EXAMPLE PROBLEMS

Problem 7.1

An RAS flow meter provides a pulse output, each pulse representing 100 ml. The signal is connected to a PLC high-speed counter input. The input PLC displays the pulse frequency as 600 Hz.

(a) What is the flow rate in mgd?

(b) What is the conversion formula for use in the PLC program to convert from Hz to mgd?

Problem 7.2

A loop powered transmitter catalog provides the information shown in Figure 7.11. A designer plans to use a 24 VDC power supply and connect the transmitter output to a PLC input with an impedance of 250 Ω.

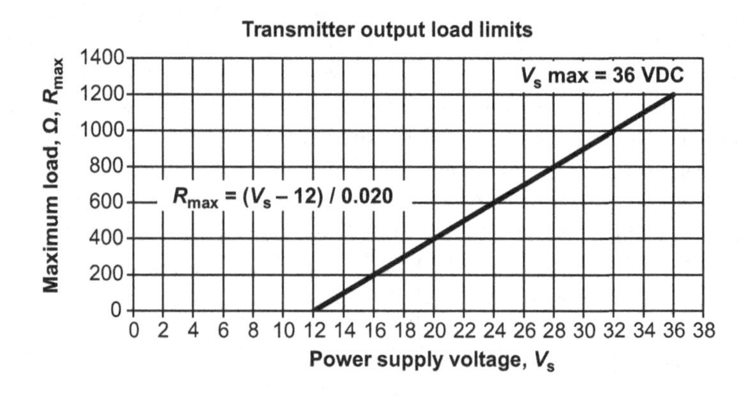

FIGURE 7.11 *Problem 7.1 sample transmitter load limits*

(a) Will this be within the transmitter load limits?

(b) The owner would like to insert an existing chart recorder with a 0–10 VDC input into the loop. Can this be accomplished using a dropping resistor?

(c) Name at least two design options to accommodate the owner's request.

Problem 7.3

A loop powered digital indicator is to be installed on the control panel to display the value of the process variable measured by a loop transmitter with characteristics shown in Figure 7.11. The indicator data sheet shows a "Loop Voltage Drop" value of 5.6 V. The signal will also be connected in series to a PLC input with an impedance of 250 Ω. Will 24 VDC loop power be adequate?

Problem 7.4

A submersible loop powered pressure transmitter with a range of 0–15 psi and a 4–20 mA output is planned for use as a level transmitter. The transmitter has a specified accuracy of ±0.5% FS and will be connected to a $4\frac{1}{2}$ digit indicator with a specified accuracy of ±0.05% FS, ±1 Count.

(a) What will be the range of level measured if the transmitter is used without scaling?

(b) What will be the output signal if the level above the transmitter is 12 ft of water?

(c) What will be the theoretical maximum error in the displayed value?

Problem 7.5

A venturi meter is stamped with "6.22 in. H_2O 2500 CFM." As part of a fine-pore diffuser replacement, a new differential pressure transmitter has been installed. The new transmitter has a range of 4–20 mA at 0–5 in. H_2O.

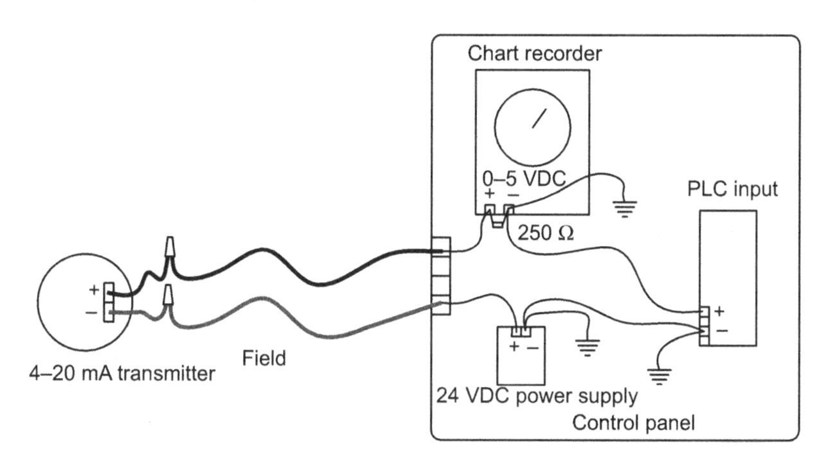

FIGURE 7.12 *Problem 7.6 transmitter wiring*

(a) The actual transmitter output is 11 mA. What is the air flow rate?

(b) What is the maximum air flow rate for the replacement transmitter?

Problem 7.6

Refer to Figure 7.12. Identify at least three problems with the wiring in this sketch.

Chapter 8

Final Control Elements

The final control elements interact with and modify the operation of the process. They convert the controller's output signals into physical actions that cause the process performance to change in a way that the instrumentation can detect and verify.

For most aeration control systems the process parameter directly controlled is flow. Other controlled parameters, such as pressure, level, or power, are indirectly manipulated by controlling flow. For example, the parameter of primary concern in aeration system control is DO concentration, which is modulated by controlling the air flow to the aeration basin.

The final control elements are less varied than the instrumentation. Most of them involve control of an electric motor, and the motor in turn affects a process parameter by manipulating the process equipment. Some pneumatic operators are still in use, but most new systems employ electric operators exclusively. The motor is not usually connected directly to the control system. Instead some type of control device is used to control the motor based on output signals from the control system. However, it is important for the designer to understand motor basics, since this will affect the control device selection and the performance of the control system.

A comprehensive analysis of electric motors is beyond the scope of this text, and beyond the requirements of control system design. The characteristics and factors most significant for control applications must be identified during the design process. As always, equipment manufacturer's data should be consulted for details.

Aeration Control System Design: A Practical Guide to Energy and Process Optimization, First Edition. Thomas E. Jenkins.

Unlike instrumentation, many final control elements operate at high energy levels and high voltages. Safety is a concern with any electrical equipment, but particular attention must be paid to safety when working with motor control and valve operators. Lockout/tagout (LOTO) procedures must be scrupulously followed. Arc flash hazards can be a significant threat to health and safety. Failure to observe the necessary safety procedures can be deadly. The required personal protection equipment (PPE) must be available and used. The designer should ensure that all required labeling is included and displayed on the control equipment.

System design can improve the safety of diagnostics and maintenance operations. Conveniently located and properly sized disconnects should be provided and a padlocking means should be included to facilitate LOTO procedures. Short-circuit protection must meet code requirements, but additional protection should be included if it enhances safety. The presence of foreign voltages (power not supplied within the panel) should be clearly indicated by wire color and labeling. Control devices and instrumentation should be located in different panels from those containing high-voltage devices and motor control systems.

8.1 VALVE OPERATORS

Valve operators can be variously categorized by the power source, the control signal, or the mechanical operation:

- Quarter-turn or multiturn operation
- Pneumatic or electric operators
- Single- or three-phase electric operators
- Analog or discrete position control
- Modulating or on/off service

Most valves used in aeration control are butterfly valves that rotate 90°, one quarter-turn, to provide full-closed to full-open position control. Ball and plug valves also require quarter-turn operators. Some types of valves, such as globe valves and gate valves, operate by moving the control element perpendicular to the pipe. The stem may be directly operated by a cylinder or diaphragm, but more often the stem has a screw that is moved by a multiturn electric motor operator.

In the very early days of automation, control systems operated pneumatically. This included not only valve actuators but sensors and the controllers themselves. A very large installed base of pneumatic operators was created, and many of them are still in service. As technology evolved, the pneumatic controllers were displaced by electronics and then by digital controls. It became necessary to provide the digital controllers with the capacity for operating the existing valve operators. Further, some owners prefer pneumatic operators for new systems in order to maintain compatibility with existing equipment. They are also comfortable with the accuracy and reliability of the proven technology.

There are two types of interface between pneumatic operators and current design controllers. For simple open/close service solenoid valves are used to control air to the operators. Interlocks should be provided to ensure that only one direction of travel is possible at any time. In low pressure service the solenoid may need to be direct acting.

Modulating valve service generally requires an analog signal for position control. The standard modulating signal for pneumatic operators is 3–15 psig air pressure. This may be directly applied to the valve, or it may be used for controlling a higher pressure pneumatic or hydraulic cylinder. The cylinder uses a mechanical linkage to provide 90° operation of the valve stem and disk.

The second type of interface is a current to pneumatic (I/P) signal conditioner. This is the preferred method for interfacing an electronic or digital controller to a pneumatic positioner. A 4–20 mA or voltage signal is supplied to the signal conditioner, which is mounted on the valve operator. The I/P signal conditioner is supplied with pressurized air and provides an output of 3–15 psig to the pneumatic positioner.

Pneumatic operators require a clean, dry, regulated air supply. This adds to the power and maintenance demands of the facility. In cold climates the air supply to the pneumatic valve positioner is subject to frost formations that may block the air stream. Condensate in the pneumatic tubing is another potential source of problems.

Electric motor operators have increased in accuracy and decreased in cost. They are the dominant technology in new applications. The motor operates through a gear reducer to provide the desired operating speed and supply the torque needed to move the valve stem. Quarter-turn and multiturn units are available. In some specialty applications, such as blower inlet guide vane (IGV) control, linear operators are used. These employ a ball screw or similar mechanism to convert the motor rotation to linear travel for extending or retracting the operator.

Limit switches and mechanical stops provide control of the minimum and maximum valve positions. The operator should be provided with two sets of limit switches. One should be set for full open and closed positions to cut power to the operator motor at the travel limits. The other set should be set to trip at the end of the normal modulating range of the valve to indicate status to the controller.

For modulating service, a positioner is provided to control the valve travel proportional to a 4–20 mA or voltage analog command. Most positioners also include local pushbuttons for manual adjustment of the valve and have a feedback signal for monitoring actual valve position. Status contacts may be provided on the manual/auto switch (or local/remote switch for some models) to indicate if the valve can be controlled automatically.

The analog signal does not directly power or control the valve operator. In most operators the motor is a constant speed single- or three-phase alternating current (AC) unit and is either off or on. The position is controlled by comparing the analog signal to a feedback signal—generally provided by a potentiometer geared to the valve stem. If the feedback signal from the valve doesn't match the command, the comparator closes a contact that causes the motor to move the valve until the signals match. To prevent hunting, a deadband adjustment is provided. When the difference between the two signals is within the deadband, there is no movement. Zero and span adjustments are also provided to allow calibration of the position limits (see

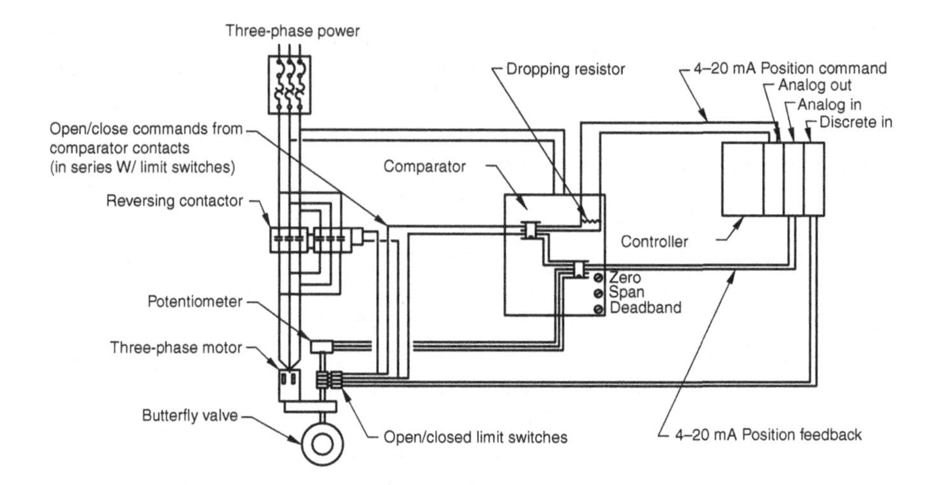

FIGURE 8.1 *Typical valve positioner schematic*

Figure 8.1). Most motor operators can be calibrated to control position within ±1% or better.

Note that analog positioners are not the only method for controlling valve position and modulating air flow. With the appropriate control logic, a timed contact closure can incrementally adjust the valve opening. This provides a simpler and more robust control method with equal precision (see Figure 8.2). It is also possible to eliminate the positioner by using position feedback from a potentiometer and performing the

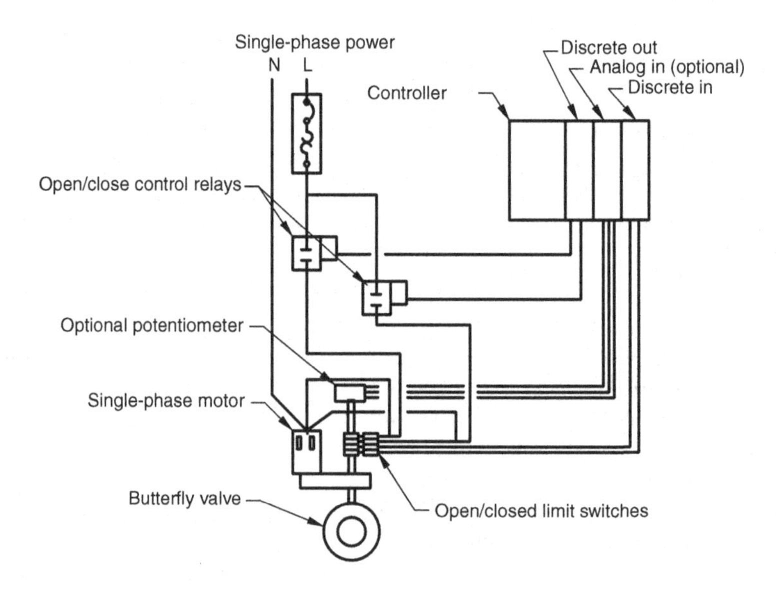

FIGURE 8.2 *Typical direct valve control schematic*

comparator functions in the controller itself. Either method eliminates field wiring and reduces the cost of the valve operator.

The open and close switches for manual positioning may be momentary or maintained. The maintained type mechanically or electrically latch to continue motor travel until a stop pushbutton is pressed, the switch is physically disengaged, or the valve hits the travel limit. This style is useful for full open and close service, but it is difficult to control if the valve is used for manually modulating flow. Momentary pushbuttons only move the valve while being pressed. They allow jogging the valve for fine position control and are generally preferred for aeration systems.

Sizing the valve operator requires consideration of the operating torque of the valve. In liquid service the aerodynamic forces on the valve may induce considerable torque on the disk and valve stem and this may dictate the motor size. For air service in aeration control, the dominant torque is the dry breakaway torque when moving the disk from the full closed position. The friction between the disk and the seat tends to increase with time as the seat elastomer deteriorates.

The service factor (SF) for the motor operator must include an allowance for the heat generated by frequent starting. The motor must be adequate for continuous modulating duty. The frequent starting and reversal of motor direction results in heating of the motor. Insulation and heat transfer characteristics must be capable of dissipating the heat load at the highest anticipated ambient temperature. Good practice requires inclusion of a thermal switch in the operator to stop the motor when unsafe temperatures are detected.

In many aeration systems the motor operators are mounted outdoors and exposed to variations in temperature and humidity. The enclosure should be rated NEMA 4 (National Electrical Manufacturers Association) or 4X to accommodate exposure to rain and snow. Even NEMA 4X enclosures are not airtight, however, and they "breathe" during temperature fluctuations. In order to prevent condensation inside the enclosure, operators mounted outdoors should be equipped with thermostatically controlled space heaters. This will prevent most moisture accumulation and corrosion inside the operator.

The gear reducer in the operator performs several functions. It multiplies the output torque of the motor to overcome friction in moving the valve disk. The gear ratio increases the time needed for the valve stem to rotate 90°. This in turn improves the ability of the control system to provide precise flow control. In some designs, particularly those employing high ratio worm gear reductions, the gear also prevents the aerodynamic forces on the disk from back-driving the motor when power is removed. This is referred to as a self-locking reducer.

Self-locking reducers do not eliminate the need for brakes on the operator motor. In air service the aerodynamic forces are so low that back-driving is seldom an issue. However, the rotating motor acts as a flywheel and it will continue to rotate after power is removed. This causes the valve to coast past the desired position, resulting in hunting or a large deadband in the control. Both of these reduce control precision. Spring applied electric release brakes should always be provided on the operator motor so that the valve travel stops as soon as power to the motor is cut.

The travel time of the valve operator is a critical factor in determining the degree of control achievable. A valve moving slowly can be positioned with greater accuracy than one traveling rapidly. This in turn results in more precise flow control. Experience has shown that 60 s for 90° rotation is the optimum for most air flow control applications. Slower travel may be needed to accommodate unusual conditions, such as oversized valves.

Note that some applications require rapid valve movement. Blow-off valves for blowers, for example, must respond rapidly to impending surge to relieve excess pressure before equipment damage results. A compromise between controllability and rapid response may be necessary. For some conditions two valves in parallel may be needed, although the economics of this solution are unfavorable. Another option may be a fast traveling blow-off valve for each blower and a slower modulating valve on a common header that will gradually close to restore normal operation if the discharge pressure is above limits.

The high gear ratios associated with slow moving valve operators also increases the torque applied to the valve stem for a given sized motor. This allows the use of less expensive single-phase operators on much larger valves than the typical practice in the wastewater industry. In general, butterfly valves up to 18 in. in diameter may be controlled with single-phase 60 s operators. This reduces equipment and wiring cost without penalty in operating life or reliability.

Other methods for controlling valve speed are occasionally employed. Some manufacturers use an internal power supply and direct current (DC) motor. This has the ability to adjust the speed of the motor and provide the required travel time. Note that the motor speed is fixed once the adjustments are completed. Another technique employs an electronic timing unit to provide controlled pulses of power to the operator motor. This method may not give satisfactory control since the motor operates at full speed when powered. If direct valve control using timed contact closure from the controller is used, the positioning accuracy may be compromised.

There have been advances in valve control technology in recent years that offer greater precision and economies than traditional operators. Digital control is available from many manufacturers. The traditional potentiometer has been replaced with an encoder, providing greater precision and simplifying calibration. Communications capabilities are available allowing direct control of the valve from the controller without the need for analog or discrete I/O and can provide advanced diagnostics. A networked system also eliminates separate control wiring for each valve. As with instrumentation, protocol compatibility and a need for redundancy in critical functions must be considered during the design.

In some applications multiturn motor operators are used to operate slide gates. These are almost exclusively three-phase operators. The position control requirements are the same as for a valve operator.

8.2 GUIDE VANES

Large single-stage centrifugal blowers are seldom controlled by throttling with butterfly valves. Instead IGVs and variable discharge diffuser vanes (VDVs) are used for capacity control (Chapter 5).

The guide vanes are typically controlled by operators provided by the blower manufacturer. There are two types of operators in common use. One is a conventional quarter-turn valve operator. The operator is mounted on the blower skid or the floor and connected to the external vane control lever by a mechanical linkage. These are usually, but not necessarily, provided with positioners and feedback to the controller. Because of the linkages involved the relationship between operator position and guide vane position may not be linear.

Another option is a linear operator mounted directly on the blower case. These are usually controlled directly by timed contacts in the controller. An analog positioner is not common in this case. A feedback potentiometer is generally employed to verify the position command is being properly implemented.

Guide vane operators do not require special construction or design features. The sizing—including torque or thrust requirements—and operating times must be closely coordinated with the blower manufacturer if the operator is not provided with the blower package.

8.3 MOTOR BASICS

The operating characteristics of electric motors vary significantly. The system designer must be aware of the critical characteristics in order to correctly size and apply the final control elements. The most critical characteristics include the following:

- Voltage type, alternating current or direct current
- Operating voltage
- Power and torque
- Operating speed
- Duty cycle: continuous, intermittent, and so on
- Ambient temperature limits
- Insulation class

The integral horsepower three-phase induction motor, often referred to as a "squirrel cage" motor, is the workhorse of industry. It is estimated that 90% of all the electricity used in a typical wastewater treatment plant is consumed by electric motors. Most process equipment, including blowers, is powered by this type of electric motor.

In new installations the blower motor selection should be performed by the blower manufacturer. If motor replacement is part of an ECM, the characteristics of the existing motor should be checked and, if appropriate, matched by the new motor. There may be advantages to the owner if the replacement motor is not identical in all respects to the existing one. There are a number of factors that must be considered relative to energy savings, control system design, and control strategies when evaluating motor replacement. These include the following:

- Nominal power, service factor, and torque characteristics
- Operating speed

- Rated supply voltage
- Enclosure type and frame size
- Insulation class and rated duty cycle
- Efficiency, power factor, and starting current

Motor power should be matched to the anticipated load created by the driven equipment. The inclination in replacing an existing motor is to provide a replacement motor with the same nominal power rating as the existing motor. This is usually safe, but not always optimum. Many motors are oversized because of excess conservatism by the original designer or because actual operating conditions differ from the original design conditions. Using a smaller replacement motor may result in improved efficiency and power factor without any reduction in motor life or equipment capability. The smaller motor will generally cost less. It can reduce the cost of replacement motor starters and variable frequency drives (VFDs) that may be part of the upgrade. Utility rebates may be available if a new high-efficiency motor replaces an existing standard efficiency one.

Conversely, blowers and pumps may have flow and pressure capabilities beyond the specified design points, but performance is limited by the motor power rating. Increasing the motor size may allow the operators to take advantage of the additional capacity and operate one blower instead of two. In some cases, using a VFD to slightly overspeed the motor will increase capacity. An example would be a PD blower that is operating below its rated capacity and speed. Obviously both blower and motor manufacturer's limits should be investigated before implementing an overspeed control strategy.

Service factor represents the amount of overload the motor can handle continuously without overheating. For most wastewater applications the motor service factor is 1.15. The service factor is intended to allow for unforeseen operating conditions and should never be included in the sizing of the motor during design. The service factor is based on the heat dissipation capabilities of the motor. Altitude and ambient temperature may reduce the service factor. It should be noted that the harmonic and transient voltages and currents on the load side of a VFD also reduce the effective service factor from 1.15 to 1.0.

The full load torque of a motor, the rated torque, is a function of the power and rotating speed:

Equation 8.1

$$T = \frac{P \cdot 5252}{N}$$

where

T = torque, lb ft
P = motor power, hp
N = rotating speed, rpm

This equation can be applied to any rotating machinery for the relationship between torque, speed, and power (Equation 5.33).

Electric motors also produce torque when they are not rotating—referred to as locked rotor torque or starting torque. The ratio of locked rotor torque to rated load torque is defined by NEMA design code letters with NEMA Design B being overwhelmingly the most common. A NEMA Design B induction motor has a locked rotor torque equal to 120–250% of rated torque. Design C and D motors have higher starting torques and are only used for special applications such as conveyors.

There are other torques ratings of interest in motor applications (Figure 8.3). Breakdown torque is the highest torque available. Load torque exceeding this value will cause the motor to stall. Pull-up torque is the minimum torque developed by the electrical motor when accelerating a load.

Many blowers are high inertia loads and the ability of the motor to accelerate the blower to operating speed before the overload protection trips can be a concern. The motor must supply the torque needed to overcome the load resistance from friction, flow, and pressure as well as torque needed for overcoming rotor inertia. For very large blowers a detailed starting toque analysis by the blower manufacturer may be a requirement. This analysis takes into account the variable motor and load torque as the speed changes. For most applications, however, a reasonable approximation can be made by assuming the available motor torque for acceleration is 166% of the rated torque. The blower inertia, expressed as WR^2 or WK^2, can be provided by the blower manufacturer. The time for acceleration can be approximated:

Equation 8.2

$$t = \frac{WK^2 \cdot \Delta N}{308 \cdot (T_m - T_l)}$$

where

t = time to accelerate the load, s
WK^2 = rotational inertia of the load at the motor shaft, lb ft^2
ΔN = change in speed, rpm
T_m = average motor torque, lb ft
T_l = average load torque, lb ft

If the motor is connected to the blower rotating assembly by a belt or gear drive, the effective inertia at the motor shaft must be corrected for the speed ratio:

Equation 8.3

$$WK_m^2 = WK_l^2 \cdot \left(\frac{N_l}{N_m}\right)^2$$

where

$WK_{m,l}^2$ = rotational inertia at the motor shaft and the load, lb ft^2
$N_{m.l}$ = speed of the motor and load, rpm

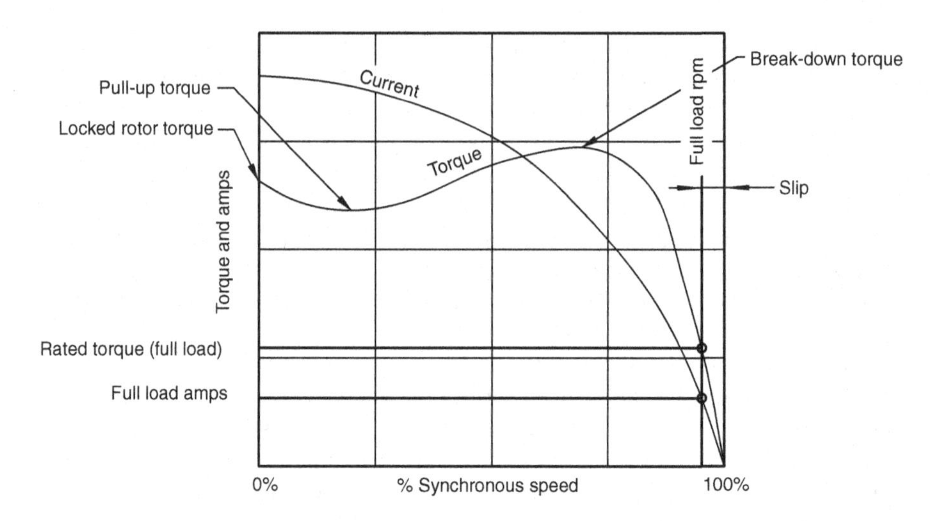

FIGURE 8.3 *Motor torque and current variation with speed*

In addition to the torque encountered during normal starting and operation, a potentially catastrophic torque can be developed by the motor while stopping a high inertia load such as a blower. It may take several minutes for the blower to coast to a stop. If voltage is applied to the motor while the blower is still rotating near full speed, a torque equal to several times the locked rotor torque may develop. This is sufficient to cause mechanical failure by breaking shafts or snapping couplings. The control system should include spin timers or interlocks with rotation detectors to prevent trying to restart a rotating blower.

The rotating speed of an electric motor is a function of the number of poles and the frequency of the alternating current:

Equation 8.4

$$N_s = \frac{120 \cdot f}{n_p}$$

where

N_s = synchronous speed of the motor, rpm
f = frequency of AC voltage, Hz (cycles/s)
n_p = number of poles, dimensionless

For blower applications the most common motors are 2-pole (3600 rpm), 4-pole (1800 rpm), and 6-pole (1200 rpm). The relationship expressed in Equation 8.4 represents the basic principle for using variable frequency drives in motor speed control.

Induction motors do not operate at their synchronous speeds, but always operate slightly slower. The difference between synchronous and actual operating speed is slip:

Equation 8.5

$$s = \frac{N_s - N_a}{N_s} \cdot 100$$

where

s = slip, %
$N_{s,a}$ = synchronous and actual speed of the motor, rpm

For a NEMA Design B motor the slip at rated load is typically 3%. Typical full load speed for a 2-pole motor, for example, is 3500 rpm. Slip for a given motor is not constant. Rather, it increases as the load on the motor increases.

Synchronous motors are constructed with separate magnetic fields on the rotor. For small motors these may be permanent magnets. Larger motors have slip rings supplying direct current to the rotor windings. Synchronous motors, as the name implies, rotate at their synchronous speed with 0% slip. They generally have higher efficiencies than comparable induction motors, but at a higher cost. This has led to their frequent application to large single-stage blowers. Synchronous motors with DC excitation also have the ability to adjust their power factor to greater than 1.0. They are often used to reduce the power factor of an entire facility.

In the United States, the most common supply voltage for motors under approximately 500 hp is 480 VAC three-phase 60 Hz. In Canada, 600 VAC motors are common. Smaller motors may be 240 VAC. These are referred to as "low-voltage" motors. Large motors may be supplied by 4160 VAC three-phase 60 Hz or, occasionally, 2400 VAC. These are referred to as "medium voltage." The selection of supply voltage is dependent on availability, of course, but in large horsepower motors the lower operating current, reduced wire size, and lower cost of controls makes medium voltage economical.

The nominal motor voltages are often referred to as "480," "240," "440," or "220" V. This may cause confusion in motor selection. The nominal utility voltage is 480 or 240 VAC, which is the rated "supply voltage." An allowance for voltage drop between the supply and the motor is considered to reduce the voltage at the motor terminals to lower voltage—such as 460 or 230 VAC, which are the "utilization" voltages.

Motors are designed to accommodate variations between rated voltage and actual voltage. NEMA standards require the motor to accommodate voltage fluctuations of ±10%. Operating with greater fluctuations will cause performance to decrease and may result in motor damage. Voltage imbalance between phases, often caused by unbalanced loads for lighting and HVAC equipment, will also cause damage. Phase monitoring equipment is recommended for large motors. Phase imbalance detection is usually an integral part of advanced motor controls such as solid-state starters and VFDs.

The physical construction of the motor is often a compromise between heat dissipation and environmental protection. The two enclosure types commonly encountered for blower systems are open drip proof (ODP) and totally enclosed fan cooled (TEFC).

The ODP motor has louvered slots in the motor body to allow cooling air to flow through the motor. As the name implies, the louvers are arranged to prevent normal dripping or splashing water from entering the motor. A TEFC motor eliminates the openings and is suitable for wash-down, a frequent occurrence in many wastewater treatment plants. The motor body is equipped with fins for heat dissipation and a fan mounted on a shaft extension opposite the drive end of the motor forces cooling air axially along the motor. TEFC motors are routinely specified in many blower applications. Other motor enclosure types, such as totally enclosed nonventilated (TENV), weatherproof (WP1 and WP2), and explosion proof enclosures (typically Class I Group 2) are available but seldom specified for aeration applications. One enclosure that is occasionally used for variable speed applications is the totally enclosed blower cooled (TEBC), which may be useful if high torque at very low speed is required.

Note that the TEFC enclosure does not eliminate breathing of ambient air into the motor as it heats and cools. If the motor is operating intermittently, this may result in internal moisture condensation and potential motor damage. Large motors are often specified with internal heaters to prevent condensation. The heaters are connected to single-phase 120 VAC power when the motor is not operating.

Another consideration for TEFC motors in variable speed applications is the reduction in cooling air flow as the motor speed is reduced. This is a particular concern in constant torque systems, such as PD blowers, which cause the motor to generate heat at reduced speed. In many cases the blower limitations restrict minimum speed to approximately 50% of nominal. Usually this results in adequate motor cooling. However, high ambient temperature, lower speeds, and marginal motor size will all increase the potential for overheating. In questionable applications or with large motors (typically greater than 200 hp), RTDs or thermal switches embedded in the stator for direct temperature monitoring are a cost-effective precaution.

The most critical physical dimensions for motor mounting have been standardized. The NEMA frame size defines the dimensions of motor shaft height, shaft length, shaft diameter, and the size and locations of mounting holes in the motor base or flange. For example, any motor with a 324T frame will have identical dimensions for these features. Charts showing the standard dimensions for various frame sizes are readily available. However, it should be noted that overall length of the motor is not a standardized dimension. If the axial space for motor mounting is restricted, caution should be used in obtaining a replacement motor.

Smaller motors are designed with adequate bearings to allow for side loads from belt drives. The allowable side loads are standardized by NEMA. Frames with a "TS" designation are not intended for side loads and should be direct-coupled to the load. In general-frame size increases with increasing hp but there is not a one-to-one correlation. The same hp may be found in several frame sizes and the same frame size may be used for several hp ratings. Speed, enclosure type, and manufacturer's preference all have a bearing on frame size.

Although the motor enclosure is designed to dissipate heat, some allowance for an increase in motor temperature is required. Different insulation classes are available which are rated for safe operation at different temperatures (see Table 7.1). Most motors for blower applications are provided with Class F insulation.

The transient voltages on the output (load) side of VFDs can put additional stress on motor winding insulation. The rapid changes in voltage—the "d*V*/d*t*" of the square wave output of most VFDs—stresses insulation. NEMA has developed a standard specifically for motors used with variable frequency drives: NEMA MG1 Part 31.4.4.2 for "Inverter Duty." Existing motors with Class F insulation and a 1.15 SF are suitable for operation with VFDs but inverter duty motors should be specified for replacements and upgrades.

Duty cycle is not generally a concern for blower applications which tend to run for long periods of time. Motors applications are considered continuous duty if operation is at full load for 60 min or more in a 24 h period. Motor starting creates the greatest thermal load on a motor because of the inrush current and high torque during load acceleration. Short cycling, the frequent starting and stopping of a motor, should be avoided—particularly with across-the-line starting. If a motor is started, stopped before the motor cooling system can dissipate the heat created, and then started again before the motor can cool to near ambient temperature, the temperature can rise to damaging levels. Motor manufacturers estimate that each occurrence of an 18 °F (10 °C) increase above the motor's insulation rating will cut the motor's life in half.

The allowable frequency of across-the-line starting varies with motor size. A conservative approximation for a 25 hp motor would be six starts per hour. For a 500 hp motor one start every 12 h would be the conservative maximum. Reduced voltage starting and VFD control greatly reduce the thermal load during starting and allow more frequent starting. Similarly, starting large motors with winding temperature monitoring may be based on actual temperature and not on cycle times.

Inrush current when a motor is connected across-the-line to start it is a very significant consideration. The inrush current is typically 600% of full load amps (FLA). Unless the motor stalls, the duration of this current is only a few seconds at most, but it represents a very high thermal load for the motor. Overloads, circuit breakers, and thermal protection devices must accommodate the inrush current. Limiting motor current draw during starting is one of the most important functions of reduced voltage starters and a significant benefit of VFD control for motors.

The locked rotor current for motors is defined by NEMA Code Letters (Table 8.1).

TABLE 8.1 Partial Listing of Three-Phase Motor Locked Rotor Current

NEMA Code Letter	KVA/hp	Typical Motor Power (Verify with Mfg.)
E	4.5–4.99	
F	5.0–5.59	15 hp and up
G	5.6–6.29	7.5–10 hp
H	6.3–7.09	5 hp
J	7.1–7.99	3 hp
K	8.0–8.99	1.5–2 hp
L	9.0–9.99	1 hp
M	10.0–11.19	

The locked rotor current can be calculated from the KVA/hp defined by the Code Letter:

Equation 8.6

$$I_{LR} = \frac{P_m \cdot \text{KVA/hp} \cdot 1000}{1.73 \cdot V}$$

where

I_{LR} = locked rotor current, A
P_m = rated motor power, hp
KVA/hp = maximum value from table, KVA/hp
V = utilization voltage at motor, V

Motor full load current should always be verified from the motor nameplate or motor data sheet during final design. For an approximation during preliminary design, the ratio of 1.25 FLA/hp may be used for 460 V motors.

Motor current is a reasonable measure of the thermal load on a motor and the motor wiring. It is the basis for most overload protection systems. However, motor current alone is not an accurate indicator of the driven load of a motor (Figure 8.4). That is because power factor decreases significantly below approximately 50% of rated load (Equation 1.6). During initial system assessment the temptation is to measure actual amps, calculate the percentage of the reading to full load amps, and use that to determine the output power as a percentage of rated power. This method will give incorrect power values throughout the motor operating range—particularly at less than half load.

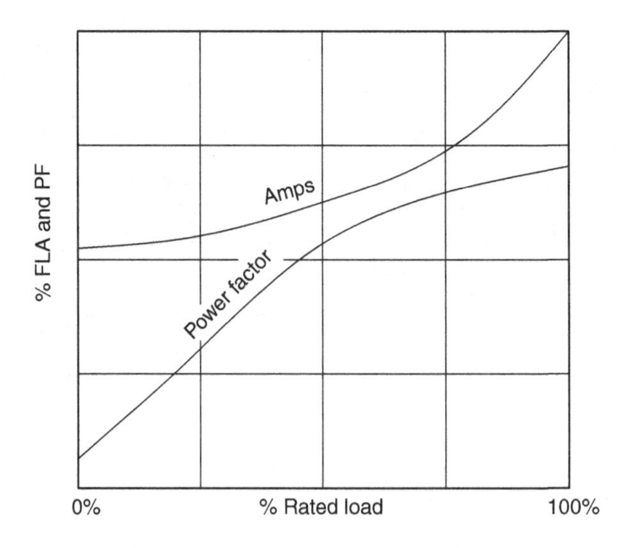

FIGURE 8.4 Variation of power factor and amps with motor load

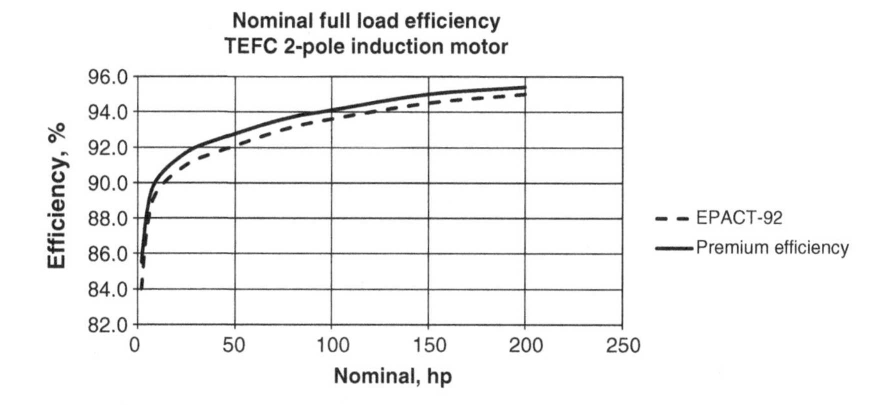

FIGURE 8.5 *Example induction motor efficiencies*

Motor efficiency also drops as the load is decreased, but not nearly so dramatically. The efficiency value for premium efficiency motors is generally quite constant until below 50%.

Motor efficiency is an important parameter in selecting blower motors in new or retrofit systems. Because blowers generally operate more than 2000 h per year at more than half load, the cost penalty of higher efficiency motors is usually recovered quickly. Motor efficiency has improved in recent years and standards have been developed to identify motors that have premium efficiency. Nominal motor efficiency and power factor generally increase with motor size (Figures 8.5 and 8.6). The nominal rpm and type of enclosure will also affect motor efficiency.

8.4 MOTOR CONTROL

Blower capacity control is an essential part of implementing aeration control and energy optimization. In some applications, for example jet aeration, pump control is also part of the process, but in most cases blowers represent a greater challenge and greater opportunity for optimization. Both pump and blower controls of necessity involve motor control. This includes starting and stopping the motors. If the blower is operated at constant speed, this is accomplished through a motor starter.

The basis of motor starting is simply connecting the motor to and disconnecting it from the power source. Since blowers and most aerators are not reversible, the starters employed are nonreversing. There is considerable variety in motor starting equipment, both in characteristics and physical construction, but the systems of interest fall into two categories: full voltage (often called "across-the-line" or "direct-on-line") and reduced voltage.

The basic full voltage nonreversing (FVNR) starter is essentially a 3-pole contactor with the addition of an overload device to shut the motor down in case of excess current draw. The overloads do not operate instantaneously—this allows for the high inrush current and provides time for the motor to accelerate the load. The required time delay is generally 10 s (Class 10) or 20 s (Class 20). The overload must also provide rapid tripping in the event of an extreme overcurrent condition such as a short to ground.

Older electro-mechanical starters used heaters to provide overload protection. These responded to the motor current draw by heating in a manner that emulated the motor winding heating. The heaters were selected based on the nominal full load amps of the motor. When the thermal capacity was exceeded the heaters would trip and generally could not be reset until sufficient time had elapsed for them to cool. The heater selection could be modified for special conditions, such as a motor installed in a location that had much higher ambient temperatures than the starter location.

Recent practice has been to replace the heater-based overload with electronic overload protection. These can provide more functions, such as phase imbalance protection and field adjustable current limiting. In some designs the electronic overloads measure motor current directly and in others a current transformer (C/T) is used to reduce the actual current passing through the overloads. The overloads have contacts wired in series with the contactor coil. When the overloads trip, the contactor drops out. This opens the circuit powering the motor and stops it.

Both the starters and the overloads should be equipped with auxiliary contacts. These are available in adder decks with various numbers of contacts and in NO and NC arrangements. The control system should utilize these as dry contacts to indicate run and fault status to the controller. Most starters will accept multiple auxiliary contact blocks and adding contacts to existing starters in retrofit applications is not usually a problem.

Motor starters are sized according to current, but for user convenience many manufacturers also provide starter ratings based on the motor hp rating at a specific voltage. Traditional U.S. designs are rated in NEMA sizes, beginning with 00 (Table 8.2).

NEMA starters are being displaced by IEC (International Electrotechnical Commission) starters in many new applications. IEC style starters are physically smaller, they are adapted to DIN rail mounting, and they are generally less expensive than their NEMA counterparts. IEC starters generally have adjustable overloads and protection for loss of phase. The sizes are listed by letter, beginning with A. There are more sizes used for IEC starters than NEMA, with a smaller range of hp ratings for each (Table 8.3).

TABLE 8.2 Partial Listing of NEMA Starter Sizes

NEMA Size	Continuous Amps	Full Voltage Starting, hp @ 460 VAC
00	9	2
0	18	5
1	27	10
2	45	25
3	90	50
4	135	100
5	270	200
6	540	400
7	810	600

TABLE 8.3 Partial Listing of IEC Starter Sizes

Size	Amp Rating (Induction Motors)	Maximum hp @ 460 VAC
A	7	3
B	10	5
C	12	7.5
D	18	10
E	25	15
F	32	20
G	40	25
H	50	30
J	63	40

The selection of IEC starters is more complex than NEMA starters. It is based on mechanical and electrical life in number of operations at maximum rated current and on motor duty cycle. Some designers prefer NEMA starters because they are more robust mechanically and electrically. However, if the selection is made correctly IEC starters will provide satisfactory operation for the life of the system.

A combination starter is simply a motor starter with the addition of a disconnect device in the same enclosure. The disconnect is most often a motor protection circuit breaker (MPCB) with trip characteristics tailored for motor starting service. However, fusible (fused) disconnects are also common—especially in smaller sizes.

Motor starters are frequently installed in motor control centers (MCC). The starters and disconnects are identical to those used with separate enclosures. The MCC provides a method of mounting multiple starters in a common enclosure which reduces the space requirements and eliminates the need to wire the supply power to each starter individually. The utility power is wired to the MCC's bus bar system, which distributes it horizontally to each section and vertically in each section to individual buckets. Each bucket is equipped with stabs, which clamp to the bus bars when the unit is slid into place. This connects the motor control devices to the power supply without wiring.

The high inrush current when starting a large motor will not only cause motor heating, it will also result in a voltage drop in the electrical supply system. This may result in lights dimming in adjacent locations and sensitive equipment may be damaged or drop out. In order to prevent this voltage drop, motors above a certain size are started with reduced voltage nonreversing (RVNR) starters. Many electric utilities mandate that motors above a certain rating—typically 40–50 hp at 460 VAC—use RVNR starters.

Older RVNR systems used an autotransformer or similar devices for reducing the voltage during starting. These transformers are equipped with several taps which provide various voltage ratios. The motor is initially started at the lowest voltage to minimize inrush current, then time delays and interlocks are used to switch to higher ratio transformer taps as the motor accelerates. When sufficient time had elapsed for the motor to reach full speed, it is switched to across-the-line operation.

This type of starter is no longer commonly used for new installations. The reduced voltage solid-state starter (RVSS), commonly referred to as a "soft starter," provides better performance at lower cost and is recommended for all new installations. The RVSS utilizes thyristors or similar solid-state switching devices to control the current and voltage applied to the motor. The level of current limiting and current ramping times are generally adjustable. A bypass contactor should be included so that once the motor is at rated speed it is connected directly to line voltage. This shunts the motor running current around the solid-state components and prevents their heating, thereby extending the life of the RVSS.

Most RVSS systems incorporate adjustable overload protection. Additional protection for phase loss or phase imbalance is included. Dry contacts are generally available to indicate run and fault status. Fault indicators on the RVSS provide diagnostic information for trip causes. Some units are equipped with communications modules to permit control, status indication and diagnostics over a digital communications network. Note, however, that a hardwired emergency stop (E-stop) should always be provided.

Soft starters were originally limited to low-voltage applications (<600 VAC). As technology has advanced, medium-voltage RVSS have become available for standard induction motors. Units are also available for retrofit applications to existing synchronous motors which include the necessary interlocks and DC excitation systems.

The application of motor starters, motor wiring, and controls are complex topics that affect safety. In the United States the National Electric Code (NEC) defines the requirements for proper installation of motors and other electric devices. Underwriters Laboratories (UL) also has standards for industrial controls. These have been adopted into many local codes and are referenced by many design specifications. In all cases, the final design must be reviewed for compliance with the NEC or other applicable codes.

Regardless of type, the function of all motor starters is to accelerate a motor to a constant operating speed and provide overcurrent protection for the motor. Optimizing power consumption for the blower system must be accomplished with throttling or guide vanes. As shown in Chapter 5, however, variable speed operation can

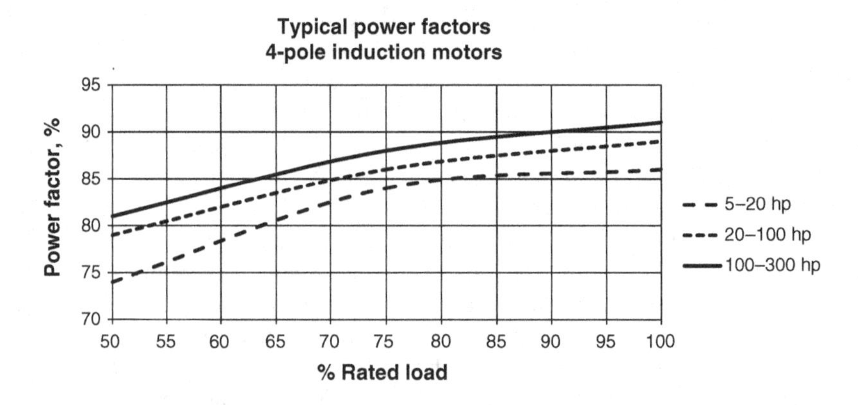

FIGURE 8.6 Example induction motor power factors

provide dramatic reductions in blower energy requirements. The benefits of reduced energy demand, improved process control, and extended equipment life have resulted in variable speed systems becoming the standard for the wastewater treatment industry.

8.5 VARIABLE FREQUENCY DRIVES

The advantages of using variable speed for blower and pump control have been recognized for many years. Earlier variable speed technology—such as wound rotor motors, eddy current clutches, variable pitch sheaves, and mechanical variable speed drives—were not generally satisfactory for aeration blower control. They were expensive, inefficient, had limited horsepower range, and suffered from high maintenance requirements. The industry norm was constant speed blower operation and control was by guide vanes and throttling.

The development of VFD technology has made most of these systems obsolete. The permanent magnet adjustable speed clutch has applications in lower horsepower and lower speed blower systems, particularly where drive harmonics are a consideration. This device actually controls the torque transmitted to the driven equipment by varying the gap between two sets of magnets, thereby controlling the speed. This is more efficient than older eddy current clutch technology—particularly at low speed.

In general, however, the variable speed system of choice is the VFD. VFDs are also variously called "adjustable speed drives," "adjustable frequency drives," "inverters," and simply "drives." Regardless of nomenclature, the VFD operates by converting incoming AC power to DC and then back to AC at a controlled frequency. For a given motor, the rotational speed is directly proportional to frequency (Equation 8.4).

There have been a variety of technologies developed for VFDs. Most state-of-the-art VFDs use pulse-width modulated (PWM) control of the output voltage (see Figure 8.7). The VFD output is a series of DC pulses of varying polarity and

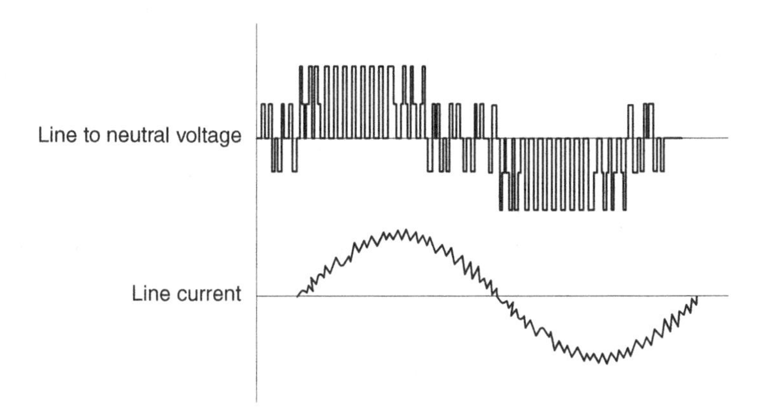

FIGURE 8.7 *PWM output voltage and motor current*

frequency. Because of the inductance in the motor, the DC pulses are attenuated and the current seen at the motor is a close approximation of sinusoidal current.

As with any competitive technology, there is a wide variation in the design details between VFD manufacturers. Most, however, have fundamentally similar operating principles (Figure 8.8). The input power has a sinusoidal waveform. For blower applications the input power is almost exclusively three phase. A rectifier section converts the AC to DC and filters the voltage. In the inverter section, the DC is switched on and off by high-power semiconductors. The frequency and duration of the pulses is controlled to match the frequency command from the local keypad or a controller.

Most blower VFD applications operate in "volts–hertz" mode. This maintains a constant ratio between the effective output voltage and the output frequency. For both U.S. 60 Hz and foreign 50 Hz power systems this ratio is 7.66:1. That is:

$$\frac{460\ \mathrm{V}}{60\ \mathrm{Hz}} = 7.66$$

$$\frac{383\ \mathrm{V}}{50\ \mathrm{Hz}} = 7.66$$

Maintaining this ratio is necessary to maintain the necessary magnetizing current at the motor fields. In general, at 50% speed the effective output voltage of a VFD will be 50% of the input voltage.

In blower control applications, open loop control based on maintaining a set output frequency is adequate. The blower speed is a secondary concern, with the primary control parameter being air flow to the process. Even in PD applications, the error in calculated flow rate introduced by ignoring motor slip is insignificant compared to other errors inherent in the control strategy.

There are more precise speed control techniques available in most VFDs. For example, closed loop vector control relies on feedback from a tachometer or encoder mounted on the motor. In sensorless vector control, the speed is calculated and controlled by the VFD's internal model of the motor. These and other techniques increase the complexity of the VFD setup and add nothing to ultimate process performance. Simple open loop frequency control is recommended for blower control applications.

There are a number of VFD parameters that routinely need adjustment during initial VFD setup. The setup can usually be performed from the local keypad. Many manufacturers also provide personal computer software for configuring the drives. The parameters generally configured at commissioning include the following:

- Maximum output current or motor FLA
- Minimum and maximum frequency
- Acceleration and deceleration times
- I/O configuration

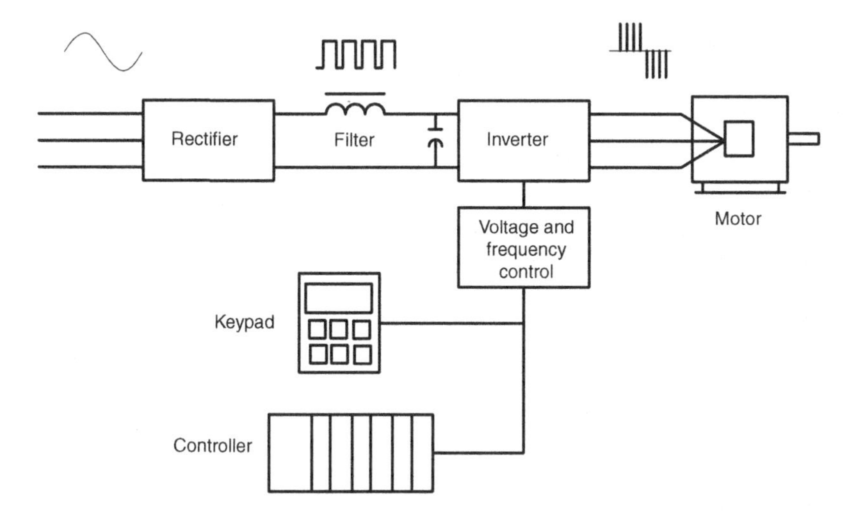

FIGURE 8.8 *Example PWM VFD configuration*

Most VFDs include two types of current protection for the motor. The first is simple overcurrent protection, where the VFD shuts off if the set allowable current for either the VFD or the motor is exceeded. This eliminates the need for separate overload devices such as those employed with motor starters. The second level of protection is current limiting. This clamps or reduces the motor speed if the preset maximum current is exceeded. This is a very useful function. It greatly reduces nuisance trips for overload. It also allows using of the full capacity of the blower up to the limit of the motor power. If discharge pressure or ambient temperature changes cause the motor to draw excess current, the VFD will slow the blower until the load is below the maximum motor capacity. When conditions change, the motor will accelerate but only up to the maximum current draw. The current limiting function is particularly useful for PD applications that may be manually controlled at the VFD.

VFDs also limit motor current during starting. This provides controlled acceleration, reduced inrush and starting currents, and minimal motor heating. Motor starts per hour can be increased in VFD applications above the limits associated with conventional starters without risking motor damage.

The minimum output frequency setting should be based on consideration of the characteristics of the blower and the motor. For PD blowers the minimum speed should be based on overheating of the blower or the motor, whichever results in the higher, more conservative, setting. For centrifugal blowers, the minimum speed is determined by the capability of the blower to provide sufficient pressure to supply air to the basins at the highest anticipated ambient temperature or by the minimum speed that will provide a safe margin above the surge point at worst-case conditions. The minimum speed setting should be a backup to the other surge prevention capabilities of the control system therefore excessive conservatism should be avoided. Setting the minimum speed at an unnecessarily high value will limit the operating range and blower turndown without improving equipment protection functions.

TABLE 8.4 Typical Speed Limits for Induction Motors

Frame Size	Maximum rpm, 2-Pole Motors	Maximum rpm, 4- and 6-Pole Motors
56–184	7200	5400
213–256	5400	4200
284–286	5400	3600
324–326	4500	3600
364–365	4500	2700
404–449	3600	2700

Maximum speed settings should also consider the blower system. The normal inclination is to set the maximum speed to match the nominal supply frequency—60 Hz in the United States. However, most VFDs will provide maximum output frequencies of 90 or 120 Hz, provided the current capability of the VFD is not exceeded. Most motors are also capable of overspeed (see Table 8.4). By taking advantage of the higher speed, it is possible to extend the operating range of the blowers—particularly at higher ambient temperature. Above 60 Hz the current and torque are limited, so the system will provide constant horsepower. Since the load's power demand increases with speed, a point is reached where the available motor power limits speed. Since many blower motors are oversized for normal operating conditions, however, this leaves a considerable margin of extra capacity. This can be used to limit blower starts and increase operating range without increasing system cost. The blower critical speed and maximum speed—as well as the maximum speed for a specific motor—must always be verified when using this technique to avoid mechanical damage from vibration or excess centrifugal forces.

Most blowers are high inertia loads. Acceleration and deceleration can result in problems with VFD control. This is true both during starting and stopping the blower and during normal speed changes encountered in blower flow control. If the acceleration time is too short, the VFD will experience overcurrent trips and shutdown. When the motor speed is being reduced it acts as a generator, creating higher than normal voltages in the DC conversion section of the VFD. This will result in overvoltage trips. It is usually necessary to change the default settings during commissioning to increase the acceleration and deceleration times. The final values depend on the WK^2 of the blower system and must be field determined. It is normal to have 30–60 s times for both. Recommended practice is to start at 30 s and increase from there as required.

Thirty seconds appears to be a long acceleration time but in a start from zero speed it is an acceptable time delay and doesn't cause operational problems. During normal control action the change in speed is only a few hertz and the delay in flow and blower performance is generally imperceptible.

Most VFDs have the ability to use a braking resistor or regenerative braking to shorten deceleration times. Braking resistors dissipate the energy stored in a rotating blower and motor as heat. Regenerative breaking converts the energy to electricity that can be fed into the power supply to the drive. Both options add cost and complexity to the VFD system and for blower applications there is only marginal benefit. Neither of these options is recommended nor required for blower control.

A related item is the availability of DC injection braking when a blower is stopped. During a controlled deceleration (as opposed to a coast-to-stop after elimination of motor power) the output power is cut off at a low frequency—usually around 6 Hz. DC voltage is applied to the motor to lock the rotor and prevent coasting. This can be beneficial in shortening the coast-down time if a blower needs to be restarted shortly after shutdown. DC injection braking is standard on most VFDs.

In some installations, leaking discharge check valves result in the blower back-spinning when it is off. The flow of air through the blower makes it act like a turbine and rotate in a reverse direction. Most VFDs have the ability to detect the reverse rotation, match output frequency to the speed, decelerate the motor, and then accelerate it in the correct direction. This eliminates the possibility of high current and torque damaging the motor, which can occur in across-the-line starting.

Most VFDs are equipped with a limited number of analog and discrete I/O points. The function of these points can be field configured to match the application requirements. For example, analog outputs may be set to retransmit motor current, motor kilowatt, or actual motor speed. Discrete inputs can be configured as run, stop, fault reset, run reverse, and so on. During system design, the available functions should be verified and matched to the control strategy. During start-up the configuration should be adjusted from factory default settings as required.

Output voltage for a VFD cannot exceed the input voltage. VFDs are classified as medium- and low-voltage based on the input power rating. For many years VFDs were economical only in low-voltage ranges for blower control applications. The cost of medium-voltage VFDs was simply too high to make the payback acceptable for all but the largest motors. Improvements in technology and a more competitive marketplace have changed this, however. Medium-voltage VFDs are increasingly applied on medium-voltage blower motors. This is particularly true for retrofit applications with large centrifugal blowers, since 2400 and 4160 VAC motors are common for this configuration.

A technology that has improved the economic justification for medium-voltage VFDs is synchronous transfer. The VFD itself is essentially a standard unit, installed with one or more across-the-line motor starters in parallel to the drive. Two or more blowers are connected each system and contactors are provided to switch any blower motor to the VFD or to a motor starter. The blower is started using the VFD and flow is modulated with speed control. When additional capacity demand requires starting another blower, the operating unit is adjusted so that the voltage output is synchronized with the utility supply. The motor is transferred to across-the-line operation and disconnected from the VFD. The VFD is set to zero output and the second blower is then connected to the VFD. The second blower can then be started and modulated using the VFD. Several blowers can be controlled by a single VFD in this manner.

A manual transfer switch is a similar technique used to reduce equipment cost in medium-voltage installations. As with synchronous transfer, multiple blowers can be grouped to operate from a single VFD. Unlike synchronous transfer, only one blower in the group can operate at a time. The blower to run must be selected when all blowers are off. Then interlocked contactors are used to connect the selected blower to the VFD.

These arrangements can reduce the cost of medium-voltage applications by eliminating the need for a VFD for every blower. In order to provide redundancy for reliability and to enable a bumpless increase in flow rate, it is recommended that at least two VFDs be provided. The operating blowers should be split between the VFDs, with two sets of synchronous or manual transfer systems.

Synchronous and manual transfer should not be confused with bypass contactors. In early applications, concern over VFD reliability resulted in the practice of providing conventional motor starters in parallel with the VFD to permit starting and operating pumps and blowers at constant speed in case of VFD failure. Improvements in VFD reliability make this an unnecessary precaution—particularly if multiple VFDs are provided. This is especially the case in large power applications where across-the-line starting is not possible. With large medium-voltage VFDs, providing reduced voltage starters in addition to the VFD may become prohibitively expensive.

VFD efficiency is obviously a concern in energy conservation applications. Full load efficiency data is available from most manufacturers and is generally 95% or higher. However, the internal losses for the VFD are not proportional to output power. The drive efficiency decreases at loads less than design and at speeds lower than 100%. Motor efficiency also decreases at lower speeds and loads which results in system efficiency that drops with decreasing load. This efficiency data is not readily available.

Fortunately, the lack of precise information does not have a dramatic impact on savings calculations. First, at lower loads the inefficiency reflects the losses of a lower total power requirement. Second, experience has shown that efficiency changes are not dramatic until speed is reduced to less than 50% and the load torque and current is reduced to less than 50% of rated values. Blower applications operate above these points in virtually all cases.

The power factor of VFDs is essentially constant and generally quite high. If more specific manufacturer's data isn't available then 95% can be used as a conservative estimate for the VFD power factor. The power factor of the VFD is the one measured by the utility metering. Power factor variations for the controlled motor won't affect power costs. The VFD will generally improve the plant's overall power factor and a reduction in penalties should be included in the savings analysis if the billing structure includes power factor.

As with most systems, the inefficiency of a VFD creates unwanted heat. Cooling for standard VFDs below approximately 1000 hp is generally provided by forced air and internal ventilating fans. Internal temperature monitors are usually provided to shut down the VFD if the temperature could potentially damage the drive. Larger VFDs are often liquid cooled, particularly for medium-voltage drives. A closed loop cooling system removes heat from the VFD and another external heat exchanger uses city water or forced air for ultimate heat removal.

Enclosures for VFDs are generally rated NEMA 1. This is to provide free air movement through the VFD for cooling. NEMA 12 enclosures with suitably louvered ventilation openings are available without significant cost penalties. However, NEMA 4 and 4X enclosure requirements significantly impede ventilation for air cooled enclosures. The designer should have an overwhelming reason for specifying VFDs with more stringent requirements than NEMA 1 or 12. It is worth noting that most VFD installations are adjacent or similar to MCC installations and most MCCs are rated NEMA 1.

Harmonics are an inevitable concern in VFD applications for blower control. The AC/DC conversion process creates electrical harmonics that are reflected into the power supply on the line side of the VFD. These can be a problem for other utility customers and can occasionally create problems within the treatment facility's internal power distribution system. Standby generators are particularly sensitive to influences from harmonic disturbances. Since blowers are a significant part of the total plant electrical load, a harmonic analysis is recommended to verify the extent of mitigation requirements needed for each installation. Most VFD suppliers can provide this at no cost.

So-called "clean power" VFDs have been developed to minimize the harmonics impact on the power supply. The most commonly specified harmonics standard is IEEE-519. This is a system standard and defines the allowable distortion at the point of common coupling (PCC), that is, the point where a facility's power source is connected to other utility customers. The PCC is generally the incoming power transformer. It has, however, become common practice in the wastewater industry to apply the IEEE-519 limits as an equipment standard, specifying the limits of harmonic distortion at the input power termination for each individual VFD.

A variety of technologies have been developed to mitigate the harmonics generated by a VFD. Early designs relied on input line reactors and 12- or 18-pulse transformers on the line side of the VFD to provide harmonics cancellation. Newer systems, particularly medium-voltage ones, use active front ends to provide dynamic cancellation of harmonics. Active front ends are less expensive, require less space, and are more efficient than the transformers used with most 18-pulse VFDs. Active front ends are gradually displacing the older technologies in new installations.

It is important to note that the "clean power" VFDs are concerned only with line side harmonics and have no bearing on the life or performance of the driven motor. In a successful application the impact of power disturbances at the load side of a VFD must also be considered. There are three principal areas of concern:

- Load side harmonics
- Potential dV/dt damage to motor insulation
- Bearing fluting

Harmonics are created by the VFD as an inherent part of the power conversion process. In extreme cases resonance can develop that generate heat and destroy the motor. The output pulses of a PWM drive can create very sharp increases in voltage in extremely short times; in other words, high dV/dt. This can stress motor insulation, and ultimately cause breakdown of the dielectric properties of the insulation and short-circuiting of the motor windings.

Bearing fluting is a rare mechanical problem caused by an electrical phenomenon. Capacitive coupling between the motor rotor and stator creates a voltage. Often the only path for the resulting current is through the bearings. This current results in what is essentially electrostatic discharge machining of the bearing races. The current erodes the bearing race and creates grooves or "flutes" in the race that ultimately result in bearing failure.

There are two precautions that minimize all three of these potential problems. Keeping the length of wiring between the VFD and the motor as short as possible minimizes the creation and impact of the phenomena. The acceptable length of cable depends on motor size, voltage, VFD design, and carrier frequency (a characteristic of the VFD output). Most VFD suppliers have cable length recommendations and these should be followed. Proper grounding of the power distribution system, the VFD, and the motor are also essential for trouble-free performance. Particular attention must be paid to ensuring proper grounding in retrofit systems. The impact of poor grounding on system operation is always deleterious and usually serious!

Reflective wave traps are essentially filters at the motor. These can reduce or eliminate the effect of harmonics and dV/dt transients on the motor.

Bearing fluting can be reduced or eliminated by short wire length and good grounding. Additional modifications to the motor may be required in demanding installations. Insulated bearings are available for new inverter duty motors to prevent currents passing through the motor bearings. If the blower and motor are direct coupled with a conductive coupling, however, the currents may pass through the blower bearings instead simply relocating but not eliminating the damage. Grounding brushes or slip rings are also used to provide a low resistance electrical path between the rotor and stator. Reducing the carrier frequency of the VFD output will also reduce the potential for fluting.

The VFDs used in turbo blowers are specifically designed for that service. They provide extremely high output frequencies (often several hundred Hertz) which are beyond the capabilities of standard VFDs. The performance characteristics of these VFDs are tailored to the high-efficiency permanent magnet synchronous motors used in most turbo blowers. The VFD is always a part of the turbo blower package.

VFDs inherently include significant microprocessor capabilities and fall into the "smart device" category. Many have built-in PID control capabilities and will provide flow or pressure control based on an analog input from a remote process transmitter. Some VFDs have more sophisticated capabilities and can include custom programming using ladder logic or high-level languages.

It is increasingly common to use the communications capabilities of VFDs for control and monitoring. Both serial and Ethernet communications are routinely available. Normal control functions, such as start/stop and speed control, can be implemented over the communications network with multiple VFDs being controlled using a single network. As with any smart device, using communications greatly increases the amount of information available for monitoring, decreases wiring cost, and simplifies diagnostics. A hardwired emergency stop, readily accessible to the operator, should be provided whether or not control is through a communications link. This hardwired stop circuit should also include major equipment fault shutdown contacts such as blower surge. The communications link should never be the only way to implement critical equipment and operator protection.

The function of the final control elements is to implement the control system strategy based on inputs from the instrumentation. The strategy and control logic provide process optimization and energy minimization based on the process needs. The control elements depend on the logic and control algorithms to determine the amount and direction of corrections to process operation.

EXAMPLE PROBLEMS

Problem 8.1

A butterfly valve is to be controlled by a directly controlled motor operator as shown in Figure 8.2 with 60 s per 90° travel. The controller output can provide timed outputs of ½ s duration.

(a) What will be the nominal increment in valve position with this system?

(b) Refer to the solution of Problem 6.4. If an 8 in. nominal valve is used, what would the approximate change in air flow be if the valve is opened one increment from an initial position of 25°?

(c) What would be the approximate change in air flow if the valve is opened one increment from an initial position of 40°?

Problem 8.2

A multistage centrifugal blower is direct coupled to a 40 hp induction motor operating at 3500 full load rpm and started across the line. The motor starter has Class 10 overload protection. The manufacturer's data sheet indicates the rotor WK^2 is 58 lb ft^2. The blower is rated at 850 SCFM and 6.5 psig discharge, with a nominal efficiency of 62%. Ambient conditions will range from 0 to 100 °F with a barometric pressure of 14.3 psia.

(a) Make a preliminary calculation to determine if the motor can accelerate the blower to operating speed without back pressure.

(b) Make a preliminary calculation to determine the ability to accelerate the blower if it started into a common header with other blowers.

(c) What design changes can be made to improve the starting capabilities of the blower?

Problem 8.3

Refer to Problem 8.2. The plant has an 80 hp portable generator that they wish to use as standby power for running one blower. The generator specifications indicate an output of 38 kW at 480 VAC, 53 A continuous, 48 kVA. The blower motor is marked NEMA Code Letter F and has a nominal efficiency of 93% and a full load power factor of 89%. Will the generator be able to start and run the blower?

Problem 8.4

Refer to Problem 8.2. The blower motor has NEMA frame size 324TS. Observation of the blower operation shows that the actual discharge pressure is only 5.5 psig. The operator has proposed running the blower at higher than nominal speed to achieve additional capacity and eliminate starting a second blower. The blower's first critical speed is 5800 rpm. Perform preliminary calculations to determine the feasibility of this design.

Chapter 9

Control Loops and Algorithms

Feedback control is essentially quite simple. You read the process parameter from the instrumentation, and if it is not acceptable you decide how much and in which direction to move the final control element. Any child that can regulate the temperature of their bathwater has mastered these fundamentals.

In the real world of automatic control systems, feedback control is more complex. This is particularly true when dealing with multivariable processes that have a high level of interaction between various pieces of process equipment. Aeration control certainly fits this category.

Most control loops are based on feedback control—the output of the controller is based on the error in a measured process variable (PV) (see Figure 9.1). Adjusting pump flow to maintain constant wet well level is an example. There are, however, also many feedforward control loops in use. A feedforward control requires a known relationship between the process and the controlled parameter. A classic example of this would be pacing the dosage of chlorine to the flow rate of well water.

It is important in developing and programming automatic control systems to keep the ultimate objective—optimizing process and energy performance—in mind. It is very easy for the designer or programmer to become absorbed in the minutiae of system operation and forget the ultimate objective. In every step of the initial evaluation—system design, performance analysis, and commissioning—one question should be applied: "How does this advance toward the objective and help optimize performance?"

Aeration Control System Design: A Practical Guide to Energy and Process Optimization,
First Edition. Thomas E. Jenkins.

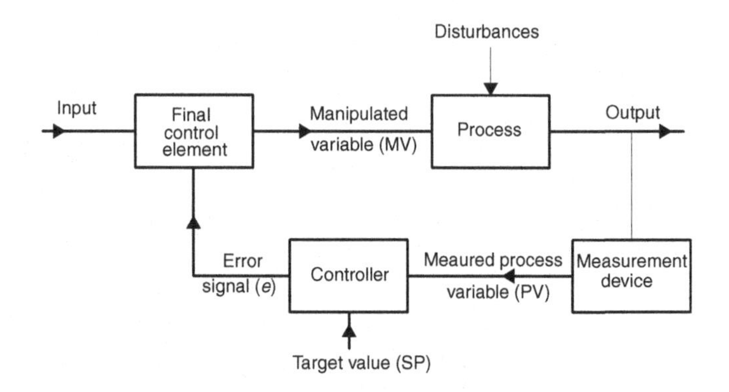

FIGURE 9.1 *Feedback control*

It is useful to recognize the historical basis for some of the automation concepts and algorithms in current use. This knowledge will help in deciding the best type of control strategy for each situation. In many cases "standard practice" is cited as the basis for a given control strategy. If the basis for the standard is understood it is easier to decide if it is the best solution, or even an appropriate solution, for meeting the specific objectives of the control system.

Manual process control is adequate for many simple processes, and if the process operates at steady-state conditions it may be the most cost-effective approach. Simple on/off automatic controls were developed for cases where manual control was inadequate or not feasible. Controlling wet well level with float switches and using pressure switches to cycle air compressors are examples. It is important to note that successful on/off control usually requires a deadband between starting and stopping values of the controlled variable to prevent short-cycling the system.

The earliest feedback controls in the modern sense were based on pneumatic sensors and actuators. Surprisingly sophisticated and complex control systems were based on this technology including proportional-integral-derivative (PID) control, limit alarms, and cascaded control loops. The early PID controls were often referred to as "three-mode" controllers. This was a reference to the "proportional," "reset," and "rate" adjustments used to increase control system accuracy. As technology progressed, the three-mode controller was emulated by single loop electronic controllers. Eventually the PID algorithm was incorporated in distributed control systems and programmable controllers as mathematical equations. Pneumatic controllers are, for all practical purposes, obsolete.

There are a variety of definitions that are applicable to all control systems. Nomenclature and mathematical symbols vary from one designer or manufacturer to the next, but the fundamental concepts are the same.

Process time lags are characteristics of the process that create a delay between a change in process inputs and a change in the measured process variable. For example, the delay between turning on the hot water tap and warm water reaching the showerhead is a process time lag. Process time lags may be further classified as resistance/capacity (RC) lags or dead time. RC lags are sometimes called transfer lags. They are created by the ability of the process to store energy or mass (capacitance lag) and by the resistance of the process to the transfer of energy or mass (resistance lag). Dead time, also referred to as transport lag, is a delay created by the time required to move energy or mass from one point to another in the system. Process time lags always increase the difficulty of achieving precise control.

The *process variable* (PV) is the measured process parameter being controlled.

The *setpoint* (SP) is the target or desired value of the process variable. It may be entered by the operator or it may be the output of another control loop.

The *error* (e) is the difference between the process variable and the setpoint.

Equation 9.1

$$e = \text{SP} - \text{PV}$$

where

e = error, in units of the PV
SP = setpoint, in units of the PV
PV = process variable, appropriate units

The *manipulated variable* (MV) is often called the *controller output* (CO). It is the command or signal sent to the final control element to change the input to the process so that the process variable approaches the target. The final control element may be a device, such as a blower or valve, or it may be another controller or control loop.

Tuning is the adjustment of the parameters in a control loop or algorithm that modifies the magnitude of the control system response to setpoint, load, or other process changes.

Cascade control is a series of two or more controllers used in controlling a single process variable. If the output (MV) of one control loop or algorithm becomes the setpoint of another control loop, the combination is referred to as cascade control.

A *direct acting* loop is one where increasing the controller output (the manipulated variable) increases the process variable. An example would be a well pump that must increase in speed to increase the water level in the discharge side water tower.

A *reverse acting* loop is one where increasing the manipulated variable decreases the process variable. An example would be a pump that must increase in speed to decrease the water level in the suction side wet well.

A *deadband*, also called a neutral zone, is an area of signal range where no action occurs. For example, the difference in temperature between the point where a heating thermostat sets (closes contacts) to start a furnace on falling temperature and where it resets (opens contacts) to stop a furnace on rising temperature is the deadband.

Hunting is a cyclic instability caused by overshooting the setpoint, with the process value being alternately above and below the desired value. Operators are justifiably reluctant to allow a system that exhibits hunting to remain in automatic control, and they will generally switch to manual control if hunting occurs. Operators are, in general, more willing to accept lack of precision in control than they are to accept hunting.

9.1 CONTROL FUNDAMENTALS

A full mathematical treatment of controllers and control theory is beyond the scope of this text. The discussions will be narrowly focused on control strategies specific to aeration control and the considerations involved in developing and optimizing aeration systems and aeration energy demand.

9.1.1 Discrete Controls

Alarming is one of the most basic types of discrete control. Alarms can be categorized by importance and by action:

- Equipment protection alarms shut down process equipment operating outside the normal range. These alarms are intended to prevent catastrophic equipment failure. Affected equipment will be stopped without operator intervention, but operator action may be required to correct process excursions or start standby equipment.
- Impending failure alarms notify the operator that the process or equipment operation is approaching normal limits and operator intervention may be required. Impending failure alarms usually precede equipment protection alarms and shutdown. They may be tripped by a different process value than the protection alarm. They may also indicate the protection process value has been reached but the condition has not been of sufficient duration to trip the protection alarm.
- Warning alarms indicate an abnormal process or equipment conditions that may indicate a problem but will not lead to equipment failure. Operator intervention may be required or advisable.
- Informational indicators show that a change in status has occurred as part of normal operation. No operator action is required.

This terminology is common but far from universal. For example, some designers and programmers classify alarms as most critical, less critical, and noncritical depending on the level of influence the condition will have on maintaining process operation. Regardless of classification, the design specifications and operator instructions should clearly identify the function of each alarm and the appropriate operator response to it.

In today's networked control environments it is possible to make the operator response to an alarm as simple as a mouse or keyboard click from a remote office. It is possible to make clearing an alarm too easy. For many equipment alarms the operator may need to look at the affected equipment and verify that

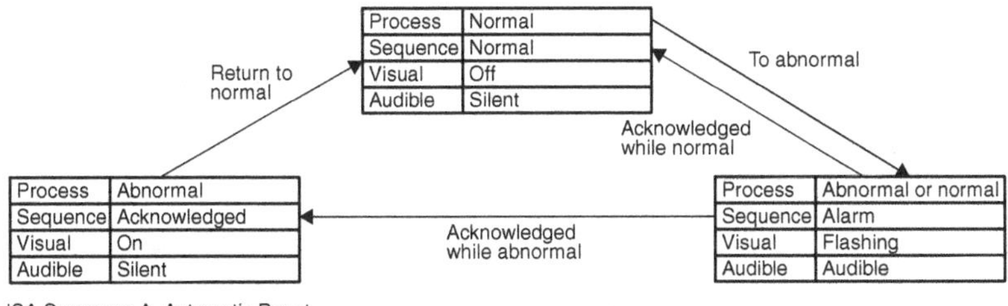

FIGURE 9.2 *ISA automatic reset alarm sequence*

the alarm condition is no longer present before the alarm should be cleared. Most operators are extremely conscientious in this regard, but in the midst of multiple alarm occurrences, repeated alarms, and the distraction of other duties, the temptation exists to simply clear the alarm and restart the equipment. If it is important to have an operator observe the equipment to inspect and approve restarting, it may be necessary to require local alarm acknowledgement in lieu of supervisory control and data acquisition (SCADA) or other remote alarm clearing.

Alarm indication was traditionally provided by an annunciator—an array of indicating lights engraved with alarm messages and providing one light (window) for each hardwired alarm. The annunciator has been made obsolete by the human/machine interface (HMI) and PC-based SCADA systems. These systems not only indicate alarms, but will time and date stamp them and may archive them for future retrieval and reference. Alarm indication may also include audible horns or buzzers, flashing beacons or stack lights, and autodialers to notify remote personnel by phone message. It is important to have a silence capability for alarms to eliminate operator distraction while considering the impact of an alarm event.

Although the annunciators are obsolete, the alarm sequences developed by the Instrument Society of America (ISA) for annunciators still provide useful guidance for alarm indication and acknowledgment. There are many sequences for indication and annunciation identified by the ISA, but the automatic and manual reset sequences are the most common (see Figures 9.2 and 9.3).

There is no universal or standard color code for alarm messages. A common practice is to use red for equipment protection or other "most critical" alarms and amber or yellow for impending failure or warning alarms. Informational alarms are usually green or gray. An alarm indicator in the off state should be white or gray. Many plants have established their own alarm color code and when it exists it should be used.

There is also no universal color code for normal equipment status indication. As with alarms, operator preference should be given priority. Two schemes are in common use:

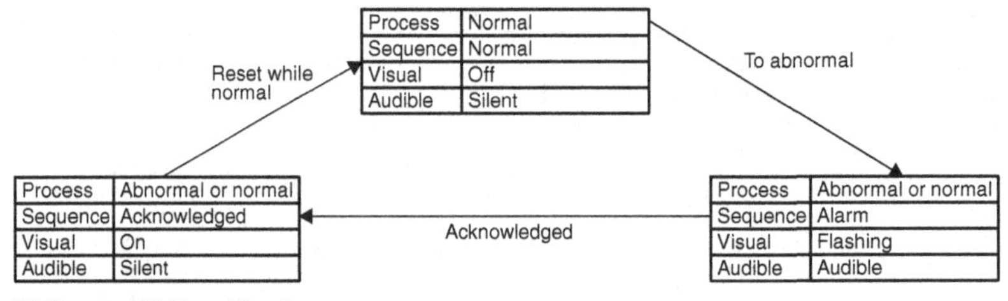

ISA Sequence M, Manual Reset

1) Includes acknowledge, reset, test, silence buttons
2) Audible alarm silenced with silence button
3) Alarms locked on until acknowledged
4) Audible and flashing stop with acknowledge button
5) Manual reset of acknowledged alarm when process returns to normal
6) Includes operational test of alarms

FIGURE 9.3 *ISA manual reset alarm sequence*

- *Traffic Light Emulation*:
  - Green for motor run and valve open
  - Amber for valve closed
  - Red for equipment or motor fault
  - Gray or background color for off status
- *Power Plant Norm*:
  - Red for running motor or open valve, indicating a potentially hazardous condition
  - Green for off status motors or closed valves, indicating a probable safe condition
  - Amber or yellow for alarm or fault

The programmer and designer should be judicious in the use of colors. A code that seems perfectly obvious to the programmer in the office may be useless, or worse—confusing—to the operator on site coping with a process emergency. Crowded screens and cute animations should be avoided. The purpose of the SCADA and HMI screens is to convey useful information to the operators for assisting them in operating the process. There is a difference between data and information. Data is numbers or status indication in random or unprioritized order. Information is data that has been culled for useful context and displayed in a fashion that clearly and quickly conveys the high-priority facts to the operator.

Similarly, only true alarms should be displayed obtrusively and audibly highlighted. Status indication and other informational indicators can be displayed and logged for operator convenience, but they should not be intermixed with true alarms. Not only is the constant interruption and audible indication annoying, excess use of the alarm function tends to make operators disregard alarm indicators—which may lead to their ignoring critical alarms.

Frequent "nuisance" alarms are another occurrence that risks causing an attitude of indifference on the part of operators. Time delays should be used for equipment shutdown conditions so that shutdown does not occur as a result of transient process conditions or random signal fluctuations. Judgment and knowledge of the equipment limitations is required in establishing appropriate delays. The time delay for alarm implementation should be field adjustable and each type of alarm for each piece of process equipment should have its own time delay setting. They may all end up with the same setting but having the flexibility to adjust alarm occurrences to field conditions will enhance performance and operator satisfaction.

Debounce timers are useful for inhibiting nuisance alarms and also for verifying functions such as equipment run status. A debounce timer is simply a short time delay, often a fraction of a second, applied to the controller input before echoing the status in the control logic. They get their name from the tendency of some mechanical contacts to physically bounce—making and breaking continuity to a controller input—and causing premature loss of status. Some controller inputs have a debounce function built into the controller I/O card and others must have it added to the control logic programming.

Loss of analog signal is an alarm condition that requires operator attention since the control strategy depends on reliable process measurement. If the analog signal is 4–20 mA, a value lower than approximately 3.6 mA should initiate a time delay, and the loss of signal alarm should be initiated when the time delay has elapsed. Many controllers include a loss of signal bit in the analog input cards. If the input signal does not have a live zero then an invariant zero input signal can be used to initiate the alarm timer. Although this is not as reliable as a 4–20 mA loss — and requires a longer delay — it is an appropriate strategy for 0–10 VDC or similar zero-based analog signals.

The program response and required operator intervention for a loss of signal alarm should be addressed in operator training. The program should set a warning alarm and lock out process control logic for affected zones on loss of some signals, such as DO concentration. For other signals, such as blower motor current, an equipment protection alarm and immediate shutdown should be initiated.

When most designers and operators think of control systems they think in terms of analog control—a continuously changing measurement being modulated by a continuously changing output. A great deal of aeration control, however, operates with discrete inputs and discrete outputs—in other words, simple on/off control. Many discrete control systems consist of equipment protection and alarming. In small systems discrete feedback control can provide cost effective control when analog control is too expensive (Section 9.1.4).

9.1.2 Analog Control

Analog control is more flexible and powerful than discrete control for most wastewater treatment process control applications. Most aeration process control uses analog inputs from the field instrumentation. Analog data is based on a continuous signal from the process transmitter. The value of the signal changes in proportion to the process value. The control logic interprets this continuous signal and executes necessary control actions on the final control elements. The control output may be an analog signal or a discrete signal, depending on the field device and the control strategy.

It is recommended that the analog control logic, HMI displays, and controller operations utilize floating-point values in engineering units. The expansion of microprocessor speed, controller memory capacity, and math function availability in modern controllers makes this implementation the most reasonable approach. The convenience and intuitive grasp of process status during programming and debugging offset any minor increase in execution speed and programming effort involved in the conversion to engineering units.

The process value data used in the controller must pass through a number of transformations before becoming useful information in engineering units. The process is sensed by a transducer which generates a low-level electrical signal. The transmitter amplifies and scales the signal so it is proportional to the calibrated range. This signal, typically 4–20 mA, is transmitted to the controller analog input channel. The signal is converted to raw data from 0 to maximum, based on the resolution of the input card's A/D (analog to digital) processor. Finally, the controller converts the signal to an engineering unit that reflects the process value (see Figure 9.4).

Scaling analog values as engineering units can be accomplished in a variety of ways, depending on controller capabilities. Some I/O cards have scaling built into the I/O firmware and — if the signal level's high and low values are entered and the minimum and maximum engineering units are entered — the I/O card will pass the engineering units directly to the program. In other systems, the input signal is directly available to the logic — as 4.00–20.00mA, for example. In this case, the scaling can be performed in the logic using Equation 7.1. Some controllers have built-in instructions to scale the process variable from the I/O card data register. If minimum

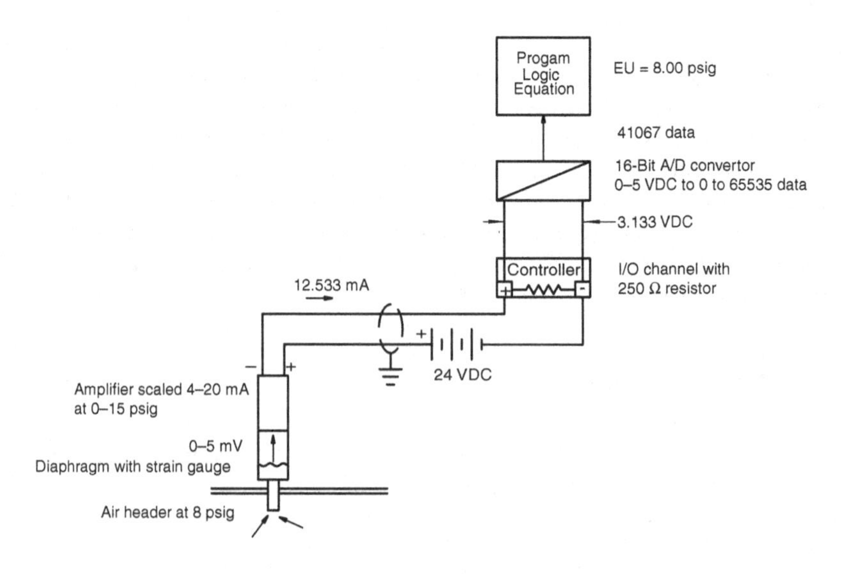

***FIGURE 9.4** Process measurement transformations*

and maximum engineering units are entered into a function block or standard command, the function block will provide the engineering units to the logic.

Regardless of format or method, virtually all scaling ultimately relies on a linear relationship between the measured value and the engineering units:

Equation 9.2

$$EU_{Actual} = \left[\frac{EU_{Span} \cdot (Data_{Actual} - Data_{Min})}{Data_{Span}}\right] + EU_{Min}$$

where

EU_{Actual} = actual process value, engineering units
$Data_{Actual}$ = actual data in register from I/O card A/D conversion
$Data_{Min}$ = data in register at minimum analog signal from transmitter
EU_{Min} = engineering units at minimum analog signal from transmitter

Equation 9.3

$$EU_{Span} = EU_{Max} - EU_{Min}$$

where

EU_{Max} = engineering units at maximum analog signal from transmitter

Equation 9.4

$$Data_{Span} = Data_{Max} - Data_{Min}$$

where

$Data_{Max}$ = data in register at maximum analog signal from transmitter

The values for scaling engineering units should be adjustable from the HMI or SCADA without requiring program changes. Over the life of a system it is reasonable to expect replacement of some transmitters and the range of the new units may not match the old. For many process measurements, such as motor amps or gauge pressure, the transmitter is inherently zero-based and an adjustable full scale is sufficient. For other measurements, such as temperature, the minimum engineering units may not be zero and providing an easily adjustable minimum value is helpful.

In an ideal system the actual and measured process values would be stable signals, varying gradually in a uniform progression as the process changes. In reality, of course, the analog input signals exhibit fluctuations. Sometimes the fluctuations are negligible, but often they are large enough and/or rapid enough to interfere with stable and accurate control logic execution.

The signal fluctuations may be caused by EMI/RFI from adjacent AC power wiring. The creation of this type of interference can be reduced by good wiring practice—shielded cable for analog signals, separate conduit and wire duct for AC and DC wiring, and so on. Signal noise of 50 and 60 Hz is usually eliminated by hardware and/or software filtering on the controller's analog input card.

Other types of signal fluctuations are more problematic. Periodic fluctuations in the actual process value are inherent in some systems. For example, the discharge pressure of a PD blower will exhibit sinusoidal fluctuations in pressure as the blower lobes rotate. Other process measurements fluctuate more randomly such as the aeration basin DO concentration.

Regardless of the source, these constantly fluctuating signals are annoying to operators, because they make it difficult to read and interpret process behavior. The fluctuations can also cause unnecessary control actions—particularly if the logic is based on a "snapshot" of the process measurement. If the reading is caught at a momentary maximum or minimum value the control output may be changed, adding to process instability.

The Nyquist rate is the theoretical minimum sample frequency necessary for control. It is equal to twice the highest frequency of fluctuations contained in the signal. Experience shows that a much higher sampling frequency is actually required for accurate control. Achieving a fast enough sample is not a problem in most systems. Issues are more likely to result from sampling the analog signal too frequently, resulting in rapidly fluctuating readings that are difficult to control and that may produce erroneous results.

Digital filtering on the analog signal can result in a reading that converges on the mean process value to improve process stability. This technique is based on a running average of current and past readings:

Equation 9.5

$$Y_n = Y_{n-1} + \left(\frac{1}{k} \cdot (X_n - Y_{n-1})\right)$$

where

Y_n = new filtered value of measurement
Y_{n-1} = filtered value of measurement from previous calculation
k = the number of samples over which the data is to be filtered
X_n = current unfiltered measurement

The frequency and level of filtering is a balancing act. If the filtering is excessive, the value may not respond to rapid changes in the process. If the filtering is insufficient then reading stability won't be obtained (see Figure 9.5). The filtering calculation may be performed at controlled time intervals initiated by a program timer or it may be executed every scan. With a very fast controller and program scans less than 100 ms it may be necessary to use a very high value for "k" to achieve stable readings. If filtering is controlled by a timer, the frequency must

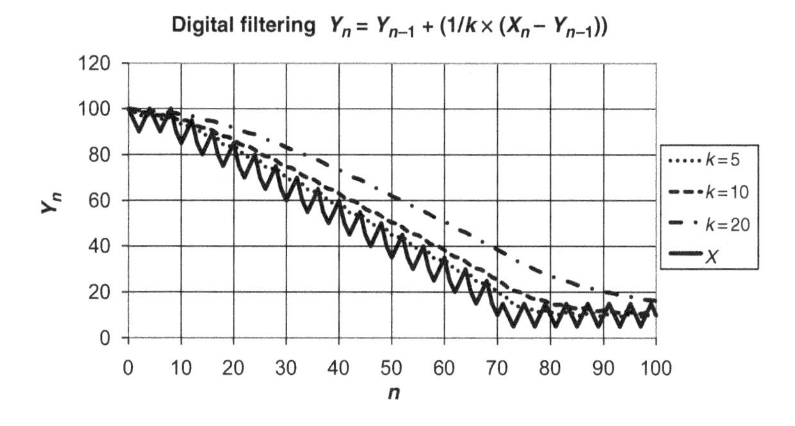

FIGURE 9.5 Effect of digital filtering

be sufficient to reflect normal process fluctuations. Experience has shown that executing the calculation every 300 ms and averaging over 10 readings ($k = 10$) is adequate for most processes. Testing and adjustment of the filtering values during commissioning is recommended.

Process response time is another consideration in executing control algorithms and logic. Wastewater aeration systems inherently include very long process time lags. Continuously executing changes to the manipulated variable tends to cause hunting and reverting to manual control.

There are many occurrences of mixed discrete and analog control in most systems. For example, a programmed process alarm based on internal logic responding to an analog input would be a mixed control strategy. A feedback control based on process measurements with the control output being timed operation of valve motor contacts would be another example. To some extent the classification of a control strategy as discrete, analog, or mixed is immaterial—what matters is how well the strategy implements the control objectives of the operator and optimizes process performance and energy consumption.

9.1.3 Proportional-Integral-Derivative

The most commonly implemented algorithm in general industrial process control practice is the PID algorithm. PID control has a long history, a large installed base, and an interesting mathematical expression. Most control engineers are taught PID algorithms in college. It is the default method of analog control for many system designers and programmers.

As it evolved from earlier mechanical and electrical three-mode controllers, the PID algorithm became expressed mathematically and implemented in programmable controllers. A common expression for the algorithm is:

Equation 9.6

$$\mathrm{MV}(t) = K_\mathrm{p} \cdot e(t) + K_\mathrm{i} \cdot \int_0^t e(t) \cdot \mathrm{d}t + K_\mathrm{d} \cdot \frac{1}{\mathrm{d}t} \cdot e(t)$$

where

$\mathrm{MV}(t)$ = manipulated variable, the control output at time t
K_p = proportional gain, dimensionless
$e(t)$ = error at time t (see Equation 9.1)
K_i = integral gain, dimensionless
K_d = derivative gain, dimensionless

There are many variations of the basic PID equation. In one of the most frequent variations, the proportional gain is common to all three control actions, so changing it will affect the response of integral and derivative action:

Equation 9.7

$$\mathrm{MV}(t) = K_\mathrm{p} \cdot \left(e(t) + \frac{1}{T_\mathrm{i}} \cdot \int_0^t e(t) \cdot \mathrm{d}t + T_\mathrm{d} \cdot \frac{\mathrm{d}}{\mathrm{d}t} \cdot e(t) \right)$$

where

T_i = integral time (often expressed as resets per second or minute)
T_d = derivative time, seconds or minutes

Most PLCs and all DCS systems have built in PID algorithms, and some controllers allow the programmer to choose between multiple forms and variations.

The three PID terms have different objectives, but the ultimate intent of combining the three control actions is to make the process variable match the setpoint.

Proportional control is the most basic function and the most intuitive. Essentially the proportional control term looks at the existing error and changes the output linearly in response to the magnitude of the error. This action is intuitive, as one would expect a greater control effort would be required to correct a greater error. The control effort, the change in the manipulated variable, is controlled by the proportional gain. Note that some controllers use "proportional band" instead of proportional gain. The two are related as follows:

Equation 9.8

$$\mathrm{PB} = \frac{1}{K_\mathrm{p}}$$

where

PB = proportional band
K_p = proportional gain

Proportional control requires a change in error to produce a change in output. Proportional control can produce an exact correction for one, and only one, load condition. At all other conditions there will be some uncorrected error, a phenomenon referred to as proportional droop or offset. The purpose of integral control action is to eliminate this offset. In earlier controllers it was referred to as reset action because it had the same effect as resetting the setpoint and driving the output to a new value. Integral action in effect looks backward, basing its effect on the control output on the accumulation of error over time.

Because of the impact of cumulative error, integral action can create significant overshoot—particularly during process or control system start-up. This effect is known as reset windup, and can be very unsettling when equipment is brought online or recovering from a power failure. Many controllers have "anti-reset windup" logic available to minimize this effect.

If the process variable is changing rapidly in response to controller output, there is a tendency to overshoot. The derivative term, referred to as rate action in earlier controllers, looks at the rate of change of the error. When the process variable (and hence the error) is moving rapidly in the correct direction, the derivative term reduces the change in the manipulated variable to reduce overshooting the setpoint. If the process variable is moving further from the setpoint, the change to the manipulated variable is increased. The faster the rate of change, the greater the impact of the derivative term is on the output. The derivative term has a tendency to increase hunting and is often eliminated from control logic by setting the derivative time or derivative gain to zero. This is called a "PI" loop.

Tuning PID loops involves both art and science. Tuning a PID control loop requires the adjustment of gain, integral gain (reset), and derivative gain (rate) to achieve the desired control system response (see Figure 9.6). The goals of loop tuning are stable control with a rapid and exact response to setpoint and load changes. Unfortunately, these two goals are often mutually exclusive. Aggressive tuning promotes rapid response, but it also tends to induce instability and hunting.

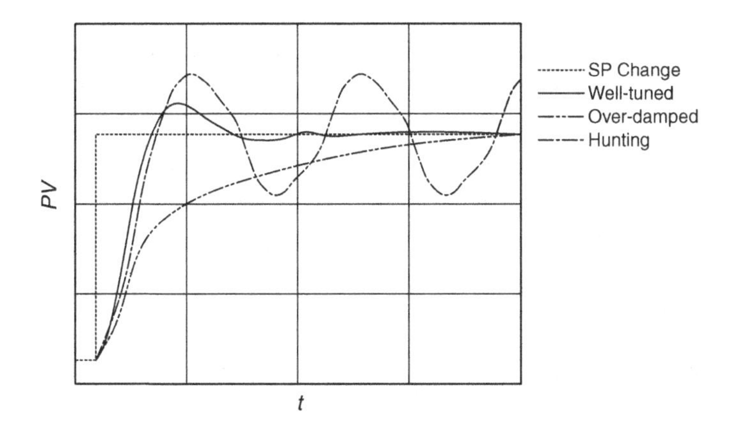

FIGURE 9.6 *Example PID tuning*

The art of tuning comes with experience in the dynamic response characteristics of both the process and the control algorithm. Very few operators have developed the "feel" for tuning that experienced programmers, technicians, and engineers acquire. This makes tuning PID loops at best challenging and at worst impossible for the staff at most treatment facilities.

The more scientific approach involves calculating initial tuning parameters, which are then modified to optimize performance. One common method is the Ziegler–Nichols method. The proportional gain is adjusted until oscillation occurs, the result being the so-called ultimate gain. The proportional gain and the other tuning parameters are then calculated as ratios of the ultimate gain. There are many other tuning methods based on a combination of test and calculation. There are also a variety of self-tuning algorithms. Most self-tuning involves a perturbation of the process by setpoint or load change and then calculating gains from the resulting response.

A dazzling number of magazine articles, books, and software packages intended to simplify and optimize tuning PID loops are introduced every year. This seems to indicate that PID tuning is both problematic and lacking in a definitive solution!

Tuning difficulties are not the only problems in applying PID to aeration control. The algorithm is best suited to linear response systems, but most aspects of aeration the process are nonlinear in nature. The response of DO concentration to air flow rate changes at low loading is very different from the response at high loading. Air flow rate response to valve movement at 20% open is very different from the response at 60% open. The change in centrifugal blower flow with pressure variation varies widely from minimum to maximum flow. These nonlinearities compound the PID tuning problem since tuning that might be satisfactory at high loads and flows may create hunting and instability at low loads.

The PID implementation in most controllers executes on an essentially continuous basis, since the scan time of the controller is orders of magnitude faster than the response time of the aeration process to changes in operating parameters. The PID loop is inherently unsuited to processes with long process time lags. Tuning that creates a stable response to normal process changes, for example, will respond slowly to sudden load changes from industrial slugs or internal plant sidestreams.

Despite these known limitations, many designers and programmers use the PID algorithm as the default technique for analog process control. Some seem to consider it the only acceptable technique and treat the equations as if they were a law of physics rather than a convenient expression for one of many approaches to process control. There are many cases where PID is an excellent choice for the control strategy—aeration control is not one of them.

9.1.4 Deadband Controllers

The simple deadband controller can be very effective and stable in applications that do not require precise control of the process variable. The operating principle is quite simple: when the process variable is in the deadband the current control status is maintained. If the process variable falls outside of the deadband, the appropriate

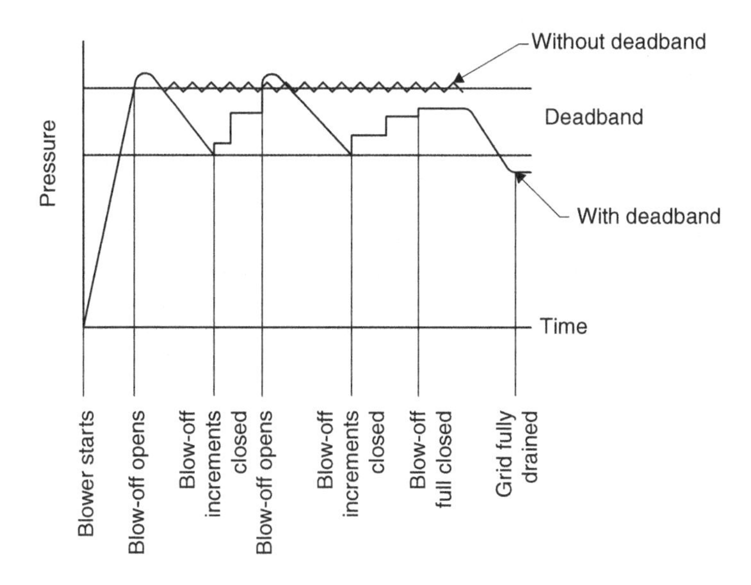

FIGURE 9.7 *Deadband pressure control for blow-off valve*

control action is implemented. The function of the deadband is to prevent rapid cycling of the manipulated variable by inhibiting control changes immediately after process variable changes.

A common application for a deadband controller is opening a main header blow-off valve to avoid blower surge when the piping system is blocked—for example, when entrained water in the diffuser grid restricts air flow after a power failure (see Figure 9.7). When the blower starts, the pressure in the main header builds as air in the system is compressed. Flow through the diffusers is blocked by the grid piping, which filled with water when the blower was off. When the pressure reaches the high setpoint, the blow-off opens to provide a path for the blower discharge air flow. The high setpoint is above the normal maximum system pressure but lower than the maximum pressure capacity of the blowers.

When the pressure in the header rises, the air begins to force the water out of the open drains in the grid piping. (The drains may require operator intervention to open.) The pressure begins to drop. If the deadband didn't exist, the pressure would immediately rise since insufficient time has elapsed to drain the grids. This would begin a cycle of the blow-off opening and closing as soon as the system pressure changed.

The deadband is established by a lower setpoint, which is above normal maximum operating pressure. While the pressure is inside the deadband, the blow-off position doesn't change. This maintains air pressure to force the draining of the header but maintains safe blower operation by providing an air flow path to atmosphere. When the pressure drops below the deadband, the blow-off closes to force the blower air flow into the aeration system. The blow-off operation may cycle several times but in a slow and controlled manner that eliminates hunting.

Aeration basins with mechanical aerators have long employed deadband controllers with discrete I/O for DO control. Most DO transmitters have high and low alarm contacts. When the aeration basin DO level rises to the high setting, the alarm contacts switch off the aerator(s). When the DO concentration drops to the low level, the aerator(s) are turned back on. For small package plants with diffused aeration the same strategy can be used to cycle blowers on and off. The deadband between the high and low settings will minimize short cycling. The resulting control won't provide a constant DO concentration in the aeration basin and aeration energy won't be optimized, but this system can provide significant energy savings with very low equipment cost. That may make it cost-effective in small facilities where the available energy savings won't justify more sophisticated and costly systems.

The on/off DO control should include timers to limit the off time to a maximum value and to limit the run time to a minimum value. The timers are recommended to ensure adequate mixing and prevent short-cycling motors. The off time limit prevents excessive settling and the run time ensures that enough aeration occurs to provide resuspension of the mixed liquor solids. The required functions can be very easily included in a micro-PLC costing just a few hundred dollars. Additional features like blower alternation, alarming, and communications interface to SCADA systems can be included with only minor additional programming.

9.1.5 Floating Control

Floating control is a flexible technique that combines responsiveness and stability. It is common in HVAC control, but has not yet found wide acceptance in general process control. Many features of other control algorithms, such as deadband control and proportional control, can be incorporated into the floating control algorithm as necessary. Time delays may be used to match control actions to system response characteristics.

Floating control provides a predetermined relationship between the error and the manipulated variable. The changes in the manipulated variable are intentionally very slow. The change may be incremental or continuous—depending on the process requirements and the nature of the final control element.

In aeration control systems floating control usually incorporates two enhancements:

- A deadband in which no control action takes place (also referred to as a tolerance zone or a neutral zone)
- Time delays between each successive execution of the control algorithm to allow the process variable to stabilize in response to the manipulated variable change

The most common floating control implementation uses a single incremental change in the manipulated variable after each control calculation. It is often referred to as "single-speed" floating control. The sequence consists of taking a snapshot of the process, determining if a correction is required, making the predetermined correction, and then implementing a time delay to allow the process to respond after

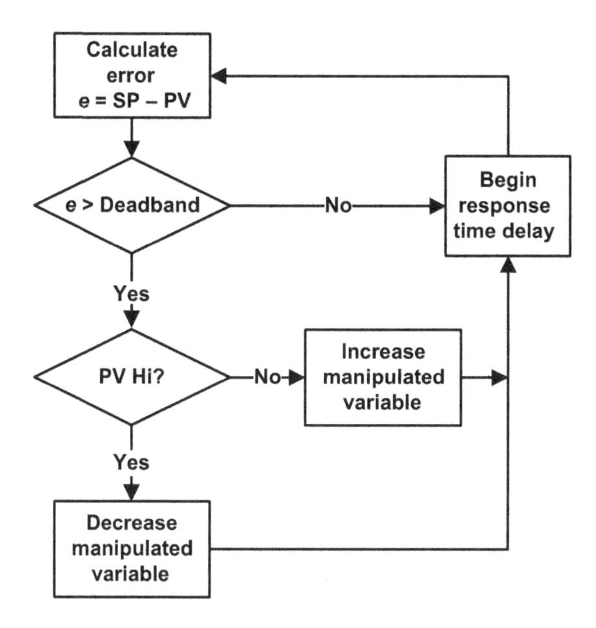

FIGURE 9.8 *Floating control with deadband logic*

the correction. If the process variable is within the deadband, no action is taken (see Figure 9.8). Note that the deadband is twice the tolerance. The logic can readily incorporate direct acting or reverse acting control.

The basic floating control algorithm is analogous to adjusting water temperature while taking a shower. The bather checks the temperature, taps the hot water valve open or closed a little, and then waits for the temperature to stabilize. The process is repeated until the shower temperature is within the comfort zone.

Refer to Figure 9.9 for an illustration of the control method. In this example, the process variable is air flow and the manipulated variable is valve position.

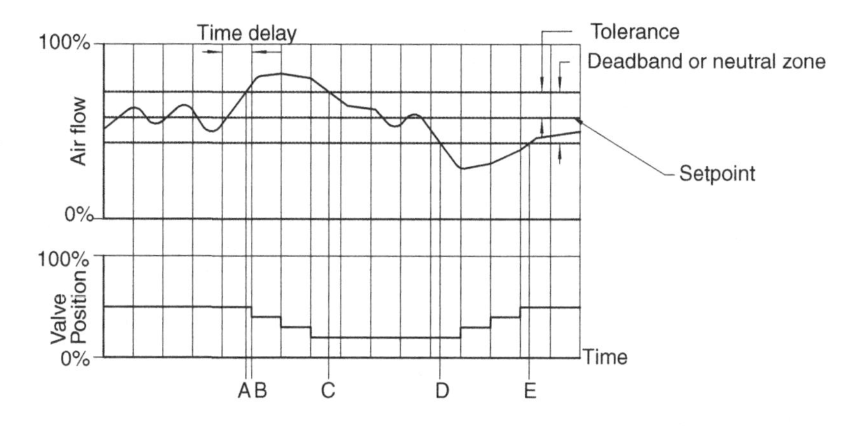

FIGURE 9.9 *Sample floating control with deadband*

From the beginning of the data until point "A", the flow is fluctuating within the deadband and no control action is required. At point "A" the flow is higher than the upper deadband limit. However, no immediate action takes place until the time delay has fully elapsed. Then, at point "B", the valve closes incrementally. After the time delay elapses again the flow is still high and the valve closes incrementally a second time. After the next time delay the valve closes incrementally a third time and at point "C" the flow falls inside the deadband. The flow remains within the set limits until point "D", when other process equipment causes it to drop below the lower limit. The control system begins a sequence of incrementally opening the valve (including response time delays) until point "E". At point "E" the flow again falls within the allowable error tolerance and no control action is required.

Floating control has many advantages over PID and other analog control methods:

- *Inherent Stability*: There is no reset windup or similar parameters that lead to hunting and the manipulated variable only changes during active correction.
- *Simplified Tuning*: The parameters of increment, time delay, and deadband are intuitive.
- *No Proportional Droop*: The control action will always act to move the process variable to within the deadband (within the capabilities of the process equipment).
- *Effective with Nonlinear Processes*: As long as the incremental change in the manipulated variable is small, the process will always move toward setpoint whether the process changes from the manipulated variable changes are rapid or slow.
- *Effective with Long Process Time Lags*: By adjusting the time delay between control actions to correspond to the process lag time, transport lags and dead time do not affect control accuracy.

The incremental change in the manipulated variable can be as simple as a timed contact closure for a valve. This eliminates the need for analog valve positioners and the accompanying expense, wiring, and calibration (Figure 8.2). In retrofit applications where analog positioners are already installed, the increment can be an adjustable percentage of valve position rather than a timed contact closure.

There are a variety of other modifications that can be made to the floating control algorithm if process and hardware considerations warrant them. Instead of a fixed increment for changing the manipulated variable, the change can be proportional to the error (Figure 9.10). This is known as proportional speed floating control. It is often used when the final control element has analog signal tracking built into it. An example is controlling blower air flow rate with a VFD. In order to minimize hunting and overshoot, the logic can include limits on the maximum change made at any one time. The calculation of the change in manipulated variable is straightforward:

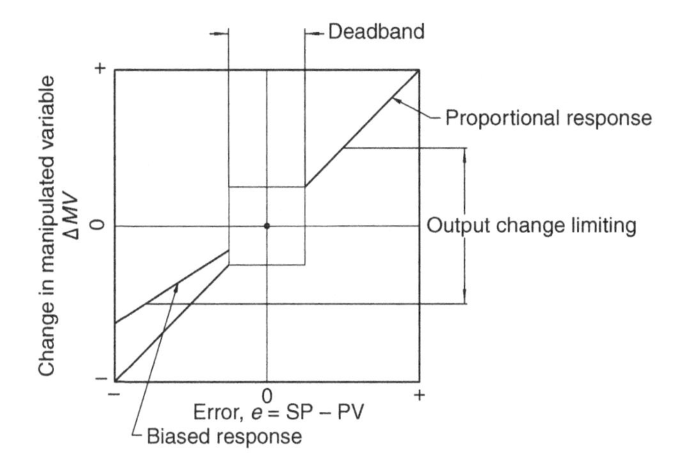

FIGURE 9.10 *Proportional speed floating control*

Equation 9.9

$$\Delta\text{MV} = G \cdot (\text{SP} - \text{PV}) = G \cdot e$$

where

ΔMV = change in the manipulated variable, units per controlled device
G = gain, units of MV/units of PV

A similar modification is the "two-speed" floating control action. In this case, two deadbands are created—an inner one and an outer one. When the process variable is within the inner deadband, the manipulated variable is unchanged. When the process variable is between the inner and outer deadbands, a small incremental change in the manipulated variable is made after each cycle (identical with the operation of the basic floating control). If the process variable is outside the outer deadband, a large incremental change or a continuous change in the manipulated variable is made after each cycle. The outer deadband is usually a multiple of the inner one. This modification allows rapid recovery from large errors in the process but eliminates hunting by making smaller corrections as the process variable approaches setpoint.

Another technique useful in eliminating hunting is a biased manipulated variable change. The incremental change in control element movements in one direction is a fraction of the incremental change when moving in the other direction. The magnitude of the bias is typically 60–75%. Biased control action eliminates hunting in two ways. First, it minimizes overshooting the setpoint in processes where the change in the process variable is skewed. An example is DO control, where the oxygen demand helps decrease the DO concentration when the control reduces air flow. Second, the difference in upward and downward increments corrects overshooting so that the process variable eventually settles to a value inside the deadband.

Using the rate of change of the process variable to inhibit control actions can also serve to reduce hunting. This differs from the derivative term in PID control, which often increases hunting and instability. The rate inhibit modification to the floating control logic essentially mimics a skilled operator's actions. If the operator determines by inspection that the process is being corrected they will wait before making additional corrections. The method is straightforward:

Equation 9.10

$$r = \frac{PV_{n-1} - PV_n}{\Delta t} = \frac{\Delta PV}{\Delta t}$$

where

r = rate of change, units of PV/s
Δt = time delay between successive control calculations, s

If the process variable is moving toward the setpoint and the rate of change exceeds an operator adjustable minimum threshold, no change is made to the manipulated variable until another time delay has elapsed.

9.2 DISSOLVED OXYGEN CONTROL

The most fundamental control strategy in aeration control is maintaining a set dissolved oxygen concentration in the aeration basins. This is also one of the most difficult control loops to implement since DO control is nonlinear, has significant process time lags, is a multivariable process, and has extreme variations in loading. These factors make the PID algorithm poorly suited for DO control. The recommended control strategy is modified floating control.

In diffused aeration systems, DO control is a cascaded control loop (see Figure 9.11). The output of the logic, the manipulated variable, is the air flow rate for the blowers and the aeration tank. The output air flow represents the

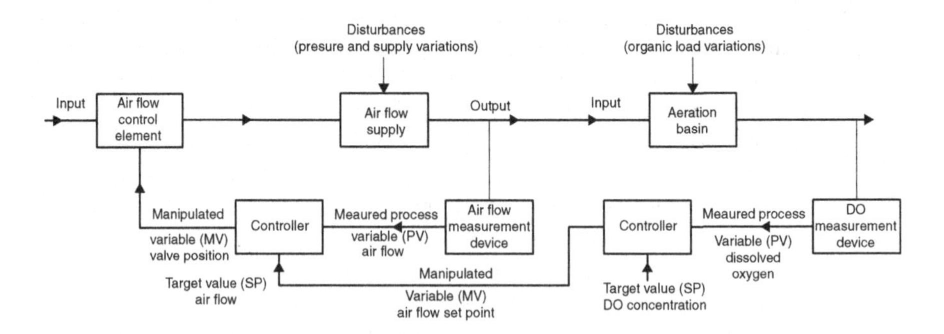

FIGURE 9.11 *Cascaded DO and air flow control loops*

estimated air flow change calculated to correct the DO error. Under ideal circumstances this will be the actual air flow at the controlled zone. However, other air flow considerations may prevent achieving the theoretical demand air flow.

Modifications to improve the performance of the basic floating control algorithm for DO control include the following:

- Timed response with the initial delay between corrections equal to 3–5 min.
- Direct acting logic—the DO control output (air flow) is increased to increase the process variable, DO concentration
- Proportional speed control with the change in air flow proportional to the error in DO concentration (The gain, CFM/ppm, varies widely with tank size, diffuser OTE, and so on, but a reasonable starting point is 1–2% of the average air flow to the controlled zone)
- Deadband—with a typical initial deadband of 0.2 to 0.3 ppm (tolerance equals ± 0.1 to ± 0.15 ppm)
- Biased response—with initial air flow changes on decreasing air flow (high DO) about 60% of the changes on increasing air flow (low DO)
- Rate of change inhibiting the control response—with the minimum rate of change about 0.1–0.15 ppm/minimum.

Note that these initial approximations for tuning must be modified during commissioning (Chapter 12).

The output from the DO control logic is the change in the theoretical air flow demand of the controlled zone. In Equation 9.9, ΔMV is the change in air flow, CFM, and G is the gain in CFM/ppm DO error. The theoretical air flow demand of the controlled zone is calculated from the change and the existing theoretical demands:

Equation 9.11

$$Q_{\mathrm{T}n} = Q_{\mathrm{T}\ n-1} \pm \Delta Q_{\mathrm{T}}$$

where

$Q_{\mathrm{T}n}$	= new theoretical demand for the controlled zone, SCFM or ACFM
$Q_{\mathrm{T}n-1}$	= previous theoretical demand for the controlled zone, SCFM or ACFM
ΔQ_{T}	= increase or decrease in air flow demand at the tank, SCFM or ACFM (calculated from equation 9.9)

The new air flow demand may not equal the actual flow setpoint used by the flow control loop for several reasons. The demand flow must be compared to and limited by the minimum mixing air flow and the maximum allowable flow per diffuser. Both of these parameters override the DO control requirements. If the limits of blower capacity have been reached, the actual flow setpoint may not equal the demand because the needed air supply is not achievable.

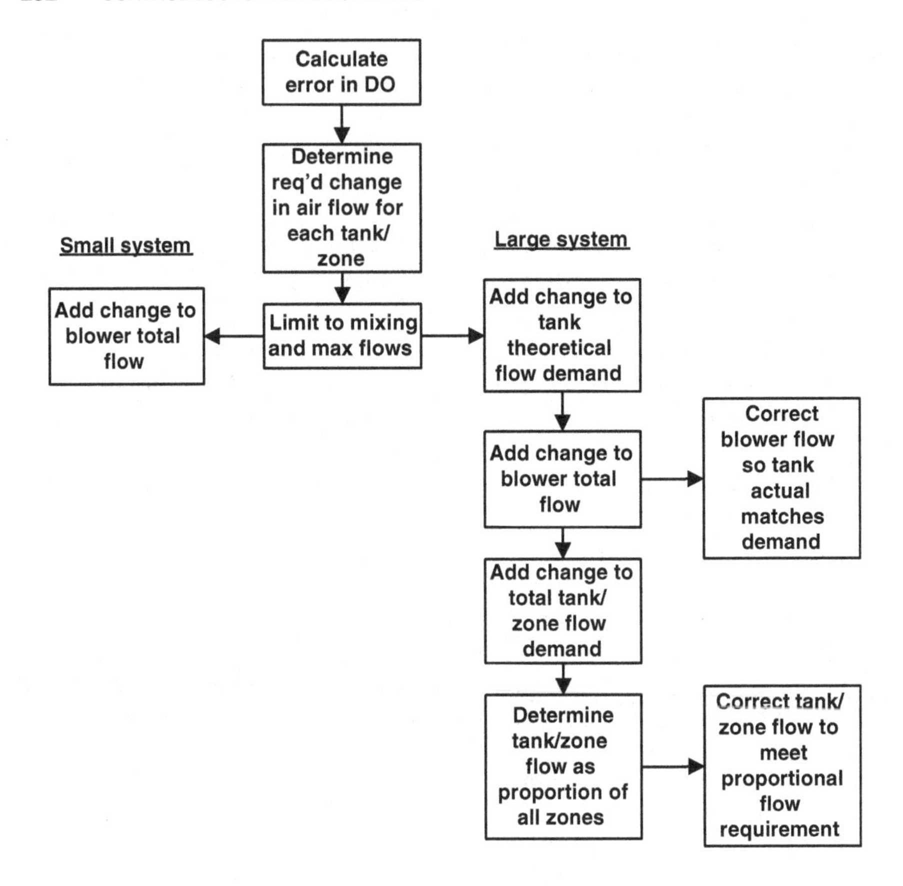

FIGURE 9.12 *DO control logic sequence*

The output of a floating control algorithm is an increment or decrement in air flow—not an absolute air flow setpoint. This is one of the many factors that add stability to the DO control system. By incrementally changing the air flow demand, the need to have the blower air flow rate match the basin air flow is eliminated. This allows for unregulated and unmeasured air flow to channel aeration, sludge holding tanks, and so on.

Small plants without automatic individual aeration basin control will use the change in air flow demand directly to increase or decrease the total blower flow rate after each tank's DO control logic executes (see Figure 9.12). In small facilities the distribution of the air flow between the aeration tanks is performed manually. The manual control is intended to balance the actual air flow and the process demand. Automatically controlling only blower flow makes DO control cost-effective for small facilities that cannot justify the cost of valves and flow meters at each aeration basin. In these systems the DO control logic and the total air flow determination is included in a master controller. Some plants may have individual blower local control panels to implement the blower flow control.

Equation 9.12

$$Q_{Bn} = Q_{B\ n-1} + \Delta Q_{Tn}$$

where

Q_{Bn} = new total blower air flow setpoint, SCFM or ICFM
Q_{Bn-1} = total blower air flow setpoint after previous iteration of logic, SCFM or ICFM
ΔQ_{Tn} = increase or decrease in air flow demand at the tank, SCFM or ACFM

One concern with small facilities with multiple tanks and manual air control is selection of the DO signal to use for determining blower air flow changes. Each tank is equipped with a DO probe and the logic must select the error signal used for calculating the air flow change. One technique is to totalize the error for all tanks using Equations 9.9, 9.11, and 9.12. In general, however, operators prefer to establish how the blowers are modulated. The most common technique is to allow the operators to select the error signal used to change the blower output. The choices include selecting the average of all DO errors, the worst-case DO error, or designating a specific tank or DO control point for the error calculation. The worst-case error is the one most below setpoint or the least above setpoint. The error is used instead of the actual DO reading to allow for a different setpoint in each tank. This may be required for some process variations, such as contact stabilization.

In larger plants with air flow control valves and flow meters at individual tanks (or individual diffuser grids within tanks), the process is slightly more complicated. Each tank or grid equipped with an air flow meter and flow control valve is a separate control zone. The air flow demand of the control zone must increase on falling DO and this occurs by changing the process demand flow using the output of the DO floating control logic. In order for the tank air flow to increase, the total air flow supplied by the blowers must increase as well. The most expeditious way to accomplish that is by adding the net change in demand for all tanks directly to the blowers—as indicated by Equation 9.12.

The flow measured by the basin flow meters may be in units of SCFM or ACFM. Blower air flow may be measured in units of SCFM or ICFM. Even if the nominal measurement units match, the total measurement at the tanks is unlikely to match the total measurement at the blowers because of normal measurement inaccuracy. Further mismatches may result if air flow to channels and so on, is unregulated. In order to bring the flow at the tanks as close to the theoretical demand flow as possible, a total flow correction routine is applied to the blowers. This compares the actual flow at the basins to the demand and uses the difference to increase or decrease the blower system total air flow.

Equation 9.13

$$Q_{B\ \text{Corrected}} = Q_{B\ \text{Current}} + F \cdot \left(\sum Q_{T\ \text{Demand}} - \sum Q_{T\ \text{Actual}}\right)$$

where

$Q_{\text{B Corrected}}$ = corrected total blower system air flow demand, SCFM or ICFM
$Q_{\text{B Current}}$ = actual total blower system air flow demand, SCFM or ICFM
F = damping factor to limit hunting, typically 0.5
$\sum Q_{\text{T Demand}}$ = total theoretical process demand air flow for all tanks, SCFM or ACFM
$\sum Q_{\text{T Actual}}$ = total measured air flow for all tanks, SCFM or ACFM

Determining the ultimate flow setpoint for the blowers is usually a two-step process—as shown in Figure 9.12. First, the net change in blower system flow is determined using Equation 9.12. If more than one tank is changing simultaneously, some may have increasing flow demand while others may have decreasing flow demand. Then the blower setpoint is "trimmed" using Equation 9.13. The trimming serves to counteract the effect of unmetered air flows and differences or inaccuracies in the various flow measurements in the system. The damping factor, typically 50%, is applied to the difference between demand and actual aeration flow.

If the flow correction in Equation 9.13 is applied without first making the adjustment to total capacity shown in Equation 9.12, the blower system air flow will still incrementally move toward matching the basin's demand air flow, and the DO concentration will still move toward the setpoint. However, by first directly adding the change in basin demand to the blower demand using Equation 9.12, the time required to achieve the target DO is decreased.

As with all of the logic associated with DO control, a time delay is used between successive applications of the blower total flow correction routine. The typical time delay between successive adjustments is 2 or 3 min. Both the damping and the time delay serve to avoid hunting by making small incremental changes in the blower output and allowing the aeration tank flow control to stabilize between successive control actions.

There are times when the blowers are operating at the upper or lower limits of the available blower capacity. In this case the air flow available must be divided between tanks or control zones in proportion to the process demand of each zone. If the distribution of flow between the various control zones is not coordinated in a logical manner, the result will be loss of control and extreme excursions in DO at some basins.

Maintaining proportional distribution is not a major concern in small systems that only automatically control blower flow and have manual control valves at the aeration basins. When the blowers reach upper or lower limits of capacity, the manually adjusted valve positions will be unchanged. This will maintain a reasonable proportion of air flow to each tank or control point.

The case is entirely different in large systems with individual tanks or grids controlled automatically. Consider the example of a system with automatically controlled blowers and aeration basins when the blowers are operating at their maximum capacity and the organic load to the aeration basins is increasing. As the

DO concentration in the aeration basins drops, the air demand increases. However, the total air flow to the basins cannot increase since blower capacity is limiting the available air flow. If the zone's air flow control used the theoretical demand as the zone air flow setpoint, the position of all the valves would move further open. Each valve control loop would try to achieve its own air flow demand setpoint. Eventually all of the valves would be at maximum opening and the distribution of air between zones would be, for all practical purposes, uncontrolled.

The opposite, and potentially more undesirable, situation occurs if the blowers are at their minimum capacity and the DO is high. In this case, all zone valves will cycle toward the closed position. The increase in restriction to flow will drive the main header pressure upward. With a PD blower the result will be increased power requirements—potentially accompanied by opening of the pressure relief valve. The blowers may shut down on high pressure or temperature. With a centrifugal blower the probable result will be blower surge and complete blower shutdown.

The solution to eliminating these extreme excursions is to override the theoretical demand air flow and coordinate all of the control zones. This is accomplished by controlling the zone valves based on an air flow setpoint. This is achieved by dividing the available air flow proportionally between the basins.

The proportionality could be based on equal air flows to all zones or air flow proportional to the number of diffusers in each zone. However, the most reasonable basis for the proportional division is to make the actual flow setpoint proportional to the process demand. This will cause all zones to have DO concentrations somewhat elevated or depressed, but will avoid any one zone being severely over-aerated or under-aerated.

Equation 9.14

$$Q_{\text{TSP}} = \frac{Q_{\text{TDemand}}}{\sum Q_{\text{TDemand}}} \cdot \sum Q_{\text{TActual}}$$

where

Q_{TSP} = actual setpoint used by the zone or tank flow control, SCFM or ICFM

Q_{TDemand} = theoretical process demand air flow for this tank, SCFM or ACFM

$\sum Q_{\text{TDemand}}$ = total theoretical process demand air flow for all tanks, SCFM or ACFM

$\sum Q_{\text{TActual}}$ = total measured air flow for all tanks, SCFM or ACFM

It is essential that the DO control (and schedule control if provided) include the ability to designate each tank as in service or out of service. If the tank is taken out of service, it should essentially be ignored by all control algorithms. DO control, basin flow control and flow proportioning, and total blower air demand should ignore any tank that is not in service.

The control logic must distinguish between the capacity of the blowers actually running and the capacity of the blowers available to run. A minimum and maximum flow rate for each blower must be provided. When the total blower air flow required (as calculated by Equation 9.13) exceeds the maximum capacity of the running blowers but is less than the capacity of the available blowers, additional blowers are required.

The logic must be slightly more complex than simply totalizing flow demand to properly control the number of operating blowers. Blowers should not be short cycled—that is, they should not be started and stopped with excessive frequency. This avoids equipment damage (Chapter 8). When additional blowers are required, the logic should initiate a time delay. This delay is typically 20–30 min. If the air flow demand is consistently greater than the capacity of the operating blowers for this period, the delay will time out and a status indication provided to indicate that additional blowers are needed.

Blowers may be started automatically, which is preferred, or manually. If the blower start/stop control is automatic, the next blower would be started after the delay times out. In some plants the operators prefer to manually start blowers. In this case, after the time delay elapses, the control system must provide an indication to the operators that an additional blower should be started.

When the total blower demand is less than the minimum capacity of the running blowers, some of the running additional blowers may be stopped. This may also be automatic or manual. If blower stopping is automatically controlled, a short time delay may be used before stopping the blower, and a minimum blower run time is generally part of the logic to prevent short cycling.

If the capacity of the running blowers is insufficient to meet demand, the calculated theoretical demand will continue to increase until the blower starting time delay elapses. If unchecked, this will result in a large difference between the running capacity of the blowers and the demand air flow of the blowers. The result of this accumulation is that when blowers are started or stopped there is a very large step in the total system air flow. (This is analogous to reset windup in a PID loop.) The dramatic change can result in process instability, as well as a short term and unnecessary jump in power consumption.

The sudden jump in flow can be reduced by limiting the accumulation of blower target air flow on increasing demand to approximately 120% of the maximum capacity of the running blowers. To avoid an analogous problem on decreasing demand, the blower target can be limited to approximately 80% of the minimum running blower air flow. Because the delay on stopping blowers is typically shorter, the accumulation is not as problematic on decreasing demand as on increasing demand.

Normally the theoretical demands of the aeration tanks are allowed to continue calculating during the time delay preceding blower starting and stopping. This allows the actual flow control setpoint to continue to be proportional to the process demand. The minimum and maximum air flow settings for each tank or control zone will normally prevent excessive demand changes for the tanks when the blowers are cycled. Some field adjustment of the minimum and maximum zone flows may be needed to make sure this is the case.

The complexities resulting from the process and control interactions of DO control can seem overwhelming. However, if the logic is viewed one step at a time it becomes manageable:

- The deviations of DO concentration are used to calculate the increments in the theoretical air flow demand for the control zone.
- The net changes in theoretical demand are used to adjust the total blower air supply to the system.
- The total theoretical demand of the control zones is used to calculate the proportion of total actual system flow to be supplied to each zone.
- The deviations between the total theoretical demand and the total actual air supply are used to trim the blower air supply to the system.
- The deviations between the proportional air flow setpoint for each zone and the current air flow to each zone are used to modulate each zone's flow control valve.

9.3 AERATION BASIN AIR FLOW CONTROL

Once the basin air flow setpoint is established by the DO control logic, controlling the air flow to each tank becomes a fairly straightforward task. The air flow could be controlled by a PID feedback loop, with air flow the process variable and valve position the manipulated variable. Generally, though, better stability and tighter control is obtained by using floating control logic. Several enhancements to the basic floating control are beneficial:

- Timed response with the initial delay between corrections set at 5–15 s.
- Direct acting logic—the control output (valve position) is increased to increase the process variable (the measured air flow rate).
- Deadband—with a typical initial deadband setting equal 2–5% average air flow (in some instances two-speed floating control action may be used to improve response—with the outer deadband on the order of 5–10 times the inner deadband).
- Biased response—with valve position changes on decreasing air flow initially set at about 60% of the changes on increasing air flow
- Depending on the configuration of the valve operator, one of three modes may used for controlling the valve position:
  - Proportional speed control: the change in valve position is proportional to the error in air flow rate. The reciprocal of the gain, CFM per % travel, varies widely, but an initial gain of 50–100 CFM per % is acceptable. (Note that the reciprocal of the actual gain is used to make the engineering units more convenient and the tuning more intuitive. The actual gain for use in Equation 9.9 would be 0.02–0.01% per CFM and the program should calculate this reciprocal.)

- Incremental position control: used when analog positioners are provided on the control valves but the simplicity and stability of the basic single-speed floating control is preferred. Each time the error exceeds setpoint the change in valve position is 0.5–2% when opening and some percentage of that on closing, depending on the bias.
- Timed valve travel control: when using single-speed or two-speed floating control. The valve operator motor is directly operated by a timed contact closure. The nominal closure time for the "creep" or slower travel is typically 0.5–2 s and for the faster travel it may range from 2 s to continuous travel—with the appropriate bias applied to both adjustments when closing.

There are three air flow rates associated with every control zone. The first, the theoretical demand flow, is calculated from the DO error, and is used to correct blower flow as well as establish the proportion of each zone's demand to the total demand. The second air flow rate is the actual air flow setpoint calculated using Equation 9.14. This setpoint is used by the zone's air flow control logic to establish the magnitude and direction of valve position changes. The final air flow is the measured air flow for the control zone, which is the process feedback for the control logic.

In order to avoid excessive pressure drops in the air distribution system or prevent the complete loss of air flow to a control zone, a minimum position may be set for each valve. A minimum opening of 15–20% is common. With analog positioners, the minimum position may be directly determined by the analog output of the controller. If timed contacts are used to control the valves, the minimum position status limit switch should be set to indicate the lower limit of valve travel has been reached.

Maximum position settings are also used but their primary benefit is to improve response time. Almost all air flow control valves are butterfly valves, and once they are more than 70% open they are near the maximum air flow rate for a given header pressure. Additional valve opening has little impact on the amount of air delivered to the basin. The response time of the air flow control loop to closing the valve is reduced by setting the maximum valve position at approximately 70% open.

The position of the zone flow control valves, whether they are automatically or manually controlled, has an impact on the pressure needed to deliver a given volume of air to the tank or zone. The pressure in the header and the resulting pressure drop across the valve also impact the valve position needed to maintain a set air flow. Both of these factors may influence the air flow rate delivered by the blower system—particularly for centrifugal blowers. The complex and interconnected relationships require integrating basin air flow control into the overall system control logic if precision in tank air flow control is to be achieved. Furthermore, without accurate control of the air flow, DO control is not possible.

9.4 PRESSURE CONTROL

The control of air flow using the methodology defined above requires coordination of all control loops and recognition of the status of each control element. A change in the process demand at one aeration zone necessitates a change in the total air flow delivery of the blower system. A change in the position of one basin air flow control valve changes the total system restriction (Equations 6.14 and 6.15). The change in system restriction in turn changes the pressure in the main distribution header, which in turn changes the air flow rate through all other control valves tapped off that header.

This coordination isn't a problem for current state-of-the-art controllers. In many cases, the DO and air flow control logic is executed in a central master controller, and the data is common to all control loops. Even if separate controllers are used for various functions, the communications capabilities of modern control systems make the necessary data available in essentially real time to all controllers.

This data availability wasn't the case when DO and aeration control systems were first applied in the late 1960s and early 1970s. Almost all controllers were single loop PID devices with no communications capability. The only "data" transfer was through a 3–15 psig pneumatic or a 4–20 mA electrical signal. In effect every control loop was totally independent of every other control loop in the system, regardless of process relationships.

This loop independence created two problems for early control system designers. First, the independent loops could not convey required changes in demand to the blower controllers, leaving the system without a way to match total air flow supply to total air flow demand. Second, as each zone's control valve adjusted position the change in back pressure would disturb the air flow to all other tanks, causing the DO in these control zones to move off setpoint.

The early system designers addressed both of these problems by employing a pressure control loop for the main distribution header. This control logic was generally implemented as a cascaded control loop. It used header pressure as the process variable for the first loop, with total system air flow as the manipulated variable. The air flow from the first loop was the setpoint for the secondary loop. This secondary loop used blower air flow as the process variable and inlet valve (or IGV) position as the manipulated variable (see Figure 9.13).

When the pressure control loop was properly tuned and operating with stability, it resolved both issues. If the basin flow control valve opened to increase air flow to a zone, the pressure in the main header would decrease. The pressure control loop would increase the blower air flow setpoint, and the blower control loop would increase the discharge air flow. If the system flow restriction remained relatively constant then the increased air flow would increase header pressure back to the original value.

The basin air flow rates would remain constant if the pressure in the header and the valve position remained constant. If a zone flow control valve changed position, the air flow would correspondingly increase or decrease — again assuming a constant header pressure.

There are several problems with the constant pressure control approach. Because of the dynamic interaction between the various loops, it is very easy to introduce instability into the process. The constant flow through unchanged valves is predicated on maintaining constant pressure, but maintaining constant pressure is

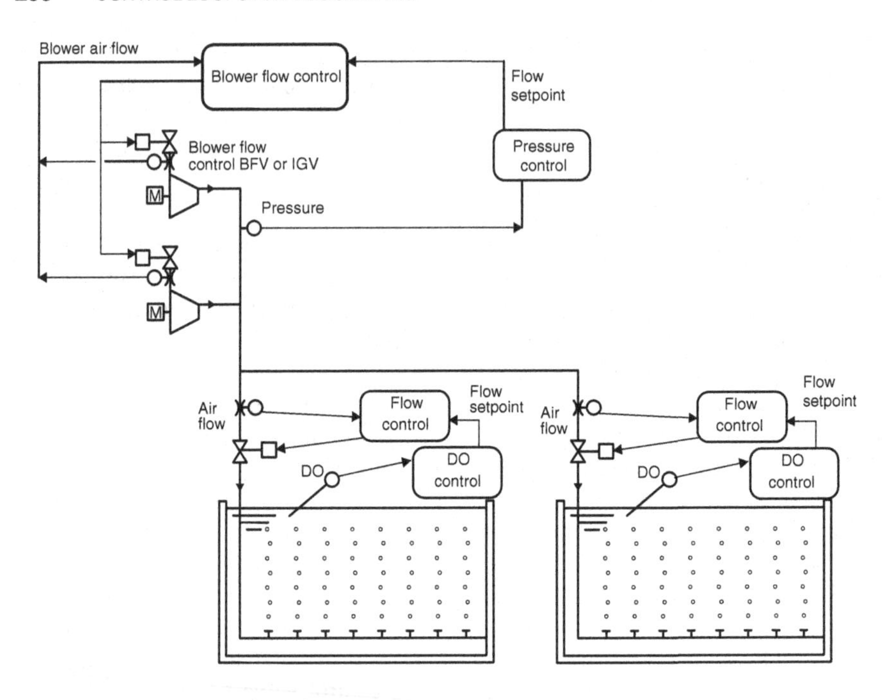

FIGURE 9.13 *Typical pressure control system*

dependent on limited changes in valve position. If the tank flow control valves open to compensate for a decrease in flow, the pressure will drop and the blower flow will increase. If the result is higher flow then the tank valves will close, resulting in higher pressure and a decrease in blower flow. This can quickly result in oscillation in the system, creating instability and hunting.

Avoiding hunting requires a very delicate balance in tuning. If the pressure and blower control loops are tuned too aggressively, the pressure swings will cause instability in the zone flow control valves, which in turn causes instability in the pressure control. If the pressure control loop's tuning results in too slow a response then the tank valves will overcompensate—also causing instability in the pressure control.

It is important to note that if conventional PID control is used then the loops all operate independently, even if they are contained in a single controller. The issues with tuning and hunting persist because the control logic does not integrate the response of the various independent loops. Integrated floating control based on flow avoids this issue. Further, the DO concentration in the aeration basins is a function of flow rate and is not dependent on pressure. Constant pressure control is essentially obsolete technology, developed to solve the problems created by independent PID control logic.

Maintaining constant pressure has a negative effect on system power requirements even when tuning is perfect. Ideally the system pressure will drop as air flow decreases. Constant pressure across the flow spectrum eliminates this decrease, and

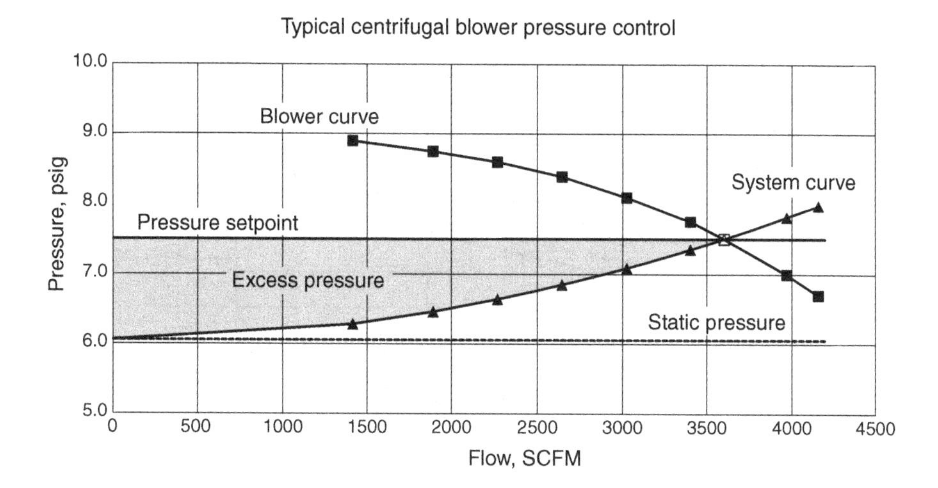

FIGURE 9.14 *Excess pressure with pressure control*

excess pressure above that needed to supply air to the tanks results (see Figure 9.14). With variable speed or IGV controlled blowers, this extra pressure creates an excess power demand. Elimination of excess pressure is the impetus behind most-open-valve (MOV) control.

9.5 MOST-OPEN-VALVE CONTROL

MOV logic is a technique developed to eliminate the excess power demand created by constant pressure control. It is not associated with maintaining DO and is not part of the blower flow control logic. It is a modification to the other control loops designed to reduce discharge pressure. When operating properly it results in a system pressure equal to the minimum value necessary to overcome static pressure and friction loss. The blower discharge pressure tracks the system curve. The result is reduced power consumption (see Figure 9.15).

The term "most-open-valve" is slightly confusing. The fundamental principle is that at least one of the aeration flow control valves will be at the maximum position at all times. This valve is referred to as "the" most open valve. By maximizing valve opening, the pressure drop across all valves and the resulting main header pressure is minimized. There are two common techniques employed with the selection depending on the other strategies employed in the system. One technique is used in conjunction with pressure control of the blowers and the other (newer) technique is used in conjunction with integrated direct flow control.

The initial MOV logic was developed to work with pressure control-based systems. It functions by readjusting the pressure setpoint based on valve position. In the most common embodiment the valves throttle at the basins based only on achieving the set air flow rate for that valve. When one valve opens to the maximum

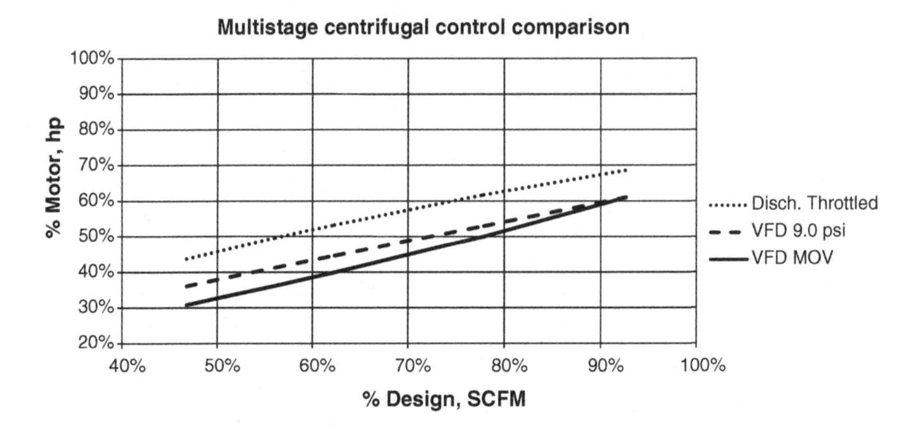

FIGURE 9.15 *Example savings with MOV control*

position (typically around 75% open) it indicates that the aeration system is calling for more air. At this point the pressure setpoint for the main header is adjusted upward, typically 0.05–0.1 psig. The increased pressure setpoint causes the blower system air flow to increase.

In this system, the reduction in pressure takes place when the basins have more air than needed. When the most-open-valve position is reduced to a preset minimum value—for example, 50% open—the pressure setpoint is decreased. This causes the blower air flow to decrease and in turn the basin valves open to decrease the back pressure. If the decrease in pressure setpoint and decrease in blower flow match the system requirements, the blower and aeration basin air flow balance out at a reduced pressure level. It may require several iterations of the MOV logic to achieve this balance. Note that with this system significant instability can easily result from errors in tuning.

The pressure-based MOV logic will not function properly if the blowers are at the minimum or maximum limits of capacity. The changing pressure setpoint can't affect the blower air flow rate, and the result is that the valves at the aeration basins cycle to the extreme position with a resulting loss in air flow control accuracy. Another disadvantage of this system is that the discharge pressure is slightly elevated while the most open valve is between the positions that trigger a change in pressure setpoint.

When floating control is used in conjunction with proportional distribution of air flow, a different approach is required. In this technique, a close permissive status in the logic must be set for each zone control valve before it is allowed to close—regardless of the air flow error. A valve that is the most open will be indicated by either analog position feedback or limit switch status. The first valve to be at the maximum position and designated most open will not have the close permissive set. Even if the air flow at this control point is above setpoint, this valve will not be allowed to close.

Because of the proportional distribution of actual air flow, if one control point is above its actual setpoint then another control point will necessarily be below its actual setpoint. The control valve for that zone will therefore cycle open. This reduces system back pressure and also reduces air flow through the most open valve. After a few iterations either the flows will be correct or the second valve will reach the open position. At that time the first valve will have the close permissive status set, and it will close to further reduce air flow and reach its setpoint.

While the valve positions are adjusting, the blower control loop will be modulating the blowers to maintain the correct total system air flow.

This system is inherently more stable than the pressure control-based logic. Further, because blower control is flow based, and because one aeration basin flow control valve is always at the maximum position, the system pressure is always at the minimum needed to provide the required flow. Time delays are used between successive iterations of the basin flow control logic. This ensures that the system stabilizes between each valve movement.

If operating properly, MOV logic typically results in average discharge pressures 0.5–1.0 psig lower than the conventional pressure control setting. That results in a decrease in blower power.

9.6 BLOWER CONTROL AND COORDINATION

Blower control requirements vary with the type of blower, the size of the blower, and the manufacturer's requirements (see Figure 9.16). Larger blowers are more expensive and this justifies more sophisticated controls. Blower control logic falls into one of three actions:

- Start/stop control and sequencing
- Air flow control
- Equipment protection (often referred to as equipment health monitoring)

The demand for blower starting and stopping is determined by the DO control logic and the resulting air flow demand. Once the blower start or stop is initiated, the proper sequencing is determined by the specific manufacturer's requirements. In many cases a manufacturer-provided local control panel is used for each blower, and the aeration control system needs only provide a dry contact or status bit to initiate the start/stop sequence. In other cases, particularly in the case of smaller systems, the blower control during starting and stopping is handled by the aeration control system. Regardless of the control source, a status should be provided to verify the successful completion of the start sequence and verify that the blower has reached normal operation. When the blower is stopped, a status should be provided confirming completion of that operation.

Each manufacturer and each installation will have specific requirements for starting the blowers. The manufacturer's operation manuals should be referenced and their recommendations followed. In addition, the plant staff will have preferences and experiences that must be considered in establishing the sequence for starting the blowers. For small constant speed PD blowers the start sequence may be

Blower Type	Positive Displacement	Multistage Centrifugal	Single-stage Centrifugal	Turbo Blower
Flow control method	Variable speed	Variable speed or inlet throttled	Variable speed or inlet/disch. vanes	Variable speed
Surge protection	N/A	Yes	Yes	Yes
Overload protection	Yes	Yes	Yes	Yes
Bearing temperature protection	Large units only	Typically yes	Yes	Sometimes
Bearing vibration protection	No	Typically yes — accelerometer	Yes — proximity probe	Sometimes
Blower case vibration protection	Large units only	Not usually	Not usually	Sometimes
Discharge air temperature protection	Yes	Typically yes	Yes	Typically yes
Discharge air pressure protection	Yes	Typically yes	Yes	Typically yes
Lube oil pressure	Large units only	Not usually	Yes	N/A
Lube oil temperature	Large units only	Not usually	Yes	N/A

FIGURE 9.16 *Blower control and protection requirements*

as simple as closing a run contact to a motor starter. For a large single-stage centrifugal blower, the start sequence may incorporate a variety of ancillary equipment and may take several minutes to complete.

When the blower is stopped, the same ancillary equipment must be sequenced. The normal stop sequence may differ from emergency stop (E-stop) sequencing. In the normal stop sequence an orderly shutdown may require several minutes to execute. With an emergency stop situation, whether initiated by an operator pressing an emergency stop pushbutton or by a major equipment fault condition, it may be more important to remove power from the blower motor immediately and then sequence ancillary devices.

Not all blower systems, even with the same type of blower employing identical control strategies, have the same configuration. The ancillary and peripheral equipment that may be part of the blower system may include the following:

- Discharge and inlet isolation valves—which must be opened for the blower to run. These are often manual, and may or may not have open status limit switches.

- Blower protection devices such as RTDs and vibration transmitters. The controller must have confirmed valid signals from all for the blower to run.
- Blower monitoring devices such as flow, pressure, and temperature transmitters. The controller must have confirmed valid signals from all for the blower to run.
- Blow-off or bypass valves are normally closed (NC) if the blower is not running, and these are generally cycled open prior to starting the blower or as part of the start sequence. Status confirmation from analog feedback or limit switches should be received.
- Inlet throttling valves or inlet and discharge guide vanes should be cycled to initial starting position and confirmed before starting the blower. The initial position may be full closed, full open, or a predefined intermediate position depending on manufacturer and operator preference.
- Lube systems, which may include external auxiliary lube pumps, must be started and pressure or flow confirmed prior to starting the blower. Integral lube pumps require confirmation of pressure or flow within a set time after blower starts.
- Lube oil cooling, which includes external cooling water or fans for heat exchangers, should be started before starting the blower.
- Variable speed drives or starters must include run status and normal speed confirmation. The controller must receive confirmation of status within a set time after starting the motor.
- The emergency stop pushbutton must be in normal position before starting. E-Stops should be maintained type, hardwired to the VFD or starter with a separate status contact to controller.
- External fault systems should be hardwired to VFDs or starters, with a separate status contact to controller.

To the extent possible, the external connections and wiring should be "fail safe." For example, NC contacts should be used for external faults and emergency stop functions. If the contact doesn't close properly or if there is a loose wire, the controller and any hardwired interlocks will interpret the loss of status as a fault and prevent running the blowers.

Very few plants require the full design capacity of the blower system for more than short periods of time. The designation of lead, first lag, second lag, and so on, establishes the order in which additional blowers will be started when additional capacity is needed to meet process demand. The sequence or priority of operating blowers should be readily adjustable by the operators—generally at the master control panel or from the SCADA system. This encourages alternating equipment. The blower control logic should enable the operator to change the sequence without stopping the blowers that will remain running in the new sequence. One way of achieving this is to have "new" sequence settings transferred from the HMI or SCADA to the control system as a block after the operator completes changes.

Equipment left idle for extended periods of time has a tendency to develop problems that can lead to catastrophic failure when operation is resumed. It is good practice to alternate the operating blowers periodically to avoid bearing brinnelling and to keep a lubricant coating on bearings and gears. The tendency for operators when alternating equipment is to start the new blowers before stopping ones already operating. This allows them to verify starting of the next blower. This procedure also, unfortunately, adds directly to demand charges and operating cost. Unless the blowers experience unusual problems that frequently result in failure to start, the running blowers should be stopped before the next blowers in the sequence are started. If the blowers are started automatically this can be accomplished simply while changing lead and lag designations.

There are many methods and potential locations for controlling blower starting and modulating air flow. These include central SCADA computers, a local control panel HMI, local panel pushbuttons and selector switches, MCC pushbuttons and selector switches, and so on. It is tempting to the system designer and programmer to include all of these options in the control scheme with the intent to provide operator flexibility.

This temptation should be avoided! Multiple points for setting Manual, Local, Auto, Remote and similar selections can quickly become confusing and unmanageable for the operators. The meaning of "Auto/Auto" may be obvious to a programmer sitting in the office, but it will be hard for the operator to decode when he is dragged out of bed at 2:00 A.M. to get the plant running after a power failure. Good design keeps the levels of redundancy to a minimum and makes labeling of the various control strategies intuitive. All functions should be kept as simple as possible and the options worked out in conjunction with the plant operators.

The primary variable indicating a blower is at the upper or lower limit of capacity is air flow rate. However, other conditions may prevent increasing or decreasing the blower flow rate even though the nominal limits of maximum or minimum air flow have not been reached. The minimum capacity may be clamped by minimum IGV position, minimum speed setting at a VFD, or minimum throttling valve position. Normally these settings should be coordinated with the flow capacity, but excursions in ambient conditions may create a disparity. If excess blowers are operating, any factor that indicates the blower is operating at the minimum available capacity can be used to initiate stopping a blower.

The maximum capacity may be limited by factors other than air flow rate as well. The IGV, VFD, or BFV may have upper limits on their operating range. Motor amps or kW may also define the upper limit of capacity at a value lower than the set maximum flow rate. Rather than rely on flow rate only, maximum and minimum capacity status should be indicated by any of these parameters. The starting of additional blowers in response to increasing demand should be initiated by any condition that indicates the maximum capacity has been reached.

Starting additional blowers often occurs when other blowers are operating near maximum capacity and the common air header is under pressure. This is not generally an issue with PD or screw blowers, but centrifugal blowers with flat capacity curves can experience surge when they are started. There are several techniques employed to alleviate the difficulty of starting a centrifugal blower against system pressure:

- Open a blow-off valve until the blower is up to speed, and then gradually close it to build pressure and bring the blower discharge flow into the main header. This allows the blower to stabilize and gradually increase its discharge pressure as it is brought online.
- Reduce the flow rate of all operating blowers to a point below 50% of operating range—typically to 15–25% of range. This reduces the pressure in the header. Once all blowers are operating properly, the flow control is released and paced from process air flow demand. Note that DO and basin air flow control may need to be locked out during this operation.
- If automatically controlled valves are provided at the aeration basins, they may all be moved to maximum position while the additional blower is started. Once all blowers are operating properly, the basin air flow is returned to automatic control. This procedure is often combined with one of the two methods described above. Besides decreasing system pressure to ease blower starting, this procedure brings all basin valves to a known position so that MOV logic is not disturbed by temporary high flows during blower starting.

It is possible for blower motor power to increase even if the aeration control is operating at steady state. As the ambient temperature changes, power for the blower will change too. A deadband controller for limiting blower power demand to a safe level is recommended. This logic is similar to the deadband controller shown in Figure 9.7 and will prevent overload conditions even when the air flow is below the set maximum value.

When a centrifugal blower is started, it necessarily passes through the surge point as it comes up to speed. This may result in the check valve cycling open and closed with attendant noise as the disk hits the seat. This is particularly common with multistage centrifugals, since both single-stage and turbo blowers generally have a blow-off valve open during start-up. Although some operators find the noise objectionable, this phenomenon doesn't represent a problem or operational concern.

Each blower's control logic should include a coast down or spin timer to prevent trying to start a blower that is still rotating above a minimal speed. Because of the high blower inertia, the time delay before a restart should be in the area of 5–10 min.The setting should be confirmed by field observation during commissioning.

The process demand is generally calculated in SCFM and the blower performance is defined by ICFM. Either parameter may be measured and used for blower flow control. The programmer must be careful, however, to keep track of the units of measurement being provided by the DO control and flow correction logic. Necessary conversions should be made to maintain consistency in the blower control.

One problem frequently encountered in blower control is the inability to achieve accurate measurement of the blower air flow. The piping in the blower room has a lot of fittings, bends, and valves in a limited space. It is difficult to obtain adequate straight pipe for proper flow transmitter installation. The volumetric flow rate for various types of blowers and controls may be calculated with sufficient accuracy for control by using the appropriate formulas. For PD blowers:

Equation 9.15

$$Q_{iPD} = CFR \cdot (N_a - N_s)$$

where

Q_{iPD} = volumetric flow rate of PD blower, ICFM
CFR = blower displacement, cubic feet per revolution
N_a = actual speed required to provide flow rate, rpm
N_s = actual blower slip, rpm

Multistage centrifugal blowers operating at constant speed and controlled with inlet throttling commonly monitor motor Amps and correlate that to mass air flow rate. This is the basis for the "calibrated ammeter" commonly used to indicate air flow for this type of blower. The manufacturer's curve provides the power at minimum and maximum air flow rate. The power in turn can be used to calculate current draw, and from these two points a linear correlation to air flow can be developed. The linear correlation is sufficiently accurate for control and blower protection in inlet throttled constant speed applications:

Equation 9.16

$$I = \frac{P_b \cdot 746}{V \cdot 1.73 \cdot \eta_m \cdot PF}$$

where

I = three-phase motor current, A
P_b = blower shaft power from blower curve, hp
V = motor volts, V
η_m = motor efficiency, decimal
PF = motor power factor, decimal

In retrofit applications it may be necessary to pick the two points for flow and amperage off a calibrated ammeter instead of from the characteristic curve:

Equation 9.17

$$m_{MS} = \frac{Q_{max} - Q_{min}}{I_{max} - I_{min}}$$

where

m_{MS} = slope of linear correlation, SCFM/A

$Q_{\text{max,min}}$ = max and min flow rate curve points, SCFM
$I_{\text{max,min}}$ = motor current draw at max and min curve points, A

Equation 9.18

$$b_{\text{MS}} = I_{\text{min}} - \frac{Q_{\text{min}}}{m_{\text{MS}}}$$

where

b_{MS} = intercept of linear correlation, A

Equation 9.19

$$Q_{\text{MS}} = m_{\text{MS}} \cdot (I_{\text{Act}} - b_{\text{MS}})$$

where

Q_{MS} = mass air flow rate, SCFM

The above relationships assume a fairly constant pressure ratio across the blower. For variable speed centrifugal blowers or those controlled with inlet or discharge guide vanes, the linear correlation of SCFM and current isn't correct because the pressure ratio varies significantly. If flow meters cannot be used, the air flow rate can be accurately calculated by going back to the thermodynamic performance characteristics of the blower. To perform the calculation requires monitoring inlet temperature, inlet pressure, discharge temperature, discharge pressure, and blower shaft power. In most applications there is no direct measurement of blower shaft power and the motor current is monitored instead. The shaft power can be calculated by rearranging Equation 9.16 and correcting for mechanical losses:

Equation 9.20

$$P_{\text{B}} = \frac{I \cdot V \cdot 1.73 \cdot \eta_{\text{m}} \cdot \text{PF}}{746} - P_{\text{L}}$$

where

P_{B} = calculated blower shaft power, bhp
P_{L} = mechanical losses for bearings, gears, lube system, and so on, bhp (typically 1–3% of motor power, depending on blower design)

If the electrical power draw of the motor is measured directly with a motor protection relay or other device, the conversion is simplified:

Equation 9.21

$$P_{\text{B}} = \frac{P_{\text{E}} \cdot \eta_{\text{m}}}{0.746} - P_{\text{L}}$$

where

P_E = measured electrical power draw of motor, kW

The actual blower efficiency can be calculated from the inlet and discharge pressures and inlet and discharge temperatures:

Equation 9.22

$$X_{AD} = \left(\frac{p_d}{p_i}\right)^{0.283} - 1$$

where

X_{AD} = adiabatic factor, dimensionless
$p_{d,i}$ = discharge and inlet pressure, psia

Equation 9.23

$$\eta_B = \frac{X_{AD} \cdot T_i}{T_d - T_i}$$

where

η_B = blower adiabatic efficiency, decimal
$T_{i,d}$ = inlet and discharge air temperature, °R

From this data, the blower volumetric air flow rate can be calculated:

Equation 9.24

$$Q_i = \frac{\eta_B \cdot P_B}{p_i \cdot X_{AD} \cdot 0.01542}$$

where

Q_i = inlet air flow rate, ICFM

This calculated air flow "measurement" is quite accurate and can be used for any type of blower and any type of control method. If the mass flow rate is required, it can be calculated from the volumetric flow rate using Equation 5.8.

Blower air flow may be controlled by a conventional PID algorithm. However, the centrifugal blower response to the manipulated variable is nonlinear, so tuning and stability are matters for concern. This is true for inlet throttling, IGV, or variable speed. Better results are usually obtained with floating control and a 3–5 s response delay.

A PD blower's air flow is linear with speed. Because the relationship is known and defined, there is no need for a feedback loop of any kind with this type of blower.

Surge control for centrifugal blowers is based on maintaining a "surge margin" between the actual surge point and a set minimum inlet flow rate. The surge point is usually taken from the blower performance curves. If the blower is controlled by inlet guide vanes, the surge point varies with guide vane position. The variation can be derived from the performance curves. An equation developed by linear regression can be used to correct the surge point based on IGV position. If the blower is variable speed, the surge point at nominal operating speed can be determined from the performance curve and corrected using Equation 5.40. The surge margin is typically set at a flow 5–10% higher than the surge point.

When the blower air flow falls below the surge margin, the control system should modulate the blower to increase the flow. Inlet guide vanes or throttling valves should be opened or the blower speed should be increased. If the blower has a blow-off valve and modulation doesn't sufficiently increase the air flow, the blow-off valve should be opened and then modulated using a deadband controller to restore normal operation. If none of the corrective measures is sufficient to bring the blower into the safe operating range within a short time, the blower should be stopped and an alarm set.

If more than one blower is operating, the process air flow demand must be shared between them. There are two methods commonly used for determining the flow rate of each blower—cascade or parallel control.

In cascade control, when two blowers are running both are initially operated at the minimum flow rate. As demand increases, the lead blower's flow rate is increased and the second blower is held at the minimum flow rate. If the lead blower reaches maximum flow rate and the process requires additional air flow, the lag blower is then modulated to increase its air flow. On decreasing air demand the process is reversed, with the lead blower kept at maximum air flow rate until the lag blower reaches its minimum flow setting.

Some designers feel the cascade approach minimizes power consumption since the lead blower is operated closer to its best efficiency point (BEP). In practice the improvement in power consumption is negligible. Although the lead blower is closer to its BEP, the lag blower is further from its BEP, and the two effects tend to cancel each other out.

In parallel control both blowers are operated at the same degree of turndown. If the blowers are identical, they will both be operating at the same air flow rate. If the blowers are of different capacities, they will both be operating at the same percentage of their operating range. The operating air flow for each blower can be calculated from the total flow range of all operating blowers:

Equation 9.25

$$Q_{1\ \mathrm{set}} = \left[\left(\frac{Q_{\mathrm{Total\ set}} - \sum Q_{\mathrm{min}}}{\sum Q_{\mathrm{max}} - \sum Q_{\mathrm{min}}}\right) \cdot (Q_{1\ \mathrm{max}} - Q_{1\ \mathrm{min}})\right] + Q_{1\ \mathrm{min}}$$

where

$Q_{1\ \mathrm{set}}$ = flow setpoint for first blower, ICFM or SCFM

$Q_{1\ min,max}$ = minimum and maximum flow for first blower, ICFM or SCFM
$Q_{Total\ set}$ = total flow setpoint for all running blowers, ICFM or SCFM
$\Sigma Q_{min,max}$ = total minimum and maximum flow setpoint for all running blowers, ICFM or SCFM

Parallel operation is preferred over the cascade method by many designers. This method is more stable and easily controlled, since both blowers are moved away from the surge line as the total process flow increases. As the blowers move away from surge, the performance curve generally becomes steeper—which in turn improves stability and makes flow rate less sensitive to disturbances. The improvement in stability more than offsets any minor difference in power consumption for most applications.

9.7 CONTROL LOOP TIMING CONSIDERATIONS

All of the control logic for DO, aeration basin air flow, and blower air flow exhibit inherent delays between the control action and the process response. The stability of the floating control method is enhanced by accommodating this response delay in the programming. The stability and accuracy are also improved by staggering the control actions so that the system has an opportunity to stabilize at a new steady-state condition before it is disturbed by another change induced by other process equipment.

The slowest responding segment of the process control system is DO control. There is a lag of several minutes between changing air flow and the DO stabilizing at the new steady-state value.

After the change in air flow demand, the basin air flow control may have to go through several iterations before the air flow settles at the new setpoint. If the response delay for the air flow logic is set to several seconds, the zone air flow has time to stabilize before the next execution of the DO control logic. Since the air flow stabilizes much faster than the DO can respond, there is adequate time for the DO to reach a new steady-state value at the new air flow rate.

In small systems, with manual control of the air flow to individual aeration tanks, the same principle applies. The blower response delay is typically set to a few seconds, so there is more than enough time for the blower system to stabilize before the next change in demand from the DO control logic executes.

The adjustments in the aeration basin flow control valves change the system pressure and may disturb the blower air flow rate. The blower response to control changes is quite rapid. The delay of a few seconds between blower modulations provides time for the blower performance to stabilize between successive basin flow control valve movements.

The use of "nested" time delays for the various control logic segments is a simple programming technique for modern controllers, but was not available in independent single loop controllers or most PID implementations. This technique provides a significant enhancement to the stability and precision of the aeration control system.

9.8 MISCELLANEOUS CONTROLS

One of the challenges in DO control is establishing the correct DO setpoint. For most facilities, establishing a constant DO setpoint that meets process performance objectives is sufficient. Starting at 2.0 ppm and decreasing the setpoint in small increments provides energy savings compared to manual control and optimizes treatment. The process performance must be monitored during this procedure. If the process performance or the population of microorganisms begins to deteriorate, the DO setpoint should be increased. A delay of several days between each increment of the DO setpoint is required to allow the process to stabilize at the new operating level.

In large systems with nutrient removal permits, it may be possible to further optimize power consumption by making the DO setpoint a function of ammonia (NH_3) by using feedback control. It is the norm for most large treatment facilities to be permitted on both BOD and nitrogen. As indicated in Chapter 3, the rate of biological nitrification is affected by the DO concentration up to approximately 3.0 ppm. A feedback control system using an ammonia concentration transmitter can reduce the DO concentration while maintaining nitrification.

In most applications that use ammonia feedback control, the ammonia concentration is monitored at the effluent end of the aeration basins. The control logic is typically deadband control. If the ammonia concentration rises above an operator adjustable threshold, the setpoint of the DO control system is increased. A higher DO concentration increases the rate of nitrification and reduces the ammonia concentration. If the ammonia level falls below a low setpoint value, it indicates excess aeration, and the DO setpoint is decreased. Between the two values the DO setpoint is held constant. Time delays are used to prevent DO setpoint changes after short-term ammonia level changes.

This is relatively new technology, since reliable nutrient monitoring equipment has only recently become commercially available. For large installations, where the energy cost justifies the additional equipment cost and maintenance, the system shows promise of further optimizing energy consumption.

Feedforward control can provide accurate process control by predicting process requirements based on measured parameters and known relationships between the process variables. Unlike feedback control, feedforward systems do not require the process to go off setpoint before corrective action is taken. Feedforward systems can be very effective in optimizing process and energy performance. Either alone or coupled with a feedback loop for trimming response to minor disturbances, a feedforward control can minimize process deviations.

Pacing air flow from influent wastewater flow is one feedforward technique that has been implemented with, at best, mixed results. The logic is based on the assumption that organic load and process demand are proportional to the influent flow rate. This is correct in a generalized way, but it is not an adequate process model for optimizing control.

Flow paced control doesn't account for variations in organic pollutant concentration in the wastewater. These exist even under dry weather conditions. In rain events the concentration can vary significantly. In the first flush at the beginning of a rain event, the pollutant concentration may rise dramatically as the high flow rate cleans out debris and organics that have settled in the sewer system. If the rain event persists, the pollutant concentration begins to fall dramatically as ground water dilutes the normal domestic and industrial flows. The air demand can be further decreased because the ground water is typically very high in DO.

The flow paced control also ignores the impact of internal sidestreams. Filtrate from belt presses, decant from digesters, and loads from filter backwashing are not monitored in the influent flow rate. These sidestreams can contribute significantly to both BOD and ammonia loading in the aeration basins—in some cases doubling the process demand. By nature the sidestreams are intermittent. Flow pacing the air supplied to the aeration basins based on the influent flow rate will not provide satisfactory results for most treatment plants.

A more effective feedforward strategy is based on measuring the actual process loads in the aeration basins and modeling the oxygen transfer characteristics of the aeration systems. The offgas analysis system measures both the process demand (OUR) and the actual oxygen transfer capacity (AOTE) of the aeration system (Chapter 7).

Organic load variations are captured in real time by the offgas OUR measurement. The OUR will determine the mass flow rate of oxygen needed for BOD removal and nitrification. The offgas system alone cannot distinguish between carbonaceous (BOD) and nitrogenous (NH_3) demand, but from the standpoint of the air supplied by the blowers this distinction is not important. The offgas system is unable to distinguish the impact of the fouling factor on diffuser performance from variations in α, but this is also not important for establishing air flow demand. The exact mass air flow required in the aeration zone to meet the process demand can be calculated using Equation 3.4.

The process oxygen demand can be provided across a spectrum of dissolved oxygen concentrations in the mixed liquor (Figure 4.19). The lowest DO concentration will provide the required mass of oxygen to the process with the lowest energy requirement. However, it is necessary to have sufficient DO in the aeration basins to prevent the development of filamentous organisms and other undesirable biology. The feedforward control is, therefore, usually trimmed with a feedback control to maintain the desired DO concentration at the effluent of the aeration basin.

Once the basin air demand is established, the basin air flow and blower air flow are controlled using the techniques described above. The combination of feedforward and feedback has proven to be very effective in providing process control with minimum energy demand. Small plants may not have sufficient energy cost savings available to justify the additional equipment, but in large facilities a very favorable payback can be achieved.

Schedule control is usually implemented in aeration control systems for use as a temporary alternate to DO control. If storm events or system maintenance makes DO control infeasible, the schedule can provide flow variations that track the typical diurnal fluctuations. Using a schedule to control blower and aeration basin air flow is also a very handy tool during commissioning procedures because it modulates blower and aeration basin air flows in a predictable and controllable manner.

The schedule typically divides the day into four time periods, with the length of each period operator adjustable. The start of the first period is generally midnight. The starting times of the other periods are set to match changes in flow according to the plant's typical diurnal flow pattern. The schedule establishes the total blower air flow rate. Once the total air flow is established, the same logic for blower starting and stopping is used for both DO control and schedule. If individual basins or zones have flow control valves, the proportion of air flow to each is based on the relative size of the tanks—which in turn is based on mixing air flow setpoints.

Establishing the control strategy is the most critical part of system development. It is the logic selection and the control algorithms that establish the effectiveness for both process performance and energy optimization. Once the logic is established a competent designer and programmer can implement it in a variety of controller platforms. The controller selection and system architecture dictate the details of the programming and its structure. These must be considered in the overall design process.

EXAMPLE PROBLEMS

Problem 9.1

Refer to Problems 3.3 and 4.2. The owner is interested in using a DO transmitter with Hi/Lo alarm contacts to operate the aerator intermittently. The measured power draw at the aerator is 15 kW. The aerator motor manufacturer recommends 6 starts per hour maximum. The desired average DO concentration is 2.0 ppm and the proposed on/off setpoints are 1.0 and 3.0 ppm. The installed cost of the DO transmitter is \$3500, a 25 hp RVSS costs \$1800 installed, and electric energy costs \$0.10/kWh. Is this a feasible ECM?

Problem 9.2

Refer to Figure 6.8, Problems 6.4 and 8.1. Each aeration grid is controlled as a separate control zone. The DO control logic is based on proportional speed floating control with bias and bandwidth. The DO tolerance for all grids is ± 0.15 ppm and the bias is 60%. The current process data is as follows:

Grid	Actual DO, ppm	Set DO, ppm	Demand ACFM	DO Gain, CFM/ppm
C	2.10	2.0	1080	125
A	1.85	1.5	1320	240
F	1.60	2.0	920	125
E	2.40	1.5	1400	200

The current blower air flow demand is 4600 SCFM.

(a) Calculate the error and change in demand air flow for each grid.

(b) Calculate the total blower air flow change and the new blower demand.

Problem 9.3

Refer to Problem 9.2. The actual air flows are as follows:

Grid	Actual ACFM	BFV % Open
C	940	55
A	1115	70
F	875	60
E	1205	48

Calculate the actual air flow control setpoint for each zone and determine the direction of valve travel.

Problem 9.4

Refer to Problem 9.3. After completion of the adjustments indicated: the actual blower air flow is 4414 SCFM, per Problem 9.2 the total blower demand is 4386 SCFM, and the zone actual air flows and demands are as follows:

Grid	Actual ACFM	CFM Demand
C	1010	1080
A	1085	1136
F	895	970
E	1040	1120

If the damping factor is 50%, determine the new blower air flow demand.

Problem 9.5

Two existing blowers have calibrated ammeters. The first is marked with 4000 CFM at 398 A and 2000 CFM at 273 A. The second is marked with 7500 CFM at 750 A and 3800 CFM at 480 A. The blowers are to be controlled by inlet throttling using a PLC. What is the equation for flow versus Amps for each blower?

Problem 9.6

The two blowers in Problem 9.5 are operating in parallel mode. The target system air flow is 9500 CFM. What is the target flow for each blower?

Problem 9.7

Refer to Problems 8.1 and 9.3. The flow control valve for zone "C" is controlled by timed contact closure with a 66% bias. The air flow response is 90 CFM/s of travel. The timer for valve opening is set to $\frac{3}{4}$ s. The control air flow setpoint is 940 SCFM $\pm$ 20 SCFM and the actual air flow is 1054 SCFM.

(a) Can the control achieve the required air flow and if so—how many iterations of valve movement will be required to do so?

(b) If the response time delay is 10 s, how long does it take for the air flow to be within tolerance?

Problem 9.8

An aeration tank with a volume of 69,000 ft^3 is equipped with an offgas analysis system. The measured AOTE is 12.9% and the OUR is 36.6 mg/(l h). After a belt press is started, the measured OUR increases to 45.8 mg/(l h). How much must the air flow increase to maintain process performance?

Chapter 10

Control Components

Developing the system design incorporates many different engineering disciplines. The successful ECM program requires defining the objectives, evaluating cost-effectiveness, and identifying the appropriate strategies. Once these are established, the required field devices for monitoring and control can be selected, and the appropriate logic and control functions established.

After the strategy is established, the control platform and interface devices can be specified, and the control system design blocked out. Control panel locations and functions can be defined, and general wiring and communications systems laid out. The programming can begin while the procurement process is under way and continue through panel building and installation.

There are several control platforms available for implementing the aeration control system. Computer control, distributed control systems (DCSs), and programmable logic controllers (PLCs) have all been used successfully. Certainly, any of the three can provide the necessary functions. In fact, recent technology advances have blurred and blended these platforms. Many systems have all three types of components linked in a single control system.

The choice of platform, therefore, comes down to economics, compatibility with installed equipment, and operator preference. There is no single "right" choice that fits all plant sizes and all owner requirements. As with any design project, cooperation with the operators and recognition of their requirements is critical to the ultimate success of the project.

Aeration Control System Design: A Practical Guide to Energy and Process Optimization, First Edition. Thomas E. Jenkins.

As discussed in Chapter 9, the earliest automation systems were pneumatic. The earliest applications were designed to meet simple control requirements. Controlling building HVAC systems was one of the most successful applications of the pneumatic control technology. Although a wide variety of ingenious devices were developed using pneumatic control systems, there were obviously limits to the technology.

The possibilities of computer control were quickly appreciated by both the process industries and the manufacturing industries as computer technology advanced. The earliest systems were programmed in high-level languages that were inaccessible to all except a few highly trained specialists. The need to make these systems applicable to the operators of the process industries and manufacturing facilities led to the development of the DCS for process industries and the PLC for discrete parts manufacturers.

The designer should recognize that the controller cost is usually a small part of the total system cost. Engineering, programming, instrumentation, installation, wiring, and field commissioning generally overwhelm the control hardware cost. The most cost-effective system is usually the one that minimizes these ancillary expenses, even if the cost of the controller itself is higher.

10.1 PROGRAMMABLE LOGIC CONTROLLERS

The PLC was originally developed to meet the requirements of the discrete component manufacturing processes associated with the automotive industry. They were intended to replace large relay logic panels. The programming mimicked the ladder diagrams used for the electrical schematics of these panels and was called relay ladder logic (RLL) or simply ladder diagram (LD). Over time, a variety of other programming languages were developed and adapted to the PLC. The ones defined by IEC 61131-3 include the following:

- Ladder diagram (LD)
- Sequential function charts (SFC)
- Function block diagram (FBD)
- Structured text (ST)
- Instruction list (IL)

Each of these languages has its own strengths and weaknesses, and each is best suited to certain kinds of tasks. Selection is, to a large extent, dependent on the programmer's preference. Most PLC systems are programmed principally in LD (see Figure 10.1). Special functions in other languages are used for various tasks and are called from the main LD-based program. Instruction in programming PLCs is beyond the scope of this text, but it is useful for the system designer to be familiar with the basics and the nomenclature involved.

Each PLC brand, and many times each model within a given brand, has a unique programming package. The program is not transportable from one brand or model

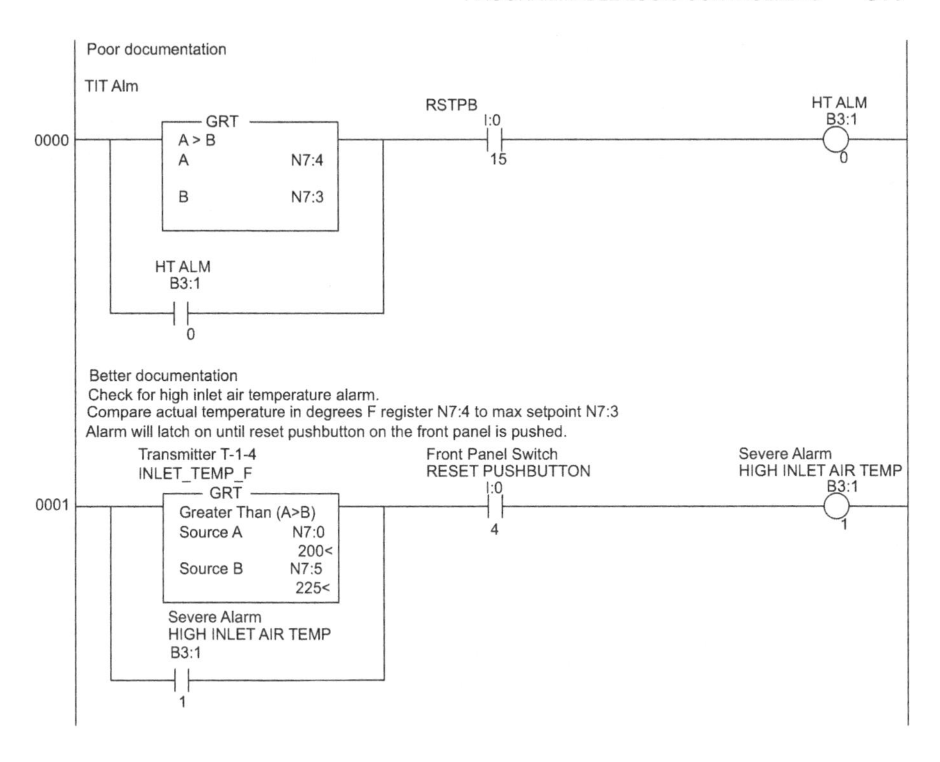

FIGURE 10.1 *Sample of ladder logic*

to another without at least some programming revisions. The basic functions of all PLCs are similar and the LD programs have similar looks. A qualified programmer can usually move from one platform to another without too much difficulty, but changing brands almost always involves some learning and expense. Most end users specify a single type of PLC for their systems to maintain compatibility of programming and reduce spare parts inventories. It is important to determine the options in PLC selection while evaluating system cost during the design process.

Program memory access can generally be categorized as register-based or tag-based. In register-based programming, a proprietary scheme is used to define the address or location of each type of data memory used in the PLC. Boolean values (also referred to as discrete status, coils, contacts, or bits), input and output (I/O) values, and internal results of logic execution and computations are assigned to memory locations. All manufacturers have developed their own nomenclature and techniques for using the data and accessing memory locations. Most programming packages also include the ability to assign descriptive nicknames to the specific registers used in a program.

Newer systems generally use tag-based memory addressing. The actual memory location and address for the program and process data is transparent to the user. Instead of using pre-defined registers and addresses, the logic and calculation results are saved and called using descriptive tags such as "BFV_Tank_1_Percent_Open". Register-based and tag-based programming each have advantages and adherents. Either can be used by a good programmer to develop efficient and understandable code. Either can be used by a less proficient programmer to develop "spaghetti code" that is almost impossible to decipher and debug. The choice of which addressing method to use is usually not up to the programmer or the designer. Once a PLC platform and a programming package are selected, the memory access method is fixed.

The early cassette loaders used for PLC programming are obsolete. Programming is now universally performed on personal computers (PCs) using proprietary software from the PLC manufacturer. The software for programming current design PLCs is overwhelmingly Windows® based. Programming software provides documentation and diagnostic functions that simplify the debugging process.

It should be noted that for most platforms the program documentation—such as register nicknames and rung comments—is not stored in the PLC. It is saved and stored in the PC. A program uploaded from most PLCs will only have logic—with undefined and undifferentiated symbols that are almost impossible to decipher. The PC package's files are the location of the register and tag names and descriptive comments. Proper program version control is essential. The end user and operators must keep the system running and they cannot be dependent on access to the original programmer. The end user should be given a copy of the PC "source code" with the latest version after commissioning. Protection of intellectual property is a concern, and non-disclosure agreements and software licenses should be used.

Version control cannot be taken casually. It is incredibly frustrating to have "identical" machines with different program versions in each! Any time a program is modified, the documentation should: include time and date, identify the programmer, and describe the changes made and their locations in the program. A comment on the first rung of the main program is an appropriate place for this. It is also incumbent on the programmer to make sure that the changes are applied to all the controllers in a system as appropriate.

Documentation of the programming is an extremely important part of the programmer's responsibilities. This step is often neglected by programmers in the rush to complete projects or get systems running in the field. Not providing adequate and complete documentation is always a mistake, and short-term pressures should not lead to poor documentation. Even the ablest programmer will need good documentation if he is called on to decipher logic years after the commissioning is completed.

Documentation should include tag names or register names that are descriptive. Cryptic shorthand such as "V1O" won't mean much during a 3:00 A.M. debugging session. A more descriptive "Tank_1_Flow_Control_Valve_Open_Limit_Switch" may take longer to enter, but the time will be more than offset by simplifying the debugging process.

Complete and descriptive rung comments are another essential tool for program documentation. It isn't necessary to comment on every rung, but every group of rungs associated with a specific function should include a comment that describes in plain language the function of those rungs. This is not just an aid in debugging. The creation of the rung comment clarifies the programmer's thinking and provides an aid in development of the logic.

The original PLCs were intended for implementing discrete logic and the original I/O was almost exclusively discrete. It was developed for accepting status signals from switches installed in the field. Analog I/O, when it became available, was very expensive and often limited in resolution. This limited the application of PLCs to process control because the majority of field devices in process control provide analog signals. As the PLC products matured, analog I/O became more available and less costly. It is now possible to obtain micro-PLCs costing a few hundred dollars with a variety of analog I/O capabilities.

The early PLCs were similarly limited in calculation capabilities. Most of them were limited to four-function math—add, subtract, multiply, and divide. As the products matured and memory and microprocessor capabilities expanded, higher level math functions were added to the PLC capabilities. Most platforms now offer functions such as square roots and logarithms, and many have the ability to enter algebraic formulas directly into ladder logic or other languages.

Early PLCs were also limited in numerical formats, with four-digit integers being a common limitation. The modern PLC offers floating-point data and scientific notation. There is virtually no function or mathematical operation required for process controls that cannot be implemented in state-of-the-art PLCs.

Older programs often used raw data or tricks such as implied decimal places to work around PLC limitations. This would often complicate both logic development and data display. With the expanded capabilities for calculation and floating-point numbers, these practices are obsolete. The recommended practice is to convert all analog data to engineering units with floating-point numbers and use these values for comparisons and data manipulations throughout the program.

PLCs were originally stand-alone units—each PLC was its own island of automation. PLC manufacturers were pushed by end users to develop communications abilities so that PLCs could exchange data with each other and with supervisory computer networks. The original communications systems developed for PLCs were serial networks. They were slow and the protocols tended to be proprietary to each supplier. The messaging was cumbersome and often required extensive programming to implement in ladder logic. PLC communications abilities have evolved and improved, and Ethernet networks have largely displaced the older serial systems. Fiber optic networks and radio transmission are becoming common and are particularly useful in noisy or physically remote applications. Although the proprietary nature of most PLC communications persists, there are now widely available public domain protocols and cost-effective communications bridges that can link new units to other brands or legacy networks.

The flexibility, capability, and cost-effectiveness of PLCs have made them the default platform for most aeration control systems. Some large municipalities that were early adopters of automation have an installed base of DCS control, but even in these applications most additions and upgrades use PLCs with advanced communications capabilities linked to the DCS systems.

10.1.1 System Architecture

The most important consideration in developing the system architecture for an aeration control application is reliability. Cost is a secondary concern. Capability is not really an issue, since almost any system architecture using state-of-the-art devices will have the ability to execute any needed control logic.

There is no universal definition of what constitutes system architecture. In aeration control applications we will use the term to describe the arrangement and configuration of the controllers and human machine interface (HMI) devices, the communications system connecting them, and the connection of field devices to the controller I/O hardware.

The simplest system architecture—and the least costly from a controller standpoint—is a single large PLC with "home run" wiring from all field devices to the central control panel I/O. This is usually not the lowest installed cost system. The amount of wiring and the difficulty in getting it from widely spaced field devices to a central location will generally offset the controller savings. Unless the necessary wiring is already in place, the single centralized unit is not cost-effective.

A single centralized controller also presents reliability issues. One central PLC represents a single point of failure—if it fails, the entire system is out of commission. This architecture may be acceptable for a small system with PD blowers that can be readily operated manually. Indeed, for this type of system, the need to provide a cost-effective solution may mandate it. For a large plant with many tanks and complex single-stage blowers, a single point of failure is not acceptable.

A more reasonable architecture for mid-sized and large systems is multiple PLCs with a communications link between them. Each PLC provides I/O and control logic for separate processes or process equipment. A central master controller provides coordination of the control logic for the individual processes. In this architecture, the failure of one controller does not stop the entire system. Even if the master PLC faults, the other units can be relied on to maintain operation of their processes based on the last commands transmitted by the master.

Physically, the individual controllers may be mounted in one or two common enclosures. Most often, however, they are each mounted in individual control panels installed physically close to the process equipment they control. Blower control is one of the most common functions of the system to be broken out from the master PLC functions. In many cases, the blower controls are provided by the blower manufacturer. This is particularly true for single-stage and turbo blowers since complex algorithms may be needed for control and protection.

Control for the aeration basins is also frequently separated from the master PLC in individual control panels. In this case, the impetus is as often reduced wiring cost as it is reliability. By installing local control panels at the basins, only a communications link needs to be connected to the master controller. The wiring between the field devices and the control system is only required at the tank itself, and this may often be installed in surface-mounted conduit.

It is tempting to use remote I/O for these functions and keep all of the control logic in the master PLC. However, the availability of low-cost micro-PLCs—compact units with fixed I/O or limited I/O expansion capability—makes this an economically marginal architecture. From a reliability standpoint, using micro-PLCs or separate PLCs at the remote locations makes more sense. The ability to provide local control and alarming is achieved with little or no additional cost. Commissioning is also simplified with this architecture, since the individual process equipment logic can be debugged and then duplicated at multiple locations.

The local PLCs and control panels are normally provided with a small HMI dedicated to the local PLC's functions. This makes operation simpler for the plant staff because the displays can be less cluttered and fewer alarms displayed at each location. Trending of the local process variables can be provided for tuning and performance monitoring. The most critical process parameters, such as DO, blower air flow, and general alarm messages, can be read directly by a SCADA link or transmitted to the master PLC for SCADA transmission.

Most PLCs, even small micro-PLCs, are typically provided with two separate communications ports. This allows the HMI communications to use one dedicated link and communications to the master PLC to use a separate link. This split communications reduces the data load on the main network, which in turn reduces latency. When the HMI is on the same communications link as multiple other devices, a perceptible delay in data display and controller response to key presses often results. This can be extremely annoying to the operators—particularly during alarm events.

Another benefit of multiple control panels is the simplicity of incorporating manual overrides for critical devices and equipment. Since reliability is an overwhelming concern for aeration systems, critical devices should have the ability to bypass the control system and operate manually in case of breakdown. If the system is based on home run wiring to a master panel, it certainly increases the amount of wiring involved in manual overrides. Perhaps more importantly, it results in a congested array of manual override switches and indicating lights on the master control front panel. Not only is this inconvenient, but the confusion created can actually result in operator error during manual operation.

10.1.2 Program Structure

When multiple PLCs are used in a distributed system architecture, it is fairly straightforward to determine which functions should be programmed in each location. The logic functions should be performed in the controller that is closest to the critical I/O for the process equipment being controlled.

As an example, some systems specify a local control panel for each blower which includes safety shutdown logic. The control of blower flow rate is often specified to be provided by an inlet valve or IGV position signal transmitted directly from a master control panel. This arrangement doesn't allow the local control panel to take action to prevent surge or overload, and if there is any latency in the data transmission to the master panel it risks unnecessary alarms and equipment shutdown. This arrangement had some merit when the only way to get performance data and commands from one controller to another was hardwired signals, but this is no longer a valid consideration.

With a distributed architecture of current design, it is more appropriate to have the master controller transmit a run command and a flow rate command to the local panel. The local panel is able to provide direct control of the flow rate and maintain operation in the safe range. If the local control panel is provided by the blower manufacturer, it may even include advanced algorithms that optimize the operating range. The local panel provides feedback of actual flow rate to the master panel, and also provides run and fail status contacts. The local panel can provide status information indicating that the blower is at the maximum or minimum air flow rate. Calculating the upper and lower limits of flow rate can be a complex undertaking with a lot of variables related to ambient conditions, specific blower capabilities, and control method. This logic is best provided by a specialist.

Two program structures that make the program debugging faster and simpler are subroutines and function blocks. They are the foundation of the modular programming concept (where proven groups of logic are developed for specific tasks or recurring equipment configurations). Using these two structures allows creation of "standard" logic.

Once the fundamental operation of the subroutine or function block has been proven, it may be copied to as many locations as appropriate. In some cases, a subroutine's logic may need "tweaking" to meet special conditions. In other cases, only register addresses or tag names need to be modified. An example might be a deadband control. Essentially identical logic can be used for a main header blow-off valve relieving pressure from flooded diffuser piping and for an individual blower blow-off valve used to prevent surge. There may be slight differences between the two operations, but the basic strategy is the same.

Function blocks are more suited for invariant logic—where the essential algorithms and logic stay the same in every application. Once the operation has been tested and verified, the function block can be reused in any program as often as that particular equipment configuration and control function are encountered. An example of this would be operation of a basin flow control valve. The standard logic incorporates: a floating control algorithm with direct control of the motor operator, timed contact closure, and limit switches for status indication. Identical logic may be needed for every control zone in a system as well as for blower inlet throttling. The same function block can be called as needed, using the specific input and output registers for each flow control location. The internal logic of the function block may be programmed in ladder logic or another supported language, depending on the platform. Once the function has been thoroughly tested and proven, the internal structure is not important.

Subroutines and function blocks do present some challenges. In some platforms, timers and counters may not execute properly if they are contained in a subroutine or function block that isn't executed every scan. It is up to the programmer to ensure the logic executes properly and if there are any precautions that need to be considered, they should be noted clearly in comments.

Programmers and engineers whose role is primarily programming tend to emphasize the creation of "efficient" code. Their objective is to reduce the length of the program, optimize memory usage, and reduce program execution time.

This is the wrong emphasis.

Owners don't want their process to run fast—they want it to run right. The important result of a "good" aeration control program is process performance that meets treatment objectives at the lowest possible energy cost. Nothing else really matters to the operators and owners. Process understanding is fundamental and more important than program technique. Less elegant programming that incorporates the performance characteristics of the treatment equipment is better than the most sophisticated programming that ignores aspects of process optimization.

One of the best ways to achieve this objective is to program to simplify the debugging process. In this context, the preferred form of "efficient" code is that which makes it fast and easy for the designer and the operator to verify the system is meeting the process objectives. Field time during commissioning is the most valuable commodity in the entire development and implementation procedure for an automation system. Deadlines loom, time is limited, and operator expectations are high. Errors are inevitable, but structuring the program and HMI configuration to make finding them efficient is the best (though not the most common) programming approach.

The program should be constructed in such a way as to make its logic operation obvious. Extensive comments and descriptive tag names enhance this. Rungs with long strings of contacts in series should be avoided. A common occurrence, for example, is a rung containing several contacts needed to provide a motor run permissive status. This construction can be very difficult to follow and can't be quickly visualized. It is better technique to group several related contacts into common status coils and use these groups in the permissive.

Some program functions are inherently difficult to monitor and their effect on logic execution may even more difficult to follow. Examples are "one shot" bits and "set/reset" functions. These execute so rapidly that the PC programming software often misses them entirely. Substituting holding contacts or other conventional logic functions can simplify tracking the logic.

A great deal of a program's time is spent waiting for response delays to time out. The program scan time is generally orders of magnitude faster than the process response. It is pointless to worry about execution times—faster scans add no value for the operators.

Sometimes it is useful to insert counters or status bits into the logic for the sole purpose of debugging. For example, it can be useful to determine if a subroutine is being called at the proper frequency. Temporary bits may be used to bypass or force some sections of logic during debugging. This code should always be clearly labeled as intended for debugging only. In general, it should be removed after testing is complete. Never leave I/O forces active in a program when the controller is not being actively monitored by the programmer.

The programming packages operating on PCs have a variety of tools available to expedite programming. The ability to cut and paste logic and documentation from one section or program to another is one of these tools. Like any tool, this has the potential to produce unwanted effects. Errors can be copied as easily as good code. The programmer must track all instances of errors when they are discovered. It is very easy to overlook some of the address changes required when subroutines or sections of code are copied. These errors can cause very interesting results on the process.

The modern PLC is a very powerful tool for general purpose process control with capabilities far beyond the original relay replacement functions. Any logic that can be described can be programmed. Distributed control architecture can enhance the cost-effectiveness of the system. The communications capabilities allow process and equipment performance information to be available to any point in the system where it might be useful for the operators or other controllers.

10.1.3 Communications Networks

Communications between controllers and field devices are a powerful tool. The ability to exchange data creates the opportunity for distributed architecture and its related advantages. It allows coordination of process equipment to optimize energy and treatment performance.

Communications networks can be difficult to set up and test. The communications medium, speed, protocol, data requirements, and similar parameters must be compatible for all of the devices in the network. Industrial communications is a field of study in itself, but it will benefit the system designer to identify some of the more common concerns.

Serial communications were the earliest implementations of PLC communications. The EIA RS-232C standard is used for short-distance (< 50 ft at 19.2 kBaud) communications—for example, between a laptop and a PLC programming port or between an HMI and a PLC in the same enclosure. Because the reference for bits is voltage to ground, it isn't sufficiently immune to EMI/RFI interference to provide reliability in many industrial applications.

The RS-422 (single point) and RS-485 (multipoint) standards were developed to extend the communications distance up to 4000 ft. The voltage changes used to signify bits are referenced to the other signal wires rather than ground. This enhances noise immunity because induced voltages are present on both conductors, so there are no false bits from voltage differences. Systems using RS-485 can include multidrop and peer-to-peer communications. Single- and dual-twisted-pair shielded wiring is used to enhance immunity to EMI/RFI. (This configuration includes the communications cable often referred to as "blue hose.")

Fiber optic systems were developed to provide immunity from EMI/RFI and to avoid equipment damage from surges and near-field lightning strikes. Because the communications connections are nonconductive they are immune to induced voltage.

Ethernet was originally developed for office communications. Ethernet is a standard for both the physical communications medium and the composition of the data packets exchanged between devices. Ethernet was not originally considered sufficiently noise immune to provide reliable communications in an industrial or process control environment. As technology improved and experience grew, it moved inexorably into PLC communications networks. It has now largely displaced serial communications in new systems.

Ethernet hardware is available using traditional twisted-pair wiring such as CAT5e and CAT6 cable. Most PLC Ethernet networks are limited to 100 meters (328 ft). Wired Ethernet can support both 10 and 100 Mbps (mega bits per second) data rates. Fiber optic communications with Ethernet have greatly extended the applications in wastewater treatment and provided a higher level of confidence in their reliability.

Radio communications can be used for remote or difficult-to-access locations. The original systems used licensed frequency radios and serial communications, and these are still used for long-distance data communications and telemetry systems. However, the licensing procedure can be cumbersome, and they are subject to interference from unlicensed "gypsy" transmitters using the same frequency.

Spread spectrum radio systems dominate in short-distance line-of-sight applications. Most applications use frequency hopping—a technique to avoid interference from other radio signals by periodically changing the transmission frequency. These radios can provide very reliable communications in noisy process control applications. Many units are limited to "short" communications distances of a few hundred feet. Others can extend the communications link up to 10 miles if high gain antennas are used. The range can be further extended by repeaters—radios that receive a signal from one device and pass it on to another.

The spread spectrum patterns of frequency and error correction are proprietary. Therefore, the same brand and model spread spectrum radio is used on all points in a communications network. Some radios are designed for Ethernet communications and others for serial networks. Two common categories, 900 MHz and 2.4 GHz, are available. The antennas are available with varying levels of gain to increase transmission distance. Most systems can use either omnidirectional antennas at the base or master radio and directional antennas (e.g., Yagi antennas) at remote "slave" locations.

Industrial communications systems have proven to be robust and reliable. Initially establishing a communications link can be one of the most difficult parts of the start-up process. Once the commissioning process is complete and the communications established, however, they continue trouble free unless interrupted by component failure. Communications protocols used with control systems incorporate some error-checking scheme. The link may fail (rarely) and no data may be transmitted, but it is not likely that incorrect data will be transmitted. Most systems incorporate some type of watchdog so that an extended interruption of the communications link will provide an alarm to the operators.

Most communications protocols are proprietary. The PLC manufacturers that developed the protocol want to encourage the use of their products exclusively and restricting communications access is one way of accomplishing this. Some proprietary protocols are published, but most are not. An exception to both of these factors is the Modbus protocol, originally developed by Modicon (now Schneider Electric).

The serial versions (Modbus RTU being the most popular) and the Ethernet version (Modbus/TCP) are both published and public domain protocols. These protocols are supported by the largest variety of devices of any single protocol. Most controllers and field devices that provide communications capabilities will support one of these.

Communications bridges are devices that convert one transmission medium to another and/or one protocol to another. Many bridges also allow mapping of discontinuous blocks of data in one controller into contiguous blocks within the bridge. This reduces communications messaging requirements. A common application would be connecting a local PLC that supports only one proprietary protocol to a Modbus/TCP network. Some programming of the bridge is needed and because of licensing and intellectual property protection, not all protocols are available. However, bridges can be extremely useful—particularly in connecting new equipment to existing networks with legacy equipment.

The communications media defines the physical connection or hardware used to link various control devices and sensors. Communications protocols define the configuration of the data packet being transmitted so that both ends of the transmission are coordinated. The protocol includes: an address to identify the communications target, a command structure to identify the type and locations of the data, and error checking to ensure the data integrity.

Protocols can be master/slave or peer-to-peer. In master/slave communications, as the name implies, one unit initiates all of the data communications requests. The master polls the individual sensors and controllers to collect data. If information from one controller is needed at another, the master must retransmit the required information. Many serial communications systems use master/slave arrangements—including Modbus RTU.

Peer-to-peer networks allow any device on the network to initiate a data transmission. If more than one device at a time is communicating the data becomes scrambled. A way to control communications must be used so that only one transmission can be active at a time. One way to control access is token passing in which a device passes the right to initiate communications to the next device after its data transaction is complete. Many serial networks use this system in their protocols.

A more common system for controlling communications is carrier sense multiple access/collision detect (CSMA/CD). In this method, each device "listens" for traffic and initiates communications only when the link is free. If two devices try to use the link simultaneously, the scrambled data is detected. Each device then waits for an empty link before trying again. This is the system used by Ethernet which is also known as TCP/IP. TCP/IP stands for transmission control protocol/Internet protocol. TCP/IP is becoming the standard communications network for process control because of its speed and reliability.

Multiple devices can be connected on the same network using TCP/IP. Hubs are multiport repeaters that join communications lines together in a star topology. They are being supplanted by Ethernet switches which provide faster communications between devices. Switches essentially connect two communication devices directly, bypassing other devices on the same network (see Figure 10.2). Strategic placement of Ethernet switches can divide the system into subnets for improved performance and reliability.

Using Ethernet does not guarantee universal communications capabilities. The physical medium is compatible between different devices and manufacturers, and the data packets can be transmitted and received. However, each data packet must

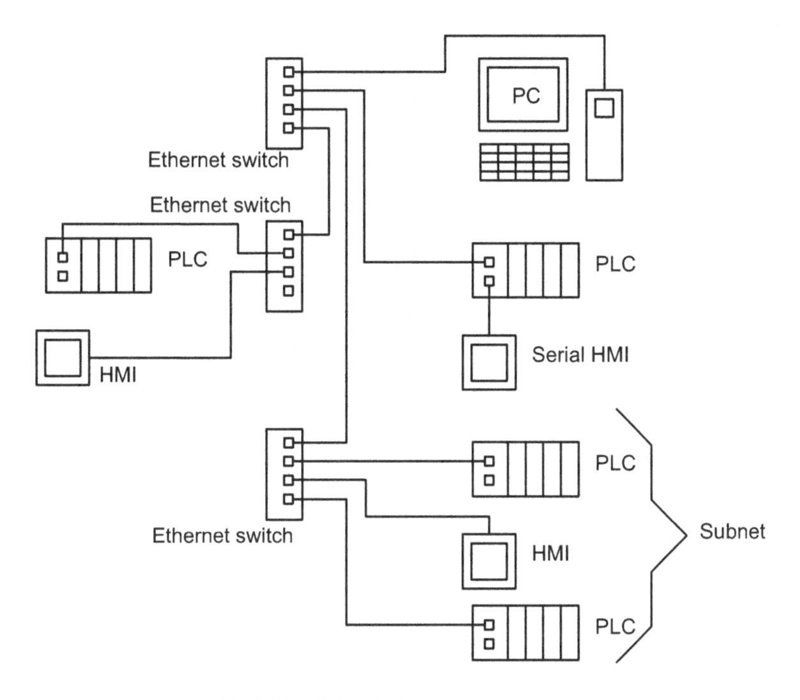

FIGURE 10.2 *Typical ethernet system*

contain commands and internal register or tag addressing that identifies the memory locations of concern. The packets also identify if the transaction is a read or a write. A controller protocol must be included in the Ethernet link compatible with both devices included in a given transaction.

One advantage of using industrial Ethernet for the control system network is accessibility to all devices in the network from any location. With many serial networks, programming changes require direct connection of the PC or laptop to the individual PLC or HMI. With properly configured Ethernet systems, the programmer can connect to a local switch and monitor multiple devices simultaneously. If the programming software supports multiple instances running simultaneously, the programmer can make changes in logic without disconnecting and reconnecting. Updates made to one PLC can be cut and pasted and downloaded into others running similar programs. This simplifies and speeds debugging and makes version control less challenging.

An exciting possibility available with advanced communications technology is remote connection to PLC and DCS networks. By using telephone connections or the Internet, it is possible to connect to a PLC in another continent. A wide and growing variety of tools make this possible. These include remote control software, virtual private networks (VPNs), and direct modem connections to PLCs.

These remote connection technologies allow an engineer or programmer to monitor system operation without traveling to the site. If control operation isn't conforming to expectations, the logic can be examined and revised if necessary.

Experienced and expert assistance for tuning can be provided to the operators. The remote connection can be used to upload current and archived data for analysis.

Security and safety are obvious concerns with remote connections. Viruses and malware are rare (but not impossible) in the PLC world, but SCADA and other PC-based systems are potentially vulnerable. It is recommended that direct and deliberate action by operators on site be required before a remote connection can be made to the control system. The operator action may include physically plugging in a cable and/or starting required access software at a local PC connection. This limits the opportunity for unauthorized access. Even more important, it provides personnel on site with warning that equipment may be operated remotely so that potential hazards can be eliminated.

Despite the potential problems, the remote access capability offers such significant opportunities that its use will increase. Reasonable precautions can be implemented to make this a safe and effective strategy.

10.1.4 Accommodating Instrument Inaccuracy and Failure

Communications links sometimes fail and alarm messages should be generated to alert the operators when that occurs. Communications devices aren't the only point of failure, of course. Field instruments and control devices also fail, and the appropriate alarm messages can be useful diagnostic tools that shorten troubleshooting time.

Some of the field failure indications are obvious—overload contacts on motor starters, or an analog signal of 0 mA, for example. Other cases of failure are not so readily recognized. Program logic may be necessary to identify real or potential problems.

A good example would be a valve travel failure alarm. If the valve is sent a command to go to a position and the feedback confirming it isn't received within a set period of time, a valve fault alarm can be logged. If a run status isn't received from a VFD or motor starter within the allotted time, a fail-to-start alarm would be appropriate.

For many field devices, calibration errors leading to inaccuracy are more common than actual failure. Periodic verification of accuracy and recalibration should be part of the normal preventive maintenance procedures for a treatment facility. The frequency of calibration checks varies with the type of instrument. Dissolved oxygen probes may require calibration checks monthly, pressure transmitters annually, and RTDs only on failure. Plant experience will determine the requirements, but it is important to note that inaccuracy may cost more in wasted energy than the expense of the most rigorous preventive maintenance program.

Operator judgment is also useful in identifying instrumentation problems. Trending parameters over time is particularly useful in this regard. For example, if header pressures are steadily increasing over a period of months, it may indicate a drifting transmitter or it may indicate fouled diffusers. If the operator compares individual blower pressure and finds they are unchanged, then the main header transmitter is suspect. If the blower pressures are also rising, then the diffusers are suspect. If the air flow required to maintain DO settings is increasing, the operator may look to the plant loadings. If they have remained constant, then the DO transmitters may need recalibration. If both pressure and air flow rate have increased, it is further indication that diffuser fouling is the most likely problem.

Some level of inaccuracy must be tolerated by the control system. One of the objectives of the recommended MOV and blower flow correction logic (Chapter 9) is accommodation of variations and inaccuracies in flow measurement. The aeration process tolerates quite significant fluctuations in process parameters without process failure. This is particularly true for short-term variations. Increased accuracy improves optimization and that is certainly the goal of the automation system. However, pursuit of arbitrary accuracy criterion shouldn't cause excess maintenance expense. It is possible to create instability in operation by striving for unrealistic or unnecessarily tight control.

10.2 DISTRIBUTED CONTROL SYSTEMS

The original DCS hardware was developed in process industries such as chemical and food processing. It provided centralized data collection capabilities and remote I/O functions. Graphical interfaces were provided at a central computer. The DCS is generally optimized for process control and PID algorithms, although current designs have the capability to provide any control strategy required.

Initially DCS communications and programming were proprietary and there was no opportunity to incorporate devices from other manufacturers. As technology has changed, the DCS has become more open. Modbus and other communications protocols are generally supported. Many DCS installations incorporate PLCs and other devices into the system. The distinctions between the two technologies have diminished.

10.3 HUMAN MACHINE INTERFACES

There are two broad categories of human machine interface used in most aeration control systems. One is the local control panel interface communicating with a single PLC and displaying data for a single piece of equipment or a small group of process equipment. This is the type of device most commonly referred to as an HMI.

The other HMI category is the supervisory control and data acquisition (SCADA) system. This is usually PC based and is connected to multiple processes throughout the plant in addition to the aeration controls. SCADA systems are usually provided by a local system integrator and are not part of the aeration controls.

The HMI in either category shares common features. One is an alarm log with date and time stamps. This is a powerful tool for diagnostics. Therefore alarm messages should be clear and precise. Means for setting the HMI internal clock should be readily accessible to the operators to make it easy to keep the alarm logging accurate.

Most current design HMIs (and virtually all SCADA systems) feature color graphical displays. Many also have animation capabilities. Both of these features should be used sparingly. The purpose of the HMI is to provide process performance data to the operator quickly and clearly. Animations and flashy colors can easily become more distracting than informative. Colors should be high contrast—primary colors are best. Color alone should not be relied on for communicating status. Approximately 7% of men are color-blind so descriptive text should be included as well. Background colors should be light and selected for high contrast with the graphics and data displays. If at all possible, the graphics should reflect the general arrangement and layout of the monitored equipment.

Process performance data should always be scaled in units that the operators are most comfortable with. The engineering units should always be displayed with the value and the equipment or process location indicated clearly. Cryptic device tags and codes should be avoided—or at least supplemented with labels in plain language.

One of the most useful capabilities of HMIs is trending. The display of process data over time indicates how well the system is performing and how the process responds to control changes. This is particularly helpful during tuning. If the control responds too aggressively and overshoots setpoint (or if the response is too slow and the process variable lags load or setpoint changes), it is apparent from the trend.

Plotting related process parameters together, or on the same screen, is helpful (see Figure 10.3). For example, if DO concentration and air flow are shown on the same trend, the operator can see the relationship. If the air flow isn't increasing, then both the logic and/or the tuning need to be verified. If the air flow changes but the DO doesn't then the gain of the control loop needs to increase. For most HMIs it is feasible to have the same process variable show on several trends. For example, the air flow and DO for one tank could show on a screen devoted to that tank. The DO for all tanks, or a group of tanks, could be displayed together on another trend screen.

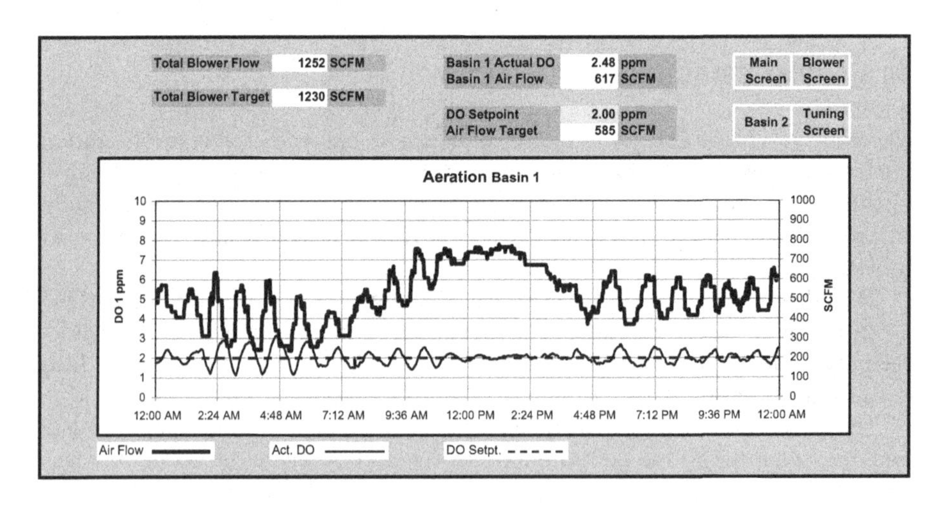

FIGURE 10.3 *Trend with multiple parameters*

The time base for trending should be compatible with the speed of process changes. It is tempting to have very rapid updates for trends simply because the HMI is capable of it. However, in most cases, it is more useful to have a longer time period visible so that the process response is apparent. For example, trending DO data at 1 second intervals and only having 15 minutes viewable at a time isn't particularly useful. Because DO changes slowly compared to other values, it is more valuable to have the DO polled once per minute and have several hours of data viewable on screen simultaneously.

Multiple trends for the same variable with different timing are useful. A rapidly scanned short-term trend of blower speed and air flow is useful for tuning the blower control loop. A 12 or 24 h trend is useful for tracking process loading changes and verifying the control is following expected diurnal fluctuations. PID "faceplates" that emulate older technology should be used sparingly. In most cases, the data can be displayed more clearly as digital values in conjunction with a trend.

Tuning parameters for the control logic should be accessible to the operators. The description of each parameter should be understandable and, if necessary, a description of the function provided as well. Text is cheap—operator confusion isn't. The tuning parameters should be grouped logically. There is no excuse for having an operator scan from screen to screen trying to find pertinent tuning. In many cases, having arrays of similar parameters for parallel equipment is convenient—having all of the DO setpoints for multiple tanks in one location is a good example.

Keys for screen navigation should be in the same spot for all screens and descriptions should be consistent. Use alarm reset keys at remote locations with caution—particularly on SCADA systems. In many instances it is preferable to have the operators go to the equipment location and observe the actual status before resetting the alarm. Warning and informational alarms may be acceptable to reset remotely, but equipment faults should generally be reset only from a local panel.

Passwords should be used when necessary. However, using too many levels of passwords is merely cumbersome and adds little to security. In a great many cases, the password can be found taped to the side of the control panel! Avoid passwords that time out after extremely brief periods of inactivity. It is very annoying to operators to have to log in after a short interruption because the password timed out.

The cost of touchscreens and monitors has continually decreased, making ever larger units available and affordable. The tendency toward "bigger is better" thinking should be limited. More data on a screen doesn't mean more information for the operator. In fact, screens which are too large limit the operator's ability to see and grasp more than a portion of the display at one time. This is particularly true of touchscreens—where the distance from screen to eye is limited by the length of the operator's arm. The upper limit on touchscreen size should be approximately 10–12 inches. Beyond that point little value is added to operator usability.

10.3.1 Supervisory Control and Data Acquisition

SCADA system capabilities have advanced with the technology of PC operating systems and hardware. Graphics and data displays can provide a "snapshot" of the overall process (see Figure 10.4). There are very few core functions that aren't

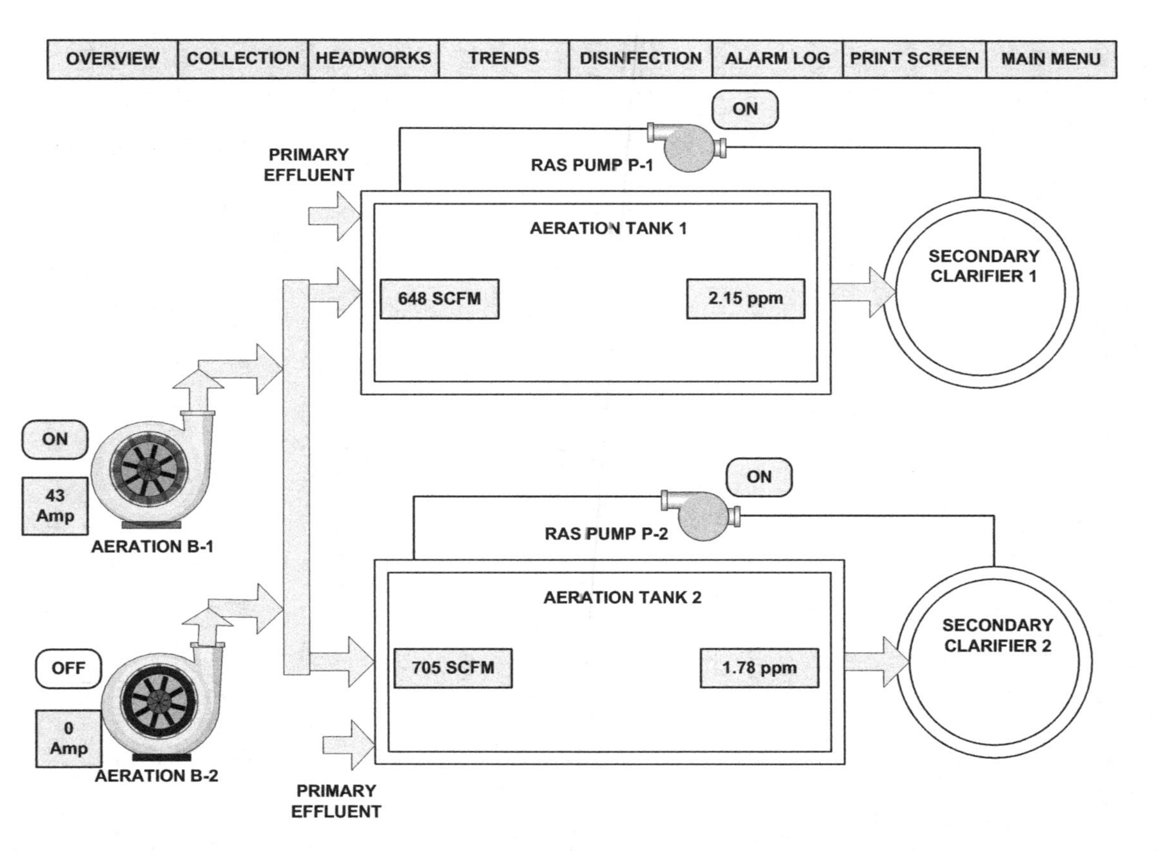

FIGURE 10.4 *Sample SCADA screen shot*

available in any commercially available SCADA software. Not only are the display and archiving functions powerful and flexible, but with new software screen configuration has been simplified and accelerated.

Data and alarm archiving is one of the most useful functions of SCADA. By saving trend data and alarm logs for future review, the operators are able to look at long-term performance. This is useful for analyzing process performance and essential in verifying success in energy conservation. The ability to look at archived data can provide the baseline for analysis of savings from control additions and modifications. The actual performance can be compared to projected performance for analysis and payback verification.

Exporting data from the archived format to spreadsheets is a useful tool. Averages, minimums, and maximums can then be calculated. Charts of historical data can be arranged and grouped to assist in the analysis. For example, hydraulic and organic loading can be plotted with air flow and DO concentration to determine if diffuser performance is degrading. Excess frequency of storing data should be avoided. It is possible to overwhelm the capabilities of software packages with long-term data logged every few seconds.

Communications latency (a time lag in the data communications between devices) is not usually a problem for current state-of-the-art controllers. The communications response is much faster than the process response. Latency and communications and data integrity verification can be minimized, however, by minimizing the number of communications transactions. This is usually accomplished by keeping data in contiguous blocks of memory. There should be at most a single read and write for analog data and a single read and write for discrete data. By combining discrete data into words or bytes and "bit stripping" the status at the controllers it is possible to eliminate separate communications for discrete data.

10.3.2 Touchscreens

Local control panels are normally provided with local operator interfaces. Current technology and economics have made the color touchscreen the predominant choice for this function.

Local HMI panels are intended to provide current operating data for the operators. The detail available generally extends beyond the level needed except at a central SCADA system. Realistically, all I/O points for the local controller should be displayed on the touchscreen. Graphics that reflect the physical arrangement of the individual equipment and instruments can be used to enhance operator awareness of the operation. A large overview screen should display the most critical and frequently required data. Multiple screens can provide any level of detail needed.

Locally displayed trends are particularly useful in tuning control response. All of the adjustments pertinent to the operation of the controls associated with a given control panel should be accessible and adjustable at the local panel—not just at the SCADA system. As a rule, having access to tuning at both locations should be avoided since it can lead to confusion.

Touchscreens have some archival capabilities, including storage on removable solid-state memory devices. However, long-term data archives should generally be performed at a more central SCADA system. Larger data storage capabilities and more convenient access are available at the SCADA PC.

Local HMIs employing serial communications should be separated from the larger system-wide communications network and access the controller directly. In addition to reducing traffic and data latency, this permits operation and monitoring of the process equipment even if the SCADA system or plant-wide network is not operating. HMIs using Ethernet communications and switches mounted in the local control panel may provide satisfactory performance without a dedicated PLC communications port. The local switch can maintain communications even if the rest of the network has faulted.

A recent development is the all-in-one operator control station. This unit combines touchscreen HMI, controller, and I/O in a single unit. Most units support both I/O mounted directly on the unit and remote I/O connected by a serial or Ethernet communications link. The logic unit can be programmed in ladder logic or other programming languages. The HMI configuration software and the logic programming are integrated in a single PC-based software package. These units provide compact and cost-effective control. They are particularly suited to local control panels serving a single piece of equipment (such as a blower) or a remote panel controlling one or more aeration basins.

10.4 CONTROL PANEL DESIGN CONSIDERATIONS

Control panel design involves as much art as science. An exhaustive study of panel design could fill another book, but it is worthwhile to indicate some of the more common considerations.

One of the most common problems is overspecification of requirements. Reliability is certainly a key aspect of owner satisfaction. Reasonable precautions must be taken to design a system that meets the functional requirements of the application. It is common, unfortunately, to find items specified that exceed the needs of the installation "just in case." This increases system cost without providing a benefit to the operators. It can also lead to operational complications and add complexity to the programming.

A good example of this is the requirement for an uninterruptible power supply (UPS) in every control panel. In an office environment, a UPS provides time for staff to back up files and engage in an orderly shutdown. In a control panel hundreds of yards from the normal operator location, the UPS does nothing but delay the inevitable. Loss of communications alarms can serve to notify the operators that there is an issue at the remote location. Unfortunately, if the three-phase power to a building has failed there is nothing for the panel to control. The UPS may keep the controller running but it can't keep the process control operating!

PLCs are equipped with battery-backed retentive memory, so the setpoints and tuning are not lost if the programmer has done the proper assignments. If not, the memory will be lost once the battery in the UPS draws down anyway. Having the processor continue to run when the field devices and final control elements are out of service requires extra programming and provides no benefit. The only exception to this would be a remote pump station where a radio link identifying loss of power may be the first notification the plant receives. This is rarely the case for in-plant controls.

Another common example of overspecification is the blanket requirement for NEMA 4X stainless steel construction for all panels. Seeing an array of these panels adjacent to a NEMA 1 motor control center demonstrates wasted money, not good design. Outdoor panels in severe climates certainly require space heaters and appropriate ratings for exposure to weather. Interior panels generally need to consider heat buildup, but in most cases, forced ventilation is adequate and panel air conditioning is not needed. The ambient conditions and the rating of other components in a panel location should be considered and the panel designed accordingly.

Most jurisdictions require control panels that conform to UL-508 and National Electric Code (NEC) standards. Good design practice requires following these codes even if the destination doesn't mandate them. The designer should be aware, however, that these codes are intended to provide for operator safety—not system function. Complying with these requirements is the beginning of the designer's job, not the end of it.

Foreign voltage in the context of control panel design refers to voltage derived from outside the panel itself. This is frequently encountered when the controller output is a dry contact to indicate status or provide control of equipment provided by others. A contact to pull in a motor starter coil with 120 VAC power coming from a control transformer in the MCC is a good example. Because these circuits will remain powered when the panel itself is turned off, the foreign voltages represent a safety hazard. A separate wire color (often yellow) should be used for foreign voltage circuits. A prominent warning label indicating the presence of foreign voltage should be mounted inside the panel. A label should also be provided with the wire color code for all internal wiring identified.

Many panel specifications, which have often been developed from pump control panels, require three-phase power and contactors within the panel for valve motor operators, small motor loads, and so on. This practice is not recommended. Increasing concerns with arc flash hazards and requirements for personal protective equipment (PPE) make this practice undesirable. Troubleshooting and testing control panels is more complicated because safety requirements must be based on the highest voltage inside the panel. Separate panels for controls and three-phase power are recommended. For safety, 24 VDC for control circuits is highly recommended if at all possible.

The panel design should consider the requirements for testing, troubleshooting, and maintenance. It must provide adequate room around all components for access to terminals and wiring. For field connections, particularly analog circuits, disconnect terminals are extremely convenient. They allow insertion of test meters into the 4–20 mA circuit without requiring wire removal. Fused and surge suppression terminal strips can provide low-cost protection of panel components from short circuits and overvoltage.

Manual overrides and disconnects should be provided for equipment. The process must remain in operation in the event of controller failure. The overrides should, of course, be clearly labeled and conveniently located.

Relays are frequently used to separate a PLC's discrete I/O from field devices. They are particularly useful with discrete outputs, since they provide dry contacts for switching any voltage for field devices. Interposing relays allow a low-current capacity output to switch much higher currents as well. During testing and troubleshooting, it is very useful to have relays with visual status indication and with manual override or test capability on the relay coil. The manual overrides allow testing the function of the controlled device without going into the program or forcing outputs from the controller.

Control panels should be thoroughly tested prior to shipment. All of the internal wiring should be checked using simulators and switches at the field terminal strips. All I/O shown on the point list should be verified. It is much more efficient to find problems with continuity and wiring errors in the shop than to wait to test on site. Particular attention should be paid to the HMIs to ensure that the data displays and setpoints are accessing the correct controller registers. Trending should be verified for scaling and readability. The results of the testing should be recorded for future reference. Exhaustive specialized test reports aren't necessary for most projects, but documenting the results of the testing is required in all cases. Creating a markup of the test results on the project documentation, including data lists and drawings, with the name of the individual conducting the test and the date of the test can be sufficient.

Testing logic in the shop requires some ingenuity. Some PLC programming packages include simulation to verify the logic performance. In other cases, signal simulators, switches, and indicating pilot lights may be wired to the terminal strip and used to provide verification of logic performance as well as I/O integrity.

The subject of control elements is a broad one. Fortunately, there is a great deal of technical literature and manufacturer support available in the area. Many of the concerns and design problems associated with this area are not industry specific. A well laid out panel for aeration control is not especially unique. It is the process understanding and control strategy that goes into system design and program development that differentiate aeration control from other applications.

EXAMPLE PROBLEMS

Problem 10.1

A discharge pressure transmitter with a range of 0–15 psig and an accuracy of ±0.75% FS will be connected to a PLC analog input. The pressure will be displayed on a touchscreen HMI with two decimal places. There are two PLC analog input cards being considered. One has 12 bit resolution and an accuracy of ±0.1% FS. The other has 15 bit resolution and an accuracy of ±0.35% FS. Assuming the HMI truncates the calculated value, what will be the possible range of values in the displayed pressure for each of the two I/O cards when the nominal pressure is 7.50 psig?

Problem 10.2

Create a ladder diagram program showing a typical implementation of an input debounce timer. Use any PLC software or word processor to create the logic.

Problem 10.3

A designer is selecting a PLC for a small aeration control project and is considering two different brands. The first brand, which the designer has programmed in the past, has a hardware cost of $2500 for the PLC and HMI and an estimated programming time of 32 h. The designer has no experience with the second brand. It has a hardware cost of $1800 for the PLC and HMI and an estimated programming time of 48 h, which includes 16 h to become familiar with the programming software. Assuming the owner has no preference, the programming software for both brands is available, and programming is billed at $75 per hour, which PLC is the most cost-effective and why?

Problem 10.4

A programmer is developing a function block for a floating control algorithm with bandwidth. It will be used for flow control with BFVs having 4–20 mA positioners and 4–20 mA position feedback signals. Identify the inputs to and outputs from the function block.

Chapter 11

Documentation

System documentation is, perhaps, one of the more burdensome and time consuming tasks associated with control system development. It is no less important for that, and indeed is one of the more critical portions of completing a successful project. The documentation serves several purposes throughout the life of a project:

- Design aid, helping establish the project requirements and process relationships
- Guide to the owner, verifying material and services included in the project
- Procurement instructions, defining the characteristics of components
- Contractor instructions, providing direction for installation and wiring
- Programmer's guide, identifying controlled equipment and the control strategy
- Operation and maintenance instructions for the operators.

It is important to measure the project documentation against these needs at each stage of the project. It is also important to remember that the documentation is a means, not an end in itself. It is easy to get caught up in checklists and paperwork and lose sight of their ultimate purpose.

The level of documentation required depends on the size of the project, the delivery model, and the expectations of the owner. A design/build project for a small facility would normally not require the same level of detail as a project for a major municipality, including public bidding with an independent consulting engineer controlling the project. However, to a large extent, the same type of documentation is

Aeration Control System Design: A Practical Guide to Energy and Process Optimization, First Edition. Thomas E. Jenkins.

required for most projects. Certainly, the requirements for accuracy and clarity are no different—regardless of project size.

The project delivery model and procurement method will have more influence on the documentation required than project size. Each delivery model imposes a different burden of documentation on each party during different phases of the project:

- Design-bid-build is typically used for large projects. The detailed design and contract documents for bidding are prepared by an engineering firm retained by the owner. The winning contractor builds the system and the system is turned over to the owner's staff for operation after construction and commissioning are complete. This delivery model may not be cost-effective on upgrades to small plants because of the engineering fees. This delivery model generally requires the most extensive documentation—particularly in the preparation of bid documents and in the creation of preconstruction documentation for approval.
- Design-build is used for any size project. The owner and an engineer (who may be an independent consultant) develop performance requirements for the system. Contractors and/or suppliers prepare bids including design, equipment, and construction. The bids are submitted to the owner or his engineer. Construction and commissioning are performed by the successful bidder. In this model, the contractor may purchase equipment from the supplier or the supplier may hire a contractor for installation. A similar model is Construction Manager at Risk. A contractor is engaged by the owner during the design process to assist in preconstruction management and to act as a general contractor during construction. Initial documentation may be less extensive than for design-bid-build.
- Performance contracting is a delivery model that includes financing by the supplier. The savings obtained from the ECMs are used to pay for the equipment and installation costs. The contractor (or energy service company—ESCO) is required to measure and verify the savings and guarantee the projections will be realized. The reduced number of project participants may also reduce the level of documentation required.
- Prepurchase of major equipment may be used with any of the above delivery models. The owner purchases the equipment or negotiates the price directly with the supplier. A contractor is responsible for installation and start-up of the equipment. This model doesn't affect the scope of documentation.

In retrofit projects, the owner is required to document the existing facility and equipment and to provide accurate data on current performance and energy requirements (Chapter 2). If the initial data is not accurate, the contractor or supplier may be relieved of some responsibility for the performance and savings. Recording the original design criteria and calculations becomes, in this case, the first and perhaps most important documentation task.

It is important to recall that design is an iterative task and documenting the design is as well. As the designer moves through various phases of the design, construction, and commissioning of a system, errors, omissions, and improvements will be discerned. The documentation must be revised to reflect changes made during the project. Each phase will create both direction for the next phase and revisions to the documents prepared during previous phases.

Version control is a crucial part of the documentation process. A number of individuals in a number of organizations will be receiving and using each document. Each document must include a revision number and date of the latest revision. Once a document has been transmitted to any organization other than the originator, all changes—however slight—necessitate a new revision number and revision date. This allows all parties to verify that they are working with the latest version of any document.

The documentation usually goes through at least three phases after the initial purchase. First is a submittal process. The most critical documentation is provided to the owner or consultant for review and approval prior to procurement. The submittal information is usually referenced as "shop drawings" because in the early days of the industry the actual drawings used in the manufacturer's shop for construction were submitted. Although this is no longer the case, particularly in automation systems, the term is still in use. The submittal information also includes "catalog cuts" that identify the manufacturer and model of key purchased components such as butterfly valves. Again, in the early days of the industry, these were usually actual pages cut from the suppliers printed catalogs. In today's industry, the information is more likely to be electronically obtained and submitted, but this term is still widely used.

The second stage of documentation use is the manufacturing and construction process. The submitted information, often expanded and with additional detail, is used by the manufacturer and the contractor to purchase, build, and install the system.

The final and most important stage in documentation is for use in the owner's ongoing operation. The documentation for construction is supplemented with additional information so that the owner may properly operate and maintain the system. The installation, operation, and maintenance manuals are usually referred to as "O&M Manuals." The drawings and data lists should be updated and provided in "as-built" condition.

The benefit to the design process of creating documentation is often overlooked. Documenting the design requires the designer to clarify the concepts and define the details. Vague conceptual ideas take definite form and items that may have been overlooked are brought forward. The act of expressing the design so it can be understood by others forces the designer to improve his own understanding.

11.1 SPECIFICATION CONSIDERATIONS

The plans and specifications, in conjunction with the bid form, are collectively known as the contract documents. They form the basis for the construction contract in a conventional design-bid-build project delivery model. This is the normal format for design documents used by consulting engineering firms. It is found in many different delivery models.

The specifications in the contract documents define the level of quality, performance requirements, and potential sources for the material and services to be provided on a project. The specifications will usually have several divisions. Division 1 will typically include terms and conditions (Ts & Cs)—the so-called "boiler plate." If there is a penalty associated with failing to meet projected savings or completion schedule, it will be found in this division. The other divisions will include technical specifications for the major equipment included in the project.

If the control system is part of a major project, it will be one of many items included in the specifications. If the ECM is a project in itself, then the specifications will be abbreviated and may only have a few divisions. Within each division there are sections divided according to equipment categories, such as valves, or individual systems, such as aeration controls. Most often there will be cross-references between the various sections. Most, but certainly not all, engineering firms use the Master Format defined by the Construction Specifications Institute for organizing the specifications.

The aeration controls will typically be found in Division 11—Equipment. This division includes material most often provided by the general contractor. Placement of the aeration controls in this section minimizes the involvement (and cost) of subcontractors. Because of the intimate connection with blowers and other process equipment, it is logical to include aeration controls in Division 11. However, some consultants prefer to include aeration controls in Division 16—Electrical. This is based on the electrical nature of much of the installation work. Others prefer Division 13—Special Construction, which may group it with SCADA systems and installation by systems integrators.

The bid form usually precedes the specifications in the contract documents. This is a very important element of the contract documents. Because of the cost of most ECM projects, local jurisdictions and owners typically require that the projects be advertised and publicly bid. The bid form includes requirements to protect the owner from collusion and price fixing. The bid form requires legally binding disclosure of information about the bidders to ensure that they are qualified and that they will provide material and services in compliance with the project specifications. Bonding requirements, if any, are identified in the bid form. Bidders that do not meet the requirements or provide the documentation called for are considered nonresponsive and excluded from participation in the project regardless of their quoted price.

In order to obtain lower pricing, many owners require competitive bidding on all projects and routinely award the project to the lowest responsive bidder. This does not always provide the best value to the owner. This is particularly true for highly technical and engineering intensive systems such as aeration control.

There are options to always awarding the project to the low bidder. Implementation of these options is subject to conformance with local statutes. Alternate procedures must also incorporate appropriate measures to protect the owner from collusion between bidders or the exercise of undue influence by potential suppliers on the owner or their consultants. Among the options are the following:

- Sole source procurement: This requires that the owner or their consultant establish a process for technical justification to establish that the selected supplier is the best value for the owner. This option requires significant analysis and must withstand third party scrutiny. In the case of sole source specifications, the equipment is often prepurchased and a separate bid process is used for selecting the installation contractor.
- Prequalification of suppliers: The potential suppliers must provide to the owner two packages. The technical qualification package may be provided on the day of bid or may be required as a prebid submittal. This technical qualification package must include details on the bidder's proposed system design, controller architecture, major components, and control strategy. The prequalification often requires submittal of experience history, references from similar projects, and may include savings or payback calculations and guarantees. The second package is the commercial proposal with pricing, terms, conditions, and so on. The commercial proposal is only opened and considered if the technical proposal provides the necessary information to indicate the supplier will meet the owner's performance objectives.
- Line item bid with base bid award: This procurement alternative is used if the owner or their consultant have selected a preferred supplier, but do not want to eliminate other potential suppliers from consideration. All bidding contractors must include the preferred supplier in their bid and indicate the price using the preferred supplier as a separate item on the bid form. The project is awarded to the responsive contractor with the lowest price for the project based on using the preferred supplier. The successful contractor may provide separate line item pricing for alternate suppliers, and the owner has the option of allowing the contractor to provide the alternate equipment if—and only if—they feel that the alternate will provide better value. The owner is not obligated to use the lower priced alternate. The owner may insist on using the preferred supplier at a higher initial cost if there is insufficient confidence that the alternate supplier will provide the performance required or better long-term value.

Review by government regulators may or may not be required prior to finalizing project specifications. If the treatment process will be affected, for example, if diffusers will be used to replace mechanical aerators, then regulatory review will probably be required. If the major process equipment will be unchanged and the project consists of automating existing operations, then regulatory review may not be needed. This is a determination that the owner or their consultant will need to make.

Neither the owner nor most consulting engineers can be expected to be experts in the areas of aeration control or energy conservation. Of necessity, they are process oriented and responsible for a wide variety of process equipment and systems. A specialist or manufacturer that has concentrated in aeration control is likely to have developed particular expertise in the field. It is normal practice for the owner or engineer to solicit information from these specialists and incorporate the most appropriate supplier's recommendations into the contract documents. Suggested specifications are often provided by the supplier to the owner or consultant. The suggested specifications are usually modified to meet the special requirements of each facility. The specifications should define the characteristics and performance requirements that led to the selection of a particular supplier or system design.

Although the specifications and plans are both part of the contract documents, they are not of equal weight. If there is a discrepancy between them, the specifications take precedence over the plans.

11.2 DATA LISTS

Data lists include tabulated information regarding the control system that can be more clearly or concisely shown in this format rather than as conventional text or as engineering drawings. The data lists used in typical system documentation include the following:

- I/O point lists
- Bills of Material
- Alarm lists
- Setpoint and tuning parameter lists
- Spare parts lists
- Troubleshooting lists

The input/output (I/O) point list (see Figure 11.1) should be the first data list prepared during the detailed system design. Preparation of the point list forces the designer to organize the project and identify the instrumentation and the system architecture. When it is developed in conjunction with the process and instrumentation diagram (P&ID), it provides an overview of the project.

The I/O type and signal provide the basis for PLC or DCS I/O selection. The description and field device tag number provide input to preparation of the P&ID. Control panel designations define the system architecture. PLC address, slot, and channel data is the basis for programming. Signal and scaling information is useful in programming and instrument procurement.

The point list also provides direction for structuring the system control strategy. If a signal is in the point list, it must have a function. The presence of an input serves as a checklist for these functions. The designer is prompted to determine if the instrument is there for control, for alarming, or for operator information only.

The instruments and PLC components identified on the point list are the beginnings of the top-level bill of material (BOM) for the overall project. Bills of material are associated with specific assemblies or subassemblies, and identify the

Project Name: Anytown WRF
Designer: T J
Date: 05/17/12
Revison No: 1
Revison Date: 06/28/12

Type	Description	Field Device Tag #	Control Panel	PLC Address	PLC Slot	PLC Channel	Signal	Min Value	Max Value	Units / Add'l
AI	Blower 1-1 Amps	JT-1-1	LCP-1-1	PLC-11	2	0	4-20 mA	0	215	A
AI	Blower 1-1 Inlet Air Flow	FIT-1-1	LCP-1-1	PLC-11	2	1	4-20 mA	0	5000	ICFM
AI	Blower 1-1 Inlet Air Temp	TIT-1-1	LCP-1-1	PLC-11	2	2	4-20 mA	0	300	°F
AI	Blower 1-1 Discharge Air Temp	TIT-1-2	LCP-1-1	PLC-11	2	3	4-20 mA	0	300	°F
AI	*Main Header Air Flow*	*FIT-1-5*	*LCP-1-5*	*PLC-5*	*n/a*		*Modbus/TCP*	*0*	*10,000*	*SCFM*
AI	*Main Header Pressure*	*PIT-1-5*	*LCP-1-5*	*PLC-5*	*n/a*		*Modbus/TCP*	*0*	*15*	*psig*
AI	*Main Header Temperature*	*TT-1-5*	*LCP-1-5*	*PLC-5*	*n/a*		*Modbus/TCP*	*0*	*300*	*°F*
AI	*SPARE*	*n/a*	*LCP-1-5*	*PLC-5*	*n/a*		*Modbus/TCP*			
DI	Tank 1 Zone 1 BFV Open L.S.	ZS-2-1-1	LCP-2-1	PLC-21	3	0	24VDC			
DI	Tank 1 Zone 1 BFV Closed L.S.	ZS-2-1-2	LCP-2-1	PLC-21	3	1	24VDC			
DI	Tank 1 Zone 2 BFV Open L.S.	ZS-2-2-1	LCP-2-1	PLC-21	3	2	24VDC			
DI	Tank 1 Zone 2 BFV Closed L.S.	ZS-2-2-2	LCP-2-1	PLC-21	3	3	24VDC			

Legend:

AI = Analog Input
AO = Analog Output
DI = Discrete Input
DO = Discrete Output

Italics = I/O Point communciated over Ethernet

FIGURE 11.1 *Example I/O point list*

components provided. The level of detail should be extensive, enabling the owner or engineer to be assured that the quality and quantity of the components meet their expectations (see Figure 11.2) The bills of material must also include sufficient information for component procurement. The item manufacturer and model and any critical performance data should be included. This information may be directly

Bill of Material		Anytown WRF			page 2 of 4
Item	Qty	Part Number Mfg. Part No.	Rev	Description Mfg. Name	Tag No.
12	10	80-0174-1 Q-45D-ODO	0	DO Transmitter, Dual Sensor Input Handrail Mount, 120 VAC power NEMA 4X, Dual 4-20 mA output for DO 0-10 ppm Dual 4-20 mA output for WW Temp 0-100 C Self-cleaning Controller Analytic Technologies Inc.	AIT-1-X X = 1 to 10
13	2	80-0186-2 Q1-24-4-420-LI-4X	1	Thermal Mass Flow Meter, 24 VDC Power, 4-20 mA output Scaled 0-400 SCFM 4" Sch 10 SS NEMA 4X Enclosure with local LCD display, SS Probe, Center Insertion, with compression fitting and Teflon ferrule Flow Devices Company	FIT-1, FIT-2
14	1	80-0186-3 Q1-24-6-420-LI-4X	1	Thermal Mass Flow Meter, 24 VDC Power, 4-20 mA output Scaled 0-800 SCFM 6" Sch 10 S NEMA 4X Enclosure with local LCD display, SS Probe, Center Insertion, with compression fitting and Teflon ferrule Flow Devices Company	FIT-3

FIGURE 11.2 *Example bill of material information*

included on the bill of material, or it may be included on component drawings referenced by part number. It is usually convenient to have the information included on the project documentation. Some system suppliers try to avoid this—hoping to keep some information proprietary or restrict repair parts sourcing. However, the owner will ultimately have access to this information from nameplates and operation manuals. It is better to provide the information in a manner that makes it convenient for the approval and procurement processes as well as future system service needs.

Alarm lists should provide the owner with both the type of alarm (warning, severe, etc.) and the control action taken by the system when conditions fall outside the normal operating range (see Figure 11.3). This information is particularly useful as part of the Operation and Maintenance (O&M) Manual. It provides the operators with a reference for diagnostics when alarms occur.

When a list of setpoints and tuning parameters is provided during the initial submittal, it helps indicate the type of control algorithms being planned by the supplier. This data list is very useful for the programmer during control logic and HMI configuration development. An important but often neglected step in commissioning is to record the various setpoints after the system tuning is complete and provide them to the operators. This permits the operators to go back to "safe" values for tuning if subsequent tuning results in degraded performance (Figure 11.4).

Most O&M Manuals include lists of recommended spare parts. They should include all wearable replaceable parts—components such as DO probe membranes, pilot light bulbs, or relays—that would be expected to require replacement during normal operation over a period of years.

The following alarms will be logged and displayed on the HMI and the actions indicated will be taken. The alarms will be displayed on the alarm list with time and date stamps. Individual alarms will be provided for each blower and each aeration control zone. Alams may be acknowledged from the HMI or a pushbutton on the local control panel. The alarm log may be cleared from the HMI with entry of the proper password. An Alarm Silence is provided at the HMI.

Alarm Description	Equipment Shutdown	HMI Display	Alarm Light	Alarm Horn
Blower surge (separate alarm for each blower)	Y	Y	Y	Y
Blower high vibration (separate alarm for each blower)	Y	Y	Y	Y
Blower high discharge air temperature (separate alarm for each blower)	Y	Y	Y	Y
High DO (separate alarm for each control zone)	N	Y	Y	N
High air flow (separate alarm for each control zone)	N	Y	Y	N

FIGURE 11.3 *Example alarm list*

Setpoints and Tuning Parameters				
Name	Description	Engineering Units	Default/Initial Value	Normal Range
DO response delay	Establishes the time delay between successive calculations of DO error and corrections to control zone air flow rate.	Minutes	3.0	1.0–5.0
DO deadband	The allowable error in DO concentration before control action is taken.	ppm	0.2	0.1–0.5
DO loop gain	The ratio of the change in air flow to DO concentration error used in calculating air flow changes needed to maintain DO setpoint.	CFM/ppm	50	10–150
DO control bias	Ratio of change in air flow on high DO to the change on low DO.	%	60%	40%–80%
Min. DO rate	The rate of change in actual DO required to inhibit changes in air flow corrections.	ppm/minute	0.15	0.1–0.3

FIGURE 11.4 *Example setpoint list*

A list of common problems and a troubleshooting guide should also be included in the O&M Manual. For each anticipated problem, the guide should indicate the observable symptoms of the problem, the possible causes of the problem, and any diagnostic tests that can identify the problem's cause. The corrective actions required to eliminate the problem should also be included.

11.3 PROCESS AND INSTRUMENTATION DIAGRAMS

The process and instrumentation diagram should be the first drawing created. Like the point list, it provides an overview of the project that is useful for clarifying and coordinating the system design. The P&ID shows the process flow streams, the mechanical process equipment, and the instrumentation. The P&ID is a schematic diagram intended to show functions and relationships—not a physical layout (see Figure 11.5).

The various symbols used in P&IDs have been standardized by the Instrument Society of America (ISA) in ANSI/ISA-S5.1. This standard provides instrument and function symbols for common process equipment and instruments. Each symbol also includes identification letters and numbers that together compose the tag number for each instrument. The identification letters have been standardized for the most common functions and are usually shown in the top half of the instrument "bubble."

The additional numbers used to create the instrument tag are not standardized. They are usually in the lower half of the symbol. A common system uses the first digit to designate the control loop. A "1" for example might indicate blower control. If there are multiple identical units, a second digit would indicate which unit. A "-1" for example would indicate blower 1 and "-2" blower 2. A third digit is used to distinguish multiple instruments on the same process equipment. Therefore, "AIT-2-3-1" could designate the first DO transmitter in the DO control loop, designated as Loop 2, located in aeration tank 3. A second DO transmitter in the same tank would have the tag number "AIT-2-3-2", and so on. This system is not universal and a

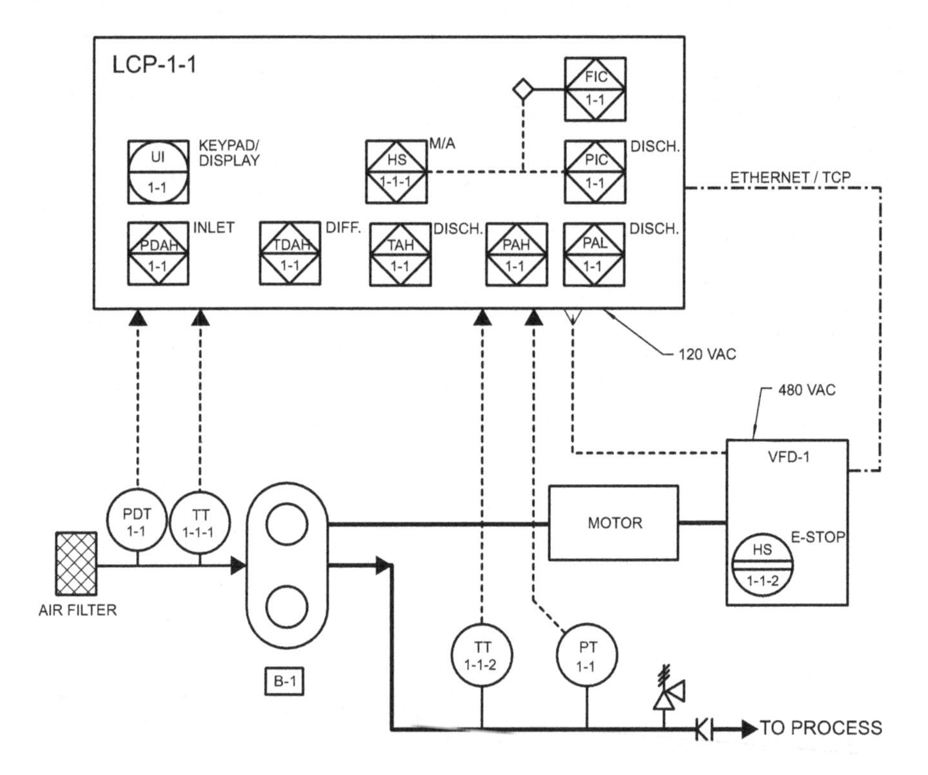

FIGURE 11.5 Simple process and instrumentation diagram

legend sheet is recommended to explain the tag number system for the project. It is not necessary to have any particular system so long as the system chosen is consistently implemented throughout the project and on the various drawings and data lists. If the owner has an existing tag number system then it should be implemented in the project, or at least cross-referenced in the point list.

All instruments used in the control system and the controlled equipment should be included on the P&IDs. It is not necessary to show all control logic in the P&ID. The most critical control loops and the key loops for energy conservation and process performance should be shown. This documents them for the submittal review and future O&M Manual use.

The P&ID establishes the basis for the design, system architecture, and the most critical functions. It is used to guide the designer through the creation of other documentation and for the programmer in creating the logic and HMI configuration.

11.4 LADDER AND LOOP DIAGRAMS

There are a variety of electrical schematic systems in use but the most common and most useful is the ladder diagram (often referred to as an elementary diagram). The symbols and practices were originally developed by the Joint Industrial Council

(JIC), and have since become part of National Fire Protection Associations (NFPA) NFPA 79 Electrical Standard for Industrial Machinery.

The ladder diagram is a schematic and not a wiring diagram or pictorial representation of the electrical components. Wiring diagrams usually show point-to-point connections and while they may be useful for construction, they are not particularly helpful for understanding how a circuit functions or for troubleshooting electrical problems. The ladder diagram, on the other hand, shows the function of each component in the system and allows tracing the power flow through the circuit for diagnostics. It is also useful in the design process to establish the required components and connections.

Most panel builders and electrical contractors understand ladder diagrams and can use them for construction and field wiring. The usefulness for construction and for diagnostics and service is enhanced by including rung numbers, wire numbers, and terminal block numbers in the ladder diagram (see Figure 11.6). Each rung and wire number should be unique for the control panel and circuit. There are a variety of systems used for designating wire numbers. As long as the system is used consistently and explained in a legend sheet, the choice of system is up to the designer.

Loop diagrams for analog circuits perform a similar function and are usually included in the system documentation. Standards for loop diagrams have been established in ISA-5.4. Loop diagrams show field devices such as transmitters,

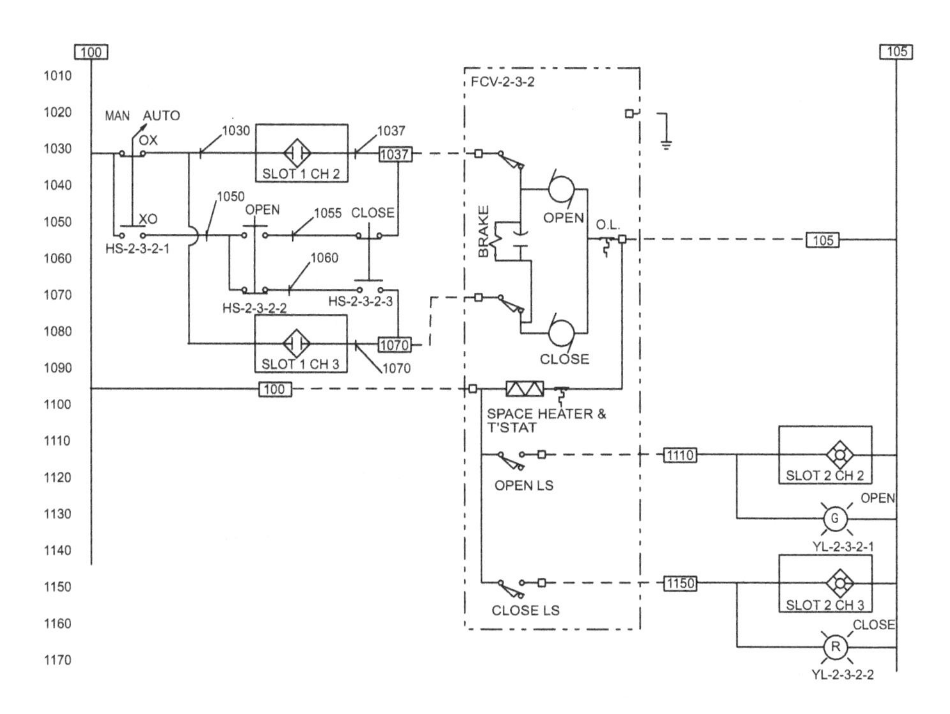

FIGURE 11.6 *Ladder diagram*

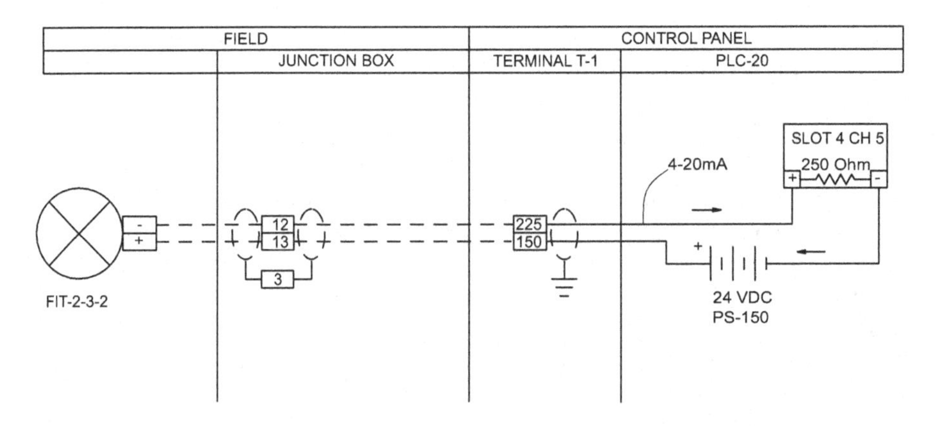

FIGURE 11.7 *Simplified loop diagram*

control panel devices such as PLC analog I/O, and interconnecting wiring and devices such as terminal junction boxes (see Figure 11.7). As with ladder diagrams, a consistent and logical system should be used to develop wire and terminal numbers which should be shown on the loop diagrams. Some items, such as field junction boxes and terminals, may be provided by the contractor and cannot be shown on the loop diagrams until the as-built drawings are prepared.

In early PLC applications, it was customary to show all of the I/O for a given input card in a single block—making the ladder diagram more like a wiring diagram. This is not particularly convenient if the I/O circuits have additional devices such as manual overrides. It is easier to diagnose the circuit functions if the individual I/O channels are shown separately.

11.5 ONE-LINE DIAGRAMS

A one-line diagram (also called a single-line diagram) is a simplified representation of a three-phase power system. One conductor is represented instead of drawing each of the three phases separately. One-line diagrams are used to show the power distribution and major components for low- and medium-voltage three-phase loads (see Figure 11.8). Control and instrumentation systems are not usually shown on the one-line diagram. As with other schematics, the one-line diagram is intended to show function instead of physical location.

In many aeration control projects, one-line diagrams are not required documentation. This is particularly true if existing blowers and three-phase loads are just being upgraded with new controls. If new blowers or blower motor control systems are included in the aeration control system scope of supply, then new one-line diagrams may be required. In this case, the existing one-line diagrams are useful references. They provide information on the capacity and utilization of the power distribution system.

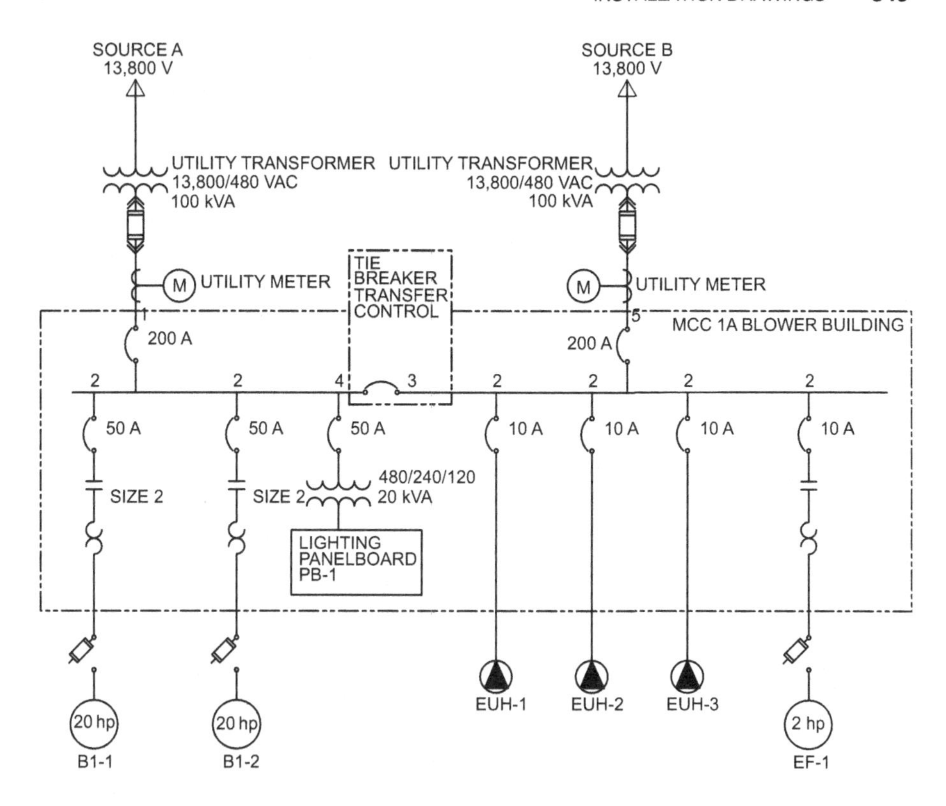

FIGURE 11.8 Simplified one-line diagram

11.6 INSTALLATION DRAWINGS

Details of construction for major process equipment, piping systems, and tanks are generally outside the scope of responsibility of the aeration control system. If the aeration controls are part of a new facility or a major upgrade to an existing one, these drawings will be provided by the owner's consultant. If the system is an ECM added to an existing facility, the owner should provide as-built construction drawings. The drawings are required for analyzing potential process and energy impacts and in designing the system.

The aeration control system documentation does require installation drawings (also known as general arrangement drawings). The contractor will need to know the locations of the control panels and field devices. The installation drawings should be physical representations of the components affected by the aeration control system. They should be drawn to scale and as accurate as possible.

It isn't generally necessary or recommended to segregate piping, electrical, and mechanical information on separate general arrangement drawings for aeration control. The purpose of the installation drawing is to guide the contractor in the placement and connection of the devices being provided as part of the aeration control system. Any pertinent feature should be shown. Tag numbers should be shown for each location to allow cross-references to the bills of material, the P&ID, and the point list. Any item included on any of these documents should be shown and identified on the installation drawing.

If modifications to existing equipment are required, the details should be defined. For example: if a new pressure transmitter is being installed, the location, size, and an acceptable method(s) of creating the threaded connection in the aeration piping should be provided.

Precautions and requirements for wiring should be provided if they are not otherwise specified. Notes specifically prohibiting wire nuts or other unacceptable wiring practices should be included on the drawing. For example, running analog and communications wiring in the same conduit as power wiring should be clearly identified as not allowed. If the contractor is to supply a particular type of shielded cable for communications connections or other purposes, then the specification or a potential source should be shown on the drawings.

Wire type and minimum size must be documented. Control and power wiring gauge must conform to all applicable codes. For three- and single-phase power, National Electric Code (NEC) requirements generally dictate the minimum size and the contractor should confirm them with local codes. Control wiring should be sized for no more than 10% voltage drop. Because control wiring is very low current, this is not usually an issue. The minimum size is often dictated by physical strength considerations for pulling the wire through conduit. Wiring that is very small is also difficult to work with while stripping and terminating. Data in Table 11.1 can be used as an initial size—to be confirmed by specific site and project requirements.

Responsibility for termination of field wiring should be defined. This is particularly true of communications systems where special end connectors may be needed. In general, the contractors would prefer that terminations be made by the system supplier but the supplier's staff is unlikely to have the appropriate local licensing. These matters should be clarified in the contract documents or the bid proposals and reiterated on the drawings.

Code compliance should typically be the contractor's responsibility. There are some generally recognized codes (such as NEC and UL-508) but the version adopted by specific jurisdictions may vary. There is also considerable variation in interpretation of codes by local inspectors. Consequently, it is prudent to put this responsibility on the local contractor. This contractor should have greater familiarity with these matters than a supplier operating from a distant location.

The planning and routing of conduit and placement of terminal junction boxes are areas where the local contractor should have responsibility and the freedom to

TABLE 11.1 Nominal Wire Sizing

Service	Minimum Size, AWG
AC power	#12
AC control	#14
DC control	#16
Analog signal	#18 twisted-pair shielded

accommodate actual site conditions. It is very difficult to optimize the conduit installations while working in the abstract. The installation drawings should simply identify the end points of field wiring and, where appropriate, the type of wiring or other technical considerations.

11.7 LOOP DESCRIPTIONS

Control loop descriptions are also known as function descriptions or control narratives. These may be included in the original specifications, or they may not be developed in full detail until completion of the programming and commissioning. The level of detail included in the loop description at various project stages will depend on the project delivery model and the expectations of the owner. By the completion of the commissioning process, fully detailed loop descriptions should be created and included in the O&M Manuals.

The final loop descriptions should include the following:

- Input signals to the loop and the related field devices
- Operator adjustments (setpoints, deadbands, gains, time delays, etc.)
- Outputs from the loop and the related field devices
- The control algorithm used by the loop (PID, deadband, floating control, etc.) and a description of operation
- Alarms and alarm indications included in the loop

There are many formats used in creating control loop descriptions. The actual format is not critical—as long as the description clearly and unambiguously describes the operation of the control strategy. One example is shown below:

Loop 4-1 Main Aeration Header Blow-off Valve Control:

Inputs: Main Header Pressure (PIT-4-1), Blow-off valve open/close status (ZS-4-1-1, ZS-4-1-2)

Operator Adjustments: Pressure setpoint, pressure deadband, valve travel increment time, pressure response time delay, valve fault alarm delay

Control Outputs: Valve open and close dry contacts (CR-1020, CR-1030) to valve operator motor (FCV-4-1)

Loop Description: The low-pressure limit will be calculated as the maximum pressure setpoint minus the deadband. If the main header pressure rises above the maximum pressure setpoint, the blow-off open contact will close and the valve will open. The valve will continuously travel open until it reaches maximum position as indicated by the open limit switch or until the header pressure drops to the low-pressure limit. When the main header pressure drops below the low-pressure limit, the blow-off valve will be incrementally closed by timed closure of the close contact as determined by the valve travel increment setpoint. After each close travel increment, the valve position will be maintained for the period established by the pressure response time delay setpoint. At the termination of the response delay, the valve close cycle will be repeated until it reaches the full closed position as indicated by the closed limit switch or the pressure increases above the maximum pressure setpoint.

Alarms: If the valve open contact closes and the pressure does not drop below the low pressure limit or the valve does not reach full open position within the time set by valve fault alarm delay, a "valve travel failure alarm" will be set. The alarm will be indicated on the master control panel HMI with time and date stamp, the alarm horn will sound, and the alarm indication pilot light will flash.

The loop descriptions can be useful for the programmer in developing the program. Creating the loop descriptions can also aid the designer by forcing him to review and rethink the control strategy. The most valuable function of the loop descriptions is to provide guidance to the operators for tuning the controls, diagnosing problems, and comparing actual to intended performance.

11.8 OPERATION AND MAINTENANCE MANUALS

The ultimate destination for the system documentation is the O&M Manuals for the system. The manuals should provide support for the operators in optimizing and maintaining the aeration control system. In order to fulfill this objective, it is important to provide as-built drawings and up-to-date data lists for the O&M Manual. It is also crucial that this information be provided to the operators at the plant.

Operators do not receive copies of the final O&M Manuals with a disturbing frequency. Too often multiple copies of the manuals are sent to the "front office" or the engineering consultants. They are reviewed, marked off on a checklist, and disappear forever into the filing system. This defeats the purpose of the documentation—it is intended for use. It should be in the hands of the contractor's on-site crew and the owner's on-site operators.

A useful tool for the operators is a printout of all HMI screen graphics. These can be created using the HMI configuration software directly or by using a print screen function. These HMI graphics should be included in the O&M Manuals.

Electronic submittal methods have reduced the problem of disseminating documentation. Electronic submittals have also simplified finding and using information in the submittals. Most plants have PCs accessible to the operators. The portable document format (PDF) is a public domain standard and PDF readers are widely available free of charge. This makes the project documentation transportable and searchable. Instead of endlessly paging through hardcopy, the user can quickly search manuals for key phrases and, if desired, print the appropriate pages for reference. Intellectual property included in the project documentation may be protected by inclusion of the appropriate copyright notices and by obtaining licensing agreements from the parties involved.

The last steps of the commissioning process should be updating the O&M Manuals and making sure the operators on site have a copy.

Once the documentation is in order and the installation is completed, the final—and sometimes most challenging step in the project—is commissioning the system.

EXAMPLE PROBLEMS

Problem 11.1

Refer to Figure 6.1. A pressure transmitter is being installed in the main aeration header immediately before it exits the blower room.

(a) Create a specification for the transmitter for use in the contract documents.

(b) Create a specification for the transmitter for use by a purchasing agent for procurement.

Problem 11.2

Refer to Problem 9.1. The decision has been made to use a micro-PLC to control the timing of the aeration on and off cycles. Create a preliminary Point List for this system, assuming only one tank.

Problem 11.3

Create a loop description for the mechanical aerator DO control in Problem 9.1.

Problem 11.4

Examine the ladder diagram in Figure 11.6.

(a) Describe the actions when the "Open" pushbutton is pressed with the circuit in the state shown.

(b) Describe the actions when the "Open" pushbutton is pressed if the selector switch in rung 1030 is moved to the "Man" position.

Chapter 12

Commissioning

The final stage of an aeration control project is, in many ways, the most difficult. It is also, in many ways, the most gratifying.

Commissioning, also known as start-up, encompasses all the activity between completion of installation and the owner taking over full responsibility for operation and maintenance of the system. There are a lot of steps in the commissioning process, and each step requires different kinds of expertise. Commissioning also demands discipline, self-control, and people skills. At this stage of the project all parties involved are experiencing anxieties over deadlines, performance, and payment. This can create a very stressful environment which is heightened by the inevitable problems that occur during commissioning.

Part of the difficulty in commissioning is the need to coordinate the activities and requirements of a variety of participants. Each of them has their own objectives and concerns, some of which conflict with the concerns of other parties. The players likely to be involved in the commissioning process include the following:

- The owner
- The contractor
- The design engineer
- The operators
- The system supplier

Aeration Control System Design: A Practical Guide to Energy and Process Optimization, First Edition. Thomas E. Jenkins.

All parties share the objective of quickly completing the commissioning process. They also want to verify that the performance of the system meets the original projections for process performance and energy cost reduction.

Unless the plant is a green field project for an entirely new facility, with no existing flow, it is necessary to maintain process performance throughout the commissioning process. This can cause conflicts between the parties regarding the staging and sequencing of the start-up. To the extent possible, the staging should be identified in the design process and appropriate levels of support to the owner identified prior to bidding. This is, unfortunately, not always done. The result is conflicting requirements for start-up.

Staging in the context of start-up means dividing the system into two or more groups of process equipment and commissioning each group separately. This usually implies a delay of weeks or months between the stages to allow for construction and installation of the equipment in the second stage.

Staging control system start-up isn't usually required if the project involves a controls upgrade without modifications to other process equipment. In this circumstance, the equipment can usually be operated manually during construction.

System modifications that generally dictate staging the commissioning are new aeration blowers, new aeration basins, and new diffusers in existing aeration basins. This equipment cannot be out of service for all tanks simultaneously. It is necessary to have some of the new process equipment installed and operating before taking the rest out of service. Full commissioning of the new controls in one step is neither needed nor possible in this circumstance.

The first set of new blowers and basins can be operating in a manual mode, while the second set is being upgraded. The first stage of the controls commissioning only requires that manual control and critical equipment protection functions are operative. The second stage of start-up would include making the controls for the second set of process equipment operable. When that has been accomplished, the overall automatic system controls (such as DO control) can be commissioned.

One factor that complicates control system commissioning is that it inevitably occurs at the end of the project. The process equipment, tank modifications, blower installation, and electrical work must be completed before the commissioning can begin. All of the mistakes and delays that have occurred during the project have accumulated. With the exception of landscaping and cleanup, the controls are usually the last item of a large project to be completed. It is a rare project that arrives at the end completely on schedule. The result is that the pressure to complete control system commissioning is very high, and a contractor exerts every effort to get the work finished quickly.

Back charges and liquidated damages (LDs) may be imminent when the control system commissioning begins. Even if the controls supplier is responsive and efficient in commissioning the system, the contractor is tempted to use back charges and LDs as a threat to the supplier to force the control system commissioning to be rushed and the project completed. Diplomacy and firmness are often required to point out that delays occurring months in advance of the call for start-up are not the responsibility of the controls supplier.

The degree of difficulty encountered in scheduling the system start-up is dependent on the scale of the project and the delivery model. If the project includes major process equipment modifications or the control system is part of a complete plant upgrade, the result is generally a long project duration. It isn't uncommon to have years elapse between the initial system assessment and the completion of commissioning for projects with broad scopes. The sizes of both the control system and the treatment plant have a bearing on scheduling. Small projects may be faster to install, although large projects with large construction crews may be installed as rapidly. Small treatment plants are usually more flexible in scheduling and staff availability. If the project is a design-build or performance contract, the control of scheduling involves fewer organizations and can be more readily controlled. In a conventional design-bid-build project, the contractor has constraints and coordination requirements with multiple suppliers to contend with.

The ability to determine the readiness of the system for commissioning can be problematic. Short notice for requesting start-up service is the norm. The contractor may request service based on his predefined schedule without verifying the actual readiness of the equipment. Many contractors do not understand the degree of integration between the controls and the process equipment, and do not understand the necessity for all the equipment and electrical installation work to be complete before the controls can be commissioned. It can be helpful to provide checklists for the contractor prior to scheduling start-up. The supplier should be aware, however, that many times the contractor bases the call for start-up on the level of work expected to be complete by a certain date rather than the actual completion.

Whether the controls are started in separate stages or in one group, there is a definite sequence that should be followed (see Figure 12.1). Some flexibility in the sequence is required, of course, and individual circumstances vary. However, it is usually most efficient to follow the suggested sequence. Even when the various steps are executed in a different sequence, it is essential to have a plan in place to establish a logical and orderly approach to the start-up. Random execution and crisis management (otherwise known as "putting out fires") will result in a longer commissioning process and reduced certainty in the outcome.

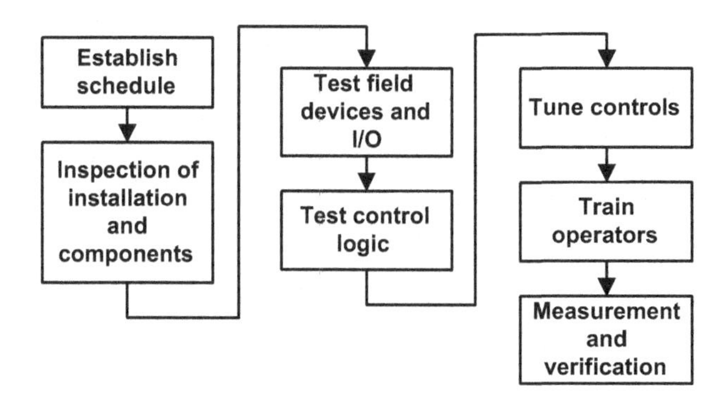

FIGURE 12.1 *Start-up activities*

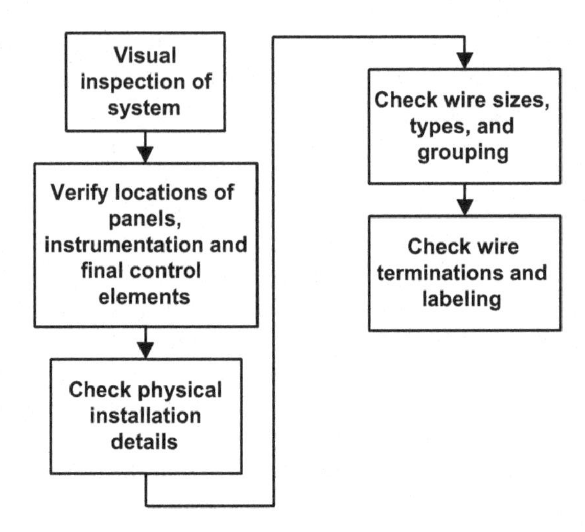

FIGURE 12.2 *Inspection sequence*

Once the scheduling has been established, the work of commissioning on site begins.

12.1 INSPECTION

You can't inspect quality into a system.

The expected performance of the system is established during the analysis and design phase of the project. The required quality of the controls and the installation should be determined during the design and preparation of the specifications. The requirements must be defined in the documentation. The purpose of the inspection process is to verify that the documented installation requirements have been met and that the components and equipment are functioning properly (Figure 12.2).

The first step in the inspection is a walk-through of the plant. The overall condition of the existing facilities and all new construction, whether or not it is associated with the aeration controls, must be examined. This can provide an indication of how well the plant is being operated, the quality level of the contractor(s), and provide a level of expectations for the installation of the controls. It will also confirm the process configuration and operating norms for the facility.

Other items that can be covered during the initial walk-through are the locations of the control panels and instrumentation. These locations should correspond to the installation drawings. The two most common installation errors are failure to locate the instrument properly relative to other process equipment and interchanging instruments that look identical but are scaled differently or otherwise different.

The probable error in location relative to other equipment varies with the type of instrument and the function. In general, flow transmitters should be upstream of valves. Suction transmitters used for dirty inlet filter detection should be upstream of any throttling valves or inlet piping. Suction transmitters that are used for calculating blower air flow rate (Equation 9.24) should be as close to the blower inlet as possible—but upstream of valves and guide vanes.

The DO transmitters should be located at points appropriate to the process control strategy. Two common problems are having them located too close to either the influent or effluent weirs. This is often the result of operators providing instruction to the contractor without being aware of the designer's intent. The DO probes are often not installed at the correct submergence or not installed at the proper angle.

Some instrumentation is interchangeable and unless tag number labels are permanently affixed locations may not be critical. DO transmitters are one example—any probe can generally be installed in any tank. Pressure and temperature transmitters are another example, contingent on either the initial scaling being identical or the scaling being easily changed in the field. The scaling of flow transmitters is generally based on the pipe inside diameter so that they may be interchanged if scaling is readily done in the field. This is one advantage of using smart transmitters—they simplify field rescaling.

Systems with distributed architecture and multiple control panels for blowers and tanks present a special problem relative to location. All of the panels for each service typically look identical. Even if the supplier takes pains to clearly identify the installation locations, the construction crew may locate them incorrectly. If the wiring has been terminated, relocation can be a significant burden. If the panels have identical components and wiring, the simplest solution is usually to change out the labels and nameplates on switches and revise the programming and addressing of the controller if possible.

Differential pressure transmitters used for suction or filter pressure drop must be inspected carefully to make sure the ports are connected properly. In this application, the high-pressure port should be open to atmospheric pressure and the low side connected to the piping. As this is counterintuitive the connections are reversed by the contractor more often than not.

Orientation of flow meters is generally critical to their correct operation. Most primary elements have an arrow or other marking to indicate the direction of flow. This is not always observed during installation and an amazing percentage of flow meters are installed backward. Orifices, pitot tubes, and flow tubes with differential pressure transmitters (for deriving the flow signal) all require close inspection to make sure that the higher pressure tap is connected to the correct port on the transmitter.

Single point flow measurement devices such as pitot tubes or thermal mass flow meters must be inserted at the correct depth into the pipe—generally, but not always, the pipe centerline. The actual insertion depth must be verified, as well as the rotation of the meter relative to the pipe centerline (sometimes referred to as the yaw).

Removal of protective caps and sleeves only installed for shipping cannot be assumed. If necessary, the probes must be removed from the piping for inspection. This is one area where observing the overall quality of workmanship can be important. If the contractor has established a confidence level, the removal for inspection may be waived unless problems become apparent later.

After the physical installation has been confirmed as correct, the wiring must be inspected. Control wiring size is not usually a problem. Minimum sizes required by specifications or code generally dictate adequate sizing. Any discrepancies between the requirements on the installation drawings and the field installation should be addressed and resolved.

There are a lot of potential issues with wiring—contractors can be ingenious in creating problems. There are some deficiencies that seem to be chronic, and the inspector should be particularly vigilant in looking for these:

- *Wire Nuts for Field Connections*: Some electricians love their wire nuts. Wire nuts especially cause problems with analog signals because the fine gauge wire makes connections mechanically unsound, and the wire nuts seem prone to picking up EMI and RFI. This causes errors in analog readings. Field connections should use terminal junction boxes. If that isn't possible, good crimp connectors (properly sized) may be acceptable.
- *Analog and Communications Wiring Sharing Conduit with AC*: This is also an EMI/RFI issue. DC power and control wiring is not a problem and may share conduit with AC or with analog wiring. Any AC wiring must be in a different conduit from analog and communications. If the wiring is run through wireway, there must be adequate distance between the AC and the other wiring to prevent interference—separate wireway is therefore preferred.
- *Reversed Polarity on Loop Powered Instruments*: For some reason, electricians have trouble with the "+ to +" connection between power supply and transmitter. The polarity seems to be reversed more than 50% of the time. The electrical schematics should show the connections clearly and the inspector should verify the polarity before powering the circuit. (Most, but not all, field instruments and PLC I/O have diodes for reverse polarity protection. However, this should not be relied on.)
- *Grounding*: This cannot be overemphasized. Running a wire to a nearby conduit is not adequate. A good earth ground must be provided and all control panels tied to it. Poor grounding leads to a multitude of problems with power, analog, and communications devices. It also exposes equipment to damage from transients since surge suppression devices depend on good grounding. Most importantly, poor grounding represents a safety hazard and will prevent GFCI and other safety devices from performing properly.
- *Improper Shield Terminations*: This is important for both analog and communications wiring. Shields should be grounded at one end—and at one end only. Terminal junction boxes should provide separate terminals for each cable's shield to provide both isolation from other signals and continuity of the shield through the box. The best practice is to ground all shields at the control panel.

Wire termination and proper labeling should be checked simultaneously. Termination should be verified at both ends. This is often made difficult by the lack of good as-built drawings for existing equipment and new equipment provided by others.

Motor starter and valve operators in particular seem to suffer from poor documentation. It may be necessary to interpret the labeling on the devices or obtain documentation from the manufacturer.

Wire numbering for field installed wires is often an issue. There are many numbering schemes in use. As long as the wires are numbered logically, consistently, and at both ends that isn't a problem. When inspecting wire numbers, the installation drawings or elementary drawings should be marked with the field wire numbers so that the as-built drawings will reflect them.

Taking good notes is an important, but often neglected, part of the inspection procedure. This obviously helps in preparing as-built drawings. Just as important is the benefit of note-taking for the balance of the commissioning process. Information comes in bursts of high intensity during most start-ups. Status, values, and data will be forgotten much faster than one imagines. Using notes to immediately record information is never a waste of time. The notes can be handwritten on paper or entered into a word processor or spreadsheet. Entry into a PC can be done directly or from handwritten notes. The advantage to using electronic notes is the ability to search by key words and the ability to transmit the notes in readable form by email.

When in doubt, err on the side of caution during the inspection process. In the long run this is less time-consuming than trying to diagnose erratic or erroneous readings.

12.2 TESTING

Testing, like any other part of the commissioning procedure, has to be approached in a systematic way. The occurrence of problems is almost inevitable. The testing should be structured so that problems are encountered as few at a time as possible, and so the source of the problem can be isolated and corrected. It is tempting to hit the big red switch and power up everything at once, but that is likely to lead to a lot of confusion.

Taking good notes during the testing process is as essential as it is during inspection. Failure to document accurately and thoroughly any problems encountered will make it difficult to determine the cause of the problem and to develop the correct solution. If deficiencies are found in equipment provided by the owner or other suppliers, it is essential to provide documentation of the problem in order to establish the responsibility for corrective measures.

Testing falls into two basic categories: hardware and software (Figure 12.3). To the extent possible, the hardware testing should be completed first and then the software should be tested. However, testing the hardware often requires using the controller to force I/O or temporary modifications of the logic. Testing the software often results in demonstrating problems with instruments or control components that weren't apparent in the hardware testing.

If forcing I/O is used to test hardware the *forces absolutely must be removed before logging off of the PLC* or other controller. Temporary program modifications should always include comments designating them as such, and they must also be removed after completing the testing.

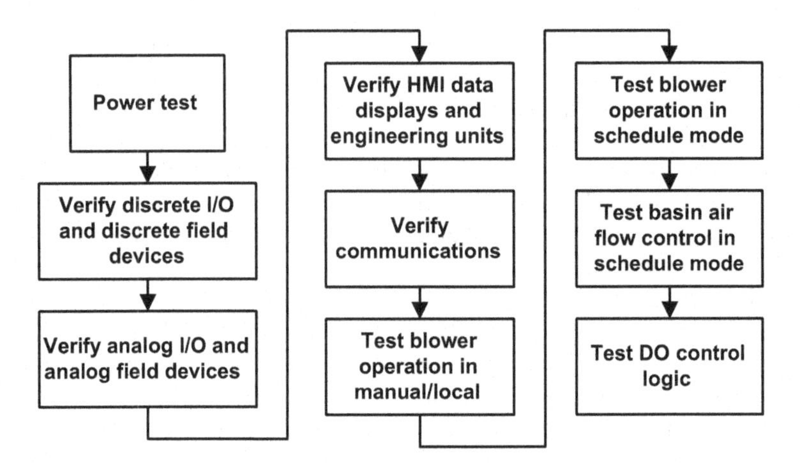

FIGURE 12.3 Testing sequence

The first test operation is the power test—sometimes referred to as a "smoke test." This requires applying power to the various system components, one at a time. If the system includes multiple panels, each panel should be tested in sequence. Normally the blower control panels are tested first, then the basin panels, and finally the master panel.

In the first step of the power test, the disconnect circuit breakers and fuses to all power supplies, other devices, and any disconnect field terminals should be opened to isolate the various circuits. The main power to each panel should be turned on first. Then the power to each circuit in the panel should be applied—one at a time—to verify proper operation.

After the panel's circuits and devices have been tested, the disconnect terminals to the field devices should be closed one at a time. The operation of each analog device should be verified by observing local indicators and measuring the 4–20 mA or voltage signals at the field terminals or the controller. It is helpful to have a laptop online with the controller so the input value can be checked and the conversion from raw data to engineering units confirmed. Discrete inputs can be tested by forcing switches or temporarily jumpering contacts if necessary. In many cases, however, if the wiring has been checked in the inspection process, separate testing of discrete I/O isn't necessary unless trouble occurs when operating the equipment.

Checking analog I/O should include verification of scaling and engineering unit displays on the HMI. Transmitters with non-zero offset in the range, such as temperature transmitters, are often not scaled correctly in the program. Smart transmitters are often supplied with factory standard scaling and require rescaling in the field. Loss of signal alarms are conveniently tested at this point—disconnect terminals make this operation simple.

If the master control panel is separate from blower or basin local control panels (LCPs), communications should be tested as each panel is powered up.

Communications with SCADA networks should be verified as connected panels are brought online. Communications are often a problem—particularly with serial networks. The required verification of communications consists of several phases for each connection:

- *Link Integrity*: Essentially verification that the physical link is functioning. Communications devices should be equipped with visual indicators such as blinking LEDs to verify the wiring is correct and transmission is occurring.
- *Data Exchange*: Verifying that data packets are being transferred between PLCs. This requires that the addressing, baud rate, and protocol in each controller match the others on the link.
- *Data Accuracy*: Confirming that the process variables and equipment status in both the local and the master controller match. This requires that the registers read and written in each controller are in the same sequence, the same data block, and have the same scaling. This task is simplified if the data is stored in contiguous blocks.

The next step in testing is to verify the operation of the process equipment. If this is existing equipment or is supplied by others, it may be necessary to have plant staff or the manufacturer's representatives on hand to verify the operation. This is not necessary for simple equipment such as flow control valves or motor starters, but is most likely a requirement for more complex equipment such as VFDs and blowers. Shaft alignment, lubrication, piping connections, and so on, should be verified before beginning tests.

In all equipment testing the objective is to verify proper operation under normal conditions and to verify the functions of equipment protection devices under reasonably anticipated abnormal conditions. This is most efficiently accomplished by limiting the number of parameters being varied at any given time. The simplest and most readily controlled operating modes should be tested first, and the more complex control strategies tested only after the simpler strategies are confirmed to be operating properly.

The first step in testing the logic is always to verify the direction of the program correction to process errors. If flow from a blower must be increased, for example, the logic must open valves or increase speed. The next step is to determine an approximate magnitude of change. This entails more than just rough tuning. If a valve can only travel from 0 to 100% open, for example, the output of the control loop should be limited to that range by the programming.

The first process equipment tested should be the aeration blowers. Mechanical checkout, including bumping to confirm the correct rotation direction, should be performed first. If there is a local manual operating mode with manual control of the flow modulation, this should be tested next. Automatic control from a manually entered setpoint should be verified. Testing protective functions and final verification of instrument performance is usually possible at this point.

The next step is to operate the blowers in remote, using schedule control or other program operating modes that allow defining blower flow demand. The objective is to verify that the blowers track the setpoint. Once that is confirmed, the starting and stopping of blowers in the proper sequence (lead first, then lag, etc.) based on system demand should be tested. This is most conveniently done based on schedule operation. If schedule operation isn't available then remote flow setpoints and start/stop commands from SCADA or a master control panel can be used for testing.

After automatic blower operation is confirmed the following step is to confirm air flow control at the basins. This is required if the system includes basin flow control strategy and the necessary hardware. The flow control zones should be brought online one at a time. Schedule control will establish a flow setpoint, and this allows tracking the direction and magnitude of flow control adjustments.

After the fundamental flow control is confirmed for the first tank or control zone, the second zone can be brought online and tested. The second zone's logic should be identical to the first, so testing the basic strategy should be simplified. Once the second tank is online testing the calculation of flow proportioning (Equation 9.14). is possible. By adjusting the mixing flows (if recommended schedule control strategies are implemented) the proportions to each tank can be modified and the calculation verified. The same technique also allows the MOV logic to be tested and confirmed.

Each successive tank should be brought online one at a time to limit the needed observations and simultaneous variations. The test of the first two tanks will have confirmed the fundamental strategy, and the successive tanks will be less difficult to test. The same rigorous method should be used for each, however, because the possibility of a typographical error or addressing mistake exists. If discipline is not used to confirm operation step by step, the errors may be missed. It is faster to use a disciplined approach initially than to rush through testing and then have to come back later to diagnose and correct problems.

DO control is the function that rouses the highest level of interest from contractors and operators. This is natural—both because it generates the highest energy cost reduction and because it is the most esoteric control loop. DO control has the most impact on process performance. The temptation to begin testing DO control immediately is strong. However, unless blower and basin air flow control is operating correctly, the DO control cannot operate correctly. The other loops must be tested and confirmed before DO control can be tested.

The testing of DO control should be done one tank or control zone at a time, usually in the sequence used for flow control testing. All control zones but one should be taken off-line and the control mode set to DO control. The setpoint should be set higher than the anticipated normal concentration setpoint and the tolerance should be set higher than the normal operational tolerance. If the recommended floating control algorithm is used, this will limit DO excursions and instability.

The initial step in testing DO control is verification of direction of changes, then the calculation of error and air flow corrections. If the calculations show a tendency to produce hunting, it may be advisable to lock the valves in a fixed manual position until the problem is corrected. Nonaggressive tuning should be used initially to reduce hunting. The calculations for changes in air flow at each zone must be confirmed and the corresponding change in total air flow provided by the blowers verified.

Once the first tank or control zone has been tested, each tank should be brought into the DO control strategy successively. As with flow control, accurate notes and discipline in testing is essential.

Operators are understandably nervous during the testing of the DO logic. They have legitimate concerns about maintaining process performance. The control system testing must include efforts to accommodate these concerns. It is important to advise the operator of each change in status as they occur and coordinate the testing to minimize the impact on the plant.

It is impossible, unfortunately, to test (or tune) a control system while maintaining steady-state operation. DO and air flow settings must be varied in order to assess the performance of the controls. The microorganisms in aerobic treatment are quite robust and they are facultative organisms that can go through extended periods without aeration and not be affected. Routinely passing through the secondary clarifiers without aeration demonstrates the robustness of the biology. The hydraulic retention time in the aeration basins is several hours. Low or fluctuating DO with durations of less than an hour will not result in loss of treatment or impair the biological performance.

If changes in logic are required to correct observed problems, it is essential that all controllers and control programming utilizing the same logic be examined and the required corrections made. Very few things are more frustrating—or unnecessary—than to have part of a system malfunctioning because known problems were not corrected across the entire system. This is one more area where taking complete and detailed notes pays off. The notes should be used to prevent the same error being programmed into future systems.

Once all of the available tanks and blowers have been tested, the system should be given time to stabilize. Then the tuning and optimization process can begin.

12.3 TUNING

Tuning is not a totally separate operation for most processes. The testing of each section of process equipment will require some tuning to provide stable operation as the testing progresses. Once all the available process equipment has been put into operation, fine-tuning for process and energy optimization becomes possible. This is the next step in the commissioning procedure (Figure 12.4).

One requirement for tuning and optimizing performance is the need to keep the system in automatic operation. Once reasonable stability has been established, this includes overnight operation in automatic. As indicated above, the plant staff may offer resistance to that, and ultimately the decision is theirs. It must be explained to them that the more data and experience that can be gained on the system, the better the final results will be. Being open about the commissioning procedures and forthcoming about problems encountered, establishes a confidence level with the operators that will increase their level of cooperation.

Although not generally considered part of the tuning process, establishing the initial setpoint for each control loop is required prior to beginning the actual tuning.

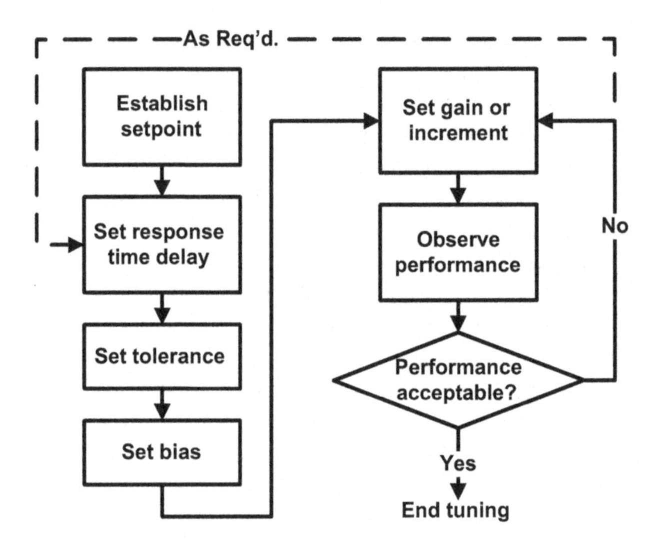

FIGURE 12.4 *Tuning floating control*

Alarm points for equipment protection should be based on manufacturer's recommendations. This includes minimum and maximum blower flows, motor full load amps, and so on. Mixing flow for basins is based on process limits and maximum flow is usually based on diffuser constraints. Setting schedule flow rates will require guidance from the operators based on their experience.

The DO setpoint is also influenced by operator experience. It is best to initially use a conservative setpoint and then gradually reduce it as experience is gained and confidence is established. An initial DO setpoint of 2.0 ppm is generally reasonable. One caution is that the control setpoint should consider the location of the DO probe. Bearing in mind that DO is only an indication that oxygen demand in the aeration process is being met, the DO concentration at the effluent end of a plug flow reactor is the most critical. Lower DO concentration at the influent end is to be expected and is not an indication of a process issue. Trying to maintain constant DO throughout the basin is usually not feasible or desirable.

One of the first tuning adjustments to be made is the digital filtering of analog inputs (Equation 9.5). The filtering interval should be less than any regular interval between fluctuations in the analog data, but significantly shorter intervals don't provide any real benefit. The number of filter samples should then be adjusted until the process value is stable at the number of significant digits used in calculations and displays. A trend of the process variable that exhibits slight fluctuations from a mean value is an indication of adequate filtering. If the process reading is fluctuating wildly, tuning the control will be difficult or impossible.

Trend charts are a valuable tool in the tuning process. They allow comparison of the response of the process variable to changes in the controlled variable. This can speed tuning by allowing some simple calculations. For example, a back calculation that consists of dividing a change in air flow by the resulting change in DO concentration can provide an initial estimate of the gain (Equation 9.10). Observing the changes in the process over time can provide guidance on the various response times for control.

There are an immense number of tuning techniques used for PID control—ranging from the simple Ziegler–Nichols to "self-tuning" controls and proprietary software. The number and variety of the software packages, books, and articles on tuning PID serves to emphasize the difficulty of tuning the algorithm. If PID algorithms are employed, reference to this literature on tuning techniques is suggested.

At some point the tuning process becomes an exercise in trial and error. Trends can be useful in tuning, regardless of the control algorithms employed. Experience is, of course, beneficial but with patience and observation tuning can be accomplished by anyone with an understanding of the process. Patience is definitely required for tuning since the effect of a change may require hours to determine. Letting the process ride through and stabilize for several load changes or setpoint changes is usually necessary.

During initial testing, coarse tuning may unfortunately require modification of several parameters at once in order to get some stability. Once the fine-tuning for process optimization begins, it is important to change only one parameter at a time. If multiple tuning values are changed simultaneously, it is impossible to know which change had the desired effect. It is even possible that the two changes counteract each other—one improving performance and one diminishing performance. Changing one value at a time and waiting to observe the effect is faster in the long term than a "shotgun" approach. Tuning is a repetitive process with a number of adjustments for each parameter needed before an acceptable value is found.

If timed floating control is the algorithm used in the program, the first parameter in the control logic to adjust is the response delay. This is not generally a sensitive parameter and a range of values will provide adequate performance. The response time setting should be slightly shorter than the time taken to reach a steady-state process value after a small load or setpoint change. The approach to the final steady-state value is usually asymptotic, and waiting to achieve that final small increment in the process value is not necessary.

The next parameter to be adjusted is the deadband or tolerance. This dictates the allowable deviation from setpoint before any control action is taken. If there are normally fluctuations in the process variable that are not removed by the digital filtering, it is probably necessary to have the tolerance zone wider than these fluctuations. This prevents the fluctuations from causing excessive control action. Otherwise the tolerance is somewhat arbitrary. Operator's opinions on what they consider acceptable agreement between setpoint and actual value should be solicited. The tolerance is normally between 1 and 5% of the setpoint or the typical operating value.

Bias adds stability to the control action by preventing the control action's repeating a specific position or value of the controlled variable if an increase is followed in the next iteration by a decrease. The value of bias isn't as critical as gain, but values of 1.0 and 0.5 should be avoided since they would result in repeating positions. Generally a bias of 0.6 or 0.7 serves to eliminate hunting.

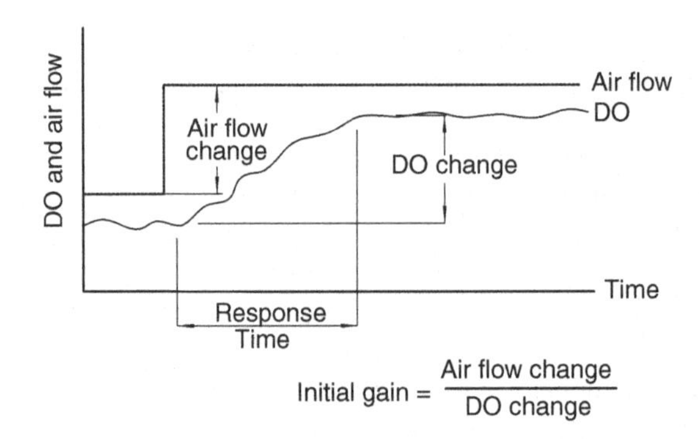

FIGURE 12.5 *Initial DO control tuning estimate*

Gain is generally the most time-consuming tuning parameter to establish. Gain represents the amount of corrective action taken by the controller in response to an error in the controlled process value. This includes both the gain constant in proportional speed floating control (Equation 9.9) and the step size for valve position in single-speed floating control (Section 9.1.5). The gain is set by assuming a value and observing the results. As indicated above, an estimate of the gain for variable speed control can be made by inducing a step change in the controlled variable and dividing it by the resulting change in the controlled variable (see Figure 12.5).

For constant speed floating control, the initial setting for the increment in the controlled variable should be about 1–2% of its range.

After getting the initial tuning parameters to provide reasonable control, another iteration of tuning will improve response and precision. The gain is the parameter that typically requires the most changes, but all parameters could be considered for modification. It is beneficial to keep track of each change in the tuning and to record if the change resulted in an improvement or decrease in performance.

The general procedure is to observe the trend of the process parameter. If the response is insufficient, then increase the gain until overshooting of the setpoint or hunting occurs. Then decrease the gain until stability is achieved. Insufficient gain is indicated if the process variable ceases to change before the set response time elapses, or if the time required and number of corrections needed to achieve setpoint is excessive. Note that "excessive" is a subjective term. Operator expectations and the physical limitations of the process play into the determination.

Response time and gain are parameters that can be interchangeable to some extent. An increased gain and slow response may provide the same control accuracy as a shorter response time and low gain. With the exception of blower air flow control, most loops will benefit from gradual corrections, since operators prefer stability to accuracy (assuming they cannot have both).

The sequence of control loop tuning is important. Blower air flow control should be the first control loop tuned and schedule control is the most convenient method for this. By changing the flow setpoint to the blowers, the blower flow control can be tuned. This loop is not affected by the source of the flow setpoint, so tuning that is stable and responsive under schedule control will also respond well when DO control is implemented. The tuning should be tested by varying the number of blowers operating and by varying the discharge pressure—if possible. Since the blower response is nonlinear, the final tuning should be set to provide acceptable response over a broad range of operating conditions.

The next loop to be tuned should be basin air flow control if it is included in the system. As with blower air flow control, this is most conveniently tuned in schedule mode. The response of flow to valve position is nonlinear and dependent on main header pressure. The tuning should be verified across a range of operating parameters, and the final tuning may be a compromise to achieve acceptable performance throughout the range.

DO control is the last fine-tuning operation. The need for patience and thorough record keeping is particularly important in tuning the DO control loop. This may require many variations before the tuning is completed. One complicating factor is that the load changes to the process vary on a diurnal basis, and the load changes cannot be controlled. Changes in setpoint can be used to induce control actions and provide data for tuning. However, the load changes from influent variations are on the order of 2:1, and the response to load changes doesn't perfectly correspond with the response to setpoint changes. In order to verify tuning, it is usually necessary to observe the system performance for several days to determine the impact of the diurnal variations. If time on site is limited, the transmission of trends from the plant can provide the necessary data, with the actual adjustments performed by remote connection to the controller or by operator entry under the supplier's guidance.

12.4 TRAINING

The training of operators in system operation and maintenance is the key to successful long-term performance and the owner's satisfaction with the system. There is a natural resistance to change by some operators and training is the best opportunity to overcome this. A successful training program should cover the following:

- Goals and objectives for the system
- Theory of operation and process performance basics
- Adjustment and tuning procedures
- Routine maintenance requirements
- Required response to emergencies and equipment failures

Scheduling training usually requires close coordination with plant staff. Many facilities have three shifts and multiple classes may be required to cover all of them. In large systems it may be advisable to separate the maintenance training from process operation training if the duties of the staff are split on that basis.

Ideally final O&Ms should be available at the time training is provided. However, the logistics of working through distribution channels and the need to provide as-built information often make this impossible. Excerpts from the manual, particularly sections dealing with tuning and troubleshooting, are useful handouts for future reference. The trainer should be careful, however, not to turn the session into a group reading of the handouts. This is never an effective way to maintain the attention of the trainees!

The most common presentation method for training is a PowerPoint™ slide show. This can be effective provided the recommended practices for any presentation are followed. As with handouts from the manual, the presentation should not be an exercise in reading the slides to the audience! The slides should be brief with only a few bullet points on each. Diagrams and text should be large enough for legibility. Terms and acronyms that may not be familiar to the staff should be defined and jargon should be avoided as much as possible. Fewer slides well-presented make a more engaging training session than a presentation that is rushed. A printed copy of the presentation slides also makes a good reference handout. More technical or obscure information can be included in supplemental handouts and may not need to be covered in the presentation.

Many design-bid-build projects have specified training based on a standardized requirement that exceeds the needs for aeration controls. Maintenance requirements for automation systems are usually minimal and do not require extensive training time. Work with the plant staff and the specifying engineer to develop a schedule and content suited to the specific project. A brief presentation with well-considered and clearly presented content will be more beneficial than a long presentation with redundant or trivial material.

Include time for periodic breaks in the agenda, and allow for questions and answers. A Q&A time at the end of the presentation is useful, but it is more important to encourage and respond to questions during the session when the context is clear.

The most important factor in successful training is a display of enthusiasm by the trainer. Participants are willing to overlook a lot if they understand that the person doing the training is personally enthusiastic about the system and its benefits. A lackluster session with the trainer just going through the motions cannot generate interest in the plant staff. Without enthusiasm and a commitment to success from the operators, the system is not likely to be a long-term success.

Many plants have a standard practice of recording training sessions. Be prepared for this and try to speak loudly and clearly for the recording equipment. It is not advisable or necessary to remain rigidly in one position, but the location and field of vision of the cameras should be borne in mind during the training.

It is very important to identify the initial goals for the system early in the training. If the operators understand the intent, and if they know that process performance will not be compromised, they are more likely to operate the system in automatic and work to improve its performance.

The theoretical basis for the control system (process functions) should be covered in the training. The connections between process operation, instrumentation, mechanical process equipment, and control logic should be explained so that operators can understand troubleshooting and diagnostic procedures. The cause and effect relationship between control actions and the resulting process changes should be addressed. The theory behind both process and control needs to be highlighted. Operators know what works and what doesn't, but may not understand why particular actions have specific results.

Changes in operating and control strategy from previous practice should be determined ahead of the training and identified and explained during the training. The reasons why the new strategy represents an improvement over the old should be emphasized.

The procedures for setpoint adjustment and tuning are very important—this will be the operators' most common interaction with the system. The various adjustments should be categorized as frequent, infrequent, and rarely required with the appropriate emphasis on each. The location of the tuning parameters in the SCADA or HMI should be indicated and the steps needed for accessing and changing them demonstrated. The effect of each parameter on the system response (increases, decreases, etc.) and indications for needed changes should be described. The need for recording changes and their results cannot be overemphasized in the training.

Maintenance requirements may be handled by operators, by technicians, or by outside contractors depending on the normal practice at a specific facility. Routine and preventive maintenance procedures should be distinguished from repair procedures. Diagnostic methods for troubleshooting problems should be included and common indications of problems identified. Recommended spare parts lists should be given to the staff.

Most electric and control components need minimal preventive maintenance. Screw terminals should be tightened on an annual basis. Flow, pressure, and similar transmitters may require calibration on an annual basis, but this may be extended based on actual plant experience. However, checking accuracy at least annually should be standard procedure.

Analytic instruments will require the most maintenance—particularly DO probes. Three distinct maintenance procedures should be covered: cleaning (typically weekly), calibration check and recalibration (typically monthly), and component replacement of caps or electrolyte and membranes (typically annually).

Alarm messages and the appropriate response to various alarms should be reviewed. The differences between informational and severe alarms should be emphasized. The types of potential catastrophic failures should be listed, along with the appropriate response. The staff will be particularly interested in the response to power outages and blower failures. The conditions indicating when manual intervention in the system operation is justified must be specifically covered along with the steps to take going to and from manual control of each type of equipment.

It is important to stress the need to leave the system in automatic unless severe equipment or process repercussions would result from continuing automatic control. The tendency for most operators is to switch to manual control as soon as the system deviates slightly from their previous experience. Expected and insignificant deviations from perfect operation, including the typical time for correction, should be explained. The training should encourage operators to leave the system in auto if at all possible, and instead of going to manual they should list the problem and the events before and during the problem. Management support and assistance in this area is invaluable.

Some items included in the aeration control system may justify specialized training by manufacturer's representatives. One example would be VFDs, which may have specialized and brand specific requirements. The special equipment training should be coordinated with the overall control system training.

The training sessions will consist of classroom instruction for most of the allotted time, but hands-on training and actual equipment operation should be included after completion of classroom instruction. It is important to have operators actually navigate through HMI screens and make setpoint entries. Actually demonstrating changing lead blowers with starting and stopping operating blowers is very effective. The hands-on sessions should include a tour of the system, including identifying all equipment that is automatically controlled. Demonstration of DO transmitter calibration and probe service is very helpful. It is even better to have the staff perform the maintenance with guidance from the trainer.

Safety should be emphasized at every step of the training process. Proper lockout and tagout procedures should be discussed. Potential hazards encountered in normal operation and during maintenance must be identified. Proper measures for ensuring safety should be emphasized. It is impossible to overstress the safety aspects of using and maintaining automatic controls.

It is recommended that a list of training attendees be obtained for future reference. It is also recommended that the owner or a representative sign-off as having had training and accepting the training as adequate. A sign-off sheet attached to the agenda makes a good record of both the completion of training and the topics covered. If appropriate, the consulting engineer's and the contractor's signatures should also be obtained.

12.5 MEASUREMENT AND VERIFICATION OF RESULTS

The final step in the commissioning process is measurement of performance and verification of the results. The results of the measurement and verification should be included in a report for submission to the owner and other appropriate parties. This step is required for several reasons:

- Establish a record of the systems process and energy performance for future reference in tuning and problem diagnostics
- Obtain acceptance by the owner and release of final payment
- Provide verification of energy savings for obtaining utility incentives and rebates and for verification of performance contract requirements
- Provide the data required for a design review—comparing projected savings during the design stage to actual savings for use in refining future estimates and designs

Measurement of system performance after implementation is not generally a difficult task. The control system design should include all of the instrumentation required for both process and energy monitoring. Total plant power information is generally readily available, but submetering of blower power must generally be provided as part of the control system. Since power (or at least motor current draw) is normally included in the blower control and protection functions, this is not an additional cost.

The measurements should include all of the key performance indicators used in the initial system assessment. This should include process data in addition to energy consumption data. In many systems the reduction in demand is at least as critical as consumption therefore the impact of the control system on demand should be included in the report.

An item-by-item comparison between actual and projected performance should be made for the design review, but this may not be required in the final report submitted to the owner and utilities. In most cases, only the total savings obtained by all the implemented ECMs is required. It is useful for the designer to determine, if possible, how much of the savings is due to improved diffuser performance and how much to DO control, for example, but this will not generally affect overall financial considerations.

It may be necessary in some cases to "normalize" the data to provide a standard basis of comparison. This is the case if organic or hydraulic loading has changed since the initial assessment was made. Adjustments for seasonal variations in wastewater temperature may be appropriate. Finally, operational changes that impact oxygen demand, such as MLSS concentration, BOD removal, or nitrification/denitrification may require adjustments to the raw energy data in order to properly asses the system.

The format of the final reports should mimic the format of the original system assessment to allow convenient comparison of the projections to the actual savings. Data showing energy consumption and process performance before and after implementation should be provided. The inclusion of graphics illustrating the performance can simplify the interpretation of the results.

Acceptance of the system and completion of the final report represents the final step in project implementation. It is a satisfying conclusion to an extended and complex task—one that provides long-term benefits to the plant management and staff.

EXAMPLE PROBLEMS

Problem 12.1

Refer to Figure 7.12.

(a) Identify at least three steps to use during troubleshooting of this circuit and the phase of the commissioning process when each step of the troubleshooting would be performed.

(b) Identify at least three corrective measures that could be implemented to correct problems with this circuit.

Problem 12.2

A system is in the test stage of commissioning and some rough tuning is required. Refer to Figure 12.6. Identify at least two possible tuning changes would the performance trend indicate? Provide an explanation for each.

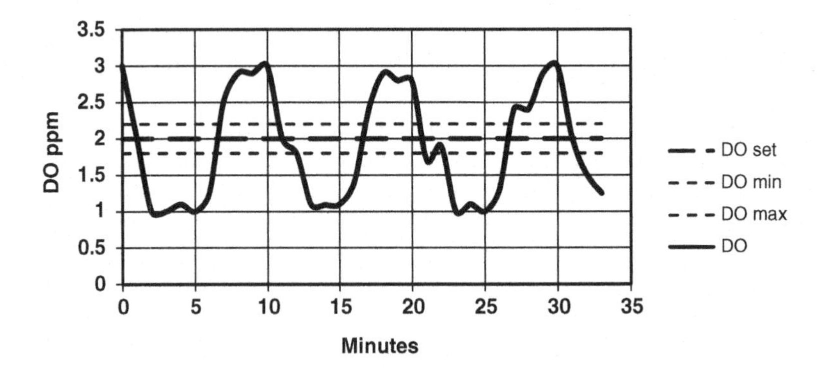

FIGURE 12.6 Problem 12.2 sample DO control performance

Problem 12.3

A facility has requested training for two separate groups of plant staff for a new system with high-efficiency blowers, new fine pore diffusers, and automatic DO and aeration control. The two groups consist of process operations and maintenance. Identify at least two areas where both groups should receive the same training, and at least two areas for each group where the training should be different.

Chapter 13

Summary

The completion of commissioning and acceptance by the owner represent the end of the procedure for implementing automatic aeration control. However, it is far from the end of the project. It should be the beginning of years of successful operation for the owner and a valuable learning experience for the designer.

The measurement and verification stage of the project should be used as a review of the successes and failures encountered. Most projects will have some of each. Lessons can be learned from both for application in future projects. The goal is to have the failures fewer and less significant in each succeeding project.

13.1 REVIEW OF INTEGRATED DESIGN PROCEDURE

The integration of all facets of the plant operation has been stressed throughout this text, with special emphasis on the various technologies encountered. Items to be integrated include:

- process and energy optimization
- mechanical equipment and process requirements
- instrumentation and control equipment
- control strategy and process performance objectives

Aeration Control System Design: A Practical Guide to Energy and Process Optimization, First Edition. Thomas E. Jenkins.

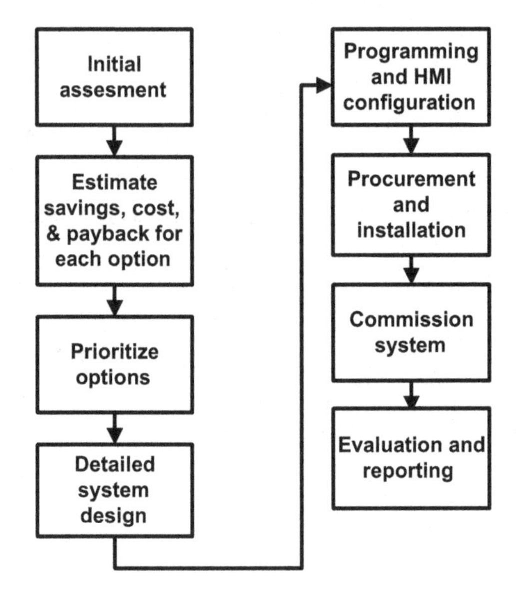

FIGURE 13.1 Design procedure outline

- control logic and equipment protection
- control hardware and HMI/SCADA.

The procedures for system evaluation and design must also integrate the various (and often conflicting) design criteria. This integration begins with the initial assessment and continues until the final report is completed (see Figure 13.1). The integration must include the various steps of the design procedure, since each step must develop data and set up the basis for the next step.

The first step is the initial system assessment (Chapter 2). The data collection must include the pertinent information on process, energy, and equipment. The analysis of the process requirements and performance, and their relationship to the mechanical equipment and energy consumption, is a critical part of the design. This incorporates the following:

- Aeration Processes (Chapter 3)
- Mechanical and Diffused Aeration Systems (Chapter 4)
- Blowers and Blower Control (Chapter 5).

Once the current baseline operation is defined, the comparison to the benchmark will help identify the opportunities for savings. Upgrading the aeration to fine-pore diffusers generally provides the greatest magnitude of savings, followed by automatic aeration controls and more efficient blowers—in that order. However, the most cost-effective ECM (in terms of payback) is typically automatic aeration controls, followed by high-efficiency blowers and fine-pore aeration. That is due to the relative capital cost of the various ECMs offsetting the savings for high-efficiency blowers and fine-pore diffusers.

To determine payback for each option requires an estimate of the design and the cost of the equipment and installation. This in turn requires an analysis of the existing equipment to determine the items needing replacement and to establish the size of the new equipment. The most attention in this analysis is usually given to the diffusers and the blowers. The aeration piping and the control valves (Chapter 6) are too often neglected in the analysis, although these can significantly affect the performance of the system. A thorough analysis should include all of these components.

It is appropriate to emphasize that the analysis and system design should be based on current and near term loads. Ultimate design loads and worst-case analysis will not provide accurate values for the savings in energy costs. Because of the long life of wastewater facilities, there is a strong temptation to base cost justification on ultimate design loads. This temptation should be resisted—more realistic short-term justification will generally lead to a higher level of owner satisfaction.

The next step is to prioritize the ECMs based on customer objectives, payback, and available capital. This may include phasing in some of the options. Controls in particular lend themselves to phased implementation. For example, the blower control and basic DO control can be implemented initially, with the addition of individual flow control for each basin or zone added later. The staging of projects may also be influenced by the financing method(s). For example, if utility rebates are available for some options they may be higher priority than others with better paybacks.

Once the project is authorized and financing obtained, the detailed design begins. This usually involves the same analysis techniques and equipment covered in the initial assessment and cost-effectiveness studies. Greater detail will be required for some calculations, and it is common to include additional site visits and data collection in this phase.

During detail design, a difficult balance must be struck between excess rigidity and chronic indecision. During the detailed design stage, there is an inevitable review of the current design in light of potential problems that come to light. There is a tendency to want to revise the design and resurrect approaches that had been rejected earlier. Sometimes that is the prudent thing to do. Most often, though, the other design has potential problems that just haven't come up. Generally, solving the problems and continuing on with the selected design are more efficient than starting over with a new concept. Experience and engineering judgment are required to decide which approach to take in each instance.

After the design is completed, the next phase is implementation. This is generally straightforward. The only complications encountered are usually during commissioning. The commissioning may require some analysis as part of the trouble-shooting procedures.

Measurement and verification requires repeating the analysis and performance calculations of the initial assessment step. The calculations will include determining the effectiveness of all of the process and energy components included in the ECM design.

13.2 POTENTIAL PROBLEM AREAS

The implementation of automatic aeration controls is, like any engineering endeavor, subject to problems. Every project offers a unique set of potential problems, but there are some challenges which are frequently encountered. The designer should be aware of them and prepared to deal with them from the beginning of the project.

It is inevitable that some assumptions must be made when calculating the performance and energy savings. One of the most common causes of the system's failure to meet expectations or deliver projected savings is the use of excessively optimistic assumptions during the initial assessment. This applies, of course, to "ballpark estimates" of savings based on experience and benchmarks. The designer must make additional assumptions as part of the analysis and calculations. It is easy to assume parameters reflecting the best case or optimal results. If these assumptions are then "buried" in an impressive set of calculations and formulas, their impact will not be obvious. It is necessary to avoid expecting a level of accuracy in the calculation results that exceeds the level of accuracy in the assumptions!

Operator resistance to change can be a tremendous barrier to system success. It is incumbent on the system designer to work with the operators throughout the ECM development process—from initial assessment to the final report. This will, of course, provide buy-in for the operators, giving them a stake in the successful implementation. It will also improve the system performance, since operators have detailed insight into the plant and process that other participants lack.

Changing loads to the process and changes in electric rates can occur between the initial assessment and the measurement and verification. It isn't uncommon to have several years elapse between project inception and completion. There is no way to eliminate these changes. Therefore, the designer must make sure to address them in the initial assessment report. This is accomplished by detailing the basis of the evaluation, itemizing changes that will impact the projected results, and normalizing the savings to allow correcting the final savings to the initial projections.

Aeration controls are frequently provided by suppliers or manufacturers specializing in the field. Most facilities rely on outside consulting engineers for technical assistance in evaluating proposed changes to the treatment plant. This is typically a civil engineering firm specializing in an environmental design. They will usually be asked by the owner to examine the supplier's proposals and projections. The engineering firm may not have expertise in energy conservation or automation that matches the capabilities of an experienced supplier. It is important that the evaluation incorporate sound engineering practice to establish a high level of confidence in the supplier's calculations. The assessment should include data and calculations demonstrating the basis of the projected savings. If the proposed system includes new techniques or technology unfamiliar to the plant staff or the consulting engineer, a detailed explanation of the operating principles, benefits, and potential problems should be provided.

Cost is always a consideration in engineering systems. The designer must avoid extreme conservatism—adding equipment or features that are of marginal benefit "just to be safe." The designer must also avoid the temptation of trying to reduce the cost of the system at the expense of reliable operation. Cost reduction is only cost-effective if the eliminations do not impair system performance.

One of the common pitfalls related to keeping costs down is the use of existing equipment in the new system. This is usually desirable if the existing equipment is operating properly. Unfortunately, the plant management may not be aware of the condition or performance limitations of the existing equipment. The designer should include an evaluation of all existing components intended for reuse in the system. This evaluation should include discussion with the operators, review of calibration records, and comparison of the performance specifications with the new project requirements.

As indicated previously, municipal projects are often awarded on the basis of low bidder price only, without considering technical merit or other factors that will influence success. The competitive bidding should be structured to incorporate all pertinent factors into the procurement decision (Chapter 11).

13.3 BENEFITS

In many engineering projects the goal is simply to solve a problem while minimizing the impact on cost or performance. Designing aeration controls provides the opportunity to produce multiple "wins" in a single effort. Operators, ratepayers, and society in general can all benefit.

A properly designed system will provide double gains in environmental benefits. The process performance will be enhanced, improving the consistency and reliability of the waste treatment. The system will also reduce energy requirements, with the attendant reductions in greenhouse gases and resource depletion.

Operators benefit from the increased process stability. A well-designed automation system will also allow them to spend more time on preventive maintenance and process optimization. This is accomplished by reducing the time and attention needed to maintain process performance.

Lower energy consumption will also result in lower energy cost, reducing a significant component of the operating expense at most facilities. This obviously has benefits for the ratepayers—the citizens that ultimately pay for the cost of maintaining and operating treatment facilities.

Designing aeration controls for wastewater treatment can be a challenging prospect. It requires integrating a variety of engineering disciplines in the analysis, design, and commissioning. When it is done properly, the results (providing long-term benefits to all of the stakeholders) are well worth the effort.

EXAMPLE PROBLEMS

None

Appendix A

Example Problem Solutions

Author's Notes:

(1) Costs for energy, equipment, installation, and so on are very dynamic due to market and technology changes. The example problems in this book are intended to demonstrate methodology, and should not be used for estimating actual costs and savings.

(2) The availability of spreadsheets, math software, and similar computer-aided engineering tools has made it both easy and tempting to carry calculations out to many decimal places. However, all engineering calculations are based on assumptions in the development of the formulas themselves. Further assumptions and approximations are involved in the selection of data and parameters entered into the formulas. The solutions to the example problems below include answers rounded to a reasonable number of digits to avoid implying a greater level of accuracy than truly exists.

Aeration Control System Design: A Practical Guide to Energy and Process Optimization,
First Edition. Thomas E. Jenkins.

CHAPTER 1: INTRODUCTION

Problem 1.1

(a) Plant staff reports a pump motor current measurement of 223 A. From the plant electric bill, it is determined that the nominal plant supply voltage is 480 VAC three phase and the average plant power factor is 83%. What is the apparent power of the motor? What is the estimated power consumption?

Solution:

From Equation 1.8:

$$S = V \cdot I \cdot \sqrt{3} = \frac{223 \cdot 480 \cdot 1.732}{1000} = 185\,\text{kVA}$$

From Equation 1.6:

$$P = V \cdot A \cdot \sqrt{3} \cdot \text{PF} = \frac{223 \cdot 480 \cdot 1.732 \cdot 0.83}{1000} = 154\,\text{kW}$$

$$154\,\text{kW} \cdot \frac{1 \cdot \text{hp}}{0.746 \cdot \text{kW}} = 206\,\text{hp}$$

(b) Subsequent measurements with a power meter at the motor terminals show the actual motor voltage to be 469 V, and the measured motor power factor is 87.8%. What is the apparent and actual motor power with the revised data? What was the percentage error between the original assumed and the actual measured data?

Solution:

From Equation 1.8:

$$\text{S} = V \cdot A \cdot \sqrt{3} = \frac{223 \cdot 469 \cdot 1.732}{1000} = 181\,\text{kVA}$$

From Equation 1.6:

$$P = V \cdot A \cdot \sqrt{3} \cdot \text{PF} = \frac{223 \cdot 469 \cdot 1.732 \cdot 0.878}{1000} = 159\,\text{kW}$$

$$159\,\text{kW} \cdot \frac{1 \cdot \text{hp}}{0.746 \cdot \text{kW}} = 213\,\text{hp}$$

$$\frac{213 - 206}{213} \cdot 100 = 3\%\ \text{error}$$

Problem 1.2

The influent lift station at a treatment plant has a design capacity of 8 mgd. The station has three identical pumps with characteristic curve shown in Figure 1.6. The average dry weather flow for the plant is 2 mgd, the reported discharge head is 80 ft, and the power consumption at the motor control center is measured as 58 kW.

(a) What is the operating flow rate of the pump?

Solution:

From the pump curve, the intersection of the flow rate and the head is 2780 gpm (see Figure A.1).

$$\frac{2780\ \cancel{\text{gpm}}}{694.4\ \cancel{\text{gpm}}/\text{mgd}} = 4.0\ \text{mgd}$$

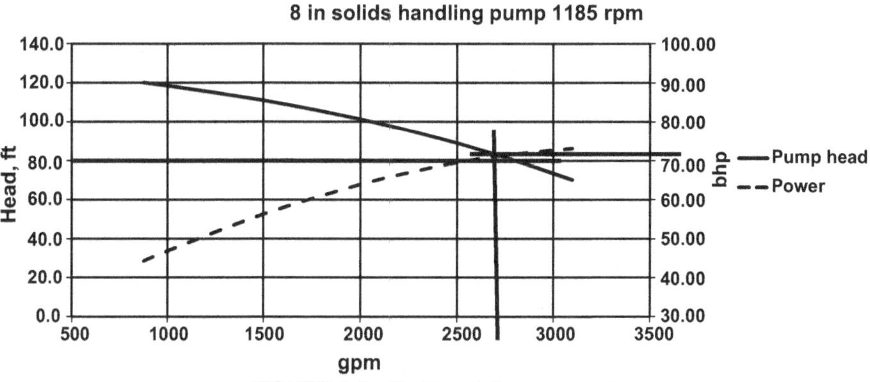

FIGURE A.1 *Problem 1.2 pump curve*

(b) What is the corresponding pump shaft power?

Solution:

From the pump curve at the flow rate of 2780 gpm, the corresponding pump shaft power is 72 bhp.

(c) What is the wire to water efficiency?

Solution:

From Equation 1.10:

$$\eta_{\text{ww}} = \frac{(Q \cdot h/3960) \cdot 0.746}{P_{\text{e}}} \cdot 100 = \frac{(2780 \cdot 80/3960) \cdot 0.746}{58} \cdot 100 = 72.2\%$$

(d) What is the pump efficiency?

Solution:

The waterpower can be determined from Equation 1.3:

$$P_{\mathrm{w}} = \frac{Q_{\mathrm{w}} \cdot h \cdot \mathrm{SG}}{3960} = \frac{2780 \cdot 80 \cdot 1.0}{3960} = 56.2\ \mathrm{hp}$$

Therefore, the pump efficiency is

$$\frac{56.2}{72} \cdot 100 = 78\%$$

Problem 1.3

What is the annual operating cost of the pump in Problem 1.2 if the cost of electricity is $0.08/kWh?

Solution:

The pump flow rate is 4 mgd, but the actual average daily flow is only 2 mgd. Therefore, the average amount of time the pump will operate must be determined:

$$24 \frac{\mathrm{h}}{\mathrm{day}} \cdot \frac{2.0\ \mathrm{mgd}}{4.0\ \mathrm{mgd}} = \frac{12\ \mathrm{h}}{\mathrm{day}}$$

From this the operating cost can be determined:

$$\frac{12\ \mathrm{h}}{\mathrm{day}} \cdot 58\ \mathrm{kW} \cdot \frac{365\ \mathrm{day}}{\mathrm{year}} \cdot \frac{\$0.08}{\mathrm{kWh}} = \frac{\$20{,}300}{\mathrm{year}}$$

Problem 1.4

It is proposed that 75 hp variable frequency drives be provided for the pump system in Problem 1.2. The installed cost of each drive is estimated at $7500 each. A preliminary estimate of cost effectiveness is requested.

(a) Is this cost-effective if the billing is based on energy consumption only at $0.08/kWh?

Solution:

No. Because the head is assumed constant, based on plant information, the total energy required to pump the required flow will not be changed and there would not be an energy cost reduction.

(b) Is this cost-effective if the billing is based on energy consumption at $0.06/kWh and a monthly demand charge of $9.00/kW?

Solution:

For a preliminary estimate, it may be assumed that the pump efficiency will remain approximately constant at the reduced flow. Therefore, although the energy will remain the same, power and demand will be reduced:

$$\frac{2\text{ mgd}}{4\text{ mgd}} \cdot 58\text{ kW} = 29\text{ kW}$$

$$(58\text{ kW} - 29\text{ kW}) \cdot \frac{12\text{ months}}{\text{year}} \cdot \frac{\$9.00}{\text{kW}} = \frac{\$3100}{\text{year}}$$

Determine payback if all three pumps are equipped with VFDs using Equation 1.11:

$$P = \frac{\$7500/\text{VFD} \cdot 3\text{VFD}}{\$3100/\text{year}} = 7.3\text{ years}$$

Determine payback if all only one pump is equipped with a VFD:

$$P = \frac{\$7500/\text{VFD}}{\$3100/\text{year}} = 2.4\text{ years}$$

Either option may be acceptable to the owner. A more detailed analysis should be performed including actual diurnal flow rates and time of day use. Note that if only one VFD is provided, a control system must include provisions for periodically cycling all pumps on and off.

Problem 1.5

List at least three benefits obtained from automating the aeration process.

Solution:

Any three of the following should be included. Additional benefits not shown may also be included based on the judgment of the reader.

- Energy cost reduction
- Reduced greenhouse gas production
- Improve sustainability
- Increased treatment performance
- Better process stability
- Longer equipment life
- Reduced operator effort
- Better process data collection

CHAPTER 2: INITIAL SYSTEM ASSESSMENT

Problem 2.1

Refer to the energy bill shown in Figure 2.2. What is the composite electric rate for this facility?

Solution:

First, determine the total power consumption:

$$190{,}000\,\text{kWh} + 320{,}000\,\text{kWh} = 510{,}000\,\text{kWh}$$

Then calculate the composite rate:

$$\frac{\$34{,}119}{510{,}000\,\text{kWh}} = \$0.067/\text{kWh}$$

Problem 2.2

The treatment plant with electric rates shown in Figure 2.2 has four 250 hp aeration blowers, but typically operates two at a time. The manufacturer's data shows the average power consumption for each blower is 180 bhp and the motor efficiency is 95%. The blowers are alternated weekly around 9:00 AM.

(a) What are the potential savings if the blowers are alternated weekly by turning off the lead before starting the lag, compared to starting the lag blower and letting it warm up for 10 min before stopping the lead?

Solution:

First estimate the kW draw of each blower from Equation 1.7:

$$P = \frac{\text{bhp} \cdot 0.746}{\eta_{\text{m}}} = \frac{180 \cdot 0.746}{0.95} = 141\ \text{kW}$$

Correct the demand reduction based on operation at 10 min out of a 15 min demand period:

$$141\ \text{kW} \cdot \frac{10}{15} = 94\,\text{kW}$$

Determine the demand savings:

$$94\,\text{kW} \cdot \frac{\$15}{\text{kW}} \cdot \frac{12\ \text{months}}{\text{year}} = \frac{\$16{,}900}{\text{year}}$$

Calculate the savings in consumption to see if they are significant:

$$141\ \text{kW} \cdot 10\ \text{min} \cdot \frac{1\ \text{h}}{60\ \text{min}} \cdot \frac{52}{\text{year}} \cdot \frac{\$0.06}{\text{kWh}} = \frac{\$73}{\text{year}}$$

The consumption savings are negligible.

(b) If the operators are concerned about changing the procedure and having reduced air flow during the alternation, what alternatives are available?

Solution:

- The operators could assist in a test of the alternate starting method to verify that there is no change in process performance
- The alternation could be changed to weekends, early morning, or late evening when demand charges are not assessed
- The warm-up period could be reduced or eliminated to reduce the time of operation with an excess blower to less than 10 min, reducing the impact of the demand charge.

Problem 2.3

A facility with the rate structure shown in Figure 2.2 has three existing constant speed positive displacement blowers with 100 hp motors. The blowers were designed for inlet pressure of 14.2 psia, barometric pressure of 14.4 psia, discharge pressure of 9.0 psig, and an air flow of 1800 ICFM. It is estimated that actual average air flow required is 1500 ICFM, and current operating pressure is 8.5 psig. The estimated installed cost of each 100 hp constant torque rated VFD is \$15,000. The estimated installed cost of each turbo blower is \$75,000 each, including drive and controls. Assume the efficiency of the existing blowers is 65%, the existing motor efficiency is 94%, and the VFD efficiency is 96%. The estimated efficiency of each turbo blower is 80%, with 98% motor efficiency and 95% VFD efficiency. Provide an initial evaluation to determine if it is more cost-effective to provide VFDs for the existing blowers or to replace them with high-efficiency variable-speed single-stage turbo blowers. Use the composite energy cost (see Problem 2.1) for the estimate.

Solution:

Estimate existing constant speed blower power consumption from Equations 2.1 and 2.2:

$$X = \left(\frac{p_d}{p_i}\right)^{0.283} - 1 = \left[\frac{(14.4 + 8.5)}{14.2}\right]^{0.283} - 1 = 0.1448$$

$$\text{kW} = 0.01151 \cdot \frac{Q_i \cdot p_i \cdot X}{\eta_b \cdot \eta_m \cdot \eta_{vfd}} = 0.01151 \cdot \frac{1800 \cdot 14.2 \cdot 0.1448}{0.65 \cdot 0.94} = 70\ \text{kW}$$

Estimate the existing blower consumption, energy cost reduction, and payback at reduced speed and air flow:

$$\text{kW} = 0.01151 \cdot \frac{Q_\text{i} \cdot p_\text{i} \cdot X}{\eta_\text{b} \cdot \eta_\text{m} \cdot \eta_\text{vfd}} = 0.01151 \cdot \frac{1500 \cdot 14.2 \cdot 0.1448}{0.65 \cdot 0.94 \cdot 0.96} = 60\ \text{kW}$$

From Equation 2.5:

$$(70\ \text{kW} - 60\ \text{kW}) \cdot \frac{\$0.067}{\text{kWh}} \cdot \frac{8760\ \text{h}}{\text{year}} = \$5900/\text{year}$$

From Equation 1.11:

$$P = \frac{C}{S} = \frac{3 \cdot \$15{,}000}{\$5900/\text{year}} = 7.6\ \text{years}$$

Estimate the turbo blower power consumption and energy cost reduction at reduced air flow:

$$\text{kW} = 0.01151 \cdot \frac{Q_\text{i} \cdot p_\text{i} \cdot X}{\eta_\text{b} \cdot \eta_\text{m} \cdot \eta_\text{vfd}} = 0.01151 \cdot \frac{1500 \cdot 14.2 \cdot 0.1448}{0.80 \cdot 0.98 \cdot 0.95} = 48\ \text{kW}$$

$$(70\ \text{kW} - 48\ \text{kW}) \cdot \frac{\$0.067}{\text{kWh}} \cdot \frac{8760\ \text{h}}{\text{year}} = \$12{,}900/\text{year}$$

$$P = \frac{C}{S} = \frac{3 \cdot \$75{,}000}{\$12{,}900/\text{year}} = 17.4\ \text{years}$$

Either option will provide savings. However, even though the turbo blower will provide twice the savings per year the payback is probably not acceptable for an existing facility.

Problem 2.4

For the blower system in Problem 2.3 the operators suggested three alternate ECMs. Estimate the payback and list advantages and disadvantages for each.

(a) Use a V-belt sheave change to reduce the air flow on one or more blowers, with an installed cost of $400 each.

Solution:

From Equation 1.11:

$$P = \frac{C}{S} = \frac{3 \cdot \$400}{\$5900/\text{year}} = 0.2\ \text{years} = 2\ \text{months}$$

Advantages: low cost; can be implemented immediately by plant staff; very good payback.

Disadvantages: does not supply step-less control of air flow; more difficult to match air flow to process demand; requires operator intervention to change air rate to the process; may not achieve potential savings because of excess aeration, so actual payback will probably be longer; more difficult to incorporate into automatic control system.

(b) Install only one VFD on the existing blowers.

Solution:

$$P = \frac{C}{S} = \frac{1 \cdot \$15{,}000}{\$5900/\text{year}} = 2.5 \text{ years}$$

Advantages: less cost than three VFDs; very good payback.

Disadvantages: if lead blower or VFD fail there will be no savings; operators must manually alternate blowers on weekends to maintain operability.

(c) Install only one new blower

Solution:

$$P = \frac{C}{S} = \frac{1 \cdot \$75{,}000}{\$12{,}900/\text{year}} = 5.8 \text{ years}$$

Advantages: less cost than three blowers

Disadvantages: payback is still marginal; if lead blower fails, there will be no savings; operators must manually operate other blowers (on weekends) to maintain operability.

Problem 2.5

A municipal utility authority operates three activated sludge wastewater treatment plants:

(a) A 0.75 mgd ADF plant with a monthly energy consumption of 180,000 kWh.
(b) A 1.5 mgd ADF plant with a monthly energy consumption of 109,000 kWh.
(c) A 2.5 mgd ADF plant with a monthly energy consumption of 220,000 kWh.

Establish the baseline, typical benchmark, and potential savings for each plant based on Table 2.1. Use the composite power rate from Problem 2.1:

Solution:

(a) $$\frac{180{,}000 \text{ kWh/month}}{0.75 \text{ mil gal/day} \cdot 30 \text{ day/month}} = 8000 \frac{\text{kWh}}{\text{mil gal}}$$

The benchmark for a typical plant of this size is 5500 kWh/mil gal.

$$\frac{8000}{5500} \cdot 100 = 145\%$$

$$\left[8000\frac{\text{kWh}}{\text{mil gal}} - 5500\frac{\text{kWh}}{\text{mil gal}}\right] \cdot 0.75\frac{\text{mil gal}}{\text{day}} \cdot 30\frac{\text{day}}{\text{month}} \cdot 12\frac{\text{months}}{\text{year}} 0.067\frac{\$}{\text{kWh}}$$

$$= \$45{,}000/\text{year probable savings}$$

(b) $$\frac{109{,}000\ \text{kWh/month}}{1.5\ \text{mil gal/day} \cdot 30\ \text{day/month}} = 2400\frac{\text{kWh}}{\text{mil gal}}$$

The benchmark for a typical plant of this size is 2500 kWh/mil gal.

$$\frac{2400}{2500} \cdot 100 = 96\%$$

This plant is operating below the typical plant benchmark. It is operating above the high-efficiency benchmark of 1700 kWh/mil gal, so there are potential savings but they will require additional effort to determine.

(c) $$\frac{220{,}000\ \text{kWh/month}}{2.5\ \text{mil gal/day} \cdot 30\ \text{day/month}} = 2900\frac{\text{kWh}}{\text{mil gal}}$$

The benchmark for a typical plant of this size is 2500 kWh/mil gal.

$$\frac{2900}{2500} \cdot 100 = 116\%$$

$$\left[2900\frac{\text{kWh}}{\text{mil gal}} - 2500\frac{\text{kWh}}{\text{mil gal}}\right] \cdot 2.5\frac{\text{mil gal}}{\text{day}} \cdot 30\frac{\text{day}}{\text{month}} \cdot 12\frac{\text{months}}{\text{year}} 0.067\frac{\$}{\text{kWh}}$$

$$= \$7200/\text{year probable savings}$$

The 0.75 mgd plant should be the highest priority because it has the greatest deviation from the typical benchmark and offers the greatest probable reduction in energy cost. The 2.5 mgd plant should receive attention next, and the 1.5 mgd plant should be studied last. All three plants use more energy than the high efficiency benchmark and offer potential savings in energy cost.

Problem 2.6

Calculate the 10 year and 20 year present worth of the turbo blower replacement in Problem 2.3. Assume a 5% discount rate and a 3% inflation rate.

Solution:

From Equation 1.14:

$$r = \frac{R-1}{1+I} = \frac{0.05-0.03}{1+0.03} = 0.019$$

From Equation 1.13:

$$\text{PWF} = \frac{(1+r)^n - 1}{r \cdot (1+r)^n} = \frac{(1+0.019)^{10} - 1}{0.019 \cdot (1+0.019)^{10}} = 9.03$$

From Equation 1.12:

$$\text{NPW} = (\text{PWF} \cdot S) - C = (9.03 \cdot \$12{,}900) - (3 \cdot \$75{,}000) = -\$108{,}500$$

The project is not justified if a 10-year term is required.
From Equation 1.13:

$$\text{PWF} = \frac{(1+0.019)^{20} - 1}{0.019 \cdot (1+0.019)^{20}} = 16.51$$

From Equation 1.12:

$$\text{NPW} = (16.51 \cdot \$12{,}900) - (3 \cdot \$75{,}000) = -\$12{,}000$$

Although the simple payback calculation indicates a 17-year payback, the net present worth method indicates that a 20-year evaluation does not justify the project based on the assumptions made. Note that different inflation and discount rate assumptions will result in a different net present worth.

Problem 2.7

A more detailed analysis of the process air demand for Problem 2.3 indicates that the average air demand from 9:00 A.M. to 9:00 P.M. is 115% of the average and the air demand for the rest of the day is 85% of the average. The maximum air demand occurs at noon and is 120% of the average. Use the actual rates shown in Figure 2.2 and recalculate the cost effectiveness of providing three VFDs for the existing blowers.

Solution:

Calculate annual power cost for the existing blower operating at a constant 70 kW using Equations 2.7–2.10:

$$\text{On Peak \$} = \frac{\$}{\text{kWh}} \cdot \text{kW} \cdot \frac{60\text{ h}}{\text{week}} \cdot \frac{52\text{ weeks}}{\text{year}} = 0.06 \cdot 70 \cdot 60 \cdot 52 = \$13{,}100$$

$$\text{Off Peak \$} = \frac{\$}{\text{kWh}} \cdot \text{kW} \cdot \frac{108\ \text{h}}{\text{week}} \cdot \frac{52\ \text{weeks}}{\text{year}} = 0.03 \cdot 70 \cdot 108 \cdot 52 = \$11{,}800$$

$$\text{Demand \$} = \frac{\$}{\text{kW}} \cdot \text{kW} \cdot \frac{12\ \text{months}}{\text{year}} = 15.00 \cdot 70 \cdot 12 = \$12{,}600$$

There is no power factor charge. The total cost per year:

$$\$13{,}100 + \$11{,}800 + \$12{,}600 = \$37{,}500$$

From Problem 2.3; the average power for the new system at 1500 ICFM average air flow rate is 60 kW. Find the power for on peak, off peak, and demand for the new system:

$$60\ \text{kW} \cdot 1.15 = 69\ \text{kW}$$

$$60\ \text{kW} \cdot 0.85 = 51\ \text{kW}$$

$$60\ \text{kW} \cdot 1.20 = 72\ \text{kW}$$

Calculate annual power cost for the new blower operating at a variable air flow rate:

$$\text{On Peak \$} = 0.06 \cdot 69 \cdot 60 \cdot 52 = \$12{,}900$$

$$\text{Off Peak \$} = 0.03 \cdot 51 \cdot 108 \cdot 52 = \$8600$$

$$\text{Demand \$} = 15.00 \cdot 72 \cdot 12 = \$12{,}900$$

$$\$12{,}900 + \$8600 + \$12{,}900 = \$34{,}400$$

$$\text{Savings} = \$37{,}500 - \$34{,}400 = \$3100\ \text{per year}$$

This is much less than the savings predicted using the composite energy cost. The demand is actually higher because the peak air flow is unchanged and the VFD losses increase the peak power demand. It should be noted that if the average process demand was lower the savings would be increased.

CHAPTER 3: AERATION PROCESSES

Problem 3.1

A plant uses airlift pumps for pumping RAS from three clarifiers to the aeration basin influent channel. An operator has proposed replacing them with centrifugal pumps and VFDs to provide better operational control and save energy. The RAS flow is 500 gpm per clarifier and the discharge point is 3 ft above the clarifier water surface with an additional 2 ft of friction head in the piping. The current pumps are 8 in.

nominal diameter, and the air entry is 12 ft below the water surface. The variable speed aeration blowers have an average efficiency of 70% and the discharge pressure is 8.0 psig with inlet pressure of 14.5 psia and barometric pressure is 14.7 psia. Assume the new pumps will have an average efficiency of 65%. Both the pump and the blower motors have a nameplate efficiency of 95% and the VFDs are 96% efficient.

(a) Are there energy savings?
(b) If the energy cost is $0.07/kWh, new pumps cost $4000 each installed and the VFDs cost $1000 each installed, what is the payback period?

Solution:

(a) Calculate the submergence from Equation 3.1:

$$\text{Sub} = \frac{S}{S+H} \cdot 100 = \frac{12}{12+(3+2)} \cdot 100 = 71\%$$

From Figure 3.8; the required air flow to pump 500 gpm at 70% submergence is 100 CFM each. Verify that the blower discharge pressure is adequate:

$$12\,\text{ft}\,\text{H}_2\text{O} \cdot \frac{0.433\,\text{psi}}{\text{ft}\,\text{H}_2\text{O}} = 5.2\,\text{psi} < 8.0\,\text{psi}$$

The blower discharge pressure is sufficient to supply the airlifts. Calculate the blower power consumption using Equations 2.1 and 2.2. The blower discharge pressure will dictate power demand:

$$X = \left(\frac{p_\text{d}}{p_\text{i}}\right)^{0.283} - 1 = \left[\frac{(14.7+8.0)}{14.5}\right]^{0.283} - 1 = 0.1352$$

$$\text{kW} = 0.01151 \cdot \frac{Q_\text{i} \cdot p_\text{i} \cdot X}{\eta_\text{b} \cdot \eta_\text{m} \cdot \eta_\text{vfd}} = 0.01151 \cdot \frac{3 \cdot 100 \cdot 14.5 \cdot 0.1352}{0.70 \cdot 0.95 \cdot 0.96} = 10.6\,\text{kW}$$

Rearrange Equation 1.10 and calculate the pump power demand:

$$P_\text{e} = \frac{(Q \cdot h/3960) \cdot 0.746}{\eta_\text{p} \cdot \eta_\text{m} \cdot \eta_\text{vfd}} = \frac{(500 \cdot 5/3960) \cdot 0.746}{0.65 \cdot 0.95 \cdot 0.96} = 0.8\,\text{kW per pump}$$

$$0.8 \frac{\text{kW}}{\text{pump}} \cdot 3\,\text{pump} = 2.4\,\text{kW total power}$$

$$\text{Savings} = 10.6\,\text{kW} - 2.4\,\text{kW} = 8.2\,\text{kW}$$

(b) Calculate the energy cost savings from Equation 2.5:

$$8.2\,\text{kW} \cdot \frac{\$0.07}{\text{kWh}} \cdot \frac{8760\,\text{h}}{\text{year}} = \$5000/\text{year}$$

Calculate the simple payback from Equation 1.11:

$$P = \frac{C}{S} = \frac{3 \cdot (\$4000 + \$1000)}{\$5000/\text{year}} = 3.0\ \text{years}$$

Even though the change might be justified on the basis of better process performance alone, it is also a cost-effective ECM and should be implemented.

Problem 3.2

The plant in Problem 3.1 has aeration basins with influent channels 80 ft long total and effluent channels 60 ft long. A small PD blower is proposed to aerate the channels for mixing instead of using air from the aeration tank blowers. The diffuser submergence will be 6 ft. If the blower-installed cost is \$10,000 and the average efficiency is 65%, will this be cost-effective?

Solution:

Assume the air required is 4 CFM per lineal foot (Section 3.1.1):

$$Q = 4\frac{\text{CFM}}{\text{ft}} \cdot (80\ \text{ft} + 60) = 560\ \text{CFM}$$

Calculate the existing aeration blower power from Equations 2.1 and 2.2:

$$X = \left(\frac{p_\text{d}}{p_\text{i}}\right)^{0.283} - 1 = \left[\frac{(14.7 + 8.0)}{14.5}\right]^{0.283} - 1 = 0.1352$$

$$\text{kW} = 0.01151 \cdot \frac{Q_\text{i} \cdot p_\text{i} \cdot X}{\eta_\text{b} \cdot \eta_\text{m} \cdot \eta_\text{vfd}} = 0.01151 \cdot \frac{560 \cdot 14.5 \cdot 0.1352}{0.70 \cdot 0.95 \cdot 0.96} = 19.8\ \text{kW}$$

Calculate the new channel aeration blower power from Equations 2.1 and 2.2. Assume 0.5 psi pressure loss through piping and diffusers:

$$6\ \text{ft H}_2\text{O} \cdot \frac{0.433\ \text{psi}}{\text{ft H}_2\text{O}} + 0.5\ \text{psi} = 3.1\ \text{psi}$$

$$X = \left(\frac{p_\text{d}}{p_\text{i}}\right)^{0.283} - 1 = \left[\frac{(14.7 + 3.1)}{14.5}\right]^{0.283} - 1 = 0.0597$$

$$\text{kW} = 0.01151 \cdot \frac{Q_\text{i} \cdot p_\text{i} \cdot X}{\eta_\text{b} \cdot \eta_\text{m}} = 0.01151 \cdot \frac{560 \cdot 14.5 \cdot 0.0597}{0.65 \cdot 0.95} = 9.0\ \text{kW}$$

$$\text{Savings} = 19.8\ \text{kW} - 9.0\ \text{kW} = 10.8\ \text{kW}$$

Calculate the energy cost savings from Equation 2.5:

$$10.8\,\text{kW} \cdot \frac{\$0.07}{\text{kWh}} \cdot \frac{8760\,\text{h}}{\text{year}} = \$6600/\text{year}$$

$$\frac{\$10{,}000}{\$6600/\text{year}} = 1.5\text{ years}$$

With less than a 2-year payback, this is a very cost-effective ECM.

Problem 3.3

A complete mix tank is 40 ft square and has a side water depth of 14 ft. It is equipped with a mechanical aerator, 25 hp with an actual site aeration efficiency (AE) of 2.0 lb O_2/hp·h. The manufacturer recommends a minimum of 0.75 hp/th ft^3 for mixing. The average daily flow rate to the tank is 0.5 mgd. Influent BOD is 165 ppm and 90% removal is required. The utilization rate is reported as 1.1 lb O_2/lb BOD removed.

(a) Will the tank be mixing limited or oxygen limited?
(b) If the tank is upgraded with fine-pore diffusers with an actual site OTE of 10%, what air rates will be required for mixing and for process performance?

Solution:

(a) Calculate the hydraulic retention time and oxygen uptake rate from Equations 3.6 and 3.7:

$$\Delta\text{BOD} = 165\text{ ppm} - 0.10 \cdot 165\text{ ppm} = 149\text{ ppm}$$

$$V_t = L \cdot W \cdot D = 40 \cdot 40 \cdot 14 = 22{,}400\text{ ft}^3$$

$$\text{HRT} = \frac{1.795 \cdot 10^{-4} \cdot V_t}{Q_{ww}} = \frac{1.795 \cdot 10^{-4} \cdot 22{,}400}{0.5} = 8\text{ h}$$

$$\text{OUR}_{\text{Req}} = \frac{\Delta\text{BOD} \cdot U_{O_2}}{\text{HRT}} = \frac{149 \cdot 1.1}{8} = 20.5\text{ mg}/(\text{l h})$$

Calculate mixing power requirement:

$$\frac{22{,}400\text{ ft}^3}{1000} \cdot 0.75 \frac{\text{hp}}{\text{th ft}^3} = 17\text{ hp}$$

Calculate average power to satisfy oxygen demand from Equation 3.5:

$$P_a = \frac{\text{OUR} \cdot V_t}{16{,}000 \cdot \text{AE}} = \frac{20.5 \cdot 22{,}400}{16{,}000 \cdot 2.0} = 14.4\text{ hp}$$

At average conditions, the tank is limited by mixing requirements. Note that the aerator may be slightly oversized.

(b) Calculate mixing air requirement from Equation 3.13:

$$Q_{\text{mix}} = L \cdot W \cdot Q_{\text{req}} = 40\,\text{ft} \cdot 40\,\text{ft} \cdot 0.12\frac{\text{CFM}}{\text{ft}^2} = 192\,\text{CFM}$$

Calculate average air flow to meet process demand:

$$Q_{\text{s}} = \frac{\text{OUR} \cdot V_{\text{t}}}{16{,}700 \cdot \text{AOTE}} = \frac{20.5 \cdot 22{,}400}{16{,}700 \cdot 0.10} = 275\,\text{SCFM}$$

With diffused aeration at average conditions, the tank would be limited by process demand.

Problem 3.4

In the spring, the mixed liquor temperature at a WWTP increases from 50 °F to 60 °F. List at least three effects that this may have on the process.

Solution:

Any three of the following should be included. Additional effects not shown may also be included based on the judgment of the reader.

- The DO control may need to be retuned
- The air flow to the aeration tanks may need to be increased
- The distribution of air within the tanks may need to be modified
- Sludge wasting may need to be increased
- Additional tanks may need to be brought on or off-line to accommodate different permit requirements
- Clarifier and sludge operation may need to be adjusted to accommodate denitrification in the sludge blanket if nitrification rates increase

CHAPTER 4: MECHANICAL AND DIFFUSED AERATION SYSTEMS

Problem 4.1

Refer to Problem 2.7. Additional testing has determined that the average aeration system effluent DO concentration is 6.0 ppm at the current operating air rate of 1800 CFM.

(a) Recalculate savings based on reducing air flow to achieve 2.0 ppm DO. Use a mixed liquor temperature of 50 °F.
(b) If the DO control system installed cost is \$30,000, what is the revised payback for the VFDs plus controls?
(c) If the blowers have 60% turndown, can they meet the reduced flow requirement?

Solution:

(a) Determine the approximate saturation concentration from Table 4.1:

$C^*_{\infty f} \cong 11.1\ \text{ppm}$

Calculate the ratio of actual to required air flow using equation 4.2:

$$\frac{Q_{as}}{Q_{2.0}} = \frac{C^*_{\infty f} - 2.0}{C^*_{\infty f} - C_a} = \frac{11.1 - 2.0}{11.1 - 6.0} = 1.8$$

Calculate the new average air flow:

$$Q_{2.0} = \frac{1800\ \text{CFM}}{1.8} = 1000\ \text{CFM}$$

From Problem 2.7; the total cost per year for the existing operating mode is = \$37,500

Calculate the average power at 1000 ICFM:

$$X = \left(\frac{p_d}{p_i}\right)^{0.283} - 1 = \left[\frac{(14.4 + 8.5)}{14.2}\right]^{0.283} - 1 = 0.1448$$

$$\text{kW} = 0.01151 \cdot \frac{Q_i \cdot p_i \cdot X}{\eta_b \cdot \eta_m \cdot \eta_{vfd}} = 0.01151 \cdot \frac{1000 \cdot 14.2 \cdot 0.1448}{0.65 \cdot 0.94 \cdot 0.96} = 40\ \text{kW}$$

Find the power for on peak, off peak, and demand for the system operating at 2.0 ppm DO:

$40\ \text{kW} \cdot 1.15 = 46\ \text{kW}$

$40\ \text{kW} \cdot .85 = 34\ \text{kW}$

$40\ \text{kW} \cdot 1.20 = 48\ \text{kW}$

Calculate annual power cost for the new blower operating at a variable air flow rate:

On Peak \$ = 0.06 · 46 · 60 · 52 = \$8600

Off Peak \$ = 0.03 · 34 · 108 · 52 = \$5700

Demand \$ = 15.00 · 48 · 12 = \$8600

\$8600 + \$5700 + \$8600 = \$22,900

Savings = \$37,500 − \$22,900 = \$14,600/year

(b) $$P = \frac{C}{S} = \frac{3 \cdot \$15{,}000 + \$30{,}000}{\$14{,}600/\text{year}} = 5.1\ \text{years}$$

The additional cost of the DO control system improves the payback significantly.

(c) Rearrange Equation 2.3 to determine the minimum air flow:

$$Q_{\text{min}} = Q_{\text{max}} \cdot \left(1 - \frac{\text{Turndown}}{100}\right) = 1800\,\text{CFM} \cdot \left(1 - \frac{60}{100}\right) = 720\,\text{CFM}$$

The minimum expected flow from Table 4.2 is 70% of the average.

$$1000\,\text{CFM} \cdot 70\% = 700\,\text{CFM}$$

The minimum blower flow rate is approximately equal to the minimum process demand. The blower system is adequate and the savings are achievable.

Problem 4.2

Refer to Problem 3.3. Calculate the standard AE for the existing mechanical aerators and the proposed diffused aeration system. The current DO concentration is 3.5 ppm and the mixed liquor temperature is 60 °F. The plant is located at 400 ft above sea level (ASL) and the barometric pressure is 14.5 psia. Assume an average annual air temperature of 55 °F and a blower efficiency of 70%.

Solution:

Rearrange Equation 4.8 to establish the relationship between standard and actual transfer efficiency:

$$\text{SAE} = \frac{\text{AE}_{\text{f}}}{\alpha \cdot F} \cdot \frac{1}{\theta^{(T-20)}} \cdot \frac{C^*_{\infty 20}}{\left(\tau \cdot \beta \cdot \Omega \cdot C^*_{\infty 20} - C\right)}$$

Note that the reported field aerator efficiency of 2.0 lb O_2/hp·h for the mechanical aerator will include the effects of α and that for a mechanical aerator $F = 1.0$.

For the surface aerator, $C^*_{\infty 20} = 9.09$ ppm.

$$\Omega = \frac{p_{\text{b}}}{p_{\text{s}}} = \frac{14.5}{14.7} = 0.99$$

$$\frac{60\,^\circ\text{F} - 32\,^\circ\text{F}}{1.8} = 15.6\,^\circ\text{C}$$

$C^*_{20} = 9.09$ ppm. Calculate C^*_{sT} at 15.6 °C by linear interpolation of Table 4.1:

$$C^*_{\text{sT}} = 11.29 + \frac{9.09 - 11.29}{20 - 10} \cdot (15.6 - 10) = 10.06\,\text{ppm}$$

$$\tau = \frac{C^*_{\text{sT}}}{C^*_{\text{s}20}} = \frac{10.06}{9.09} = 1.11$$

Assume $\beta = 0.95$.

$$\mathrm{SAE} = \frac{\mathrm{AE_f}}{\alpha \cdot F} \cdot \frac{1}{\theta^{(T-20)}} \cdot \frac{C^*_{\infty 20}}{\left(\tau \cdot \beta \cdot \Omega \cdot C^*_{\infty 20} - C\right)}$$

$$= 2.0 \cdot \frac{1}{1.024^{(15.6-20)}} \cdot \frac{9.09}{(1.11 \cdot 0.95 \cdot 0.99 \cdot 9.09 - 3.5)} = 3.4 \frac{\mathrm{lb\,O_2}}{\mathrm{hp \cdot h}}$$

Calculate the aerator efficiency for the diffused aeration system. The reported field $\mathrm{OTE_f}$ of 10% includes α and F.

Since the manufacturer's data is not available, calculate an approximate $C^*_{\infty 20}$. Assume the top of the diffusers is 9 in. above the tank floor. The side water depth was given as 14 ft:

$$\mathrm{Sub} = 14 - \frac{9}{12} = 13.25\ \mathrm{ft}$$

$$C^*_{\infty 20} = 9.09 + 0.1 \cdot \mathrm{Sub} = 9.09 + 0.1 \cdot 13.25 = 10.5$$

Calculate SOTE:

$$\mathrm{SOTE} = \frac{\mathrm{OTE_f}}{\alpha \cdot F} \cdot \frac{1}{\theta^{(T-20)}} \cdot \frac{C^*_{\infty 20}}{\left(\tau \cdot \beta \cdot \Omega \cdot C^*_{\infty 20} - C\right)}$$

$$= 0.10 \cdot \frac{1}{1.024^{(15.6-20)}} \cdot \frac{10.5}{(1.11 \cdot 0.95 \cdot 0.99 \cdot 10.5 - 3.5)} = 0.16 = 16\%$$

Calculate inlet and discharge pressures, assuming 0.2 psi filter losses and 1.0 psi friction loss in piping and diffusers:

$$p_\mathrm{i} = 14.5\ \mathrm{psia} - 0.2\ \mathrm{psia} = 14.3\ \mathrm{psia}$$

$$p_\mathrm{d} = 14.5\ \mathrm{psia} + 13.25\ \mathrm{ft\,H_2O} \cdot \frac{0.433\ \mathrm{psia}}{\mathrm{ft\,H_2O}} + 1.0\ \mathrm{psia} = 21.2\ \mathrm{psia}$$

Calculate the standard aerator efficiency for the diffused aeration system:

$$\mathrm{SAE_d} = \frac{2418 \cdot \mathrm{SOTE} \cdot \eta_\mathrm{b}}{T_\mathrm{i} \cdot \left[(p_\mathrm{d}/p_\mathrm{i})^{0.283} - 1\right]} = \frac{2418 \cdot 0.16 \cdot 0.70}{(55 + 460) \cdot \left[(21.2/14.3)^{0.283} - 1\right]}$$

$$= 4.5 \frac{\mathrm{lb\,O_2}}{\mathrm{hp \cdot h}}$$

Problem 4.3

A treatment plant is permitted for effluent of 10 ppm BOD_5 and 5 ppm ammonia at 1.8 mgd ADF. The plant has five aeration basins, each 15 ft wide × 100 ft long × 18 ft SWD. The diffusers have the characteristics shown in Figures 4.16 and are installed in three equal grids per tank with 100 diffusers in each grid. The diffusers are distributed equally along the length of each aeration basin. The operators indicate that a minimum retention time of 12 h at a DO concentration of 2.0 ppm is required to meet the treatment requirements. The plant elevation is 100 ft ASL, average annual air temperature is 60 °F, and the average blower efficiency is 70%, inlet losses are 0.1 psi, and discharge pressure equals static pressure plus 1.0 psi. Actual ADF is 0.9 mgd and the primary clarifier effluent is 130 ppm BOD_5 and 45 ppm NH_3. The mixed liquor temperature is 68 °F and the average aeration basin effluent DO concentration is currently much higher than 2.0 ppm. Assume the fouling factor $F = 0.9$, $\alpha = 0.6$, and $\beta = 0.95$. The composite power rate is reported as $0.12/kWh.

(a) Are the tanks mixing limited?
(b) If so, how many tanks can be taken out of service?
(c) What energy cost savings can be expected by taking these tanks out of service?

Solution:

(a) Calculate tank volumes and average retention time using Equation 3.7:

$$V_t = L \cdot W \cdot \text{SWD} = 100\,\text{ft} \cdot 15\,\text{ft} \cdot 18\,\text{ft} = 27{,}000\,\text{ft}^3 \text{ each}$$

$$\text{HRT} = \frac{1.795 \cdot 10^{-4} \cdot V_t}{Q_{ww}} = \frac{1.795 \cdot 10^{-4} \cdot (27{,}000 \cdot 5)}{0.9} = 27\,\text{h}$$

This is well above the required 12 h detention time.

Calculate the mixing air flow requirement using Equation 3.13. Use a conservative 0.12 SCFM/ft^2:

$$Q_{mix} = L \cdot W \cdot Q_{mix} = 100\,\text{ft} \cdot 15\,\text{ft} \cdot 5\,\text{tanks} \cdot 0.12\frac{\text{SCFM}}{\text{ft}^3} = 900\,\text{SCFM}$$

Calculate the air flow required to meet the process demand. The first step is to determine the field OTE for the diffusers based on mixing air flow and a nominal 2.0 ppm DO:

$$\text{Number of diffusers} = 5\,\text{tanks} \cdot \frac{3\,\text{grids}}{\text{tank}} \cdot 100\frac{\text{diffusers}}{\text{grid}} = 1500\,\text{diffusers}$$

$$Q_{dif} = \frac{900\,\text{SCFM}}{1500\,\text{diffusers}} = 0.6\frac{\text{SCFM}}{\text{diffuser}}$$

From Figure 4.16 at 0.6 SCFM/diffuser the SOTE is 2.5%/foot submergence. Assume the diffuser surface is 9 in. above the tank bottom:

$$\text{Depth} = 18 - \frac{9}{12} = 17.25\ \text{ft}$$

$$\text{SOTE}_{0.6} = 2.5\frac{\%}{\text{ft}} \cdot 17.25\ \text{ft} = 43.1\%$$

The system is essentially at sea level and the wastewater temperature is 68 °F = 20 °C. Therefore, $\Omega = 1.0$ and $\tau = 1.0$. Calculate the field oxygen saturation concentration and the field OTE:

$$C^*_{\infty 20} = 9.09 + 0.1 \cdot \text{Depth} = 9.09 + 0.1 \cdot 17.25 = 10.8\ \text{ppm}$$

$$C^*_{\infty \text{f}} = \tau \cdot \beta \cdot \Omega \cdot C^*_{\infty 20} = 1.0 \cdot 0.95 \cdot 1.0 \cdot 10.8 = 10.3$$

$$\text{OTE}_\text{f} = \alpha \cdot F \cdot \text{SOTE} \cdot \theta^{(T-20)} \cdot \frac{C^*_{\infty \text{f}} - C}{C^*_{\infty 20}} = 0.6 \cdot 0.9 \cdot 0.431 \cdot 1.024^{(20-20)} \cdot \frac{10.3 - 2}{10.8}$$

$$\text{OTE}_\text{f} = \text{AOTE} = 0.179 = 17.9\%$$

Calculate the ratio of field to standard OTE:

$$\frac{\text{OTE}_\text{f}}{\text{SOTE}} = \frac{17.9\%}{43.1\%} = 0.42$$

Estimate the air flow required for BOD removal and nitrification assuming this OTE and 1.1 lb O_2/lb BOD. Use Equations 3.8, 3.9, and 3.11:

$$w_{\text{BOD}} = C \cdot Q_{\text{ww}} \cdot 8.34 = (130 - 10) \cdot 0.9 \cdot 8.34 = 900\frac{\text{lb}}{\text{day}}\ \text{BOD removed}$$

$$w_{\text{NH}_3} = C \cdot Q_{\text{ww}} \cdot 8.34 = (45 - 5) \cdot 0.9 \cdot 8.34 = 300\frac{\text{lb}}{\text{day}}\ \text{NH}_3\ \text{nitrified}$$

$$Q_\text{s} = \frac{w_{\text{BOD}} \cdot U_{\text{O}_2} + w_{\text{NH}_3} \cdot 4.6}{\text{AOTE} \cdot 24.84} = \frac{900 \cdot 1.1 + 300 \cdot 4.6}{0.179 \cdot 24.84} = 530\ \text{SCFM}$$

The process demand is less than the mixing requirement. The basins are mixing limited. If the blowers have sufficient turndown, then the air requirement and energy cost can be reduced by taking some basins out of service.

For reference, estimate the actual DO concentration in the aeration basins using Equation 4.2, the process demand, and the actual air flow based on mixing:

$$C_\text{a} = C^*_{\infty \text{f}} - \frac{Q_{2.0} \cdot \left(C^*_{\infty \text{f}} - 2.0\right)}{Q_{\text{as}}} = \frac{530 \cdot (10.3 - 2.0)}{900} = 4.9\ \text{ppm}$$

This exceeds the 2.0 ppm required by the operators.

(b) Determine how many basins are required based on the required hydraulic retention time. Rearranging Equation 3.7:

$$V_t = \frac{\text{HRT} \cdot Q_{\text{ww}}}{1.795 \cdot 10^{-4}} = \frac{12 \cdot 0.9}{1.795 \cdot 10^{-4}} = 60{,}200\ \text{ft}^3$$

$$\frac{60{,}000\ \text{ft}^3}{27{,}000\ \text{ft}^3/\text{tank}} = 2.2\ \text{tanks required}$$

Keep three tanks in service.

$$V_t = 27{,}000\ \frac{\text{ft}^3}{\text{tank}} \cdot 3\ \text{tanks} = 81{,}000\ \text{ft}^3$$

(c) With fewer tanks in service, the loading to each tank will increase. This may increase the air flow per diffuser, which would decrease the AOTE. If the aeration basins operate at 2.0 ppm instead of 4.9 ppm, the OTE will increase. It is necessary to calculate the new mixing air flow and the new process demand air flow.

$$Q_{\text{mix}} = 100\ \text{ft} \cdot 15\ \text{ft} \cdot 3\ \text{tanks} \cdot 0.12\ \ \text{SCFM}/\text{ft}^3 = 540\ \text{SCFM}$$

Calculate the oxygen uptake rate based on the loading and the HRT with three tanks in service using Equations 3.7 and 3.6:

$$\text{HRT} = \frac{1.795 \cdot 10^{-4} \cdot 81{,}000}{0.9} = 16.2\ \text{h}$$

$$\text{OUR}_{\text{Req}} = \frac{\Delta\text{BOD} \cdot U_{\text{O}_2} + \Delta\text{NH}_3 \cdot 4.6}{\text{HRT}} = \frac{120 \cdot 1.1 + 40 \cdot 4.6}{16.2} = 19.5\ \text{mg}/(\text{l h})$$

Calculate the required oxygen transfer rate from Equation 4.9:

$$\text{ROTR} = \frac{\text{OUR}_{\text{Req}} \cdot V_t}{16{,}000} = \frac{19.5 \cdot 81{,}000}{16{,}000} = 99\ \text{lb}/\text{h}$$

At steady state actual OTR (AOTR) must equal ROTR. Calculate the AOTR at no less than two different flow rates and two different DO concentrations and graph them to determine the air flow per diffuser required using Equation 4.11. First calculate the OTE and AOTR at 2.0 mg/l DO and then correct to other DO concentrations.

From Figure 4.16, at 0.5 CFM/diffuser SOTE = 2.5 %/ft

At 1.0 mg/l and at 8.0 mg/l DO:

$$\text{SOTE}_{0.5,1} = (2.5\ \%/\text{ft} \cdot 17.25\ \text{ft}) \cdot \frac{\text{OTE}_\text{f}}{\text{SOTE}} \cdot \frac{C^*_{\infty\text{f}} - C_\text{a}}{C^*_{\infty\text{f}} - 2.0}$$

$$= (0.025 \cdot 17.25) \cdot 0.42 \cdot \frac{10.3 - 1.0}{10.3 - 2.0}$$

$$SOTE_{0.5,1} = 0.20 = 20\%$$

$$SOTE_{0.5,8} = (0.025 \cdot 17.25) \cdot 0.42 \cdot \frac{10.3 - 8.0}{10.3 - 2.0} = 0.050 = 5.0\%$$

From Figure 4.16, at 1.0 CFM/diffuser SOTE = 2.35 %/ft
At 1.0 mg/l and at 8.0 mg/l DO:

$$SOTE_{1.0,1} = (0.0235 \cdot 17.25) \cdot 0.42 \cdot \frac{10.3 - 1.0}{10.3 - 2.0} = 0.19 = 19\%$$

$$SOTE_{1.0,8} = (0.0235 \cdot 17.25) \cdot 0.42 \cdot \frac{10.3 - 8.0}{10.3 - 2.0} = 0.047 = 4.7\%$$

Calculate the number of diffusers and total air flows:

$$\text{Number of diffusers} = 3\ \text{tanks} \cdot \frac{3\ \text{grids}}{\text{tank}} \cdot 100 \frac{\text{diffusers}}{\text{tank}} = 900\ \text{diffusers}$$

Use Equation 4.5 and calculate the AOTR for each point:

$$AOTR_{0.5,1} = Q_s \cdot AOTE \cdot 1.035 = (0.5 \cdot 900) \cdot 0.20 \cdot 1.035 = 93\ \text{lb}O_2/\text{h}$$

$$AOTR_{0.5,8} = (0.5 \cdot 900) \cdot 0.05 \cdot 1.035 = 23\ \text{lb}\ O_2/\text{h}$$

$$AOTR_{1.0,1} = (1.0 \cdot 900) \cdot 0.19 \cdot 1.035 = 177\ \text{lb}\ O_2/\text{h}$$

$$AOTR_{1.0,8} = (1.0 \cdot 900) \cdot 0.047 \cdot 1.035 = 44\ \text{lb}\ O_2/\text{h}$$

The results are graphed in Figure A.2. From the graph, at an OTR of 99 lb O_2/h the air flow per diffuser will be approximately 0.6 SCFM/diffuser.

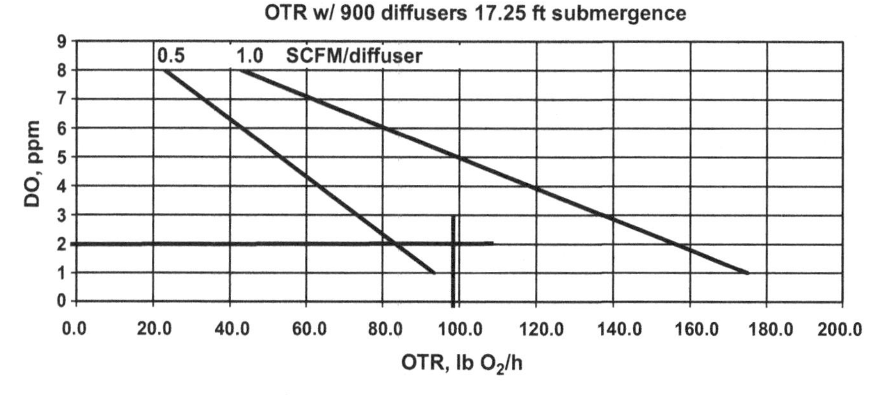

FIGURE A.2 *Problem 4.3 OTR vs. DO*

As an alternate procedure, the new air flow rate can be determined by iteration. An air flow is estimated, the SOTE, AOTE, and the AOTR are calculated using the procedure above. The air flow is modified until the calculated AOTR is approximately equal to the ROTR. For example, at 0.59 SCFM/diffuser, the SOTE is 2.5 %/ft submergence. At the target 2.0 pm DO:

$$\mathrm{SOTE}_{0.6,2} = (0.025 \cdot 17.25) \cdot 0.42 = 0.18 = 18.0\%$$

$$\mathrm{AOTR}_{0.6,2} = (0.59 \cdot 900) \cdot 0.18 \cdot 1.035 = 99\ \mathrm{lb\ O_2/h}$$

The calculation can be repeated until the required level of accuracy is obtained. Calculate the new total air flow rate:

$$\mathrm{Q} = 0.6\ \mathrm{CFM/diffuser} \cdot 900\ \mathrm{diffuser} = 540\ \mathrm{SCFM}$$

The nominal air flow savings are

$$900\ \mathrm{SCFM} - 540\ \mathrm{SCFM} = 360\ \mathrm{SCFM}$$

Calculate the blower inlet pressure, discharge pressure, and reduction blower horsepower from Equations 4.12 and 4.6:

$$p_{\mathrm{i}} = 14.7 - 0.1 = 14.6\ \mathrm{psia}$$

$$p_{\mathrm{d}} = 14.7 + 17.25 \cdot 0.433 + 1.0 = 23.2\ \mathrm{psia}$$

$$P_{\mathrm{b}} = \frac{(Q_{\mathrm{s}}/60) \cdot \gamma \cdot R \cdot T_{\mathrm{i}}}{550 \cdot (k - 1/k) \cdot \eta_{\mathrm{b}}} \cdot \left[\left(\frac{p_{\mathrm{d}}}{p_{\mathrm{i}}}\right)^{(k-1/k)} - 1\right]$$

$$P_{\mathrm{b}} = \frac{(360/60) \cdot 0.075 \cdot 53.3 \cdot (60 + 460)}{550 \cdot 0.283 \cdot 0.7} \cdot \left[\left(\frac{23.2}{14.6}\right)^{0.283} - 1\right] = 16\ \mathrm{bhp}$$

Assume 95% motor efficiency and calculate the electric power savings from Equation 1.7:

$$\mathrm{kW} = \frac{16 \cdot 0.746}{\eta_{\mathrm{m}}} = 12.5\ \mathrm{kW}$$

Calculate the theoretical energy cost savings:

$$\mathrm{S} = 12.5\ \mathrm{kW} \cdot \frac{\$0.12}{\mathrm{kWh}} \cdot \frac{8760\ \mathrm{h}}{\mathrm{year}} = \$13{,}100/\mathrm{year}$$

The assessment should note that the actual savings may be less than the theoretical savings because of diurnal flow variations. During low flow periods, the three operating tanks may still be limited by the mixing flow rate that is approximately equal to the average process air flow demand. However, this ECM has no capital cost and will not reduce process performance. Therefore it should be initiated immediately.

Problem 4.4

A plant is being designed for an ADF of 6.0 mgd and each of four aeration basins is designed to treat 1.5 mgd. The effluent permit requires BOD removal only, with an effluent concentration of 30 mg/l. Primary effluent BOD_5 is 175 mg/l and the design wastewater temperature is 50 °F. The tanks are 30 ft wide × 120 ft long × 16 ft deep. The tanks will have three equal grids of fine-pore diffusers with 100 diffusers/grid. The base design has grids manually controlled, with the air flow equally divided between them. The expected AOTE is 11% at an average DO of 2.0 mg/l. It is estimated that the first grid will remove 50% of the total BOD_5, the second grid 30% of the total, and the third grid 20% of the total. The calculated blower power consumption is 34 SCFM/kW. Barometric pressure is 14.5 psia. The installed cost of a DO transmitter, control valve, and air flow transmitter is estimated as $9000 for each grid and the electric power composite rate is $0.15/kWh. For the individual grid control, assume a DO profile of 0.75, 1.25, and 2.0 mg/l from influent to effluent grids. Will it be cost-effective to install automatic control on each grid?

Solution:

Calculate HRT of each tank and grid using Equation 3.7 and the total OUR and the OUR for each grid using Equation 3.6. Assume the utilization is 1.1 lb O_2/lb BOD.

$$V_t = L \cdot W \cdot D = 120 \cdot 30 \cdot 16 = 57{,}600\ \text{ft}^3 \text{ per tank}$$

$$V_{1,2,3} = \frac{57{,}600}{3} = 19{,}200\ \text{ft}^3/\text{grid}$$

$$\text{HRT} = \frac{1.795 \cdot 10^{-4} \cdot 57{,}600}{1.5} = 6.9\ \text{h total per tank}, 2.3\ \text{h/grid}$$

$$\Delta\text{BOD} = 175 - 30 = 145\ \text{mg/l}$$

$$\text{OUR} = \frac{\Delta\text{BOD} \cdot U_{O_2}}{\text{HRT}} = \frac{145 \cdot 1.1}{6.9} = 23.1\ \text{mg/(l h total)}$$

$$\text{OUR}_1 = \frac{0.5 \cdot 145 \cdot 1.1}{2.3} = 34.7\ \text{mg/(l h)}$$

$$\text{OUR}_2 = \frac{0.3 \cdot 145 \cdot 1.1}{2.3} = 20.8\ \text{mg/(l h)}$$

$$\text{OUR}_3 = \frac{0.2 \cdot 145 \cdot 1.1}{2.3} = 13.9\ \text{mg/(l h)}$$

Calculate total air flow per tank and grid air flows from Equation 3.4:

$$Q_{as} = \frac{OUR \cdot V_t}{16{,}700 \cdot AOTE} = \frac{23.1 \cdot 57{,}600}{16{,}700 \cdot 0.11} = 720\ \text{SCFM total per tank}$$

$$Q_{as1,2,3} = \frac{720\ \text{SCFM}}{3} = 240\ \text{SCFM/grid}$$

For reference, calculate the SCFM/diffuser:

$$Q_{dif} = \frac{240\ \text{SCFM}}{100\ \text{diffusers}} = 2.4\ \text{SCFM/diffuser}$$

This is within the normal range.

Calculate new OTE for each grid. The saturation concentration from Table 4.1 is 11.1 ppm:

$$AOTE_1 = 0.11 \cdot \frac{11.1 - 0.75}{11.1 - 2.0} = 0.125$$

$$AOTE_2 = 0.11 \cdot \frac{11.1 - 1.25}{11.1 - 2.0} = 0.120$$

$AOTE_3 = 0.11$ (this grid and standard conditions are at 2.0 ppm)

Calculate the required air flow for each grid based on process demand:

$$Q_{req1} = \frac{OUR \cdot V_t}{16{,}700 \cdot AOTE} = \frac{34.7 \cdot 19{,}200}{16{,}700 \cdot 0.125} = 320\ \text{SCFM}$$

$$Q_{req2} = \frac{OUR \cdot V_t}{16{,}700 \cdot AOTE} = \frac{20.8 \cdot 19{,}200}{16{,}700 \cdot 0.12} = 200\ \text{SCFM}$$

$$Q_{req3} = \frac{OUR \cdot V_t}{16{,}700 \cdot AOTE} = \frac{13.9 \cdot 19{,}200}{16{,}700 \cdot 0.11} = 150\ \text{SCFM}$$

Check to ensure the process demand air flow is above mixing limits:

$$Q_{mix} = (L \cdot W) \cdot Q_{req} = \frac{120}{3} \cdot 30 \cdot 0.12 = 144\ \text{SCFM per grid}$$

This is less than the process demand. Calculate the reduction in air flow:

$$Q_s = 320 + 200 + 150 = 670\ \text{SCFM}$$

$$720\ \text{SCFM} - 670\ \text{SCFM} = 50\ \text{SCFM}$$

$$\frac{50}{720} = 7\%\ \text{savings}$$

Calculate the savings and payback:

$$\text{kW saved} = \frac{50\ \text{SCFM}}{34\ \text{SCM/kW}} = 1.5\ \text{kW per tank}$$

$$\text{S} = 1.5\ \text{kW} \cdot \frac{\$0.15}{\text{kWh}} \cdot \frac{8760\ \text{h}}{\text{year}} = \$1900/\text{year}$$

$$P = \frac{C}{S} = \frac{3 \cdot \$9000}{\$1900/\text{year}} = 14\ \text{years}$$

The individual grid controls will be cost-effective. This improved design may be implemented if the owner's criterion for new construction is a 20-year payback.

CHAPTER 5: BLOWERS AND BLOWER CONTROL

Problem 5.1

The specifications for a blower system include the following data points:

SCFM	°F	%RH
3500	40	20
4000	60	36
4500	80	60
5000	100	80

The site barometric pressure is given as 14.2 psia. Convert the flow rate to ACFM and lb/h of air while,

(a) ignoring relative humidity,
(b) including relative humidity, and
(c) calculate the percentage error for each point if humidity is ignored.

Solution:

(a) Ignoring relative humidity, rearrange Equation 5.8:

$$Q_a = \frac{Q_s \cdot T_a}{p_a \cdot 35.92}$$

$$\frac{3500 \cdot (460 + 40)}{14.2 \cdot 35.92} = 3431\ \text{ACFM}$$

$$\frac{4000 \cdot (460 + 60)}{14.2 \cdot 35.92} = 4078\ \text{ACFM}$$

$$\frac{4500 \cdot (460 + 80)}{14.2 \cdot 35.92} = 4764 \text{ ACFM}$$

$$\frac{5000 \cdot (460 + 100)}{14.2 \cdot 35.92} = 5489 \text{ ACFM}$$

Modify Equation 5.9 to calculate pounds/hour:

$$q_m = Q_s \cdot \rho \cdot 60 \frac{\text{min}}{\text{h}}$$

$$3500 \text{ SCFM} \cdot \frac{0.075 \text{ lb}_m}{\text{ft}^3} \cdot \frac{60 \text{ min}}{\text{h}} = \frac{15{,}750 \text{ lb}_m}{\text{h}}$$

$$4000 \text{ SCFM} \cdot \frac{0.075 \text{ lb}_m}{\text{ft}^3} \cdot \frac{60 \text{ min}}{\text{h}} = \frac{18{,}000 \text{ lb}_m}{\text{h}}$$

$$4500 \text{ SCFM} \cdot \frac{0.075 \text{ lb}_m}{\text{ft}^3} \cdot \frac{60 \text{ min}}{\text{h}} = \frac{20{,}250 \text{ lb}_m}{\text{h}}$$

$$5000 \text{ SCFM} \cdot \frac{0.075 \text{ lb}_m}{\text{ft}^3} \cdot \frac{60 \text{ min}}{\text{h}} = \frac{22{,}500 \text{ lb}_m}{\text{h}}$$

(b) Including relative humidity use Equation 5.12. Use Table 5.2 to determine the saturation vapor pressure:

$$Q_a = Q_s \cdot \frac{14.58}{p_b - (\text{RH} \cdot p_{sat})} \cdot \frac{T_a}{528} \cdot \frac{\cancel{p_b}}{\cancel{p_i}}$$

Note that because this is actual ambient air flow and not inlet air flow $p_i = p_b$ and the last term cancels.

$$3500 \cdot \frac{14.58}{14.2 - (0.2 \cdot 0.1217)} \cdot \frac{(460 + 40)}{528} = 3409$$

$$4000 \cdot \frac{14.58}{14.2 - (0.36 \cdot 0.2561)} \cdot \frac{(460 + 60)}{528} = 4071$$

$$4500 \cdot \frac{14.58}{14.2 - (0.6 \cdot 0.5068)} \cdot \frac{(460 + 80)}{528} = 4829$$

$$5000 \cdot \frac{14.58}{14.2 - (0.8 \cdot 0.9492)} \cdot \frac{(460 + 100)}{528} = 5753$$

Because the original data was provided as SCFM, which is a mass flow rate, there is no need to recalculate the mass flow/hour using relative humidity.

(c) Tabulate results and calculate percent error:

$$\%\text{Error} = \frac{b - a}{b} \cdot 100$$

Specified SCFM	ACFM without RH	ACFM with RH	%Error
3500	3431	3409	−0.6
4000	4078	4071	−0.2
4500	4764	4829	1.3
5000	5490	5489	4.6

Note that the error increases with increasing temperature and increasing relative humidity, both of which increase the effective vapor pressure.

Problem 5.2

A lobe type PD blower is connected to a 1750 rpm 100 hp motor by a belt drive with a 1.7:1 speed increasing ratio and 3% slip. Motor efficiency is 94%. The blower has a displacement of 0.662 CFR, a 1 psi slip of 108 rpm, a maximum rated speed of 3400 rpm, and friction hp of 1.24 bhp/1000 rpm. The actual discharge pressure varies between 7.0 and 8.0 psig. Worst-case ambient conditions are 100 °F and barometric pressure is 14.4 psia.

(a) What is the current flow rate in SCFM?
(b) If the belt ratio is changed, what is the maximum flow in SCFM that can be obtained from the blower?
(c) A proposed ECM includes installing a VFD on the blower. Construct a chart for kW versus SCFM at 7.0 and 8.0 psig with the new belt ratio and variable speed.

Solution:

(a) Calculate slip at 8.0 psi from Equation 5.32. Assume 0.2 psi loss from the inlet filter and 0.2 psi loss from the inlet and discharge silencers:

$$p_\text{i} = 14.4 - 0.2 - 0.2 = 14.0\ \text{psia}$$

$$p_\text{d} = 14.4 + 0.2 + 8.0 = 22.6\ \text{psia}$$

$$N_\text{s} = N_1 \cdot \sqrt{(p_\text{d} - p_\text{i}) \cdot \frac{14.7}{p_\text{i}} \cdot \frac{T_\text{i}}{528}}$$

$$= 108 \cdot \sqrt{\left((22.6 - 14.0) \cdot \frac{14.7}{14.0} \cdot \frac{(100 + 460)}{528}\right)}$$

$$N_\text{s} = 334\ \text{rpm}$$

Find blower rpm and calculate the current ICFM and SCFM by rearranging Equation 5.30:

$$N_a = N_m \cdot \text{belt ratio} \cdot (1 - \text{belt slip}) = 1750 \cdot 1.7 \cdot 0.97 = 2886\,\text{rpm}$$

$$Q_i = (N_a - N_s) \cdot \text{CFR} = (2886 - 334) \cdot 0.662 = 1689\,\text{ICFM}$$

Calculate SCFM at inlet conditions from Equation 5.8:

$$Q_s = Q_a \cdot \frac{p_a \cdot 35.92}{T_a} = 1689 \cdot \frac{14.0 \cdot 35.92}{(100 + 460)} = 1517\,\text{SCFM}$$

(b) The maximum flow may be limited by motor hp or maximum rpm. Calculate the actual friction hp from Equation 5.34 and the maximum allowable speed at 100 hp by rearranging Equation 5.35:

$$\text{FHP}_a = \text{FHP}_{\text{nom}} \cdot \frac{N_a}{1000} = 1.24 \cdot \frac{2886}{1000} = 3.6\,\text{hp}$$

$$N_a = \frac{P_{\text{PD}} \cdot (1 - \text{belt slip}) - \text{FHP}}{0.0044 \cdot \text{CFR} \cdot (p_d - p_i)} = \frac{100 \cdot 0.97 - 3.6}{0.0044 \cdot 0.662 \cdot (22.6 - 14.0)} = 3730\,\text{rpm}$$

The maximum rated speed is 3400 rpm, so this will limit capacity and not motor power.

Calculate the new belt ratio:

$$\text{Ratio} = \frac{3400}{1750 \cdot 0.97} = 2.0 : 1$$

Calculate the ICFM and SCFM at the maximum rpm from the rearranged Equations 5.31 and 5.8:

$$Q_i = (N_a - N_s) \cdot \text{CFR} = (3400 - 334) \cdot 0.662 = 2030\,\text{ICFM}$$

$$Q_s = Q_a \cdot \frac{p_a \cdot 35.92}{T_a} = 2030 \cdot \frac{14.0 \cdot 35.92}{(100 + 460)} = 1820\,\text{SCFM}$$

(c) Assume minimum blower speed is 50% of maximum. Calculate the blower power at min and max speed for each pressure using Equation 5.35:

$$1.24 \cdot \frac{3400}{1000} = 4.2\,\text{hp}$$

$$P_{\text{PD 8 max}} = \{[0.0044 \cdot N_a \cdot \text{CFR} \cdot (p_d - p_i)] + \text{FHP}_a\} \cdot (1 + \text{belt slip})$$

$$\text{P}_{\text{PD 8 max}} = \{[0.0044 \cdot 3400 \cdot 0.662 \cdot (22.6 - 14.0)] + 4.2\} \cdot 1.03 = 92\,\text{hp}$$

Calculate friction hp at minimum rpm:

$$1.24 \cdot \frac{1700}{1000} = 2.1\,\text{hp}$$

$$P_{PD\,8\,min} = \{[0.0044 \cdot 1700 \cdot 0.662 \cdot (22.6 - 14.0)] + 2.1\} \cdot 1.03 = 46\,\text{hp}$$

$$P_{PD\,7\,max} = \{[0.0044 \cdot 3400 \cdot 0.662 \cdot (21.6 - 14.0)] + 4.2\} \cdot 1.03 = 82\,\text{hp}$$

$$P_{PD\,7\,min} = \{[0.0044 \cdot 1700 \cdot 0.662 \cdot (21.6 - 14.0)] + 2.1\} \cdot 1.03 = 41\,\text{hp}$$

Calculate kW:

$$\text{kW} = \frac{\text{bhp} \cdot 0.746}{\eta_m}$$

$$\text{kW}_{8\,max} = \frac{92 \cdot 0.746}{0.94} = 73\,\text{kW}$$

$$\text{kW}_{8\,min} = \frac{46 \cdot 0.746}{0.94} = 37\,\text{kW}$$

$$\text{kW}_{7\,max} = \frac{82 \cdot 0.746}{0.94} = 65\,\text{kW}$$

$$\text{kW}_{7\,min} = \frac{41 \cdot 0.746}{0.94} = 33\,\text{kW}$$

Calculate ICFM and SCFM:

$$Q_{i8\,max} = (N_a - N_s) \cdot \text{CFR} = (3400 - 334) \cdot 0.662 = 2030\,\text{ICFM}$$

$$Q_{s8\,max} = Q_a \cdot \frac{p_a \cdot 35.92}{T_a} = 2030 \cdot \frac{14.0 \cdot 35.92}{(100 + 460)} = 1820\,\text{SCFM}$$

$$Q_{i8\,min} = (1700 - 334) \cdot 0.662 = 900\,\text{ICFM}$$

$$Q_{s8\,min} = 900 \cdot \frac{14.0 \cdot 35.92}{(100 + 460)} = 810\,\text{SCFM}$$

Calculate slip rpm at 7.0 psi:

$$N_s = 108 \cdot \sqrt{\left((21.6 - 14.0) \cdot \frac{14.7}{14.0} \cdot \frac{(100 + 460)}{528}\right)} = 314\,\text{rpm}$$

$$Q_{i7\,max} = (3400 - 314) \cdot 0.662 = 2040\,\text{ICFM}$$

$$Q_{s7\ max} = 2040 \cdot \frac{14.0 \cdot 35.92}{(100 + 460)} = 1830\ \text{SCFM}$$

$$Q_{i7\ min} = (1700 - 314) \cdot 0.662 = 920\ \text{ICFM}$$

$$Q_{s7\ min} = 918 \cdot \frac{14.0 \cdot 35.92}{(100 + 460)} = 830\ \text{SCFM}$$

Notes: An analysis of this type would be performed using a spreadsheet or other analysis software to reduce the engineering time required. This practice is encouraged. However, it is important to use the rigorous calculation method on some of the data points to verify the accuracy of the spreadsheet formulas and data entry.

Problem 5.3

A plant has four geared single-stage blowers with inlet guide vane control. Plant SCADA trends show a pattern generally conforming to Table 4.2 with an average air flow of 28,500 SCFM and occasional excursions during storm events to 56,000 SCFM (see Figure 5.15). The blower performance characteristics and ambient conditions are shown in Figure 5.16 for 51 °F, which is the annual average temperature at the site. The motors are 2000 hp, 4160 VAC 3 Phase power, and have an average efficiency of 96%. Current controls maintain a constant discharge pressure of 10.3 psig. It is proposed that the guide vane control be replaced with 97% efficient VFD control. The energy cost is $0.06/kWh on and off peak, plus a demand charge of $5/kW for the highest demand within a 12-month period (ratchet charge). Diffuser submergence is 19.5 ft.

(a) If each VFD has an installed cost of $300,000, including control system upgrades, will the system be cost-effective?
(b) Adding a synchronous transfer system to permit controlling two blowers from one VFD adds $80,000 to the installed cost of a VFD for each blower. What are the pros and cons of this system?
(c) How is the payback and system selection affected if a utility rebate of $50/hp is available for each VFD?

Solution:

Establish the tabulated flow pattern for analysis (use a spreadsheet for data analysis):

Procedure:

- Calculate the flow rates to be evaluated based on the percentages of Table 4.2 and a discharge pressure of 10.3 psig

Example: $\text{SCFM} = \%\text{ADF} \cdot Q_{ave} = 0.7 \cdot 28{,}500 = 19{,}950\ \text{SCFM}$

- By inspection or interpolation determine the IGV position that corresponds to each flow rate on the SCFM vs. psig plot (see Figure A.3)

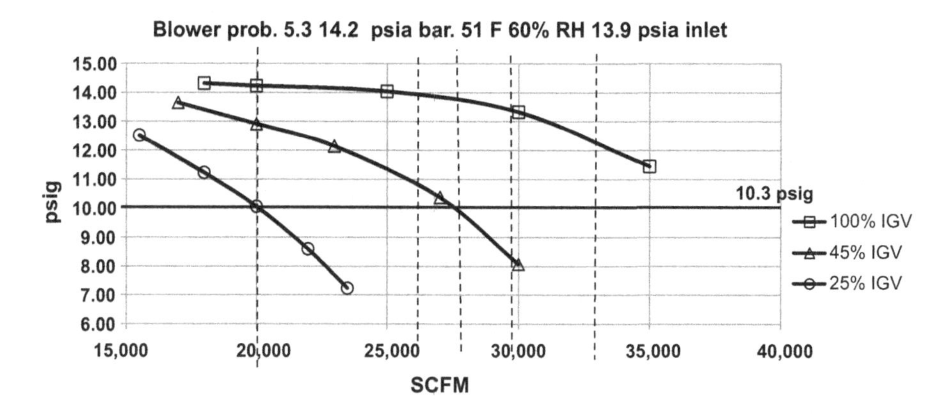

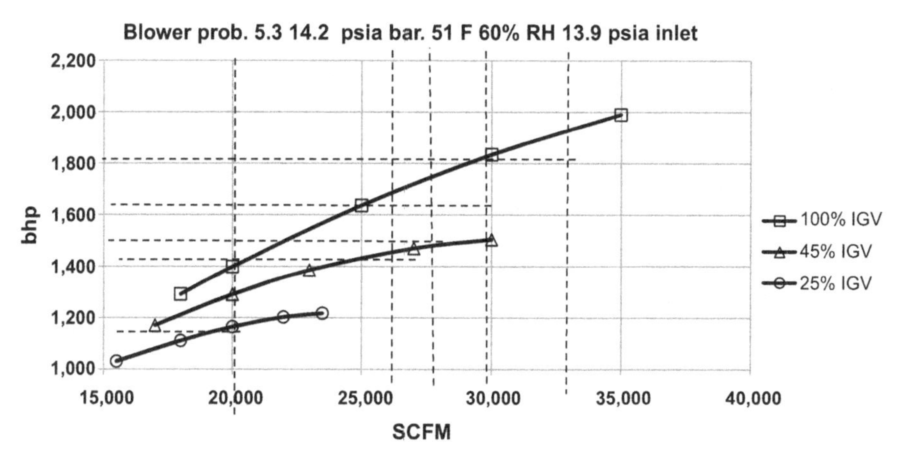

FIGURE A.3 *Problem 5.3 blower performance with IGV*

- Using the IGV position and flow rate determine the corresponding blower hp for each flow rate by inspection or interpolation (see Figure A.3)
- Calculate the corresponding kW for each flow rate

$$\text{Example:}\quad \text{kW} = \frac{\text{bhp} \cdot 0.746}{\eta_{\text{motor}}} = \frac{1180\,\text{bhp} \cdot 0.746}{0.96} = 917\,\text{kW}$$

- Prorate the kW based on % time × kW

$$\text{Example:}\quad \text{Prorated kW} = \text{kW} \cdot \%\text{Time} = 917 \cdot 0.2083 = 191\,\text{kW}$$

- Total the prorated kW values to establish the evaluated kW

% Time	%ADF	SCFM	psig	IGV	hp ea.	kW	Prorated kW
20.83	70	19,950	10.3	26	1180	917	191
12.50	90	25,650	10.3	44	1450	1,127	141
8.33	100	28,500	10.3	54	1570	1,220	102
33.33	107.50	30,638	10.3	68	1660	1,290	430
25.00	120	34,200	10.3	85	1840	1,430	357
						Total	1,221

At the peak air flow demand of 56,000 SCFM two blowers will be required, each drawing approximately 1200 kW for a total demand of 2400 kW.

Calculate the baseline energy cost per year by combining Equations 2.7 and 2.8, and Equation 2.9:

$$\text{On Peak \$} + \text{Off Peak \$} = \frac{\$}{\text{kWh}} \cdot \text{kW} \cdot \frac{168\ \text{h}}{\text{week}} \cdot \frac{52\ \text{weeks}}{\text{year}} = 0.06 \cdot 1221 \cdot 168 \cdot 52$$

$$\text{On Peak \$} + \text{Off Peak \$} = \$640{,}000\ \text{per year}$$

$$\text{Demand \$} - \text{Demand}\frac{\$}{\text{kW}} \cdot \text{kW}_{\text{peak}} \cdot \frac{12\ \text{months}}{\text{year}} = 5 \cdot 2400 \cdot 12 = \$144{,}000\ \text{demand}$$

$$\text{Total \$} = \$640{,}000 + \$144{,}000 = \$784{,}000/\text{year}$$

Establish the system curve parameters for the MOV logic evaluation using Equations 4.12, 5.14, and 5.15. Assume 10.3 psig at the max capacity of one blower, 35,000 SCFM, which is a conservative assumption:

$$p_s = 19.5 \cdot 0.433 = 8.44\ \text{psig}$$

$$k_f = \frac{(p_{\text{des}} - p_{\text{stat}})}{Q^2_{\text{des}}} = \frac{(10.3 - 8.44)}{35{,}000^2} = 1.51837 \cdot 10^{-9}$$

Repeat the evaluation, using the 100% IGV curve as the 60 Hz performance, and calculating the performance at reduced speed:

Procedure:

- Calculate the required discharge pressure at each flow rate from the table above.

  Example: $p_{\text{tot}} = p_{\text{stat}} + k_f \cdot Q^2 = 8.44 + 1.51837 \cdot 10^{-9} \cdot 19{,}950^2 = 9.04\ \text{psig}$

- By inspection or iteration determine the motor speed or VFD output frequency that corresponds to each flow rate on the variable speed SCFM versus psig plot (see Figure A.4). Use Equations 5.40 and 5.41, calculate the discharge pressure at reduced speed from the initial data at 60 Hz:

$$\text{Example:}\quad Q_a = Q_c \cdot \frac{N_a}{N_c} = 23{,}797 \cdot \frac{50.3}{60} = 19{,}950\ \text{SCFM}$$

$$\text{Example:}\quad r_p = \frac{p_d}{p_i} = \frac{(14.1 + 14.2)}{13.9} = 2.037$$

$$\text{Example:}\quad X = r_p^{(k-1/k)} - 1 = 2.037^{(0.283)} = 0.223$$

$$\text{Example:}\quad X_a = X_c \cdot \left(\frac{N_a}{N_c}\right)^2 = 0.223 \cdot \left(\frac{50.3}{60}\right)^2 = 0.156$$

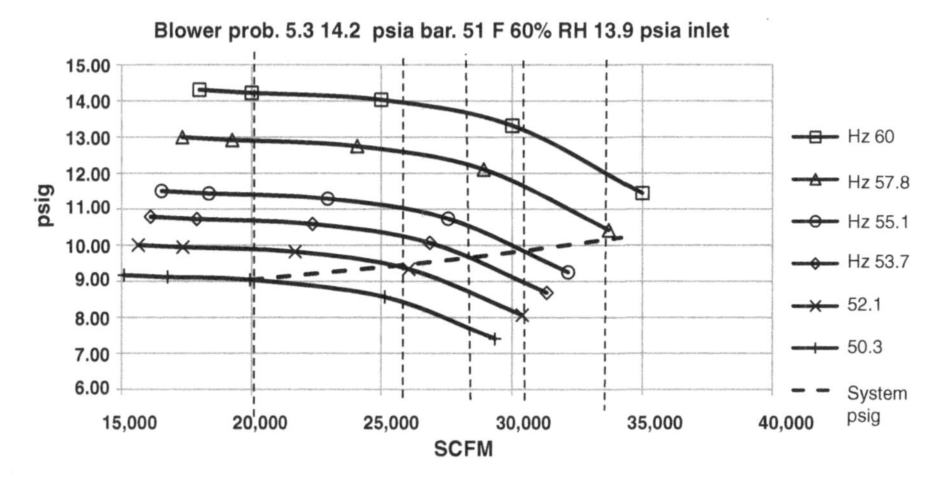

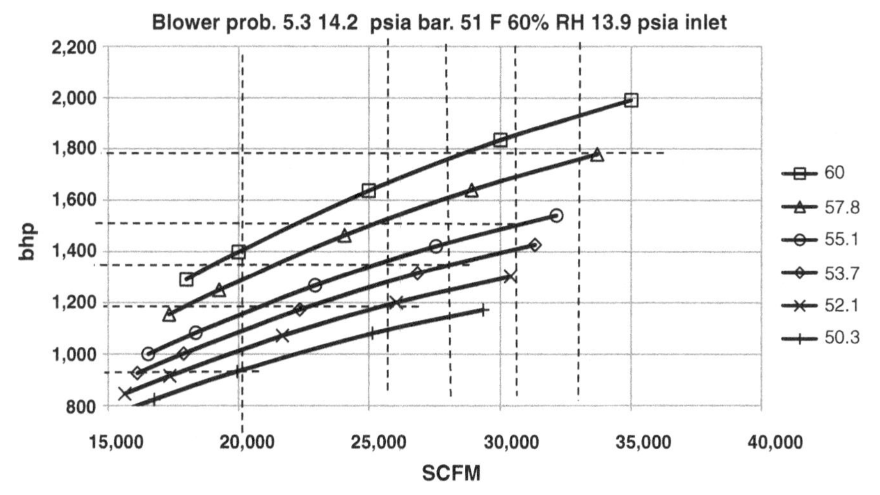

FIGURE A.4 *Problem 5.3 blower performance with VFD*

- Rearrange Equations 5.20 and 5.18 to determine the pressure:

$$\text{Example: } r_p = (X_a + 1)^{(k/k-1)} = (0.156 + 1)^{3.532} = 1.669$$

$$\text{Example: } p_d = r_p \cdot p_i - p_{bar} = 1.669 \cdot 13.9 - 14.2 = 9.00\,\text{psig}$$

This is approximately equal to the system pressure at that flow rate, so the speed is correct for this data point.

- Using the blower speed and hp corresponding to the original curve flow rate, determine the corresponding blower hp for each flow rate by inspection or interpolation (see Figure A.4). Use Equation 5.42 to calculate the new hp:

$$\text{Example: } P_a = P_c \cdot \left(\frac{N_a}{N_c}\right)^3 = 1583\,\text{bhp} \cdot \left(\frac{50.3}{60}\right)^3 = 933\,\text{bhp}$$

- Calculate the corresponding kW for each flow rate

$$\text{Example: } \text{kW} = \frac{\text{bhp} \cdot 0.746}{\eta_{motor} \cdot \eta_{VFD}} = \frac{933\,\text{hp} \cdot 0.746}{0.96 \cdot 97} = 747\,\text{kW}$$

- Prorate the kW based on % time × kW

$$\text{Example: Prorated kW} = \text{kW} \cdot \%\text{Time} = 747 \cdot 0.2083 = 156\,\text{kW}$$

- Total the prorated kW values to establish the evaluated kW system

% Time	%ADF	SCFM	psig	Hz	hp ea.	kW	Prorated kW
20.83	70	19,950	9.04	50.3	932	746	155
12.50	90	25,650	9.44	52.1	1,193	956	119
8.33	100	28,500	9.67	53.7	1,359	1,088	91
33.33	107.50	30,638	9.87	55.1	1,504	1,205	402
25.00	120	34,200	10.22	57.8	1,795	1,438	360
						Total	1,127

Calculate the annual energy cost with MOV logic and VFD blower control:

$$\text{On Peak \$} + \text{Off Peak \$} = \frac{\$}{\text{kWh}} \cdot \text{kW} \cdot \frac{168\,\text{h}}{\text{week}} \cdot \frac{52\,\text{weeks}}{\text{year}} = 0.06 \cdot 1127 \cdot 168 \cdot 52$$

$$\text{On Peak \$} + \text{Off Peak \$} = \$590{,}700/\text{year}$$

At 56,000 SCFM, two blowers will operate at approximately 1088 kW each, 2176 kW total.

$$\text{Demand\$} = \text{Demand}\frac{\$}{\text{kW}} \cdot \text{kW}_{\text{peak}} \cdot \frac{12\ \text{months}}{\text{year}} = 5 \cdot 2176 \cdot 12 = \$130{,}600\ \text{demand}$$

$$\text{Total\$} = \$590{,}700 + \$141{,}600 = \$721{,}300/\text{year}$$

Calculate annual savings:

$$\text{Savings} = \$784{,}000 - \$721{,}300 = \$62{,}600/\text{year}$$

Calculate the payback if all four blowers are retrofitted with VFDs and controls:

$$P = \frac{C}{S} = \frac{4 \cdot \$300{,}000}{\$62{,}700} = 19.9\ \text{years}$$

This is not an acceptable payback period for a retrofit project, although it might be acceptable for new construction.

(b) There are two synchronous transfer options. Two VFDs each controlling two blowers is one choice, and one VFD with synchronous transfer ability to all four blowers is another.

With two VFDs:

$$\frac{2 \cdot \$300{,}000 + 4 \cdot \$80{,}000}{\$60{,}300} = 15.3\ \text{years}$$

With one VFD the demand savings will be one half of the previous calculation, since only one blower will be on the VFD:

$$\text{S} = \$60{,}300 - 0.5 \cdot (\$144{,}000 - \$130{,}600) = \$53{,}600$$

$$\frac{1 \cdot \$300{,}000 + 4 \cdot \$80{,}000}{\$53{,}600} = 11.6\ \text{years}$$

Advantages: Improved payback.

Disadvantages: This system is more complex for the operators to control. The need to alternate blowers requires that all units be able to operate on variable speed, but not more than two at once. If only one blower at a time can be on the VFD, the demand savings are reduced. It also does not allow for energy savings if the VFD is out of service.

(c) The availability of a rebate or incentive from the utility improves the payback:

$$\frac{\$50}{\text{hp}} \cdot 2000\ \text{hp} = \$100{,}000$$

$$P = \frac{C}{S} = \frac{4 \cdot (\$300{,}000 - \$100{,}000)}{\$60{,}300} = 13.3\ \text{years}$$

Four VFDs will be a cost-effective alternate. Check the payback if two VFDs are used with synchronous transfer:

$$\frac{2 \cdot (\$300{,}000 - \$100{,}000) + 4 \cdot \$80{,}000}{\$60{,}300} = 11.9 \text{ years}$$

This will also be cost-effective, but the operational advantages of four VFDs make this a better choice.

Problem 5.4

A 75 hp screw blower with an inlet filter pressure drop of 0.2 psi and a discharge pressure of 9.0 psig has an efficiency of 70%. The blower case is 24 in. × 12 in. × 9 in. The motor efficiency is 97% and the VFD efficiency is 97%. The package has a temperature rating of 40 °C (104 °F). The ambient air temperature is 90 °F and the altitude is 800 ft above sea level.

(a) What is the discharge temperature of the process air flow?
(b) How much ventilating air will be required to keep the enclosure temperature to safe limits?

Solution:

(a) Use Equation 5.26 to determine the discharge air temperature. First the pressure ratio and the adiabatic factor X must be determined:

$$p_{\mathrm{bar}} = 14.7 - \frac{\mathrm{Alt}}{2000} = 14.7 - \frac{800}{2000} = 14.3 \text{ psia}$$

$$p_{\mathrm{i}} = p_{\mathrm{bar}} - \mathrm{Losses} = 14.3 - 0.2 = 14.1 \text{ psia}$$

$$p_{\mathrm{d}} = 14.3 + 9.0 = 23.3 \text{ psia}$$

$$r_{\mathrm{p}} = \frac{p_{\mathrm{d}}}{p_{\mathrm{i}}} = \frac{23.3}{14.1} = 1.652$$

$$X = r_{\mathrm{p}}^{(k-1/k)} - 1 = 1.652^{0.283} - 1 = 0.153$$

$$T_{\mathrm{d}} = T_{\mathrm{i}} + \frac{T_{\mathrm{i}} \cdot X}{\eta} = (90 + 460) + \frac{(90 + 460) \cdot 0.153}{0.7} = 670\,^{\circ}\mathrm{R}$$

$$670 - 460 = 210\,^{\circ}\mathrm{F}$$

(b) Determine the heat rejection for each part of the system. Assume the blower will draw full motor power, a conservative estimate. Use Equation 5.27 to determine the heat rejected by the motor and VFD:

$$H_{\mathrm{d}} = 2544 \cdot P_{\mathrm{mot}} \cdot (1 - \eta_{\mathrm{mot}} \cdot \eta_{\mathrm{VFD}}) = 2544 \cdot 75 \cdot (1 - 0.97 \cdot 0.97) = 11{,}280 \frac{\mathrm{BTU}}{\mathrm{h}}$$

Determine the blower case area and use Equation 5.28 to determine the heat rejected by the blower case:

$$A_b = \frac{24}{12} \cdot \frac{12}{12} \cdot \frac{9}{12} = 1.5\,\text{ft}^2$$

$$H_b = 2.4 \cdot 1.25 \cdot A_b \cdot (T_d - T_a) = 2.4 \cdot 1.25 \cdot 1.5 \cdot (210 - 90) = 540\frac{\text{BTU}}{\text{h}}$$

Calculate the air flow required to maintain the enclosure temperature at 100 °F from Equation 5.29:

$$Q_{\text{fan}} = \frac{H_d + H_b}{1.08 \cdot (T_r - T_o)} = \frac{11{,}280 + 540}{1.08 \cdot (100 - 90)} = 1094\,\text{CFM}$$

CHAPTER 6: PIPING SYSTEMS

Problem 6.1

Refer to the piping layout in Figure 6.8. The plant is at an altitude of 1000 ft ASL. Ignore friction losses and assume equal air flow to all diffuser drop legs. Calculate the velocity head and the static pressure at points "A" and "E" if two blowers are operating at rated capacity. Determine if the effect is significant.

Solution:

Obtain the pipe inside diameters (ID) from Table 6.4 and calculate the velocity in each section of concern by rearranging Equation 6.5:

$$A_{14} = \frac{\pi d^2}{4} = \frac{\pi (13.624/12)^2}{4} = 1.012\,\text{ft}^2$$

$$A_8 = \frac{\pi (8.329/12)^2}{4} = 0.378\,\text{ft}^2$$

Note that velocity is a function of volumetric flow rate (ACFM) and pipe size. The mass air flow rate, SCFM, must be converted to volumetric flow using Equation 5.7 and a rearranged Equation 5.8. Calculate the velocity at Point "G" using rearranged Equation 6.5:

$$p_{\text{bar}} = 14.7 - \frac{\text{Alt}}{2000} = 14.7 - \frac{1000}{2000} = 14.2\,\text{psia}$$

$$Q_a = \frac{Q_s \cdot T_a}{p_a \cdot 35.92}$$

$$Q_{\mathrm{G}} = \frac{4800 \cdot (215 + 460)}{(14.2 + 10.3) \cdot 35.92} = 3680\ \mathrm{ACFM}$$

$$V = \frac{Q_{\mathrm{a}}}{A}$$

$$V_{\mathrm{G}} = \frac{2760}{1.012} = 3640\ \mathrm{ft/min}$$

Calculate the density using Equation 6.3 and the velocity head at point "G" using Equation 6.4:

$$\rho_{\mathrm{aG}} = \frac{2.692 \cdot p_{\mathrm{a}}}{T_{\mathrm{a}}} = \frac{2.692 \cdot (14.2 + 10.3)}{(215 + 460)} = 0.098 \frac{\mathrm{lb_m}}{\mathrm{ft^3}}$$

$$p_{\mathrm{dG}} = \frac{\rho \cdot V^2}{3.335 \cdot 10^7} = \frac{0.098 \cdot 3640^2}{3.335 \cdot 10^7} = 0.039\ \mathrm{psi}$$

At points "A" and "E" the air flow is 1200 SCFM. Repeat the above calculations for each point. It is reasonable to ignore the velocity head in working with air piping calculations, but the impact may be verified after completing the calculations:

$$Q_{\mathrm{A}} = Q_E = \frac{1200 \cdot (215 + 460)}{(14.2 + 10.3) \cdot 35.92} = 920\ \mathrm{ACFM}$$

$$V_{\mathrm{A}} = \frac{920}{0.378} = 2430\ \mathrm{ft/min}$$

$$p_{\mathrm{dA}} = \frac{0.098 \cdot 2430^2}{3.335 \cdot 10^7} = 0.017\ \mathrm{psi}$$

The static pressure at each point is obtained from the original static pressure at the blower room and the difference in velocity head. Rearranging Equation 6.2:

$$p_{\mathrm{totalA}} = p_{\mathrm{totalG}} + p_{\mathrm{dG}} - p_{\mathrm{dA}} = 10.3 + 0.039 - 0.017 = 10.32\ \mathrm{psig}$$

The static pressure at point "A" is slightly (0.02 psig) higher than the static pressure at point "G", where the air flow leaves the blower building.

$$V_{\mathrm{E}} = \frac{920}{1.012} = 910\ \mathrm{ft/min}$$

$$p_{\mathrm{dE}} = \frac{0.098 \cdot 910^2}{3.335 \cdot 10^7} = 0.002\ \mathrm{psi}$$

$$p_{\mathrm{totalE}} = p_{\mathrm{totalG}} + p_{\mathrm{dG}} - p_{\mathrm{dE}} = 10.3 + 0.039 - 0.002 = 10.34\ \mathrm{psig}$$

The static pressure at point "E" is also slightly (0.04 psig) higher than the static pressure at point "G". In both locations the air velocity is well below the limits of Table 6.3. On a practical basis the difference in static pressure is not likely to be significant at either location.

Problem 6.2

An orifice is to be installed in fine-pore diffusers to compensate for errors in the installed height of a diffuser. This will be done by introducing ½ in. of additional pressure drop at minimum flow. The nominal depth is 18 ft. The flow rate through each diffuser will range from 1.0 SCFM to 4.0 SCFM and the nominal air temperature at the diffuser face is 50 °F.

(a) What is the recommended orifice diameter?

(b) What is the total pressure drop at minimum and maximum air flow rates if the pressure drop through the membrane itself is shown in Figure 4.16?

Solution:

(a) Calculate the volumetric flow rate, the static pressure, and the required pressure drop using rearranged Equations 5.8, 4.12 and 6.3. Assume barometric pressure is 14.7 psia, which will be conservative:

$$Q_{\text{a min}} = \frac{Q_{\text{s}} \cdot T_{\text{a}}}{p_{\text{a}} \cdot 35.92} = \frac{1.0 \cdot (50 + 460)}{(14.7 + 7.80) \cdot 35.92} = 0.631 \text{ ACFM}$$

$$Q_{\text{a max}} = \frac{4.0 \cdot (50 + 460)}{(14.7 + 7.80) \cdot 35.92} = 2.52 \text{ ACFM}$$

$$\Delta p_{\text{o}} = \text{Depth} \cdot 0.433 = \frac{0.5}{12} \cdot 0.433 = 0.018 \text{ psi}$$

$$p_{\text{stat}} = 18 \cdot 0.433 = 7.80 \text{ psi}$$

$$\rho_{\text{aG}} = \frac{2.692 \cdot p_{\text{a}}}{T_{\text{a}}} = \frac{2.692 \cdot (14.7 + 7.80)}{(50 + 460)} = 0.119 \frac{\text{lb}_{\text{m}}}{\text{ft}^3}$$

Use an orifice coefficient of 0.6 for the drilled hole, which corresponds to a sharp edged orifice. Solve for the diameter by rearranging Equation 6.17:

$$d = \left(\frac{\rho \cdot Q_{\text{a}}^2}{992.2 \cdot C_{\text{o}}^2 \cdot \Delta p_{\text{o}}}\right)^{\frac{1}{4}} = \left(\frac{0.119 \cdot 0.631^2}{992.2 \cdot 0.60^2 \cdot 0.018}\right)^{0.25} = 0.293 \text{ in.}$$

Use a 5/16 in. (0.3125 in.) diameter orifice:

$$\Delta p_{\text{o}} = \frac{\rho \cdot Q_{\text{a}}^2}{992.2 \cdot C_{\text{o}}^2 \cdot d^4} = \frac{0.119 \cdot 0.631^2}{992.2 \cdot 0.6^2 \cdot 0.3125^4} = 0.014 \text{ psi} = 0.39 \text{ in. } H_2O$$

This is acceptable. Check pressure drop at max flow:

$$\Delta p_o = \frac{0.119 \cdot 2.52^2}{992.2 \cdot 0.6^2 \cdot 0.3125^4} = 0.22 \text{ psi} = 6 \text{ in. } H_2O$$

This is a reasonable value.

(b) From Figure 4.18 the pressure drop through the membrane is 9 in. H_2O at 1 SCFM and 19 in. H_2O at 4 SCFM. Calculate the total pressure drop:
At 1 SCFM:

$$\frac{9}{12} \cdot 0.433 + 0.014 = 0.34 \text{ psig}$$

At 4 SCFM:

$$\frac{19}{12} \cdot 0.433 + 0.22 = 0.91 \text{ psig}$$

Problem 6.3

Refer to Problem 6.1. Calculate the pressure drop from friction between the blower room and points "A" and "E". Assume the blowers are at max capacity, all valves are full open, and the air flow is split equally among all four grids.

Solution:

Analyze the three sections with different flow rates or pipe arrangements separately. The first section will be between the blower room and point "G" where the first lateral takes off from the main header. The first step is to determine the equivalent length of 14 in. pipe using a spreadsheet, Equation 6.13, and Table 6.6.

For Schedule 10 and 14 in. nominal pipe size the ID = 13.624:

Description	Qty.	L/D	Eq. Lgth. Each	Total Lgth. ft
Horizontal pipe	1	n/a	20	20
Vertical pipe	2	n/a	15	30
90 ° Elbow	4	30	35	140
			Total	190

Assume $p_m = 10.2$ psig and calculate pressure drop using Equation 6.10:

$$\Delta p_f = 0.07 \cdot \frac{Q_S^{1.85}}{d^5 \cdot p_m} \cdot \frac{T}{528} \cdot \frac{L_e}{100}$$

$$= 0.07 \cdot \frac{4800^{1.85}}{13.624^5 \cdot (14.2 + 10.2)} \cdot \frac{(215 + 460)}{528} \cdot \frac{190}{100}$$

$$\Delta p_f = 0.10 \text{ psig}$$

Check assumed mean pressure using Equation 6.12:

$$p_{\mathrm{m}} = p_{\mathrm{initial}} - \frac{\Delta p_{\mathrm{f}}}{2} = 10.3 - \frac{0.1}{2} = 10.25$$

Recalculate:

$$0.07 \cdot \frac{4800^{1.85}}{13.624^5 \cdot (14.2 + 10.25)} \cdot \frac{(215 + 460)}{528} \cdot \frac{190}{100} = 0.10\,\mathrm{psig}$$

No further iterations are required.

$$p_{\mathrm{final\ G}} = p_{\mathrm{initial}} - \Delta p_{\mathrm{f}} = 10.3 - 0.10 = 10.2\,\mathrm{psig}$$

Calculate pressure drop from "G" to "E":

Description	Qty.	L/D	Eq. Lgth. Each	Total Lgth., ft
BFV 100	1	20	23	23
Horizontal pipe	1	n/a	60	60
Tee run	1	20	23	23
Tee branch	1	60	70	70
			Total	176

After the final iteration, the calculated pressure drop is

$$\Delta p_{\mathrm{f}} = 0.07 \cdot \frac{2400^{1.85}}{13.624^5 \cdot (14.2 + 10.19)} \cdot \frac{(215 + 460)}{528} \cdot \frac{176}{100} = 0.02\,\mathrm{psig}$$

$$p_{\mathrm{m}} = p_{\mathrm{initial}} - \frac{\Delta p_{\mathrm{f}}}{2} = 10.20 - \frac{0.02}{2} = 10.19$$

$$p_{\mathrm{final\ E}} = p_{\mathrm{initial}} - \Delta p_{\mathrm{f}} = 10.20 - 0.02 = 10.18\,\mathrm{psig}$$

Notes: Because the pressure drop is so low, the change in flow rate after the first drop leg at point "F" has been ignored. If the air velocity was higher and the pressure drop more significant, the section between "F" and "E" would be analyzed separately.

The pressure drop from "G" to "E" must be done in two steps because both the flow rate and the pipe diameter changes at the first drop leg, point "C". From point "G" to point "C":

Description	Qty.	L/D	Eq. Lgth. Each	Total Lgth., ft
Horizontal pipe	1	n/a	30	30
90 ° Elbow	1	30	35	35
BFV 100	1	20	23	23
Horizontal pipe	1	n/a	20	20
Tee run	1	20	23	23
			Total	131

After the final iteration, the calculated pressure drop for the section from "G" to "C" is:

$$\Delta p_{\mathrm{f}} = 0.07 \cdot \frac{2400^{1.85}}{13.624^5 \cdot (14.2 + 10.19)} \cdot \frac{(215 + 460)}{528} \cdot \frac{131}{100} = 0.02 \text{ psig}$$

$$p_{\mathrm{m}} = p_{\mathrm{initial}} - \frac{\Delta p_{\mathrm{f}}}{2} = 10.20 - \frac{0.02}{2} = 10.19$$

$$p_{\mathrm{final\ C}} = p_{\mathrm{initial}} - \Delta p_{\mathrm{f}} = 10.20 - 0.02 = 10.18 \text{ psig}$$

The ID of 8 in. Schedule 10 pipe is 8.329 in. The calculated pressure drop for the section from "C" to "A" is

Description	Qty.	L/D	Eq. Lgth. Each	Total Lgth. ft
Transition	1	20	13	13
Horizontal pipe	1	n/a	40	40
Tee branch	1	60	40	40
			Total	93

$$\Delta p_{\mathrm{f}} = 0.07 \cdot \frac{1200^{1.85}}{8.329^5 \cdot (14.2 + 10.15)} \cdot \frac{(215 + 460)}{528} \cdot \frac{93}{100} = 0.04 \text{ psig}$$

$$p_{\mathrm{m}} = p_{\mathrm{initial}} - \frac{\Delta p_{\mathrm{f}}}{2} = 10.18 - \frac{0.04}{2} = 10.16$$

$$p_{\mathrm{final\ A}} = p_{\mathrm{initial}} - \Delta p_{\mathrm{f}} = 10.18 - 0.04 = 10.14 \text{ psig}$$

Although the total pressure drop on the branch from "G" to "A" (0.06 psi) is triple the drop from "G" to "E" (0.02 psi), it is still negligible. If this were a new plant consideration would be given to reduce the pipe size.

Problem 6.4

The existing 8 inch butterfly valves for the diffuser drop legs in Figure 6.8 are being considered for replacement. The control valve is limited to travel between 20° and 70° open. The C_v for a 6, 8, and 10 in. nominal valve are shown in Figure 6.6 and are tabulated below:

°Open	6 in. C_v	8 in. C_v	10 in. C_v
20	45	89	151
30	95	188	320
40	205	408	694
50	366	727	1237
60	605	1202	2047
70	958	1903	3240

The upstream temperature is 215 °F and the upstream pressure is 10.1 psig. The flow will vary between 600 SCFM and 2400 SCFM. Static pressure is 7.8 psig and the pressure drop for the diffusers and grid piping is shown below:

SCFM	Δp psi
600	0.50
1200	0.77
1800	1.03
2400	1.30

Calculate the valve positions for each of the four flow rates and make a recommendation for the valve size.

Solution:

- Establish the pressure drop through the valve for each flow rate (use a spreadsheet for data analysis):

Example: $\Delta p_{600} = 10.10 - 7.80 - 0.50 = 1.80\,\text{psi}$

- Calculate the required C_v for each valve and flow rate using Equation 6.20:

$$\text{Example: } C_{v600} = \frac{Q_s}{22.66} \cdot \sqrt{\frac{SG \cdot T_u}{p_u \cdot \Delta p_v}} = \frac{600}{22.66} \cdot \sqrt{\frac{1.0 \cdot (215 + 460)}{(14.2 + 10.10) \cdot 1.80}}$$

Example: $C_{v\,600} = 104$

SCFM	Δp	p_d	Δp_v	C_v Required
600	0.50	8.30	1.80	104
1200	0.77	8.57	1.53	225
1800	1.03	8.83	1.27	372
2400	1.30	9.10	1.00	558

- Determine the corresponding valve position for each size and flow rate from inspection of Figure 6.6 or interpolation of the tabulated data.

°Open			
SCFM	6 in.	8 in.	10 in.
600	31°	22°	17°
1200	42°	32°	25°
1800	50°	39°	32°
2400	59°	45°	37°

The 10 in. valve would be unacceptable. It has a limited travel range of 20° and would be unable to throttle the flow to the minimum flow rate because the position is less than the minimum 20° setting.

Either the 6 in. or the 8 in. would be acceptable. The 6 in. would be preferred, particularly for new construction, since it has a slightly wider range of travel and the travel covers the middle of the allowable range. Since this is defined as a retrofit project the 8 in. could be retained if it is necessary to minimize expense, since keeping the 8 in. would eliminate piping modifications.

Problem 6.5

The system in Figure 6.8 has the drop leg valves controlled manually. The air flow rate at point "D" is 2400 SCFM. The grid connected to point "C" has 400 diffusers with an estimated pressure drop through the butterfly valve, drop leg, grid piping, and diffusers of 0.75 psi at 900 SCFM. The grid connected to point "A" has 600 diffusers with an estimated pressure drop (including the above plus the piping) between "C" and "A" of 0.90 psig at 1200 SCFM. The static pressure is 7.8 psig. Ignoring velocity head regain, how much air flow will go to each grid if the butterfly valves are not used for throttling?

Solution:

The two pipe sections operate in parallel from point "C." It is necessary to calculate the constant of proportionality, k_f, for each section using a rearranged form of Equation 5.14 for each branch:

$$k_{fC} = \frac{\Delta p_a}{Q_{des}^2} = \frac{0.75}{900^2} = 926 \cdot 10^{-9}$$

$$k_{fA} = \frac{0.90}{1200^2} = 625 \cdot 10^{-9}$$

The proportion of the total air flow to each grid can be calculated from Equation 6.14:

$$Q_C = \frac{Q_{total}}{1 + \sqrt{k_{f\,C}/k_{f\,A}}} = \frac{2400}{1 + \sqrt{(926 \cdot 10^{-9})/(625 \cdot 10^{-9})}} = 1080\ \text{SCFM}$$

$$Q_A = Q_{total} - Q_C = 2400 - 1080 = 1320\ \text{SCFM}$$

These values can be verified by calculating the pressure for each section using Equation 5.15. The pressures for each must be equal:

$$p_{tot\,A} = p_{stat} + k_f \cdot Q^2 = 7.80 + 625 \cdot 10^{-9} \cdot 1320^2 = 8.89\ \text{psi}$$

$$p_{tot\,C} = 7.80 + 926 \cdot 10^{-9} \cdot 1080^2 = 8.88\ \text{psi}$$

The two values agree well within the expected level of accuracy. The calculated flows are correct.

Problem 6.6

The 60 ft long pipe section from "D" to "A" in Figure 6.8 is subject to a worst case temperature change of 225 °F between being out of operation in winter and maximum discharge air temperature in the summer. The pipe is anchored at point "D".

(a) How much movement will be expected at point "A" if the pipe is stainless steel?
(b) How much movement will be expected at point "A" if the pipe is fiberglass?

Solution:

(a) Calculate the change in length using Equation 6.7 and Table 6.2:

$$\Delta L = L_0 \cdot \alpha \cdot \Delta T = 60 \cdot 12 \cdot 9.9 \cdot 10^{-6} \cdot 225 = 1.6 \text{ in.}$$

(b) $\Delta L = 60 \cdot 12 \cdot 14.0 \cdot 10^{-6} \cdot 225 = 2.3$ in.

CHAPTER 7: INSTRUMENTATION

Problem 7.1

An RAS flow meter provides a pulse output, each pulse representing 100 ml. The signal is connected to a PLC high-speed counter input. The input PLC displays the pulse frequency as 600 Hz.

(a) What is the flow rate in mgd?
(b) What is the conversion formula for use in the PLC program to convert from Hz to mgd?

Solution:

(a) $$600\text{ Hz} \cdot \frac{1\text{ pulse/s}}{\text{Hz}} \cdot \frac{100\text{ ml}}{\text{pulse}} \cdot \frac{1}{1000\text{ ml}} \cdot \frac{0.26417\text{ gal}}{1} \cdot 60\frac{\text{s}}{\text{min}} \cdot \frac{1\text{ mgd}}{694.44\text{ gal/min}}$$
$$= 1.37\text{ mgd}$$

(b) Calculate the conversion factor for this meter:

$$1\,\cancel{\text{Hz}} \cdot \frac{1\,\cancel{\text{pulse/s}}}{\cancel{\text{Hz}}} \cdot \frac{100\,\cancel{\text{ml}}}{\cancel{\text{pulse}}} \cdot \frac{\text{L}}{1000\,\cancel{\text{ml}}} \cdot \frac{0.26417\,\cancel{\text{gal}}}{\cancel{\text{l}}} \cdot 60\frac{\cancel{\text{s}}}{\cancel{\text{min}}} \cdot \frac{1\text{ mgd}}{694.44\,\cancel{\text{gal/min}}}$$

$$= 2.282 \cdot 10^{-3}\text{ mgd}$$

The conversion formula is

$$\text{Hz} \cdot 2.282 \cdot 10^{-3} \frac{\text{mgd}}{\text{Hz}} = \text{mgd}$$

Problem 7.2

A loop powered transmitter catalog provides the information shown in Figure 7.11. A designer plans to use a 24 VDC power supply and connect the transmitter output to a PLC input with an impedance of 250 Ω.

(a) Will this be within the transmitter load limits?
(b) The owner would like to insert an existing chart recorder with a 0–10 VDC input into the loop. Can this be accomplished using a dropping resistor?
(c) Name at least two design options to accommodate the owner's request.

Solution:

(a) By looking at the chart the maximum burden for 24 VDC loop power is 600 Ω. This is much higher than the 250 Ω input card impedance, and there should be no problems. This can be confirmed with the formula given:

$$R_{\text{max}} = \frac{(V_S - 12)}{0.020} = \frac{(24 - 12)}{0.020} = 600\,\Omega$$

(b) Use Ohm's law to determine the dropping resistor required:

$$R = \frac{V}{I} = \frac{10\,\text{V}}{0.020\,\text{A}} = 500\,\Omega$$

$$P = V \cdot A = 10 \cdot 0.020 = 0.20\,\text{W}$$

A 500 Ohm ¼ W precision resistor would be required. This will be acceptable if only the chart recorder is in the loop. If both the recorder and the PLC are connected in the loop the total burden is 750 Ω, which exceeds the allowable load for 24 VDC loop power.

(c) Any two of the following would be acceptable:
- Increase the power supply voltage to at least 27 VDC. The compatibility with any other devices powered from the same supply would have to be confirmed. If this option is selected, the chart recorder would have to be recalibrated to accommodate the resulting 2–10 VDC signal instead of the existing 0–10 VDC scale.
- Provide a separate power supply for this loop set to 27 VDC minimum.
- Use a signal conditioner in the loop to provide a separate 0–10 VDC signal to the recorder.

- Use a PLC analog output to provide a signal to the chart recorder and use program logic to echo the input value to the output.
- Many chart recorders have selectable input ranges. If the recorder has a 4–20 mA selection and the impedance is $\leq 350\,\Omega$, the loop signal can be used directly. If the recorder has a 0–5 VDC input selection then a 250 Ω dropping resistor could be used and the recorder recalibrated for 1–5 VDC.

For any of the above options, the existence of ground loops should be checked and eliminated if they exist.

Problem 7.3

A loop-powered digital indicator is to be installed on the control panel to display the value of the process variable measured by a loop transmitter with characteristics shown in Figure 7.11. The indicator data sheet shows a "Loop Voltage Drop" value of 5.6 V. The signal will also be connected in series to a PLC input with an impedance of 250 Ω. Will 24 VDC loop power be adequate?

Solution:

Calculate the equivalent load to drive the loop-powered indicator using Ohm's law:

$$R = \frac{V}{I} = \frac{5.6}{0.020} = 280\,\Omega$$

$$R = \frac{V}{I} = \frac{5.6}{0.004} = 1400\,\Omega$$

The equivalent load "seen" by the transmitter will range from 280 to 1400 Ω. Since the maximum load allowable for the transmitter is 1000 Ω the indicator is not compatible with the transmitter.

Problem 7.4

A submersible loop-powered pressure transmitter with a range of 0–15 psi and a 4–20 mA output is planned for use as a level transmitter. The transmitter has a specified accuracy of $\pm 0.5\%$ FS and will be connected to a 4 ½ digit indicator with a specified accuracy of $\pm 0.05\%$ FS, ± 1 count.

(a) What will be the range of level measured if the transmitter is used without scaling?

(b) What will be the output signal if the level above the transmitter is 12 ft of water?

(c) What will be the theoretical maximum error in the displayed value?

Solution:

(a) Rearrange Equation 4.12 to solve for the depth:

$$\text{Depth} = \frac{p_{\text{stat}}}{0.433} = \frac{15\,\cancel{\text{psi}}}{0.433\,\cancel{\text{psi}}/\text{ft H}_2\text{O}} = 34.64\ \text{ft H}_2\text{O}$$

(b) Use Equation 7.1 to calculate the output signal:

$$S_{\text{Out}} = \left[\frac{\text{PV} - \text{EU}_{\text{Min}}}{\text{EU}_{\text{Span}}} \cdot (S_{\text{Max}} - S_{\text{Min}})\right] + S_{\text{Min}}$$
$$= \left[\frac{12 - 0}{34.64} \cdot (20 - 4)\right] + 4 = 9.54\ \text{mA}$$

(c) The maximum error from the analog signal is obtained by multiplying the individual errors of each component:

$$\left[\left(1 + \frac{0.5}{100}\right) \cdot \left(1 + \frac{0.05}{100}\right) - 1\right] \cdot 100 = 0.55025\% \approx 0.55\%$$

The error in reading will be the analog signal error plus the error in the digits displayed. The meter will be scaled to display XX.XX, and one count will be ± 0.01 ft.

$$34.64 \cdot \frac{0.55}{100} = 0.19\ \text{ft}$$

$$0.19\ \text{ft} + 0.01\ \text{ft} = 0.20\ \text{ft}\ (= 2.4\ \text{in.})$$

Problem 7.5

A venturi meter is stamped with "6.22 in. H_2O 2500 CFM." As part of a fine-pore diffuser replacement, a new differential pressure transmitter is installed. The new transmitter has a range of 4–20 mA at 0–5 in. H_2O.

(a) The actual transmitter output is 11 mA. What is the air flow rate?
(b) What is the maximum air flow rate for the replacement transmitter?

Solution:

(a) Rearrange Equation 7.1 and solve for the differential pressure:

$$\text{PV} = \frac{S_{\text{out}} - S_{\text{min}}}{S_{\text{max}} - S_{\text{min}}} \cdot \text{EU}_{\text{span}} + \text{EU}_{\text{min}} = \frac{11 - 4}{20 - 4} \cdot 5 + 0 = 2.188\ \text{in H}_2\text{O}$$

Use Equation 7.9 and solve for the new air flow rate:

$$Q_2 = Q_1 \cdot \sqrt{\frac{\Delta p_{\text{m2}}}{\Delta p_{\text{m1}}}} = 2500\ \text{CFM} \cdot \sqrt{\frac{2.188\ \text{in. H}_2\text{O}}{6.22\ \text{in. H}_2\text{O}}} = 1480\ \text{CFM}$$

(b) Use Equation 7.9 to solve for the air flow rate at the new maximum output of 5 in. H_2O:

$$Q_{2\,\max} = 2500\,\text{CFM} \cdot \sqrt{\frac{5.0\,\text{in.}\,H_2O}{6.22\,\text{in.}\,H_2O}} = 2240\,\text{CFM}$$

Problem 7.6

Refer to Figure 7.12. Identify at least three problems with the wiring in this sketch.

Solution:

Any of three of the following would be acceptable:

- The polarity of the wiring between the power supply and the transmitter is reversed. The power supply "+" should connect to the "+" terminal on the transmitter. As wired the transmitter will not work, and unless it has reverse polarity protection it may be damaged.
- The field wiring is not shielded twisted pair and will be subject to EMI/RFI-induced signal noise.
- Wire nuts are used to splice the field wiring. This will increase signal noise and is not a secure connection.
- The chart recorder is scaled for 0–5 VDC input, but a 4–20 mA signal through the 250 Ω dropping resistor will create a 1–5 VDC signal. The chart will not be accurate unless it is recalibrated.
- The power supply, the PLC input, and the chart recorder are all grounded on the "–" terminal. This creates a ground loop. The transmitter output will not be sensed at the PLC input.

CHAPTER 8: FINAL CONTROL ELEMENTS

Problem 8.1

A butterfly valve is to be controlled by a directly controlled motor operator as shown in Figure 8.2 with 60 s per 90° travel. The controller output can provide timed outputs of ½ s duration.

(a) What will be the nominal increment in valve position with this system?

(b) Refer to the solution to Problem 6.4. If an 8 in. nominal valve is used, what would the approximate change in air flow be if the valve is opened one increment from an initial position of 25°?

(c) What would be the approximate change in air flow if the valve is opened one increment from an initial position of 40°?

Solution:

(a) Calculate the amount of travel for each ½ second increment:

$$0.5\frac{\text{s}}{\text{increment}}\cdot\frac{90°}{60\text{ s}}=\frac{0.75°}{\text{increment}}$$

(b) The change in flow could be calculated rigorously using Equation 6.19, but sufficient accuracy for the purpose of approximation can be achieved by linear interpolation of the tabulated results in Problem 6.4:

$$\frac{1200\text{ SCFM}-600\text{ SCFM}}{32°-22°}\cdot 0.75°=45\text{ SCFM}$$

(c) Calculate by interpolation:

$$\frac{2400\text{ SCFM}-1800\text{ SCFM}}{45°-39°}\cdot 0.75°=75\text{ SCFM}$$

Note that at the mid-range valve position, the change in flow rate is nearly twice the change that results from the same movement when the valve is near the closed position.

Problem 8.2

A multistage centrifugal blower is direct coupled to a 40 hp induction motor, 1.15 SF, operating at 3500 full load rpm and started across the line. The motor starter has Class 10 overload protection. The manufacturer's data sheet indicates the rotor WK^2 is 58 lb. ft.^{2}. The blower is rated at 850 SCFM and 6.5 psig discharge, with a nominal efficiency of 62%. Ambient conditions will range from 0 to 100 °F with a barometric pressure of 14.3 psia.

(a) Make a preliminary calculation to determine if the motor can accelerate the blower to operating speed without back pressure.
(b) Make a preliminary calculation to determine the ability to accelerate the blower if it started into a common header with other blowers.
(c) What design changes can be made to improve the starting capabilities of the blower?

Solution:

(a) For a preliminary estimate use an average motor torque. Calculate the motor full load torque using Equation 8.1 and assume the average acceleration torque is 166% of full load torque:

$$T_{\text{FL}}=\frac{P\cdot 5252}{N}=\frac{40\cdot 5252}{3500}=60\text{ lb ft}$$

$$T_{\text{Acc}}=T_{\text{FL}}\cdot\frac{166\%}{100}=100\text{ lb ft}$$

Use Equation 8.2 to estimate the time required to accelerate the load. If the blower is starting unloaded $T_l = 0$:

$$t = \frac{WK^2 \cdot \Delta N}{308 \cdot (T_m - T_l)} = \frac{58 \cdot 3500}{308 \cdot 100} = 7\ \text{s}$$

A Class 10 overload delays 10 s prior to tripping. The blower can be started if it is unloaded.

(b) Calculate the nominal blower power and torque and then determine the net torque available for acceleration. Assume inlet losses = 0.2 psi, and use Equation 4.6 to calculate the power:

$$P_b = \frac{(Q_s/60) \cdot \gamma \cdot R \cdot T_i}{550 \cdot (k - 1/k) \cdot \eta_b} \cdot \left[\left(\frac{p_d}{p_i} \right)^{k-1/k} - 1 \right]$$

$$P_b = \frac{(850/60) \cdot 0.075 \cdot 53.3 \cdot (460 + 100)}{550 \cdot 0.283 \cdot 0.62} \cdot \left[\left(\frac{14.3 + 6.5}{14.3 - 0.2} \right)^{0.283} - 1 \right] = 38\ \text{hp}$$

Calculate torque at full load:

$$T_l = \frac{38 \cdot 5252}{3500} = 57\ \text{lb ft}$$

The blower torque requirement will increase gradually from zero to maximum as the discharge air flow increases with speed. A conservative assumption is that the blower load torque will increase linearly and uniformly from zero torque at zero speed to max torque at max speed. The average load torque would therefore be 50% of the max torque. Recalculate the acceleration time:

$$t = \frac{WK^2 \cdot \Delta N}{308 \cdot (T_m - T_l)} = \frac{58 \cdot 3500}{308 \cdot (100 - 0.5 \cdot 57)} = 9\ \text{s}$$

The application is marginal if the blower must be started against a load.

It is possible to run a more exact analysis using integration or small time increments with actual motor torque and blower load torque curves, but based on the assumptions required this is probably not justified.

(c) Design options include:

- Use a class 20 overload.
- Provide a RVSS instead of across the line starting.
- Use a VFD instead of across the line starting. (This may have energy conservation benefits as well.)
- Provide a blow-off valve to eliminate starting the blower against a load.

Problem 8.3

Refer to Problem 8.2. The plant has an 80 hp portable generator that they wish to use as standby power for running one blower. The generator specs indicate an output of 38 kW at 480 VAC, 53 A continuous, 48 kVA. The blower motor is marked NEMA Code Letter F and has a nominal efficiency of 93% and a full load power factor of 89%. Will the generator be able to start and run the blower?

Solution:

Check the motor kW:

$$\mathrm{kW} = \frac{\mathrm{hp} \cdot 0.746}{\eta_{\mathrm{m}}} = \frac{40 \cdot 0.746}{0.93} = 32\,\mathrm{kW}$$

That is within the generator rating. Check motor full load amps by rearranging Equation 1.6:

$$I = \frac{P \cdot 1000}{V \cdot \sqrt{3} \cdot \mathrm{PF}} = \frac{32 \cdot 1000}{480 \cdot 1.73 \cdot 0.89} = 43\ \mathrm{A}$$

That is within the generator rating. Check nominal motor kVA:

$$\mathrm{kVA} = \frac{480\ \mathrm{V} \cdot 43\ \mathrm{A} \cdot \sqrt{3}}{1000} = 36\,\mathrm{kVA}$$

That is within the generator rating. Check starting kVA using the worst case value from Table 8.1:

$$40\,\mathrm{hp} \cdot 5.59\,\mathrm{kVA/hp} = 224\,\mathrm{kVA}$$

This is significantly higher than the generator kVA rating. The generator would stall out when trying to start the blower. This is not a good application if the across the line starting is maintained. Either the generator supplier should be consulted for an appropriately sized unit adequate for starting this blower or a VFD or RVSS should be used to limit the starting kVA to a level that the generator could handle. Note that VFD harmonics can impose additional loads on generators, and the supplier should be consulted before finalizing the design.

Problem 8.4

Refer to Problem 8.2. The blower motor has NEMA Frame Size 324TS. Observation of the blower operation shows that the actual discharge pressure is only 5.5 psig. The operator has proposed running the blower at higher than nominal speed to achieve additional capacity and eliminate starting a second blower. The blower first critical speed is 5800 rpm. Perform preliminary calculations to determine the feasibility of this design.

Solution:

Determine the limiting speed of the blower based on four criteria: Critical speed (vibration), VFD current capacity, motor power, and motor speed limit.

Standard practice is to avoid operating a multistage centrifugal blower faster than 90% of the first critical speed:

$$N_{\text{max Vibr}} = 0.90 \cdot 5800\,\text{rpm} = 5200\,\text{rpm}$$

Determine the corresponding VFD output frequency by rearranging Equation 8.4:

$$f = \frac{N \cdot n_{\text{p}}}{120} = \frac{5200 \cdot 2}{120} = 87\,\text{Hz}$$

Most VFDs will provide output in excess of 90 Hz, so this is not a problem. The typical drive will provide constant current and voltage above 60 Hz, effectively limiting output to the rated power. Rearrange Equation 4.6 and solve for the maximum flow rate achievable without exceeding the motor rated power:

$$Q_{\text{S max}} = 60 \cdot \left[\frac{P_{\text{b}} \cdot 550 \cdot (k - 1/k) \cdot \eta_{\text{b}}}{\gamma \cdot R \cdot T_{\text{i}}} \cdot \left(\frac{1}{(p_{\text{d}}/p_{\text{i}})^{k-1/k} - 1}\right)\right]$$

$$Q_{\text{S max}} = 60 \cdot \left[\frac{40 \cdot 550 \cdot 0.283 \cdot 0.62}{0.075 \cdot 53.3 \cdot (460 + 100)} \cdot \left(\frac{1}{((5.5 + 14.3)/(14.3 - 0.2))^{0.283} - 1}\right)\right]$$
$$= 1020\,\text{SCFM}$$

Use rearranged Equation 5.40 to determine the blower speed required for this flow rate:

$$N_{\text{a}} = \frac{Q_{\text{a}}}{Q_{\text{c}}} \cdot N_{\text{c}} = \frac{1020}{850} \cdot 3500 = 4200\,\text{rpm}$$

Determine the corresponding VFD output frequency at max flow rate by rearranging Equation 8.4:

$$f = \frac{N \cdot n_{\text{p}}}{120} = \frac{4200 \cdot 2}{120} = 70\,\text{Hz}$$

The nominal motor speed limit for a two-pole motor from Table 8.4 is 4500 rpm.

The limiting factor will be the motor power capacity, 4200 rpm at 70 Hz. The design is reasonable. More detailed calculations using the actual blower performance curve should be provided to verify the design and determine the energy savings.

CHAPTER 9: CONTROL LOOPS AND ALGORITHMS

Problem 9.1

Refer to Problems 3.3 and 4.2. The owner is interested in using a DO transmitter with Hi/Lo alarm contacts to operate the aerator intermittently. The measured power draw at the aerator is 15 kW. The aerator motor manufacturer recommends 6 starts per hour maximum. The desired average DO concentration is 2.0 ppm, and the proposed on/off setpoints are 1.0 ppm and 3.0 ppm. The installed cost of the DO transmitter is $3500, a 25 hp RVSS costs $1800 installed, and electric energy costs $0.10/kWh. Is this a feasible ECM?

Solution:

Use average conditions to estimate the feasibility. Use Equation 4.8 and the parameters calculated in Problem 4.2 to calculate the field saturation concentration:

$$C^*_{\infty f} = \tau \cdot \beta \cdot \Omega \cdot C^*_{\infty 20} = 1.11 \cdot 0.95 \cdot 0.99 \cdot 9.09 = 9.5$$

Per Problem 3.3 the current average DO concentration is 3.5 ppm. Rearrange Equation 4.11 to estimate the energy savings:

$$\mathrm{OTR}_{2.0} = \mathrm{OTR_a} \cdot \frac{C^*_{\infty f} - 2.0}{C^*_{\infty f} - C_a} - 2.0 \frac{\mathrm{lb\,O_2}}{\mathrm{hp\ h}} \frac{9.5 - 2.0}{9.5 - 3.5} - 2.5 \frac{\mathrm{lb\,O_2}}{\mathrm{hp \cdot h}}$$

$$\frac{2.0}{2.5} \cdot 100 = 80\%$$

The new power requirement would be 80% of the existing power. Verify the payback is reasonable:

$$\mathrm{S} = (\text{Existing kW} - \text{Projected kW}) \cdot \mathrm{Rate} \frac{\$}{\mathrm{kWh}} \cdot 8760 \frac{\mathrm{h}}{\mathrm{year}}$$

$$\mathrm{S} = (15 - 0.80 \cdot 15) \cdot 0.10 \cdot 8760 = \$2600/\mathrm{year}$$

$$P = \frac{C}{S} = \frac{\$3500}{\$2600/\mathrm{year}} = 1.3\ \mathrm{years}$$

The payback is quite good. Check the motor duty cycle. The OUR from Problem 3.3 is 20.5 mg/l/h. Rearrange Equation 3.3 to estimate the time required for the DO to drop from 3.0 to 1.0 ppm:

$$t_2 - t_1 = \frac{C_1 - C_2}{\mathrm{OUR}} = \frac{3.0 - 1.0}{20.5} = 0.1\ \mathrm{h} = 6\ \mathrm{min}$$

If the motor duty cycle is 80% on, 20% off:

$$\text{Cycle time} = \frac{6\ \mathrm{min}}{0.2} = 30\ \mathrm{min}$$

The motor will be on for 24 min, off for 6 min, and will have 2 starts per hour. This is well within the manufacturer's recommendation. As a precaution a RVSS could be installed if the payback isn't excessive:

$$P = \frac{C}{S} = \frac{\$3500 + \$1800}{\$2600/\text{year}} = 2.0 \text{ years}$$

Using a simple on/off DO control would be cost-effective.

Problem 9.2

Refer to Figure 6.8, Problems 6.4 and 8.1. Each aeration grid is controlled as a separate control zone. The DO control logic is based on proportional speed floating control with bias and bandwidth. The DO tolerance for all grids is ± 0.15 ppm and the bias is 60%. The current process data is

Grid	Actual DO, ppm	Set DO, ppm	Demand ACFM	DO Gain, CFM/ppm
C	2.10	2.0	1080	125
A	1.85	1.5	1220	240
F	1.60	2.0	920	125
E	2.40	1.5	1300	200

The current blower air flow demand is 4600 SCFM.

(a) Calculate the error and change in demand air flow for each grid.
(b) Calculate the total blower air flow change and the new blower demand.

Solution:

(a) Use Equation 9.1 to calculate the errors:

$$e_C = \text{SP}_C - \text{PV}_C = 2.0 - 2.1 = -0.1 \text{ ppm high}$$

$$e_A = 1.5 - 1.85 = -0.35 \text{ ppm high}$$

$$e_F = 2.0 - 1.60 = 0.40 \text{ ppm low}$$

$$e_E = 1.5 - 2.40 = -0.90 \text{ ppm high}$$

Use Equation 9.9 to calculate the air flow change for each basin and Equation 9.11 to calculate the new demand for each basin.

The error for zone "C" is less than the $\pm$ DO tolerance, and therefore there is no air flow demand change for this zone.

$$\Delta Q_A = G \cdot (\text{SP} - \text{PV}) = G_A \cdot e_A = 240 \text{ CFM/ppm} - 0.35 \text{ ppm} = -84 \text{ CFM}$$

$$Q_{\mathrm{T\,A}\,n} = Q_{\mathrm{T\,A}\,n-1} \pm \Delta Q_{\mathrm{T}} = 1220\ \mathrm{CFM} - 84\ \mathrm{CFM} = 1136\ \mathrm{CFM}$$

$$\Delta Q_{\mathrm{F}} = G \cdot (\mathrm{SP} - \mathrm{PV}) = G_{\mathrm{F}} \cdot e_{\mathrm{F}} = 125\ \mathrm{CFM/ppm} \cdot 0.40\ \mathrm{ppm} = 50\ \mathrm{CFM}$$

$$Q_{\mathrm{T\,F}\,n} = Q_{\mathrm{T\,F}\,n-1} \pm \Delta Q_{\mathrm{T}} = 920\ \mathrm{CFM} + 50\ \mathrm{CFM} = 970\ \mathrm{CFM}$$

$$\Delta Q_{\mathrm{E}} = G \cdot (\mathrm{SP} - \mathrm{PV}) = G_{\mathrm{E}} \cdot e_{\mathrm{E}} = 200\ \mathrm{CFM/ppm} - 0.90\ \mathrm{ppm} = -180\ \mathrm{CFM}$$

$$Q_{\mathrm{T\,E}\,n} = Q_{\mathrm{T\,E}\,n-1} \pm \Delta Q_{\mathrm{T}} = 1300\ \mathrm{CFM} - 180\ \mathrm{CFM} = 1120\ \mathrm{CFM}$$

(b) Calculate the total change in blower air flow from Equation 9.12:

$$\Delta Q_{\mathrm{T}n} = 0 - 84 + 50 - 180 = -214\ \mathrm{CFM}$$

$$Q_{\mathrm{B}\,n} = Q_{\mathrm{B}\,n-1} + \Delta Q_{\mathrm{T\,n}} = 4600\ \mathrm{SCFM} - 214\ \mathrm{CFM} = 4386\ \mathrm{SCFM}$$

Problem 9.3

Refer to Problem 9.2. The actual air flows are

Grid	Actual ACFM	BFV % Open
C	940	55
A	1115	70
F	875	60
E	1205	48

Calculate the actual air flow control setpoint for each zone, and determine the direction of valve travel.

Solution:

Calculate the actual flows and the proportions of flow for each zone using Equation 9.14:

$$\sum Q_{\mathrm{T\,Demand}} = 1080\ \mathrm{CFM} + 1136\ \mathrm{CFM} + 970\ \mathrm{CFM} + 1120\ \mathrm{CFM} = 4306\ \mathrm{CFM}$$

$$\sum Q_{\mathrm{T\,Actual}} = 940\ \mathrm{CFM} + 1115\ \mathrm{CFM} + 875\ \mathrm{CFM} + 1205\ \mathrm{CFM} = 4135\ \mathrm{CFM}$$

$$Q_{\mathrm{T\,SP\,C}} = \frac{Q_{\mathrm{T\,Demand\,C}}}{\sum Q_{\mathrm{T\,Demand}}} \cdot \sum Q_{\mathrm{T\,Actual}} = \frac{1080}{4306} \cdot 4135 = 1037\ \mathrm{CFM}$$

$1037 > 940$, so this valve will open

$$Q_{\mathrm{T\,SP\,A}} = \frac{1136}{4306} \cdot 4135 = 1091\ \mathrm{CFM}$$

1091 < 1115, so this valve would have to close to meet setpoint. However, this is the most open valve, so the close permissive will not be set and this valve will maintain current position.

$$Q_{\mathrm{T\ SP\ F}} = \frac{970}{4306} \cdot 4135 = 931\ \mathrm{CFM}$$

931 > 875 so this valve will open

$$Q_{\mathrm{T\ SP\ E}} = \frac{1120}{4306} \cdot 4135 = 1076\ \mathrm{CFM}$$

1076 < 1205 so this valve will close
Summarize the results in a table:

Grid	Actual ACFM	BFV % Open	Actual CFM Setpoint	Travel Direction
C	940	55	1037	Open
A	1115	70	1091	None
F	875	60	931	Open
E	1205	48	1076	Close

Problem 9.4

Refer to Problem 9.3. After completion of the adjustments indicated the actual blower air flow is 4414 SCFM, per Problem 9.2 the total blower demand is 4386 SCFM, and the zone actual air flows and demands are

Grid	Actual ACFM	CFM Demand
C	1010	1080
A	1085	1136
F	895	970
E	1040	1120

If the damping factor is 50%, determine the new blower air flow demand.

Solution:

Calculate the total actual and demand air flows and use Equation 9.13 to calculate the new blower demand:

$$\sum Q_{\mathrm{T\ Demand}} = 1080 + 1136 + 970 + 1120 = 4306\ \mathrm{CFM}$$

$$\sum Q_{\mathrm{T\ Actual}} = 1010 + 1085 + 895 + 1040 = 4030\ \mathrm{CFM}$$

$$Q_{\text{B Corrected}} = Q_{\text{B Current}} + F \cdot \left(\sum Q_{\text{T Demand}} - \sum Q_{\text{T Actual}}\right)$$

$$Q_{\text{B Corrected}} = 4386 + 0.50 \cdot (4306 - 4030) = 4524 \text{ SCFM}$$

Because the demand is higher than the actual flows at the aeration tanks the blower output must increase.

Problem 9.5

Two existing blowers have calibrated ammeters. The first is marked with 4000 CFM at 398 A and 2000 CFM at 273 A. The second is marked with 7500 CFM at 750 A and 3800 CFM at 480 A. The blowers are to be controlled by inlet throttling using a PLC. What is the equation for flow versus amps for each blower?

Solution:

Use Equations 9.17, 9.18, and 9.19 to develop the linear equations to determine air flow from motor current:

$$m_{\text{MS 1}} = \frac{Q_{\max} - Q_{\min}}{I_{\max} - I_{\min}} = \frac{4000 - 2000}{398 - 273} = 16.0 \text{ CFM/A}$$

$$b_{\text{MS 1}} = I_{\min} - \frac{Q_{\min}}{m_{\text{MS}}} = 273 - \frac{2000}{16.0} = 148 \text{ A}$$

$$Q_{\text{iMS 1}} = m_{\text{MS}} \cdot (I_{\text{Act}} - b_{\text{MS}}) = 16.0 \cdot (I_{\text{Act}} - 148)$$

$$m_{\text{MS 2}} = \frac{7500 - 3800}{750 - 480} = 13.7 \frac{\text{CFM}}{\text{A}}$$

$$b_{\text{MS 2}} = 480 - \frac{3800}{13.7} = 203 \text{ A}$$

$$Q_{\text{iMS 1}} = 13.7 \cdot (I_{\text{Act}} - 203)$$

Problem 9.6

The two blowers in Problem 9.5 are operating in parallel mode. The target system air flow is 9500 CFM. What is the target flow for each blower?

Solution:

Use Equation 9.25 to determine the required air flow target for each blower to operate at the same proportion of their range:

$$Q_{\text{1 set}} = \left[\left(\frac{Q_{\text{Total set}} - \sum Q_{\min}}{\sum Q_{\max} - \sum Q_{\min}}\right) \cdot (Q_{\text{1 max}} - Q_{\text{1 min}})\right] + Q_{\text{1 min}}$$

$$Q_{1\ \mathrm{set}} = \left[\left(\frac{9500 - (2000 + 3800)}{(4000 + 7500) - (2000 + 3800)}\right) \cdot (4000 - 2000)\right] + 2000 = 3298$$

$$Q_{1\ \mathrm{set}} = \left[\left(\frac{9500 - (2000 + 3800)}{(4000 + 7500) - (2000 + 3800)}\right) \cdot (7500 - 3800)\right] + 3800 = 6202$$

Check:

$3298 + 6202 = 9500$ OK

Problem 9.7

Refer to Problems 8.1 and 9.3. The flow control valve for zone "C" is controlled by timed contact closure with a 66% bias. The air flow response is 90 CFM per second of travel. The timer for valve opening is set to ¾ s. The control air flow setpoint is 940 SCFM ± 20 SCFM and the actual air flow is 1054 SCFM.

(a) Can the control achieve the required air flow, and if so how many iterations of valve movement will be required to do so?
(b) If the response time delay is 10 s how long does it take for the air flow to be within tolerance?

Solution:

(a) First estimate the increase in flow on each valve opening increment and the decrease in flow on each valve closing increment:

$$\Delta \mathrm{SCFM_O} = 90\frac{\mathrm{SCFM}}{\mathrm{s}} \cdot 0.75\frac{\mathrm{s}}{\mathrm{increment}} = 68\frac{\mathrm{SCFM}}{\mathrm{increment}}$$

$$\Delta \mathrm{SCFM_C} = 90\frac{\mathrm{SCFM}}{\mathrm{s}} \cdot 0.66 \cdot 0.75\frac{\mathrm{s}}{\mathrm{increment}} = 45\frac{\mathrm{SCFM}}{\mathrm{increment}}$$

Then calculate the minimum and maximum air flows that will be within the control tolerance:

$$\mathrm{SCFM_{max}} = 940 + 20 = 960\ \mathrm{SCFM}$$

$$\mathrm{SCFM_{max}} = 940 - 20 = 920\ \mathrm{SCFM}$$

Calculate the flow after the first closure:

$\mathrm{SCFM_1} = 1054 - 45 = 1009$ SCFM. This is too high.

$\mathrm{SCFM_2} = 1009 - 45 = 964$ SCFM. This is too high.

$\mathrm{SCFM_3} = 964 - 45 = 919$ SCFM. This is too low.

$\mathrm{SCFM_4} = 919 + 68 = 987$ SCFM. This is too high.

$\mathrm{SCFM_5} = 987 - 45 = 942$ SCFM.

This is within the tolerance and the valve will maintain this position until changes in the other valves in the system disturb the air flow or a new flow setpoint is calculated.

(b) There are a total of four time delays before the valve reaches the final position after the fifth movement. The total time for adjustment is 40 s. Refer to Figure A.5 for an illustration of the flow response.

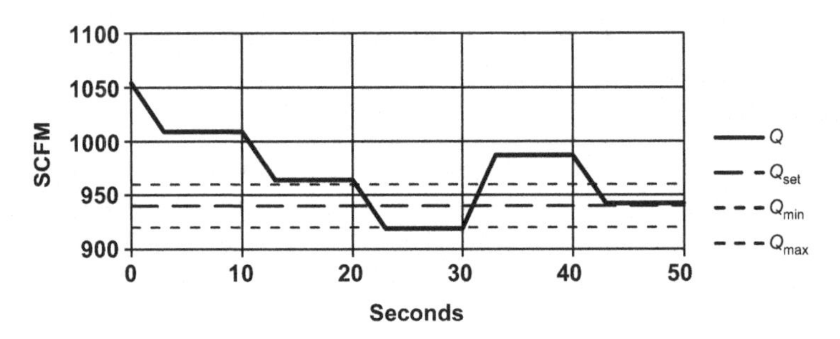

FIGURE A.5 *Problem 9.7 flow control valve response*

Problem 9.8

An aeration tank with a volume of 69,000 ft^3 is equipped with an off-gas analysis system. The measured AOTE is 12.9% and the OUR is 36.6 mg/(l h). After a belt press is started the measured OUR increases to 45.8 mg/(l h). How much must the air flow increase to maintain process performance?

Solution:

Calculate the change in process demand:

$$\Delta \text{OUR} = 45.8 - 36.6 = 9.2\,\text{mg}/(\text{l h})$$

Use Equation 3.4 to determine the change in air flow required to meet the change in process demand:

$$Q_\text{s} = \frac{\text{OUR} \cdot V_\text{t}}{16{,}700 \cdot \text{AOTE}} = \frac{9.2 \cdot 69{,}000}{16{,}700 \cdot (12.9/100)} = 295\ \text{CFM increase}$$

CHAPTER 10: CONTROL COMPONENTS

Problem 10.1

A discharge pressure transmitter with a range of 0–15 psig and an accuracy of ±0.75% FS will be connected to a PLC analog input. The pressure will be displayed on a touch screen HMI with two decimal places. There are two PLC analog input cards being considered. One has 12-bit resolution and an accuracy of ±0.1% FS. The other has 15-bit resolution and an accuracy of ±0.35% FS. Assuming the HMI truncates the calculated value, what will be the possible range of values in the displayed pressure for each of the two I/O cards when the nominal pressure is 7.50 psig?

Solution:

First determine the maximum and minimum value in the pressure transmitter signal:

$$\pm\frac{0.75}{100}\cdot 15.00 = \pm 0.1125\ \text{psig}$$

Calculate the maximum and minimum transmitted values at 7.50 psig:

$$7.50 + 0.1125 = 7.6125\ \text{psig}$$

$$7.50 - 0.1125 = 7.3875\ \text{psig}$$

Calculate the maximum and minimum values for the first card:

$$2^{12} = 4096$$

$$7.6125 + \frac{0.1}{100}\cdot 15.0 + \frac{1}{4096}\cdot 15 = 7.6312$$

$$7.3875 - \frac{0.1}{100}\cdot 15.0 - \frac{1}{4096}\cdot 15 = 7.3688$$

The first card will result in a display between 7.63 and 7.36 psig. Calculate the maximum and minimum values for the second card:

$$2^{15} = 32{,}768$$

$$7.6125 + \frac{0.35}{100}\cdot 15.0 + \frac{1}{32{,}768}\cdot 15 = 7.66$$

$$7.3875 - \frac{0.35}{100}\cdot 15.0 - \frac{1}{32{,}768}\cdot 15 = 7.33$$

The second card will result in a display between 7.66 and 7.33 psig. The 12-bit card will result in a more accurate displayed value. The error in the transmitter is much greater than the error from either I/O card.

Problem 10.2

Create a ladder diagram program showing a typical implementation of an input debounce timer. Use any PLC software or word processor to create the logic.

Solution:

See Figure A.6 for an example.

This is an example of standard de-bounce logic for a discrete input.
A 0.25 second time delay is used to eliminate losing the status if the physical contact on the motor starter bounces, alternately making and breaking continuity.

The motor starter auxiliary contact at the MCC must close and remain closed for the set delay time. The timer will then time out and turn on (pull in) an internal logic coil. Contacts asociated with that coil are then used by the rest of the program logic instead of using the input directly. In this brand of processor, the timer will stay active and the "Done" contact will remain closed, as long as the contact initiating the timer remains "on". When input to the timer opens the timer will reset, the "Done" contact will open, and the status will change to indicate the motor is not running.

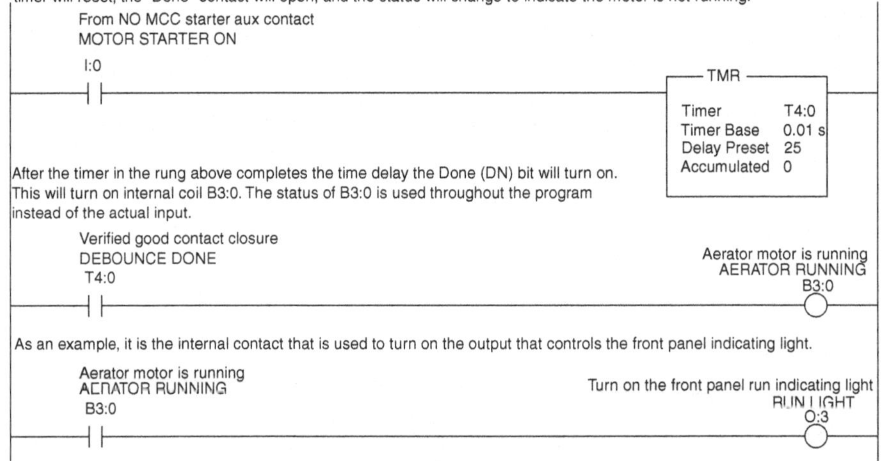

FIGURE A.6 *Problem 10.2 debounce timer*

Problem 10.3

A designer is selecting a PLC for a small aeration control project and is considering two different brands. The first brand, which the designer has programmed in the past, has a hardware cost of \$2500 for the PLC and HMI and an estimated programming time of 32 h. The designer has no experience with the second brand. It has a hardware cost of \$1800 for the PLC and HMI and an estimated programming time of 48 h, which includes 16 h to become familiar with the programming software. Assuming the owner has no preference, the programming software for both brands is available, and programming is billed at \$75/h, which PLC is the most cost-effective and why?

Solution:

Calculate the total cost for both projects, including hardware and programming time:

$$\text{Cost}_1 = \$2{,}500 + 32 \cdot 75 = \$4{,}900$$

$$\text{Cost}_2 = \$1{,}800 + 48 \cdot 75 = \$5{,}400$$

The more expensive PLC results in the most cost effective system because lower programming cost offsets the more expensive hardware.

Problem 10.4

A programmer is developing a function block for a floating control algorithm with bandwidth. It will be used for flow control with BFVs having 4–20 mA positioners and 4–20 mA position feedback signals. Identify the inputs to and outputs from the function block.

Solution:

Any or all of the following could be included. Additional data not shown may also be included based on the judgment of the reader.

Inputs:

- Flow Setpoint
- Flow Tolerance
- Open Position Increment
- Bias
- Response Delay
- Actual Flow
- Current Position Command
- Current Position Feedback

Outputs:

- At Max Position Status
- At Min Position Status
- Flow Error
- New Position Command

CHAPTER 11: DOCUMENTATION

Problem 11.1

Refer to Figure 6.1. A pressure transmitter is being installed in the main aeration header immediately before it exits the blower room.

(a) Create a specification for the transmitter for use in the contract documents.
(b) Create a specification for the transmitter for use by a purchasing agent for procurement.

Solution:

(a) As a minimum the specification should include the following:

- The range of pressure to be sensed, for example, 0–15 psig.
- The minimum acceptable accuracy, for example, ±0.5%.
- The environmental rating required, for example, NEMA 4X.
- Transmitter output, for example, 4–20 mA.
- At least one potential manufacturer.

(b) The procurement specification or Bill of Material data should include:

- The required manufacturer and vendor.
- The manufacturer's model or part number designation.
- 0–15 psig pressure range.
- Identify calibration certificate requirements if needed.
- Tag or nameplate requirements.
- The model's defined accuracy.
- NEMA 4X environmental rating required.
- Transmitter output as 4–20 mA and minimum load capabilities.
- Power requirement such as 24 VDC loop powered.
- Process connection type and size.
- Temperature ratings for process connection and electronics/ambient.
- Conduit or other electrical connection designation.

Problem 11.2

Refer to Problem 9.1. The decision has been made to use a micro-PLC to control the timing of the aeration on and off cycles. Create a preliminary point list for this system, assuming only one tank.

Solution:

As a minimum, the following information should be included:

Type	Description	Tag#	Panel	PLC Channel	Signal
DI	DO high	ASH-1-1	PLC-1	I:0.0	24 VDC
DI	DO low	ASL-1-1	PLC-1	I:0.1	24 VDC
DI	Aerator running	YS-1-1	PLC-1	I:0.2	24 VDC
DO	Start aerator	YS-1-2	PLC-1	O:0.0	Dry contact
DO	Common alarm	YS-1-3	PLC-1	O:0.1	Dry contact

Problem 11.3

Create a loop description for the mechanical aerator DO control in Problem 9.1

Solution:

Loop 1-1 Aerator DO Control:

Inputs:

High DO Contact from AIT-1 (ASH-1-1)
Low DO Contact from AIT-1 (ASL-1-1)
Motor Run Contact from MCC Starter (YS-1-1)

Outputs:

Motor Start dry contact (YS-1-2) to MCC starter
Common Alarm Contact (YS-1-3) for SCADA input

Operator Adjustments:

High DO limit and Low DO limit setpoint, both adjustable at the DO transmitter AIT-1-1

Minimum aerator run time and minimum aerator off time, adjustable from local HMI

Alarm delay time, adjustable from local HMI

Loop Description: The aerator run contact closure will initiate the minimum run time delay. High DO contact closure and completion of the minimum run time delay will cause the aerator to stop. The aerator run contact opening will initiate the minimum off time delay and the maximum off time delay. Completion of the minimum off time delay plus either Low DO contact closure OR completion of the maximum off time delay will cause the aerator to start.

Alarms: If the aerator start contact closes and a run confirmation is not received within the set time delay an alarm will initiate and be displayed on the HMI. If the aerator start contact opens and the run confirmation is not lost within the set time delay an alarm will initiate and be displayed on the HMI. On the occurrence of either alarm the alarm output will close indicating an aerator failure to SCADA.

Problem 11.4

Examine the ladder diagram in Figure 11.6.

(a) Describe the actions when the "Open" pushbutton is pressed with the circuit in the state shown.

(b) Describe the actions when the "Open" pushbutton is pressed if the selector switch in rung 1030 is moved to the "Man" position.

Solution:

(a) With the circuit as shown there is no continuity between the power rail (connected to terminal 100) and the Open pushbutton. Pressing it will result in no actions.

(b) If the selector switch is moved to Man there will be voltage at both sets of contacts on the Open pushbutton. Pressing the pushbutton will open the lower contacts and close the upper contacts. This will provide continuity through the Open and Close switch contacts to terminal 1037, through the upper limit switch, and to the valve Open motor coil. This will cause the valve to move open as long as the switch is pressed.

CHAPTER 12: COMMISSIONING

Problem 12.1

Refer to Figure 7.12.

(a) Identify at least three steps to use during troubleshooting of this circuit and the phase of the commissioning process when each step of the troubleshooting would be performed.

(b) Identify at least three corrective measures that could be implemented to correct problems with this circuit.

Solution:

(a)
- Identify the incorrect type of field wiring during inspection.
- Identify improper use of wire nuts during inspection.
- Identify incorrect polarity at transmitter during inspection or during testing if no signal is present.
- Trace the wiring in the panel and identify the ground loops during shop testing.
- Observe incorrect scaling of chart recorder during shop testing.

(b)
- Correct the polarity reversal at either the transmitter or the control panel terminal strip.
- Replace wire nuts with a terminal strip or crimp connectors.
- Replace the field wiring with shielded twisted pair and ground the shield at the control panel terminal strip.
- Eliminate the ground loops by removing the ground wires and letting the chart recorder float or install a signal isolator to eliminate the ground loop.
- Rescale the chart recorder to 4–20 mA input, install a dropping resistor and rescale to 1–5 VDC input, or use a signal conditioner that will provide 0–5 VDC output from a 4–20 mA input.

Problem 12.2

A system is in the test stage of commissioning and some rough tuning is required (refer to Figure 12.6). Identify at least two possible tuning changes would the performance trend indicate? Provide an explanation for each.

Solution:

- Reduce the response time delay. The figure indicates that the DO concentration is stabilizing in about 3 min after a change, so the response time delay should be decreased to around 2.5–3 min.
- Reduce the gain to about half of its current value. The DO is consistently overshooting the setpoint by a considerable amount and exhibiting hunting, indicating severe overcorrection.

- Add bias to the tuning. The trends show an equal DO change for both increasing and decreasing. By increasing the bias the DO will have a tendency to approach the setpoint by reducing the overshoot on decreasing DO concentration.

Problem 12.3

A facility has requested training for two separate groups of plant staff for a new system with high-efficiency blowers, new fine-pore diffusers, and automatic DO and aeration control. The two groups consist of process operations and maintenance. Identify at least two areas where both groups should receive the same training, and at least two areas for each group where the training should be different.

Solution:

Areas of commonality could include:

- Safety considerations and safe practices
- Basic goals and objectives
- Theory of operation
- Required response to equipment failures

Areas specific to Process Operations could include:

- Relationships between controlled parameters and process performance
- Tuning procedures and guidelines
- HMI displays and trending capabilities
- Alternate modes of operation: schedule, manual, and so on.

Areas specific to Maintenance could include:

- DO transmitter and probe calibration and maintenance
- Preventive maintenance procedures for all equipment
- Blower performance limits
- Blower filter replacement and lubrication procedures
- Calibration procedures for all instrumentation
- Equipment protection alarm settings and adjustments

CHAPTER 13: SUMMARY

No problems

Appendix B

List of Equations and Variables

Equation 1.1

$$E = F \cdot d$$

where

E = energy, ft lb_f
F = force, lb_f
d = distance, ft

Equation 1.2

$$\text{horsepower} = \frac{550 \cdot \text{foot} \cdot \text{lb}}{\text{s}}$$

where

s = time, seconds

Aeration Control System Design: A Practical Guide to Energy and Process Optimization, First Edition. Thomas E. Jenkins.

Equation 1.3

$$P_w = \frac{Q_w \cdot h \cdot SG}{3960}$$

where

P_w = water power, hp
h = pump head, ft
Q_w = flow rate, gpm
SG = specific gravity, water = 1.0, dimensionless

Equation 1.4

$$P = V \cdot I$$

where

P = power, W
V = voltage or electromotive force (emf), V
I – current, A

Equation 1.5 (Single-phase AC only)

$$P = V \cdot I \cdot PF$$

Equation 1.6 (Three-phase AC only)

$$P = V \cdot I \cdot \sqrt{3} \cdot PF$$

where

P = power, W
V = voltage or electromotive force (emf), V
I = current, A
PF = power factor, decimal

Equation 1.7

$$P = \frac{\text{bhp} \cdot 0.746}{\eta_m \cdot \eta_{vfd}}$$

where

P = power, kW
bhp = shaft power, hp

η_{m} = efficiency of motor, decimal
η_{vfd} = efficiency of variable frequency drive if used, decimal

Equation 1.8

$$S = V \cdot I \cdot \sqrt{3} = \frac{P}{\mathrm{PF}}$$

where

S = apparent power, VA

Equation 1.9

$$E = P \cdot t$$

where

E = energy, Wh or kWh
P = power, W or kW
t = time, h

Equation 1.10

$$\eta_{\mathrm{ww}} = \frac{(Q \cdot h/3960) \cdot 0.746}{P_{\mathrm{e}}} \cdot 100$$

where

η_{ww} = system wire to water efficiency, %
Q = flowrate, gpm
h = pump head, ft
P_{e} = total electrical input power to the pump system, kW

Equation 1.11

$$P = \frac{C}{S}$$

where

P = simple payback period, the time required for the savings to offset the investment, years
C = net total initial cost for equipment and installation
S = net annual savings (energy and maintenance), \$/year

Equation 1.12

$$\mathrm{NPW} = (\mathrm{PWF} \cdot S) - C$$

Equation 1.13

$$\mathrm{PWF} = \frac{(1+r)^n - 1}{r \cdot (1+r)^n}$$

Equation 1.14

$$r = \frac{R - I}{1 + I}$$

where

- NPW = net present worth of investment, \$, NPW > 0 means the investment is economically justified
- S = annual savings, \$
- C = capital cost for equipment and installation \$,
- n = total evaluation period, years
- I = annual inflation rate, decimal
- R = annual discount rate, decimal, typically the interest rate or rate of return on alternate investments
- r = effective annual discount rate, decimal
- PWF = present worth factor, dimensionless

Equation 2.1

$$\mathrm{kW_b} = 0.01151 \cdot \frac{Q_\mathrm{i} \cdot p_\mathrm{i} \cdot X}{\eta_\mathrm{b} \cdot \eta_\mathrm{m} \cdot \eta_\mathrm{vfd}}$$

where

- $\mathrm{kW_b}$ = blower electrical power, kilowatts
- Q_i = blower inlet flow rate, inlet cubic feet per minute (ICFM)
- p_i = blower inlet pressure, psia
- $\eta_\mathrm{b,m,vfd}$ = blower, motor, and VFD efficiency, decimal
- X = adiabatic factor (see Equation 2.2)

Equation 2.2

$$X = \left(\frac{p_\mathrm{d}}{p_\mathrm{i}}\right)^{0.283} - 1$$

where

- p_d = blower discharge pressure, psia

Equation 2.3

$$\text{Turndown\%} = \frac{Q_{\max} - Q_{\min}}{Q_{\max}} \cdot 100$$

where

$Q_{\max,\min}$ = maximum and minimum safe flow rates, SCFM

Equation 2.4

$$\text{NPW} = \sum_{1}^{n} \frac{S}{(1+r)^n} - \text{Investment}$$

where

NPW = net present worth, NPW > 0 if the project is financially justified
r = effective annual discount rate, decimal
n = total evaluation period, years
S = net annual savings (energy and maintenance), \$/year

Equation 2.5

$$\text{S} = (\text{Existing kW} - \text{Projected kW}) \cdot \text{Rate}\frac{\$}{\text{kWh}} \cdot 8760\frac{\text{h}}{\text{year}}$$

Equation 2.6

$$\text{S} = \sum \text{On Peak \$} + \text{Off Peak \$} + \text{Demand \$} + \text{Power Factor \$}$$

Equation 2.7

$$\text{On Peak \$} = \text{On Peak} \cdot \frac{\$}{\text{kWh}} \text{kW}_{\text{saved}} \cdot \frac{60\text{ h}}{\text{week}} \cdot \frac{52\text{ weeks}}{\text{year}}$$

Equation 2.8

$$\text{Off Peak \$} = \text{Off Peak}\frac{\$}{\text{kWh}} \cdot \text{kW}_{\text{saved}} \cdot \frac{108\text{ h}}{\text{week}} \cdot \frac{52\text{ weeks}}{\text{year}}$$

Equation 2.9

$$\text{Demand \$} = \text{Demand}\frac{\$}{\text{kW}} \cdot \text{kW}_{\text{peak saved}} \cdot \frac{12\text{ months}}{\text{year}}$$

Equation 2.10

$$\text{Power Factor } \$ = \%\text{Penalty} \cdot \left((\text{PF}_{\text{min}} - \text{PF}_{\text{actual}}) \cdot \text{Demand} \frac{\$}{\text{kW}} \text{kW}_{\text{peak}} \cdot \frac{12\,\text{months}}{\text{year}} \right)$$

Equation 3.1

$$\text{Sub} = \frac{S}{S+H} \cdot 100$$

where

Sub = Submergence of airlift pump, %
S = distance from source water surface to air entry point, ft
H = distance from source water surface to airlift discharge point, ft

Equation 3.2

$$Q_{\text{s}} = 0.33 \cdot \frac{Q_{\text{WW}} \cdot C_{\text{s}}}{\text{OTE} \cdot 1.024^{(T-20)}} \cdot \ln\left(\frac{C_{\text{s}} - C_{\text{i}}}{C_{\text{s}} - C_{\text{f}}}\right)$$

where

Q_{s} = theoretical air flow for postaeration, SCFM
Q_{WW} = wastewater flow, mgd
C_{s} = saturation concentration of O_2 at 20°C, mg/l (9.08 mg/l at sea level)
OTE = oxygen transfer efficiency, decimal
C_{i} = initial DO concentration, mg/l
C_{f} = final DO concentration, mg/l

Equation 3.3

$$\text{OUR} = \frac{C_1 - C_2}{t_2 - t_1}$$

where

OUR = oxygen uptake rate, mg/(l h)
C_1, C_2 = initial and final DO concentration, mg/l
t_1, t_2 = initial and final time, h

Equation 3.4

$$Q_s = \frac{\text{OUR} \cdot V_{\text{t}}}{16{,}700 \cdot \text{AOTE}}$$

where

Q_s = air flow required to meet demand, SCFM
OUR = oxygen uptake rate, mg/(l h)
V_t = aeration tank volume, cu ft
AOTE = actual site oxygen transfer efficiency, decimal

Equation 3.5

$$P_a = \frac{\text{OUR} \cdot V_t}{16{,}000 \cdot \text{AE}}$$

where

P_a = aerator shaft power required, bhp
AE = actual site aerator efficiency, lb O_2/hp h

Equation 3.6

$$\text{OUR}_{\text{Req}} = \frac{\Delta\text{BOD} \cdot U_{O_2}}{\text{HRT}}$$

where

OUR_{Req} = OUR required to meet treatment objective, mg/(l h)
ΔBOD = BOD removed from influent to effluent, mg/l
U_{O_2} = utilization of oxygen, lb O_2/lb BOD
HRT = hydraulic retention time, h

Equation 3.7

$$\text{HRT} = \frac{1.795 \cdot 10^{-4} \cdot V_t}{Q_{\text{ww}}}$$

where

Q_{ww} = wastewater flow, mgd

Equation 3.8

$$w = C \cdot Q_{\text{ww}} \cdot 8.34$$

where

w = mass loading rate of constituent, lb/day
Q_{ww} = wastewater flow, mgd
C = concentration of constituent, ppm

Equation 3.9

$$Q_s = \frac{w_{BOD} \cdot U_{O_2}}{AOTE \cdot 24.84}$$

where

Q_s = air flow required to meet demand, SCFM
w_{BOD} = mass loading rate of BOD, lb/day
U_{O_2} = utilization of oxygen, lb O_2/lb BOD
AOTE = actual site oxygen transfer efficiency, decimal

Equation 3.10

$$P_a = \frac{w_{BOD} \cdot U_{O_2}}{AE \cdot 24}$$

where

P_a = aerator shaft power required, bhp
AE = actual site aerator efficiency, lb O_2/hp h

Equation 3.11

$$Q_s = \frac{w_{NH_3} \cdot 4.6}{AOTE \cdot 24.84}$$

where

Q_s = air flow required to meet demand, SCFM
w_{NH_3} = mass loading rate of ammonia, lb/day
AOTE = actual site oxygen transfer efficiency, decimal

Equation 3.12

$$P_a = \frac{w_{NH_3} \cdot 4.6}{AE \cdot 24}$$

where

P_a = aerator shaft power required, bhp
AE = actual site aerator efficiency, lb O_2/hp h

Equation 3.13

$$Q_{mix} = (1 \cdot w) \cdot Q_{req}$$

where

Q_{mix} = total mixing air flow to basin(s), SCFM
l = tank length, ft
w = tank width, ft
Q_{req} = required mixing air rate, SCFM/ft^2

Equation 3.14

$$r_{su} = -\frac{\mu_{max} \cdot X \cdot S}{Y \cdot (K_S + S)}$$

where

r_{su} = rate of substrate concentration change, g/(m^3 d)
μ_{max} = maximum bacterial growth rate, g/(g d)
X = biomass concentration, g/m^3
S = growth limiting substrate concentration, g/m^3
Y = yield coefficient, g/g
K_s = half velocity constant, g/m^3

Equation 4.1

$$C = \frac{m_{solute}}{V_{solvent}}$$

where

C = concentration of solution, mg/l
m_{solute} = mass of solute dissolved in the solvent, mg
$V_{solvent}$ = volume of solvent, l

Equation 4.2

$$\frac{Q_{as}}{Q_{2.0}} = \frac{C^*_{\infty f} - 2.0}{C^*_{\infty f} - C_a}$$

Or

$$C_a = C^*_{\infty f} - \frac{Q_{2.0} \cdot \left(C^*_{\infty f} - 2.0\right)}{Q_{as}}$$

where

Q_{as} = actual required air flow, SCFM
$Q_{2.0}$ = required air flow to maintain 2.0 ppm DO
$C^*_{\infty f}$ = steady-state value of dissolved oxygen saturation concentration at infinite time in field process water, ppm
C_a = actual DO concentration, ppm

Equation 4.3

$$\mathrm{SAE_m} = \frac{\mathrm{SOTR_m}}{P_\mathrm{m}}$$

where

$\mathrm{SAE_m}$ = aerator standard aerator efficiency, $\mathrm{lb_m\ O_2/(hp\ h)}$
$\mathrm{SOTR_m}$ = aerator oxygen transfer rate, $\mathrm{lb_m\ O_2/h}$
P_m = aerator power draw, hp

Equation 4.4

$$\mathrm{SOTR_d} = Q_\mathrm{s} \cdot \mathrm{SOTE} \cdot \rho \cdot C_{\mathrm{O_2}} \cdot 60 \frac{\mathrm{min}}{\mathrm{h}}$$

where

$\mathrm{SOTR_d}$ = diffuser oxygen transfer rate, $\mathrm{lb_m\ O_2/h}$
Q_s = air flow rate, standard $\mathrm{ft^3/min}$ (SCFM)
SOTE = diffuser standard oxygen transfer efficiency, decimal
ρ = density of air at standard conditions, 0.075 $\mathrm{lb_m/ft^3}$
$C_{\mathrm{O_2}}$ = concentration of O_2 in air, decimal, 0.23 (23% by weight) infinite time in field process water, ppm

Equation 4.5

$$\mathrm{SOTR_d} = Q_\mathrm{s} \cdot \mathrm{SOTE} \cdot 1.035$$

Equation 4.6

$$P_\mathrm{b} = \frac{(Q_\mathrm{s}/60) \cdot \gamma \cdot R \cdot T_\mathrm{i}}{550 \cdot (k - 1/k) \cdot \eta_\mathrm{b}} \cdot \left[\left(\frac{p_\mathrm{d}}{p_\mathrm{i}} \right)^{(k-1/k)} - 1 \right]$$

where

P_b = blower shaft power, bhp
Q_s = air flow rate, standard $\mathrm{ft^3/min}$ (SCFM)
γ = specific weight of air at standard conditions, 0.075 $\mathrm{lb_f/ft^3}$
R = gas constant for air, 53.3 $\mathrm{ft\ lb_f/(lb_m\ ^\circ R)}$
T_i = blower inlet air temperature, °R (°R = °F + 459.67)
k = ratio of specific heats for air, $C_\mathrm{p}/C_\mathrm{v} = 1.395$
$k - 1/k = 0.283$, dimensionless
η_b = blower efficiency, decimal
p_d = blower discharge pressure, psia (psia = psig + barometric pressure)
p_i = blower inlet pressure, psia

Equation 4.7

$$\mathrm{SAE_d} = \frac{2418 \cdot \mathrm{SOTE} \cdot \eta_b}{T_i \cdot \left[(p_d/p_i)^{0.283} - 1\right]}$$

where

$\mathrm{SAE_d}$ = diffused aeration system standard aerator efficiency, $\mathrm{lb_m\ O_2/(hp\,h)}$

Equation 4.8

$$\mathrm{OTR_f} = \alpha \cdot F \cdot \mathrm{SOTR} \cdot \theta^{(T-20)} \cdot \frac{\left(\tau \cdot \beta \cdot \Omega \cdot C^*_{\infty 20} - C\right)}{C^*_{\infty 20}}$$

$$= \alpha \cdot F \cdot \mathrm{SOTR} \cdot \theta^{(T-20)} \cdot \frac{C^*_{\infty f} - C}{C^*_{\infty 20}}$$

where

$\mathrm{OTR_f}$ = oxygen transfer rate for the system operating under process conditions, $\mathrm{lb_m/h}$
SOTR = standard oxygen transfer rate of new aeration system, $\mathrm{lb_m/h}$
α = correction factor for basin geometry and wastewater characteristics with new aeration devices, dimensionless, typically $\alpha < 1.0$
β = correction factor for salinity and dissolved solids, = process water C^*_{st}/clean water C^*_{st}, dimensionless, typically $\beta = 0.95$
C^*_{sT} = tabular value of dissolved oxygen surface saturation concentration at actual process water temperature, barometric pressure of 14.7 psia, 100% relative humidity, ppm (see Table 4.1)
C = average actual process water dissolved oxygen concentration, ppm
θ = empirical temperature correction factor, dimensionless, typically $\theta = 1.024$
F = fouling factor, process water SOTR of a diffuser after a given time in service/SOTR of a new diffuser in the same process water, dimensionless, typically $F \leq 1.0$
$C^*_{\infty 20}$ = steady-state value of dissolved oxygen saturation at infinite time at 20 °C and a barometric pressure of 14.7 psia, ppm

- for mechanical aeration, the surface saturation concentration at 20°C is used, and therefore $C^*_{\infty 20} = C^*_{s20}$
- for diffused aeration, $C^*_{\infty 20}$ can be obtained from manufacturer's data
- for diffused aeration where manufacturer's data is not available $C^*_{\infty 20}$ may be approximated: $C^*_{\infty 20} = 9.09 + 0.1 \times$ ft. submergence

τ = temperature correction factor for dissolved oxygen saturation, dimensionless, $\tau = C^*_{sT}/C^*_{s20}$

C^*_{s20} = value of dissolved oxygen surface saturation concentration at 20°C, barometric pressure of 14.7 psia, and 100% relative humidity, ppm. $C^*_{s20} = 9.09$ ppm
Ω = p_b/p_s (for tanks less than 20 ft in depth), dimensionless
p_b = barometric pressure under field conditions, psia
p_s = standard barometric pressure, psia
$C^*_{\infty f}$ = $\tau \cdot \beta \cdot \Omega \cdot C^*_{\infty 20}$ = steady-state value of dissolved oxygen saturation concentration at infinite time in field process water, ppm

Equation 4.9

$$\mathrm{ROTR} = \frac{\mathrm{OUR} \cdot V_t}{16{,}000}$$

where

ROTR = required oxygen transfer rate, $\mathrm{lb_m/h}$
OUR = oxygen uptake rate, mg/(l h)
V_t = tank volume, ft^3

Equation 4.10

$$\mathrm{AOTR_{act}} = \mathrm{AOTR_{2.0}} \cdot \frac{C^*_{\infty f} - C_a}{C^*_{\infty f} - 2.0}$$

where

$\mathrm{AOTR_{act}}$ = actual oxygen transfer rate at alternate DO, $\mathrm{lb_m/h}$
$\mathrm{AOTR_{2.0}}$ = actual oxygen transfer rate at 2.0 mg/l, $\mathrm{lb_m/h}$
$C^*_{\infty f}$ = steady-state value of dissolved oxygen saturation concentration at infinite time in field process water, ppm
C_a = actual DO concentration, ppm

Equation 4.11

$$\frac{\mathrm{OTR_a}}{\mathrm{OTR_{2.0}}} = \frac{C^*_{\infty f} - C_a}{C^*_{\infty f} - 2.0}$$

Equation 4.12

$$p_{stat} = \mathrm{Depth} \cdot 0.433$$

where

p_{stat} = static pressure at diffuser, psig
Depth = submergence of diffuser surface, ft

Equation 4.13

$$Q_s = \frac{0.335 \cdot Q_{ww}}{\text{AOTE}} \cdot (\Delta\text{BOD} \cdot U_{O_2} + \Delta\text{NH}_3 \cdot 4.6)$$

where

Q_s = air flow required to meet demand, SCFM
Q_{ww} = wastewater flow, mgd
ΔBOD = BOD removed from influent to effluent, mg/l
U_{O_2} = utilization of oxygen, lb O_2/lb BOD
ΔNH_3 = ammonia nitrified, mg/l
AOTE = actual site oxygen transfer efficiency, decimal

Equation 4.14

$$\frac{\text{cost}}{\text{h}} = \text{kW} \cdot \text{rate}$$

where

cost/h = on peak or off peak consumption cost, \$/h
rate = appropriate on peak or off peak cost, \$/kWh

Equation 4.15

$$\text{consumption cost} = \left(\sum_{1}^{24} \frac{\text{cost}_{\text{weekday}}}{\text{h}}\right) \cdot 5 \cdot 52 + \left(\sum_{1}^{24} \frac{\text{cost}_{\text{weekend}}}{\text{h}}\right) \cdot 2 \cdot 52$$

Equation 4.16

$$\text{annual cost} = \text{consumption cost} + \text{demand charge} + \text{PF charge}$$

Equation 4.17

$$\text{annual estimated cost} = 365 \cdot \sum_{1}^{5} \text{kW}_{\text{ave}} \cdot \%\ \text{time} \cdot \%\ \text{ADF} \cdot \text{rate}$$

where

kW_{ave} = power required at average daily flow, kW
%time = estimated percent operating at a given demand, from table, decimal
%ADF = estimated proportion of average power, from table, decimal
rate = composite energy cost, \$/kWh

Equation 5.1

$$\frac{V_2}{V_1} = \frac{p_1}{p_2}$$

where

$V_{1,2}$ = volume at condition one and two
$p_{1,2}$ = absolute pressure at condition one and two

Equation 5.2

$$\frac{V_2}{V_1} = \frac{T_2}{T_1}$$

where

$V_{1,2}$ = volume at condition one and two
$T_{1,2}$ = absolute temperature at condition one and two

Equation 5.3

$$p \cdot V = n \cdot R_0 \cdot T$$

where

p = pressure, lb/ft^2
V = volume, ft^3/(lb mol)
T = absolute temperature, °R, where °R = °F + 460
R_0 = universal gas constant, 1545 ft lb$_f$/mol °R
n = number of moles, dimensionless

Equation 5.4

$$p \cdot V = m \cdot R_{air} \cdot T$$

where

p = pressure, psia
V = volume, ft^3
m = mass, lb$_m$
T = absolute temperature, °R
R$_{air}$ = gas constant for air, 0.37046 ft^3 lb$_f$/lb$_m$ °R in^2

Equation 5.5

$$Q_2 = Q_1 \cdot \frac{p_1}{p_2} \cdot \frac{T_2}{T_1}$$

where

$Q_{1,2}$ = volumetric flow rate at condition 1 and 2, cubic ft^3/min
$p_{1,2}$ = pressure at condition 1 and 2, psia
$T_{1,2}$ = temperature at condition 1 and 2, °R

Equation 5.6

$$\text{psia} = \text{psig} + \text{barometric}$$

where

psia = absolute pressure, psia
psig = gauge pressure, psig
barometric = barometric pressure, psia

Equation 5.7

$$p_{\text{bar}} = 14.7 - \frac{\text{Alt}}{2000}$$

where

p_{bar} = barometric pressure, psia
Alt = altitude, feet above sea level (FASL)

Equation 5.8

$$Q_s = Q_a \cdot \frac{p_a \cdot 35.92}{T_a}$$

where

Q_s = mass flow rate, standard ft^3/min
Q_a = volumetric flow rate, ft^3/min
p_a = actual pressure, psia
T_a = actual temperature, °R

Equation 5.9

$$q_m = Q_s \cdot \rho$$

where

q_m = mass flow rate, lb_m/min
Q_s = mass flow rate, standard ft^3/min
ρ = density of air at standard conditions, 0.075 lb_m/standard ft^3

Equation 5.10

$$p_{\text{total}} = p_{\text{a}} + p_{\text{b}} + p_{\text{c}} + \dots$$

where

$p_{\text{a,b},\dots}$ = partial pressure of gas a, b, and so on, psia

Equation 5.11

$$\%\text{RH} = \frac{p_{\text{v}}}{p_{\text{sat}}} \cdot 100$$

where

%RH = relative humidity, %
p_{sat} = saturation vapor pressure at actual dry bulb temperature, psia
p_{v} = actual vapor pressure, psia

Equation 5.12

$$Q_{\text{i}} = Q_{\text{s}} \cdot \frac{14.58}{p_{\text{b}} - (\text{RH} \cdot p_{\text{sat}})} \cdot \frac{T_{\text{i}}}{528} \cdot \frac{p_{\text{b}}}{p_{\text{i}}}$$

where

Q_{i} = inlet volumetric flow rate, ICFM
RH = relative humidity, decimal
Q_{s} = mass flow rate, SCFM
p_{b} = barometric pressure, psia
p_{sat} = saturation vapor pressure at actual dry bulb temperature, psia
p_{i} = actual inlet pressure, psia
T_{i} = actual inlet air temperature, °R

Equation 5.13

$$Q_{\text{s}} = Q_{\text{n}} \cdot 0.6386$$

where

Q_{s} = mass flow rate, SCFM, which will provide equal oxygen to the process
Q_{n} = mass flow rate, Nm^3/h, 0°C, 101.3 kPa,0% RH

Equation 5.14

$$k_{\text{f}} = \frac{(p_{\text{des}} - p_{\text{stat}})}{Q_{\text{des}}^2}$$

where

k_f = constant of proportionality for friction losses, psi/SCFM^2
p_{stat} = static pressure, psig
p_{des} = total system pressure at design flow, psig
Q_{des} = design system air flow rate, SCFM

Equation 5.15

$$p_{tot} = p_{stat} + k_f \cdot Q^2$$

where

p_{tot} = total system pressure, psig

Equation 5.16

$$k = \frac{C_p}{C_v} = \frac{C_p}{C_p - R_{air}} \approx 1.395$$

where

k = ratio of specific heats for dry air, dimensionless
C_p = specific heat at constant pressure, for dry air at 68°F ≈ 0.240 BTU/lb °R
C_v = specific heat at constant volume, for dry air at 68°F ≈ 0.172 BTU/lb °R
R_{air} = gas constant for dry air, ≈ 0.0686 BTU/lb °R

Equation 5.17

$$\frac{k-1}{k} \approx 0.283$$

Equation 5.18

$$r_p = \frac{p_d}{p_i}$$

where

r_p = pressure ratio, dimensionless
p_d = discharge pressure, psia
p_i = inlet pressure, psia

Equation 5.19

$$H_{air} = \frac{R_{air} \cdot T_i \cdot \left[r_p^{(k-1/k)} - 1\right]}{(k-1)/k}$$

where

H_{air} = adiabatic head for air, ft lb_f/lb_m
R_{air} = gas constant for air, 53.34 ft lb_f/lb_m °R
T_i = inlet air temperature, °R

Equation 5.20

$$X = r_p^{(k-1/k)} - 1$$

Equation 5.21

$$P_{gas_{ad}} = \frac{q_m \cdot T_i \cdot X}{175.1}$$

where

$P_{gas_{ad}}$ = adiabatic isentropic power in the gas stream, hp
q_m = mass flow rate, lb/min
T_i = inlet air temperature, °R
X = adiabatic factor, dimensionless

Equation 5.22

$$P_{gas_{ad}} = q_m \cdot C_p \cdot \Delta T$$

where

C_p = specific heat, 5.66 hp min/lb °R
ΔT = difference between inlet and discharge temperature, °R

Equation 5.23

$$\eta = \frac{P_{gas_{ad}}}{P_{meas}} \cdot 100$$

where

η = efficiency, %
P_{meas} = measured power, hp

Equation 5.24

$$P_{wa} = \frac{Q_s \cdot T_i}{\eta_{wa} \cdot 3131.6} \cdot X$$

where

P_{wa} = wire to air power, kW
Q_s = mass flow rate of air, SCFM
η_{wa} = wire to air system efficiency, decimal

Equation 5.25

$$P_{\mathrm{wa}} = \frac{Q_{\mathrm{i}}}{\eta_{\mathrm{wa}} \cdot 86.9} \cdot \left[\left(p_{\mathrm{i}}^{0.717} \cdot p_{\mathrm{d}}^{0.283}\right) - p_{\mathrm{i}}\right]$$

where

P_{wa} = wire to air power, kW
Q_{i} = volumetric flow rate of air, ICFM
η_{wa} = wire to air system efficiency, decimal
$p_{\mathrm{i,d}}$ = inlet and discharge pressure, psia

Equation 5.26

$$T_{\mathrm{d}} = T_{\mathrm{i}} + \frac{T_{\mathrm{i}} \cdot X}{\eta}$$

where

T_{d} = discharge temperature, °R

Equation 5.27

$$H_{\mathrm{d}} = 2544 \cdot P_{\mathrm{mot}} \cdot (1 - \eta_{\mathrm{mot}} \cdot \eta_{\mathrm{VFD}})$$

where

H_{d} = heat rejected by drive system, including motor and VFD, BTU/h
P_{mot} = motor power, hp
$\eta_{\mathrm{mot,VFD}}$ = motor and VFD efficiency, decimal

Equation 5.28

$$H_{\mathrm{b}} = 2.4 \cdot 1.25 \cdot A_{\mathrm{b}} \cdot (T_{\mathrm{d}} - T_{\mathrm{a}})$$

where

H_{b} = heat rejected by the blower case, BTU/h
A_{b} = nominal surface area of the blower case, ft^2
$T_{\mathrm{d,a}}$ = discharge and ambient air temperature, °F

Equation 5.29

$$Q_{\mathrm{fan}} = \frac{H_{\mathrm{d}} + H_{\mathrm{b}}}{1.08 \cdot (T_{\mathrm{r}} - T_{\mathrm{o}})}$$

where

Q_{fan} = required ventilating fan air flow rate, CFM
$T_{\mathrm{o,r}}$ = outside and room air temperatures, °F

Equation 5.30

$$N_a = \frac{Q_i}{\text{CFR}} + N_s$$

where

N_a = actual speed required to provide flow rate, rpm
Q_i = volumetric flow rate, ICFM
CFR = blower displacement, ft^3/revolution
N_s = actual blower slip, rpm

Equation 5.31

$$N_a = \frac{Q_i}{\text{CFR} \cdot \eta_v}$$

where

η_v = volumetric efficiency, decimal

Equation 5.32

$$N_s = N_1 \cdot \sqrt{(p_d - p_i) \cdot \frac{14.7}{p_i} \cdot \frac{T_i}{528}}$$

where

N_s = actual slip at operating conditions, rpm
N_1 = slip at 1 psig differential pressure, rpm
$P_{i,d}$ = pressure at blower inlet and discharge, psia
T_i = blower inlet air temperature, °R

Equation 5.33

$$P = \frac{T \cdot N}{5252}$$

where

P = power required, hp
T = torque, ft lb
N = rotating speed of the machine, rpm

Equation 5.34

$$\text{FHP}_a = \text{FHP}_{nom} \cdot \frac{N_a}{1000}$$

where

FHP_a = actual friction power loss, hp
FHP_{nom} = nominal friction power, hp/1000 rpm

Equation 5.35

$$P_{PD} = [0.0044 \cdot N_a \cdot CFR \cdot (p_d - p_i)] + FHP_a$$

where

P_{PD} = positive displacement blower power required, hp

Equation 5.36

$$T_d = T_i + \frac{T_i \cdot P_{PD}}{0.01542 \cdot p_i \cdot Q_i}$$

where

$T_{d,i}$ = inlet and theoretical discharge air temperature, °R

Equation 5.37

$$\eta_v = \frac{Q_i}{CFR \cdot N_a}$$

where

η_v = volumetric efficiency, decimal
Q_i = volumetric flow rate delivered to process, ICFM
CFR = blower displacement, ft^3/revolution
N_a = actual speed, rpm

Equation 5.38

$$p_{da} = p_c \cdot \frac{T_{ic}}{T_{ia}} \cdot \frac{p_{ia}}{p_{ic}}$$

where

p_{da} = actual pressure increase from inlet to discharge, psig
p_c = discharge pressure from curve, psig
$T_{ic,ia}$ = inlet temperature for curve and actual, °R
$p_{ic,ia}$ = inlet pressure for curve and actual, psia

Equation 5.39

$$P_a = P_c \cdot \frac{T_{ic}}{T_{ia}} \cdot \frac{p_{ia}}{p_{ic}}$$

where

P_a = actual blower power, hp
P_c = blower power from curve, hp

Equation 5.40

$$Q_a = Q_c \cdot \frac{N_a}{N_c}$$

where

$Q_{a,c}$ = actual and curve volumetric flow rate, ICFM
$N_{a,c}$ = actual and curve rotational speed, rpm

Equation 5.41

$$X_a = X_c \cdot \left(\frac{N_a}{N_c}\right)^2$$

where

$X_{a,c}$ = actual and curve adiabatic factor, dimensionless

Equation 5.42

$$P_a = P_c \cdot \left(\frac{N_a}{N_c}\right)^3$$

where

$P_{a,c}$ = actual and curve blower power, hp

Equation 6.1

$$\frac{p_1}{\gamma} + \frac{V_1^2}{2 \cdot g_c} + z_1 = \frac{p_2}{\gamma} + \frac{V_2^2}{2 \cdot g_c} + z_2$$

where

$p_{1,2}$ = static pressure or head, lb_f/ft^2
γ = specific weight of fluid, lb_f/ft^3
$V_{1,2}$ = velocity of fluid, ft/s
g_c = acceleration of gravity, 32.17 ft/s^2
$z_{1,2}$ = height of liquid column, ft

Equation 6.2

$$p_1 + \frac{\rho \cdot V_1^2}{3.335 \cdot 10^7} = p_2 + \frac{\rho \cdot V_2^2}{3.335 \cdot 10^7} + \Delta p_f$$

where

$p_{1,2}$ = static pressure, psi
ρ = density at air flow conditions, lb_m/ft^3
$V_{1,2}$ = air velocity, ft/min
Δp_f = pressure drop due to friction, psi

Equation 6.3

$$\rho_a = \frac{2.692 \cdot p_a}{T_a}$$

where

ρ_a = density at air flow conditions, lb_m/ft^3
p_a = actual pressure, psia
T_a = actual temperature, °R

Equation 6.4

$$p_d = \frac{\rho \cdot V^2}{3.335 \cdot 10^7}$$

where

p_d = velocity head or dynamic pressure, psi

Equation 6.5

$$Q_a = A_1 \cdot V_1 = A_2 \cdot V_2$$

where

Q_a = volumetric flow rate at actual conditions, ACFM (ft^3/min)
$A_{1,2}$ = cross-sectional area of pipe, ft^2
$V_{1,2}$ = velocity, ft/min

Equation 6.6

$$V_2 = \frac{Q_2}{Q_1} \cdot V_1$$

Equation 6.7

$$\Delta L = L_0 \cdot \alpha \cdot \Delta T$$

where

ΔL = change in length, in.
L_0 = initial length, in.
α = linear coefficient of thermal expansion, in./in. °F
ΔT = change in temperature, °F

Equation 6.8

$$\sigma = E \cdot \alpha \cdot \Delta T$$

where

σ = stress in pipe, psi
E = Young's modulus (modulus of elasticity), psi

Equation 6.9

$$A = \frac{Q_a}{V}$$

where

A = cross-sectional area of pipe, ft^2
Q_a = volumetric flow rate at actual conditions, ACFM (ft^3/min)
V = design velocity, ft/min

Equation 6.10

$$d = 24 \cdot \sqrt{\frac{Q_a}{\pi \cdot V}}$$

where

d = diameter of pipe, in.

Equation 6.11

$$\Delta p_f = 0.07 \cdot \frac{Q_S^{1.85}}{d^5 \cdot p_m} \cdot \frac{T}{528} \cdot \frac{L_e}{100}$$

where

Δp_f = pressure drop due to friction, psi
Q_S = air flow rate, SCFM
d = actual pipe inside diameter, in.
p_m = mean system pressure, psia
T = air temperature, °R
L_e = equivalent length of pipe and fittings, ft

Equation 6.12

$$p_{\mathrm{m}} = p_{\mathrm{initial}} - \frac{\Delta p_{\mathrm{f}}}{2} = p_{\mathrm{final}} + \frac{\Delta p_{\mathrm{f}}}{2}$$

where

p_{initial} = system pressure at the beginning of the pipe section, psia
p_{final} = required system pressure at the end of the pipe section, psia

Equation 6.13

$$L_{\mathrm{e}} = \mathrm{Ratio} \cdot \frac{d}{12}$$

where

L_{e} = equivalent length of fitting, ft
d = nominal pipe diameter, in.
Ratio = *L/D* from Table 6.6

Equation 6.14

$$Q_1 = Q_2 \cdot \sqrt{\frac{k_{\mathrm{f2}}}{k_{\mathrm{f1}}}} = \frac{Q_{\mathrm{total}}}{1 + \sqrt{k_{\mathrm{f1}}/k_{\mathrm{f2}}}}$$

where

$Q_{1,2,\mathrm{total}}$ = air flow rate through segment 1, 2 and total flow, CFM
$k_{\mathrm{f1,f2}}$ = constant of proportionality for friction losses in each segment, psi/ CFM^2

Equation 6.15

$$k_{\mathrm{f\ total}} = k_{\mathrm{f1}} + k_{\mathrm{f2}}$$

Equation 6.16

$$k_{\mathrm{f\ total}} = \frac{1}{\left((1/\sqrt{k_1}) + (1/\sqrt{k_2})\right)^2}$$

Equation 6.17

$$\Delta p_{\mathrm{o}} = \frac{\rho \cdot Q_{\mathrm{a}}^2}{992.2 \cdot C_{\mathrm{o}}^2 \cdot d^4}$$

where

Δp_{o} = pressure drop through orifice, psi
ρ = density at air flow conditions, lb_{m}/ft^{3}
Q_{a} = air flow rate, ACFM
C_{o} = orifice coefficient, dimensionless, = 0.60 for sharp edged orifice
= 0.80 for knife edge orifice
d = orifice diameter, in.

Equation 6.18

$$\Delta p_{v} = \left(\frac{Q_{s}}{22.66 \cdot C_{v}}\right)^{2} \cdot \frac{SG \cdot T_{u}}{p_{u}}$$

where

Δp_{v} = pressure drop across the valve, psi
Q_{s} = air flow rate, SCFM
C_{v} = valve flow coefficient from manufacturer's data, dimensionless
SG = specific gravity, dimensionless, = 1.0 for air
T_{u} = upstream absolute air temperature, °R
p_{u} = upstream absolute air pressure, psia

Equation 6.19

$$Q_{s} = 22.66 \cdot C_{v} \cdot \sqrt{\frac{p_{u} \cdot \Delta p_{v}}{SG \cdot T_{u}}}$$

Equation 6.20

$$C_{v} = \frac{Q_{s}}{22.66} \cdot \sqrt{\frac{SG \cdot T_{u}}{p_{u} \cdot \Delta p_{v}}}$$

Equation 7.1

$$S_{Out} = \left[\frac{PV - EU_{Min}}{EU_{Span}} \cdot (S_{Max} - S_{Min})\right] + S_{Min}$$

Equation 7.2

$$EU_{Span} = EU_{Max} - EU_{Min}$$

where

$S_{Out,\ Max,\ Min}$ = Transmitter actual output signal, maximum signal, and minimum signal, signal electrical units

PV $=$ Measured process variable, engineering units
$EU_{Max,\ Min}$ $=$ Maximum and minimum engineering units measured
EU_{Span} $=$ Engineering unit span

Equation 7.3

$$Q_w = 1496 \cdot 1 \cdot h^{1.5}$$

where

Q_w $=$ flow rate of water, gpm
l $=$ length of weir, ft
h $=$ height of water over the weir, ft

Equation 7.4

$$Q_w = 239 \cdot C \cdot h^2 \cdot \sqrt{64.4 \cdot h} \cdot \tan\frac{\alpha}{2}$$

where

C $=$ discharge coefficient, typically $C = 0.58$
α $=$ included angle of v-notch, degree symbol

Equation 7.5

$$Q_w = 1795 \cdot w \cdot h_a{}^n$$

where

Q_w $=$ flow rate of water, gpm
w $=$ width of flume throat, ft
h_a $=$ depth of water at flume inlet, ft
n $=$ empirical coefficient
n $=$ 1.55 for 1 ft wide flume
n $=$ $1.522 \cdot w^{0.026}$ for wider flumes

Equation 7.6

$$q_{m}' = 300 \cdot \frac{\pi}{4} \cdot d^2 \cdot C \cdot \varepsilon \cdot \sqrt{\frac{2 \cdot \rho \cdot \Delta p_m \cdot g_c}{1 - \beta^4}}$$

where

q_m $=$ mass flow rate, lb_m/h
d $=$ bore or throat diameter of the flow element, in.

C = discharge coefficient of the flow element, dimensionless
$C \approx 0.6$ for orifice plates
ε = expansion factor for compressible fluids, dimensionless
$\varepsilon \approx 1$ at typical air flow conditions
ρ = density of fluid at flow conditions, $\text{lb}_\text{m}/\text{ft}^3$
Δp_m = differential pressure of flow measurement element, $\text{lb}_\text{f}/\text{ft}^2$ (psi)
g_c = $32.174\,\text{lb}_\text{m}\,\text{ft}/\text{lb}_\text{f}\,\text{s}^2$
β = ratio of bore or throat to pipe diameters, dimensionless

Equation 7.7

$$\beta = \frac{d}{D}$$

where

d = diameter of bore or throat of the flow element, in.
D = inside diameter of the pipe, in.

Equation 7.8

$$\Delta p_\text{f} = \Delta p_\text{m} \cdot \left(1 - \beta^2\right)$$

where

Δp_f = permanent friction pressure drop, psi

Equation 7.9

$$Q_2 = Q_1 \cdot \sqrt{\frac{\Delta p_\text{m2}}{\Delta p_\text{m1}}}$$

where

$\Delta p_\text{m1,m2}$ = differential pressure at flow 1 and 2
$Q_{1,2}$ = flow rate at differential pressure 1 and 2

Equation 7.10

$$V = \sqrt{\frac{3.336 \cdot 10^7 \cdot \Delta p_\text{p}}{\gamma}}$$

where

V = velocity of air stream, ft/min
Δp_p = differential pressure at pitot tube, psi

$\Delta p_{\mathrm{p}} = p_{\mathrm{total}} - p_{\mathrm{velocity}}$
γ = specific weight of air at flow conditions, $\mathrm{lb_f/ft^3}$

Equation 7.11

$$\mathrm{AOTE} = \frac{\mathrm{O}_{2_\mathrm{i}} - \mathrm{O}_{2_\mathrm{o}}}{\mathrm{O}_{2_\mathrm{i}}} \cdot 100$$

where

AOTE = oxygen transfer efficiency, %
$\mathrm{O}_{2_{\mathrm{i,o}}}$ = concentration of oxygen into aeration (ambient) and out of aeration, %

Equation 7.12

$$\mathrm{AOTR} = Q_\mathrm{s} \cdot 60 \cdot \frac{\mathrm{AOTE}}{100} \cdot \rho \cdot C_{\mathrm{O_2}}$$

where

AOTR = actual oxygen transfer rate, $\mathrm{lb_m/h}$
Q_s = mass flow rate, SCFM
AOTE = oxygen transfer efficiency, %
ρ = density of air at standard conditions, 0.075 $\mathrm{lb_m/ft^3}$
$C_{\mathrm{O_2}}$ = concentration of $\mathrm{O_2}$ in air, decimal, 0.23 (23% by weight)

Equation 7.13

$$\mathrm{OUR} = \frac{\mathrm{AOTR} \cdot 16{,}019}{V_\mathrm{t}}$$

where

OUR = oxygen uptake rate, mg/(l h)
V_t = tank or control zone volume, $\mathrm{ft^3}$

Equation 7.14

$$d = \frac{19{,}100 \cdot v}{f}$$

Equation 7.15

$$a = \frac{f \cdot v}{3687}$$

Equation 7.16

$$a = \left(\frac{f}{8383}\right)^2 \cdot d$$

where

d = peak to peak displacement, mils (1 mil = 0.001 in.)
v = peak velocity, in./s
f = frequency, cycles/min
a = peak acceleration, g's (1 g = 386.087 in./s^2)

Equation 8.1

$$T = \frac{P \cdot 5252}{N}$$

where

T = torque, lb ft
P = motor power, hp
N = rotating speed, rpm

Equation 8.2

$$t = \frac{WK^2 \cdot \Delta N}{308 \cdot (T_m - T_l)}$$

where

t = time to accelerate the load, s
WK^2 = rotational inertia of the load at the motor shaft, lb ft^2
ΔN = change in speed, rpm
T_m = average motor torque, lb ft
T_l = average load torque, lb ft

Equation 8.3

$$WK_m^2 = WK_l^2 \cdot \left(\frac{N_l}{N_m}\right)^2$$

where

$WK_{m,l}^2$ = rotational inertia at the motor shaft and the load, lb ft^2
$N_{m.l}$ = speed of the motor and load, rpm

Equation 8.4

$$N_s = \frac{120 \cdot f}{n_p}$$

where

N_s = synchronous speed of the motor, rpm
f = frequency of AC voltage, Hz (cycles/s)
n_p = number of poles, dimensionless

Equation 8.5

$$s = \frac{N_s - N_a}{N_s} \cdot 100$$

where

s = slip, %
$N_{s,a}$ = synchronous and actual speed of the motor, rpm

Equation 8.6

$$I_{LR} = \frac{P_m \cdot \mathrm{KVA/hp} \cdot 1000}{1.73 \cdot V}$$

where

I_{LR} = locked rotor current, A
P_m = rated motor power, hp
KVA/hp = maximum value from table, KVA/hp
V = utilization voltage at motor, V

Equation 9.1

$$e = \mathrm{SP} - \mathrm{PV}$$

where

e = error, in units of the PV
SP = setpoint, in units of the PV
PV = process variable

Equation 9.2

$$\mathrm{EU}_{\mathrm{Actual}} = \left[\frac{\mathrm{EU}_{\mathrm{Span}} \cdot (\mathrm{Data}_{\mathrm{Actual}} - \mathrm{Data}_{\mathrm{Min}})}{\mathrm{Data}_{\mathrm{Span}}}\right] + \mathrm{EU}_{\mathrm{Min}}$$

where

EU_{Actual} = actual process value, engineering units
$Data_{Actual}$ = actual data in register from I/O card A/D conversion
$Data_{Min}$ = data in register at minimum analog signal from transmitter
EU_{Min} = engineering units at minimum analog signal from transmitter

Equation 9.3

$$EU_{Span} = EU_{Max} - EU_{Min}$$

where

EU_{Max} = engineering units at maximum analog signal from transmitter

Equation 9.4

$$Data_{Span} = Data_{Max} - Data_{Min}$$

where

$Data_{Max}$ = data in register at maximum analog signal from transmitter

Equation 9.5

$$Y_n = Y_{n-1} + \left(\frac{1}{k} \cdot (X_n - Y_{n-1})\right)$$

where

Y_n = new filtered value of measurement
Y_{n-1} = filtered value of measurement from previous calculation
k = the number of samples over which the data is to be filtered
X_n = current unfiltered measurement

Equation 9.6

$$\mathrm{MV}(t) = K_\mathrm{p} \cdot e(t) + K_\mathrm{i} \cdot \int_0^t e(t) \cdot \mathrm{d}t + K_\mathrm{d} \cdot \frac{1}{\mathrm{d}t} \cdot e(t)$$

where

$\mathrm{MV}(t)$ = manipulated variable, the control output at time t
K_p = proportional gain, dimensionless
$e(t)$ = error at time t (see Equation 9.1)
K_i = integral gain, dimensionless
K_d = derivative gain, dimensionless

Equation 9.7

$$\mathrm{MV}(t) = K_\mathrm{p} \cdot \left(e(t) + \frac{1}{T_\mathrm{i}} \cdot \int_0^t e(t) \cdot \mathrm{d}t + T_\mathrm{d} \cdot \frac{\mathrm{d}}{\mathrm{d}t} \cdot e(t) \right)$$

where

T_i = integral time (often expressed as resets per second or minute)
T_d = derivative time, seconds or minutes

Equation 9.8

$$\mathrm{PB} = \frac{1}{K_\mathrm{p}}$$

where

PB = proportional band
K_p = proportional gain

Equation 9.9

$$\Delta\mathrm{MV} = G \cdot (\mathrm{SP} - \mathrm{PV}) = G \cdot e$$

where

ΔMV = change in the manipulated variable, units per controlled device
G = gain, units of MV/units of PV

Equation 9.10

$$r = \frac{\mathrm{PV}_{n-1} - \mathrm{PV}_n}{\Delta t} = \frac{\Delta \mathrm{PV}}{\Delta t}$$

where

r = rate of change, units of PV/s
Δt = time delay between successive control calculations, s

Equation 9.11

$$Q_{\mathrm{T}\,n} = Q_{\mathrm{T}\,n-1} \pm \Delta Q_\mathrm{T}$$

where

$Q_{\mathrm{T}\,n}$ = new theoretical demand for the controlled zone, SCFM or ACFM
$Q_{\mathrm{T}\,n-1}$ = previous theoretical demand for the controlled zone, SCFM or ACFM

ΔQ_T = increase or decrease in air flow demand at the tank, SCFM or ACFM, calculated from Equation 9.8

Equation 9.12

$$Q_{B\,n} = Q_{B\,n-1} + \Delta Q_{T\,n}$$

where

$Q_{B\,n}$ = new total blower air flow setpoint, SCFM or ICFM
$Q_{B\,n-1}$ = total blower air flow setpoint after previous iteration of logic, SCFM or ICFM
$\Delta Q_{T\,n}$ = increase or decrease in air flow demand at the tank, SCFM or ACFM

Equation 9.13

$$Q_{B\ \text{Corrected}} = Q_{B\ \text{Current}} + F \cdot \left(\sum Q_{T\ \text{Demand}} - \sum Q_{T\ \text{Actual}}\right)$$

where

$Q_{B\ \text{Corrected}}$ = corrected total blower system air flow demand, SCFM or ICFM
$Q_{B\ \text{Current}}$ = actual total blower system air flow demand, SCFM or ICFM
F = damping factor to limit hunting, typically 0.5
$\sum Q_{T\ \text{Demand}}$ = total theoretical process demand air flow for all tanks, SCFM or ACFM
$\sum Q_{T\ \text{Actual}}$ = total measured air flow for all tanks, SCFM or ACFM

Equation 9.14

$$Q_{T\ SP} = \frac{Q_{T\ \text{Demand}}}{\sum Q_{T\ \text{Demand}}} \cdot \sum Q_{T\ \text{Actual}}$$

where

$Q_{T\ SP}$ = actual setpoint used by the zone or tank flow control, SCFM or ICFM
$Q_{T\ \text{Demand}}$ = theoretical process demand air flow for this tank, SCFM or ACFM
$\sum Q_{T\ \text{Demand}}$ = total theoretical process demand air flow for all tanks, SCFM or ACFM
$\sum Q_{T\ \text{Actual}}$ = total measured air flow for all tanks, SCFM or ACFM

Equation 9.15

$$Q_{iPD} = \text{CFR} \cdot (N_a - N_s)$$

where

Q_{iPD} = volumetric flow rate of PD blower, ICFM
CFR = blower displacement, cubic feet per revolution
N_a = actual speed required to provide flow rate, rpm
N_s = actual blower slip, rpm

Equation 9.16

$$I = \frac{P_b \cdot 746}{V \cdot 1.73 \cdot \eta_m \cdot \text{PF}}$$

where

I = three-phase motor current, A
P_b = blower shaft power from blower curve, hp
V = motor volts, V
η_m = motor efficiency, decimal
PF = motor power factor, decimal

Equation 9.17

$$m_{\text{MS}} = \frac{Q_{\max} - Q_{\min}}{I_{\max} - I_{\min}}$$

where

m_{MS} = slope of linear correlation, SCFM/A
$Q_{\max,\min}$ = max and min flow rate curve points, SCFM
$I_{\max,\min}$ = motor current draw at max and min curve points, A

Equation 9.18

$$b_{\text{MS}} = I_{\min} - \frac{Q_{\min}}{m_{\text{MS}}}$$

where

b_{MS} = intercept of linear correlation, A

Equation 9.19

$$Q_{\text{MS}} = m_{\text{MS}} \cdot (I_{\text{Act}} - b_{\text{MS}})$$

where

Q_{MS} = mass air flow rate, SCFM

Equation 9.20

$$P_B = \frac{I \cdot V \cdot 1.73 \cdot \eta_m \cdot PF}{746} - P_L$$

where

P_B = calculated blower shaft power, bhp
P_L = mechanical losses for bearings, gears, lube system, and so on, bhp (typically 1–3% of motor power, depending on blower design)

Equation 9.21

$$P_B = \frac{P_E \cdot \eta_m}{0.746} - P_L$$

where

P_E = measured electrical power draw of motor, kW

Equation 9.22

$$X_{AD} = \left(\frac{p_d}{p_i}\right)^{0.283} - 1$$

where

X_{AD} = adiabatic factor, dimensionless
$p_{d,i}$ = discharge and inlet pressure, psia

Equation 9.23

$$\eta_B = \frac{X_{AD} \cdot T_i}{T_d - T_i}$$

where

η_B = blower adiabatic efficiency, decimal
$T_{i,d}$ = inlet and discharge air temperature, °R

Equation 9.24

$$Q_i = \frac{\eta_B \cdot P_B}{p_i \cdot X_{AD} \cdot 0.01542}$$

where

Q_i = inlet air flow rate, ICFM

Equation 9.25

$$Q_{1\ \text{set}} = \left[\left(\frac{Q_{\text{Total set}} - \sum Q_{\text{min}}}{\sum Q_{\text{max}} - \sum Q_{\text{min}}}\right) \cdot (Q_{1\ \text{max}} - Q_{1\ \text{min}})\right] + Q_{1\ \text{min}}$$

where

$Q_{1\ \text{set}}$ = flow setpoint for first blower, ICFM or SCFM
$Q_{1\ \text{min, max}}$ = minimum and maximum flow for first blower, ICFM or SCFM
$Q_{\text{Total set}}$ = total flow setpoint for all running blowers, ICFM or SCFM
$\Sigma Q_{\text{min,max}}$ = minimum and maximum flow setpoint for all running blowers, ICFM or SCFM

Bibliography

American Society of Civil Engineers. *ASCE-18-96: Standard Guidelines for In-Process Oxygen Transfer Testing*. New York, NY: American Society of Civil Engineers; 1997.

American Society of Mechanical Engineers. *PTC 9-1974: Displacement Compressors, Vacuum Pumps, and Blowers*. New York, NY: American Society of Mechanical Engineers; 1997.

American Society of Mechanical Engineers. *PTC 10-1997: Performance Test Code on Compressors and Exhausters*. New York, NY: American Society of Mechanical Engineers; 1998.

American Society of Mechanical Engineers. *ASME PTC 19.5-2004: Flow Measurement Performance Test Code*. New York, NY: American Society of Mechanical Engineers; 2004.

Baumeister, T., editor. *Standard Handbook for Mechanical Engineers*, Seventh Edition. New York, NY: McGraw Hill; 1967.

Bloch, H.P. *A Practical Guide to Compressor Technology*. Hoboken, NJ: John Wiley & Sons, Inc.; 2006.

Considine, D.M. *Process Instruments and Controls Handbook*. New York, NY: McGraw Hill; 1985.

Gartmann, H., editor. *De Laval Engineering Handbook*. New York, NY: McGraw Hill; 1970.

Aeration Control System Design: A Practical Guide to Energy and Process Optimization,
First Edition. Thomas E. Jenkins.

Jenkins, T.E. Rethinking Conventional Wisdom. *Water Environment & Technology*. November 2005; 42–48.

Jenkins, T.E. *Beloit Gets Fast Payback on Medium Voltage Drives, ABB Automation and Power World, Houston, TX;* April 23–26, 2012.

Jenkins, T.E., Redmon, D.T., Hilgart, T.D., Trillo, J., Trillo, I., inventors; Advanced Aeration Control LLC assignee. *Controlling wastewater treatment processes*. US Patent 7,449,113 B2. November 11, 2008.

Menon, E.S. *Piping Calculations Manual*. New York, NY: McGraw Hill; 2005.

Metcalf & Eddy, Inc. *Wastewater Engineering: Treatment and Reuse*. New York, NY: McGraw Hill; 2004.

Moore, R.L. *Control of Centrifugal Compressors*. Research Triangle Park, NC: Instrument Society of America; 1989.

Rollins, J.P., editor. *Compressed Air and Gas Handbook*, Fourth Edition. New York, NY: Compressed Air and Gas Institute; 1973.

Stephenson, R.L., Nixon, H.E. *Centrifugal Compressor Engineering*. East Syracuse, NY: Hoffman Air & Filtration Systems; 1986.

US Department of the Interior, Bureau of Reclamation. *Water Measurement Manual*. Washington, DC: US Department of the Interior; 2001.

US Environmental Protection Agency Office of Research and Development. *Design Manual: Fine Pore Aeration Systems*. Cincinnati, OH: US Environmental Protection Agency; 1989.

US Environmental Protection Agency Office of Wastewater Management. *Evaluation of Energy Conservation Measures for Wastewater Treatment Facilities*, EPA 832-R-10-005. Washington, DC: US Environmental Protection Agency; 2010.

Vesilind, P.A. *Wastewater Treatment Plant Design*. Alexandria, VA: Water Environment Federation; 2003.

Water Environment Federation. *Manual of Practice No. 11: Operation of Wastewater Treatment Plants*. Washington, DC: Water Pollution Control Federation; 1976.

Water Environment Federation. *Manual of Practice No. 8: Design of Municipal Wastewater Treatment Plants*. Alexandria, VA: Water Environment Federation/McGraw Hill; 2010.

Water Environment Federation. *Manual of Practice No. 32: Energy Conservation in Water and Wastewater Facilities*. Alexandria, VA: Water Environment Federation/McGraw Hill; 2010.

Water Pollution Control Federation. *Manual of Practice No. FD-4: Design of Wastewater and Stormwater Pumping Stations*. Alexandria, VA: Water Pollution Control Federation/Moore & Moore Lithographers; 1981.

Index

4-20 mA, 195

AC, 6, 194
Accuracy, 198
ACFM, 115
activated sludge, 2, 49, 50
AE, 39, 79
aerobic digesters, 34, 59
air flow control, 287
air lift, 57
alarm, 264, 340, 367
altitude, 118
ammeter, 226
ammonia, 51, 66, 303
analog, 195
analytic, 216
annunciator, 265
anoxic, 72
AOTR, 93, 223
architecture, 314

ball valves, 182
baseline, 21, 24
bearing fluting, 257
belt, 137, 142
belt presses, 60
benchmark, 21, 40
benefits, 14, 375
BEP, 9, 33, 149, 301
Bernoulli, 164, 197
BFV, 163, 182, 216
bias, 279, 363
biochemical oxygen demand, 4
biosolids, 59
blower, 107, 114
blower efficiency, 86, 129
blow-off, 103, 141, 238, 297
BNR, 67
BOD, 4, 16, 50, 62
braking resistor, 254
butterfly valve, 103, 163, 182, 216

C/T, 225
calibration, 201
carbonaceous, 50
cascade, 301
cascade control, 263
centrifugal, 143

Aeration Control System Design: A Practical Guide to Energy and Process Optimization,
First Edition. Thomas E Jenkins.

ceramic, 96
channel, 59
channel aeration, 59
characteristic curve, 144
check valve, 96, 103, 150
clarifier, 56
coarse bubble, 95
COD, 51, 63
commissioning, 9, 14, 351
communications, 197, 313, 318
complete mix, 53
composite rate, 28, 44, 109
compressibility, 119
constant pressure, 107, 124
consumption, 26
contact stabilization, 53
controller output, 263
Cv, 185

data lists, 338
DC, 6, 194
DCS, 310
deadband, 263, 274, 363
demand, 27, 28, 109
demand period, 32
denitrification, 51
density, 84
diffused, 87, 95
diffuser, 39
digester, 59, 106
digester gas, 29
discharge pressure, 71, 100, 114, 123
discrete, 193
disk, 183
dissolved oxygen, 217
dissolved oxygen control, 280
diurnal, 30, 108, 110
DO control, see *dissolved oxygen control*
documentation, 312
drain, 103
duty cycle, 123
dV/dt, 245
dynamic blowers, 143

ECM, 21, 373
efficiency, 8
effluent, 36
EMI, 193, 195, 199
energy, 4
energy audit, 21, 24
energy cost, 25
EPA, 24
EPDM, 96, 177
EQ, see *equalization*
equalization, 31, 69, 106
equivalent length, 178
error, 263
Ethernet, 318
eutrophication, 51
extended aeration, 53

F:M, 57
facultative, 52
feedback, 49, 261
feedforward, 224, 261, 303, 304
fiber optic, 313, 318
fiberglass, 176
filamentous, 18, 72
filter, 102, 125
filtering, 270
filters, 61
fine pore, 95
floating aerators, 90
floating control, 276, 287, 290, 362
floc, 2, 52, 72
flocculate, 57
flow, 39, 209
flow paced, 303
flumes, 210
fouling, 101
fouling factor, 85
frame size, 244
friction loss, 101, 178
function block, 316

gain, 272, 364
grids, 99
grit, 56
ground loops, 199
guide vanes, 149, 235, 238

harmonics, 257
HART, 197
HDPE, 168, 177
head, 5, 127, 145, 147
HMI, 11, 315, 323
HRT, 52
humidity, 119

hunting, 263
HVAC, 24, 29
hydraulic loading, 36

I/O, 39, 314
ICFM, 115
IEC, 248
IFAS, 55
IGV, 149, 153, 238
induction motor, 239
influent, 35, 85
inrush current, 32, 245, 249
inspection, 354
installation drawings, 345
inverters, see *VFD*

jet aeration, 102

Kelvin, 117

ladder, 310, 343
lagoons, 55
leachate, 35
level, 226
lighting, 29
limit switches, 229
liquidated damages, 352
lobe, 134
loop descriptions, 347
loop powered transmitter, 195
LOTO, 234

magmeters, 211
manipulated variable, 263
manual transfer, 255
mass flow, 115
MBBR, 55
MBR, 61
MCC, 249
mechanical, 40, 71, 87, 88
membrane, 96, 219
mixed liquor, 2, 52
mixing, 70, 92, 96, 100
MLE, 67
MLSS, 52
Modbus, 319
motor, 42, 239
MOV, 146, 291
MPR, 226
multi-stage, 144, 150

NEC, 10, 250, 329, 346
NEMA, 193, 237, 241
nitrification, 4, 51, 66
nitrogen, 51
Nocardia, 72
non-linear, 49
normalized, 116, 122
nutrient, 51

O&M, 335, 340, 348
ODP, 244
off peak, 27
offgas, 82, 104, 221, 304
on peak, 27
one-line, 344
operator, 13
optical, 219
orifice, 101, 181, 211
ORP, 224
OSHA, 10
OTR, 79, 100
OUR, 63, 93, 104, 221
over current, 225
oxidation ditch, 53, 88, 219, 98
oxygen limited, 71
oxygen transfer, 78
oxygen utilization, 64

P&ID, 341
panel, 100, 328
parallel, 301
payback, 15, 43
PC, 11, 312
PD, 134
pH, 224
phosphorus, 51
PID, 262, 271
piping, 161
pitot, 211, 213
Plan-Do-Check-Act, 46
PLC, 11, 310
plug flow, 52
pneumatic, 235, 310
point list, 338
post aeration, 34, 61
power, 5
power factor, 7, 27, 109
precision, 198
present worth, 16, 43

pressure, 202
pressure control, 288
pressure drop, 172, 178
pressure ratio, 127, 299
primary treatment, 50
process time lags, 262
process variable, 263
properties of air, 114
proportional, 284
proportional speed, 278, 364
protocols, 319
proximity switches, 229
psia, 117
psychrometric, 120
pumping, 5, 29
PVC, 176
PWM, 251

radio, 319
rain events, see storm events
Rankine, 117
RAS, 23
register based, 311
repeatability, 198
report, 46
residence time, 52, 72, 73
resolution, 198
response time, 49
review, 371
RFI, 195, 199
RH, 119
ROTR, 93
RTD, 197, 206, 207
RVSS, 249

SAE, 83
Safety, 9, 368
saturation (O2), 78
saturation vapor pressure, 120
savings, 42
SBR, 54, 68
SCADA, 11, 30, 323, 325
scaling, 269
SCFM, 115
schedule control, 304
screw blowers, 138
secondary treatment, 50
sensor, 193
septage, 35
service factor, 142, 237, 240
setpoint, 263
sidestream, 34, 61
simple payback, 16
simulation, 73
single stage, 144, 152
slip, 135, 243
slug, 30
soft starter, 249
SOTE, 83
SOTR, 83
specific heat, 126
specific weight, 84
specification, 335
squirrel cage, 239
SRT, 55
stainless steel, 175
starter, 247
static pressure, 100
storm events, 30
straight pipe, 169, 215
submergence, 60, 101, 123
submetering, 32
subroutines, 316
substrate, 50
surge, 132, 150, 155, 297, 301
suspended, 52
synchronous motors, 243
synchronous transfer, 255
system curve, 124

tag based, 311
tapered aeration, 99
TDS, 81
TEFC, 142
temperature measurement, 205
testing, 357
thermal expansion, 167
thermal mass flow meter, 214
thermocouple, 206
throttling, 146
time of day rates, 26
touchscreens, 327
training, 365
transducer, 193
transmitter, 193
transport lag, 49, 263
trend, 362
trimming, 284, 287, 303

triple bottom line, 14
TSS, 18
tuning, 263, 340, 361
turbo blower, 144, 154
turndown, 37, 107, 132, 142
two speed floating control, 279

UL, 10, 329, 346
ultrasonic, 211, 227
units, 192
utility representatives, 26

valve, 39
valve operators, 39, 234
VDV, 149, 153, 238
velocity head, 164
velocity profile, 169
ventilation, 132
venturis, 211
verification, 368
VFD, 92, 107, 141, 153, 154, 251
vibration, 228
Viton, 177
volumetric flow, 115

WAS, 23
weirs, 209
wire nuts, 356
wire to air, 130
wire to water, 8

α, 86

Printed in the United States
By Bookmasters